Advances in Intelligent Systems and Computing

Volume 768

Series editor

Janusz Kacprzyk, Polish Academy of Sciences, Warsaw, Poland
e-mail: kacprzyk@ibspan.waw.pl

The series "Advances in Intelligent Systems and Computing" contains publications on theory, applications, and design methods of Intelligent Systems and Intelligent Computing. Virtually all disciplines such as engineering, natural sciences, computer and information science, ICT, economics, business, e-commerce, environment, healthcare, life science are covered. The list of topics spans all the areas of modern intelligent systems and computing such as: computational intelligence, soft computing including neural networks, fuzzy systems, evolutionary computing and the fusion of these paradigms, social intelligence, ambient intelligence, computational neuroscience, artificial life, virtual worlds and society, cognitive science and systems, Perception and Vision, DNA and immune based systems, self-organizing and adaptive systems, e-Learning and teaching, human-centered and human-centric computing, recommender systems, intelligent control, robotics and mechatronics including human-machine teaming, knowledge-based paradigms, learning paradigms, machine ethics, intelligent data analysis, knowledge management, intelligent agents, intelligent decision making and support, intelligent network security, trust management, interactive entertainment, Web intelligence and multimedia.

The publications within "Advances in Intelligent Systems and Computing" are primarily proceedings of important conferences, symposia and congresses. They cover significant recent developments in the field, both of a foundational and applicable character. An important characteristic feature of the series is the short publication time and world-wide distribution. This permits a rapid and broad dissemination of research results.

More information about this series at http://www.springer.com/series/11156

Pradeep Kumar Mallick · Valentina Emilia Balas
Akash Kumar Bhoi · Ahmed F. Zobaa
Editors

Cognitive Informatics and Soft Computing

Proceeding of CISC 2017

 Springer

Editors
Pradeep Kumar Mallick
Department of Computer Science
  and Engineering
Vignana Bharathi Institute of Technology
Hyderabad, Telangana
India

Valentina Emilia Balas
Faculty of Engineering
Aurel Vlaicu University of Arad
Arad
Romania

Akash Kumar Bhoi
Department of Electrical
  and Electronics Engineering,
Sikkim Manipal Institute of Technology,
  Sikkim Manipal University
Rangpo
India

Ahmed F. Zobaa
Department of Electronic
  and Computer Engineering
Brunel University London
Uxbridge, Middlesex
UK

ISSN 2194-5357   ISSN 2194-5365   (electronic)
Advances in Intelligent Systems and Computing
ISBN 978-981-13-0616-7   ISBN 978-981-13-0617-4   (eBook)
https://doi.org/10.1007/978-981-13-0617-4

Library of Congress Control Number: 2018942007

This Springer imprint is published by the registered company Springer Nature Singapore Pte Ltd.
The registered company address is: 152 Beach Road, #21-01/04 Gateway East, Singapore 189721, Singapore

Preface

Cognitive Informatics (CI) is the multi-disciplinary study of cognition and information sciences, which investigates human information processing mechanisms and processes and their engineering applications in computing. The focus of CI is on understanding the work processes and activities within the context of human cognition and the design of interventional solutions (often engineering, computing, and information technology solutions) that can improve human activities. Within the context of biomedical informatics, CI plays a key role—in terms of understanding, describing, and predicting the nature of clinical work activities of its participants (e.g., clinicians, patients, and lay public) and in terms of developing engineering and computing solutions that can improve clinical practice (e.g., a new decision-support system), patient engagement (e.g., a tool to remind patients of their medication schedule), and public health interventions (e.g., a mobile application to track the spread of an epidemic). *Soft Computing* is an emerging approach to computing which parallels the remarkable ability of the human mind to reason and learn in an environment of uncertainty and imprecision. Soft computing is based on some biologically inspired methodologies such as genetics, evolution, ant's behaviors, particles swarming, human nervous systems. Now, soft computing is the only solution when we do not have any mathematical modeling of problem-solving (i.e., algorithm) and need a solution to a complex problem in real time, and this is easy to adapt with changed scenario and can be implemented with parallel computing. It has enormous applications in many areas such as medical diagnosis, computer vision, hand-written character recondition, pattern recognition, machine intelligence, weather forecasting, network optimization, VLSI design.

The subject choice of particular parts of this edited book was determined through discussions and reflection arising during the International Conference on Cognitive Informatics & Soft Computing (CISC-2017) which was held at VBIT, Hyderabad, from December 20 to 21, 2017. This contributed book enlists selected 74 chapters from 194 presented topics during CISC-2017.

The event was organized with five different tracks, e.g.,
Track-1: Cognitive Informatics
Track-2: Cognitive Computing
Track-3: Computational Intelligence
Track-4: Advanced Computing
Track-5: Hybrid Intelligent Models and Applications

The selected 74 papers from the above tracks are arranged in this proceedings, where the chapters "Multi-tumor Detection and Analysis Based on Advance Region Quantitative Approach of Breast MRI" (U. Ravi Babu) and "Categorizing Kidney Stones Using Region Properties and Pixel Intensity Matrix" (Punal M. Arabi, Gayatri Joshi, Surekha Nigudgi) deal with the image processing techniques for breast cancer and kidney stone characterization. Chapter "Security Measures in Distributed Approach of Cloud Computing" (K. Shirisha Reddy, M. Bala Raju, Ramana Naik) deals with security issues such as mean failure cost (MFC), and multi-dimension failure cost (M^2FC) in cloud computing. Chapter "Performance of Time-Varying Particle Swarm Optimizer to Predict Cancers" (T. R. Vijaya Lakshmi) presents the cancer prediction using particle swarm optimizer. Chapter "An Analytical Review of Different Approaches for Detection and Analysis of Electrocardiographic ST Segment" (Akash Kumar Bhoi, Karma Sonam Sherpa, Bidita Khandelwal, Pradeep Kumar Mallick) presents critical analysis of ST segment. Chapter "Survey on Sentiment Analysis Methods for Reputation Evaluation" (P. Chiranjeevi, D. Teja Santosh, B. Vishnuvardhan) evaluates the reputation by surveying sentiment analysis. Chapter "A Study on the Turnout on the Use of E-Shopping Websites with Arabic Platform in Bahrain" (Yousef A. Baker El-Ebiary, Najeeb Abbas Al-Sammarraie) discusses e-commerce. Chapter "A Novel Trimet Graph Optimization (TGO) Topology for Wireless Networks" (Shanmuk Srinivas Amiripalli, Veeramallu Bobba, Sai Prasad Potharaju), and chapters "Performance Analysis of Reliability and Throughput Using an Improved Receiver—Centric MAC Protocol and Itree-MAC Protocol in Wireless Sensor Networks" (Ananda Kumar K. S., Balakrishna R.), "A Secure and Computational-Efficient Multicast Key Distribution for Wireless Networks" (B. Srinivasa Rao, P. Premchand), and "Data Mining Approaches for Correlation and Cache Replacement in Wireless Ad Hoc Networks" (Y. J. Sudha Rani, M. Seetha) discuss wireless network and protocol. Chapter "An Approach to Detect an Image as a Selfie Using Object Recognition Methods" (Madhuri A. Bhalekar, Mangesh V. Bedekar, Saba Aslam) discusses object recognition methods. Chapter "T Wave Analysis: Potential Marker of Arrhythmia and Ischemia Detection-A Review" (Akash Kumar Bhoi, Karma Sonam Sherpa, Bidita Khandelwal, Pradeep Kumar Mallick) presents a survey on T wave analysis. Chapter "Functional Link Artificial Neural Network (FLANN) Based Design of a Conditional Branch Predictor" (Abhaya Kumar Samal, Pradeep Kumar Mallick, Jitendra Pramanik, and Subhendu Kumar Pani, Rajasekhar Jelli) discusses functional link artificial neural network (FLANN). Chapter "Development of a

Model Recommender System for Agriculture Using Apriori Algorithm" (Santosh Kumar M. B., Kannan Balakrishnan) presents a development model for agriculture. The proceedings chapters (from "Emotion Speech Recognition Based on Adaptive Fractional Deep Belief Network and Reinforcement Learning" to "Steady-State Visual Evoked Potential-Based Real-Time BCI for Smart Appliance Control") focus its application on various applied areas *such as* speech recognition, electronic health record, hybrid CAT, medical images, malware, Freeman K-set model, mammograms, MANET in grid topology, sonar targets, visually impaired, deep neural network, support vector machine, survey on Big Data, IoT-aided robotics devices, relay selection in WBAN, evolutionary ANN, genetic algorithm, multiple matching algorithm, kinship verification, AI used in low-order harmonic elimination, object classification, clustering, hybrid approach for address IP traceback problem, OEFC algorithm, isolated word recognition, keyword extraction, unified power flow controller, Web-enabled Big Data, MAC optimization, epileptic EEG classification, green cloud computing, character extraction from the license plates, ranking decision rules, image segmentation on mobile devices, memetic algorithm, opinion mining, extreme learning machine variants, convolutional neural network, machine learning techniques, intelligent RST robust controller, energy audit, computer-aided diagnosis of epilepsy, cognitive radio networks, optical signal processing module, Parkinson's disease diagnosis, telemedicine, shearlet transform, multi-domain fusion methods, computational studies of CT complexes, smart toilet in healthcare units designing using cognitive approach, cardiac event coherence, and visual evoked potential-based BCI. This book will be of special value to a large variety of professionals, researchers, and students working in the areas of informatics and soft computing.

Special thanks to Honorable Chairman Dr. N. Goutam Rao and Hon. Secretary Dr. G. Manohar Reddy for their valuable support to host CISC-2017 at the Department of Computer Science and Engineering (CSE), Vignana Bharathi Institute of Technology, Hyderabad, Telangana. The editors are thankful to all the editorial board members for their contributions in various ways in the preparation of this volume.

Hyderabad, India   Pradeep Kumar Mallick
Arad, Romania   Valentina Emilia Balas
Rangpo, India   Akash Kumar Bhoi
Uxbridge, UK   Ahmed F. Zobaa

Editorial Board

Organizing Committee

Dr. Sandeep Kumar Satapathy, Department of CSE, VBIT, Hyderabad
Dr. K. Sreenivasa Rao, Department of CSE, VBIT, Hyderabad
Dr. Shruti Mishra, Department of CSE, VBIT, Hyderabad
Mrs. K. Shirisha Reddy, Department of CSE, VBIT, Hyderabad
Mr. G. Arun, Department of CSE, VBIT, Hyderabad
Mrs. N. Swapna, Department of CSE, VBIT, Hyderabad
Mr. P. Naveen Kumar, Department of CSE, VBIT, Hyderabad
Mr. Rajasekhar Jelli, Department of CSE, VBIT, Hyderabad
Mr. Palem Praveen, Department of CSE, VBIT, Hyderabad
Mrs. P. Subhadra, Department of CSE, VBIT, Hyderabad
Mr. D. Kiran Kumar, Department of CSE, VBIT, Hyderabad
Mr. Praveen P., Department of CSE, VBIT, Hyderabad
Mr. V. Sridhar Reddy, Department of IT, VBIT, Hyderabad
Mrs. Yamini Devi, Department of CSE, VBIT, Hyderabad
Ms. N. Indira Priyadarshini, Department of IT, VBIT, Hyderabad
Mrs. S. Bhagyarekha, Department of IT, VBIT, Hyderabad
Mr. N. Venketesh, Department of CSE, VBIT, Hyderabad
Mr. Rakesh Reddy, Department of CSE, VBIT, Hyderabad
Ms. Kumari Jelli, Department of CSE, VBIT, Hyderabad
Mrs. P. Laxmi, Department of CSE, VBIT, Hyderabad
Mrs. P. Swathi, Department of CSE, VBIT, Hyderabad
Mrs. G. Laxmi Deepthi, Department of CSE, VBIT, Hyderabad
Ms. Navaneeta K., Department of CSE, VBIT, Hyderabad
Mrs. M. Kalpana, Department of CSE, VBIT, Hyderabad
Mrs. Adilaxmi, Department of CSE, VBIT, Hyderabad
Mrs. K. Keerthana, Department of CSE, VBIT, Hyderabad
Ms. G. Aruna Jyothi, Department of CSE, VBIT, Hyderabad
Mrs. S. Bhavani, Department of CSE, VBIT, Hyderabad
Mr. D. Srinivas Goud, Department of CSE, VBIT, Hyderabad
Mrs. Sashi Prabha, Department of CSE, VBIT, Hyderabad
Mr. G. Suneel, Department of CSE, VBIT, Hyderabad
Mrs. Chaitanya Sri A. P., Department of CSE, VBIT, Hyderabad
Mr. G. Thirupathi, Department of CSE, VBIT, Hyderabad
Mrs. Sravani, Department of CSE, VBIT, Hyderabad
Mrs. Sowmya, Department of IT, VBIT, Hyderabad
Mrs. Manjulatha, Department of IT, VBIT, Hyderabad
Ms. Meghana, Department of IT, VBIT, Hyderabad
Mrs. Srividya, Department of IT, VBIT, Hyderabad
Mrs. Ambhica, Department of IT, VBIT, Hyderabad
Ms. B. Manga, Department of IT, VBIT, Hyderabad

Contents

About the Editors

Pradeep Kumar Mallick is Professor in the Department of Computer Science and Engineering at Vignana Bharathi Institute of Technology, Hyderabad, India. He has completed his doctorate from Siksha Ó'Anusandhan University, India, in 2016. His area of research includes algorithm design and analysis, and data mining. He has published more than 30 research papers in national and international journals and conference proceedings.

Valentina Emilia Balas is currently Associate Professor in the Department of Automatics and Applied Software at the Faculty of Engineering, "Aurel Vlaicu" University of Arad, Romania. She holds a Ph.D. in applied electronics and telecommunications from Polytechnic University of Timisoara (Romania). She is the author of more than 160 research papers in refereed journals and international conferences. Her research interests include intelligent systems, fuzzy control, soft computing, smart sensors, information fusion, modeling, and simulation. She is Editor-in-Chief of *International Journal of Advanced Intelligence Paradigms* (IJAIP), member in Editorial Board of several national and international journals, and evaluator expert for national and international projects.

Akash Kumar Bhoi completed his B.Tech. (Biomedical Engineering) from the Trident Academy of Technology (TAT), Bhubaneswar, India, and M.Tech. (biomedical instrumentation) from Karunya University, Coimbatore, India, in the years 2009 and 2011, respectively. He is pursuing his Ph.D. (in biomedical signal processing) at Sikkim Manipal University, Sikkim, India. He is currently working as Assistant Professor in the Department of Electrical and Electronics Engineering (EEE) and also Faculty Associate in the R&D Section of Sikkim Manipal Institute of Technology (SMIT), Sikkim Manipal University. He has published book chapters and several papers in national and international journals and conferences.

Ahmed F. Zobaa received his B.Sc. (Hons), M.Sc., and Ph.D. in Electrical Power and Machines from Cairo University, Egypt. Also, he received a Postgraduate Certificate in Academic Practice from the University of Exeter, UK. He was an instructor, a teaching assistant, and an assistant professor at Cairo University from 1992 to 2007. He moved to the UK as a senior lecturer in renewable energy at the University of Exeter in 2007. He was Associate Professor (on leave) at Cairo University from 2008 to 2013. He was promoted to Full Professor (on leave) at Cairo University in 2013. Currently, he is a senior lecturer (associate professor) in power systems, an M.Sc. course director in sustainable electrical power, and a full member of the Institute of Energy Futures at Brunel University London, UK. His main areas of expertise are power quality, (marine) renewable energy, smart grids, energy efficiency, and lighting applications.

Multi-tumor Detection and Analysis Based on Advance Region Quantitative Approach of Breast MRI

U. Ravi Babu

Abstract The proposed Advance Region Quantitative Approach (ARQA) method is used for Breast multi-tumor region segmentation which helps in decease detection and also detects the multi-tumors in different scenarios. The present approach uses the existing preprocessing methods and filters for effectual extraction and analysis of MRI images. The mass regions are well segmented and further classified as malignant disease by computing texture features based on vision gray-level co-occurrence matrices (VGCMs) and logistic regression method. The proposed algorithm is an easy approach for doctors and physicians to provide easy option for medical image analysis.

Keywords MRI · VGCM · ARQA · Multi-tumor · Quantitative approach
Tumor detection

1 Introduction

Breast cancer is the present disease that involves women all over the world after blood cancer [1]. It is report there are more than 1,500,000 new cases of breast cancer internationally including 226,870 new cases in USA [2]. This growth more often than not affects mature women between the ages of 40–49 years. The cause of the disease is unidentified, but one of them is supposed connected to hereditary factors. Early discovery of tumors in the breast area can be performing by mammography breast show through radiology and X-ray pictures that are known as mammograms [3]. This action is safe due to low-dose radiation. The objective of mammography is to detect lumps in the breast, even for a very small to feel myself.

The present paper mainly focuses on the analytical technology of identifying the state of tumors in the breast tissues from MRI Images. Breast Cancer jointly with other breast disease has turn out to be active dreadful deceases in women's health.

U. R. Babu (✉)
DRK College of Engineering and Technology, Hyderabad, TS, India
e-mail: uppu.ravibabu@gmail.com

© Springer Nature Singapore Pte Ltd. 2019
P. K. Mallick et al. (eds.), *Cognitive Informatics and Soft Computing*,
Advances in Intelligent Systems and Computing 768,
https://doi.org/10.1007/978-981-13-0617-4_1

The occurrence of breast cancer in India is on the go up and is fast becoming the cancer in females. Six percentage of deaths in India causes breast cancer. Up to now, there is no effective way to stop the occurrence of breast cancer [4]. As a result, a large figure of mammograms needs to be examined by an in-complete figure of radiologists, resultant in misdiagnoses due to human errors by visual fatigue [1, 3, 5]. Magnetic Resonance Imaging has developed into a widely used technique of high excellence medical imaging. MRI is a higher medical imaging technique as long as rich in order about the human soft tissue anatomy. The quantitative analyses of MRI Breast tumor allows to obtain useful key indicator of disease sequence.

According to appropriate technical research, breast cancer is an extensive field in which scientists can get their research techniques and give new solutions to breast tumor detection problems and they may find a more effective method to detect the breast cancer at a previous stage. The present study mainly focus on image processing for segmentation of breast parenchyma in order to detect and restrict the attendance of doubtful masses, and additional categorization of the masses detected as kind based on the textural in order there in their margins [6, 7].

This can occur when the nearby cancerous area are also other tissues such as the breast tissue breast milk and blood vessel. At mammography performance, the wavy agreement of tissues gives the same reply with hypothetical cancer tissue with the approximately the same image power to the harmful tissue. Due this condition, the present paper put efforts to remove the wavy structure in mammograms so that the area of the theoretical cancer can be more easily detect.

Naga et al. [8] proposed a technique for the finding of tumor in mammographic images that work Gaussian smooth and sub-example operation as preprocessing steps. The mass portion is segmented to establish strength links from the middle portion of masses into the nearby areas. The present study brings a method to analyze leaning flow-like textural in order in mammograms. Skin based on flow orientation in adaptive ribbons of pixels crossways the margins of ample is the future to categorize the region detected as true mass regions or false-positives (FPs). Setiawan et al. [9]: proposed to image preprocessing algorithms for improving the presentation of breast cancer detection algorithm that has been evaluated and their presentation is reported in this paper. Those algorithms are implementing productively in dissimilar series combination, those are (i) strength change before taking away of wavy structure and (ii) strength change after taking away of curvilinear structure. Intensity change task applied thresholding method and taking away of wavy arrangement task practical difficulty process.

To overcome the above disadvantages of existing method the present paper proposes Advance Region Quantitative Approach (ARQA) approach which ensures the overall performance of diseases detection and result comparison of multi-tumor detection in different scenarios.

Breast tumor detection and analysis is not possible in at present scenario. So the mammography is the procedure of using low influence X-rays power to look at the human breast. The goal of mammography is the early detection of breast cancer, normally from adjacent-to-adjacent discovery of characteristic masses and/or micro-calcifications. [7, 9]. The electrical signals can be understood by writing on processor screens, permit more treatment of images to hypothetically allow radiologists to more obviously view the consequences. The present approach helps in medical field to mechanize the assessment of MRI's of Breast to differentiate the tumor from non-tumor values to upgrade the therapeutic medical consideration.

The reminder of this work is organized as follows: Section four describes the Advance Region Quantitative Approach (ARQA) methodology. Section five is related to the description of region of segmentation and section six has a brief description about the noise removal. Section seven describes proposed algorithm and simulation results and analysis of the proposed method is described in section eight and finally, conclusions are included.

2 Proposed Method

The original recognition of the cancerous district in breast cells affected in tumor helps in near the beginning analysis of an unhealthy being which can decrease loss potential. Methods developed for discovery of the spiteful area in mammograms may not be able to give results productively. To conquer this restriction, it is essential to expand a move toward which can part mean regions correctly [6, 10, 11]. This enthused me to work on the difficulty of breast tumor detection in mammogram images which are able to section cancerous region along with detection. Expenditure of time in implementation is also significant to give high-quality results in real time. Having this issue in mind, the present paper expands an important method which first detects the tumor region and then segment the region enclosed by malignant tissues. In this research, the study focuses onto detect the multi-tumor tissues in breast which stands for superior strength values compared to the background in other region of the breast. However, in case of some usual opaque tissues having alike intensities to tumor region, it is essential to detect tumor area exclusive of those region productively. The diagram of the proposed approach is shown in Fig. 1a and functional diagram in shown in Fig. 1b.

(a)   **(b)**

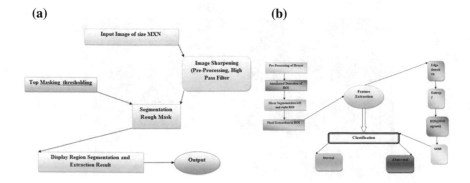

Fig. 1 **a** ARQA method overview, **b** functional diagram of ARQA

The present study concentrated on breast tumor cells detection in Advance Region Quantitative Approach (ARQA) so this proposed a technique counting detection followed by segmentation of images based on advanced image processing techniques which give good results in real time. My method consists of two main steps (1) Advanced detection and (2) Region segmentation. In the detection phase, a vision averaging sifts and thresholding process is applied on original input image which outputs malignant region area. To find the multi-tumor tissues, the present paper creates a rectangular window around the outputted region area and applies Least Max-Mean and Least-Variance technique. In segmentation stage, a tumor scrap is established by morphological final process and image incline method to find the area boundary. The present paper highlights the ensuing region boundary and detects malignant tissues on the original input image [8, 12].

2.1 Implementation AQRA and Analysis from MRI

The accuracy of diagnosis depends on how well the segmentation of the region of interest is done [6, 13, 14]. The functional diagram of the present approach is shown in Fig. 2. There are five main steps for segmentation of ROI in breast thermo grams for asymmetry analysis. The algorithm listed in Algorithm 1.1. The ARQA algorithm consists of four steps: preprocessing, edge detection, segmentation, axis in image, and region of segmentation.

Algorithm 1: ARQA implementation

Step I: **Preprocessing**: Before importing the input image and the pseudo-color pallet matching to each thermo gram into the scheme they are rehabilitated into jpeg arrangement to be process in MATLAB. The first step of preprocessing is background taking away and then resizing the image to remove the unwanted body portion. Using the color pallet a matrix is created matching to the gray value of each pixel in the image [7, 8].

(a)   **(b)**

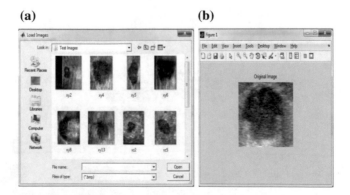

Fig. 2 **a** Image load from database, **b** database input image

Step II: **Canny edge detection**: Detecting the edges of the preprocessed image using canny edge operator. The operator needs two parameters, i.e., threshold and sigma. The present method uses 0.4 as a threshold values and 1 as considered as sigma.

Step III: **Region of Segmentation:** the present paper uses Left and Right border curves by Gradient Operator Region of MRI. The major steps included input image sharpen, identification of the backdrop image, recognition of the objects and show of the results. As it can be noticed in the block diagram, the region of interest is recognized and then extracts using both the original input image and the backdrop texture as a difference mask [1, 10].

Step IV: **Axis in image:** Generally, The body limits are easily detected by finding parabolic curves that shows the lower limits of the breasts but it very difficult. The present paper considers the breast boundaries are usually in parabolic shape. So, curves that can be distinct by the parameter, i.e., the curvature (a), X and Y coordinates of its lowly peak.

Figure 1a shows the interior arrangement of the stage where the backdrop texture is recognized and then extracted. The procedure in this stage starts with the computation of the local entropy of the picture using [9, 12]. This stage has two inputs: the high-pass filtered picture and the gray value of the threshold. The last is an input changeable of the worldwide procedure. According to the first-order indulgence talk about above, a window size of 9×9 was used to calculate the local entropy and then take out the property of the local textures based only on the neighboring (surrounding) neighbors of the present pixel of the image.

The tumor image is perusing from the evidence and preprocessed by evacuate the noise utilize Speckle Noise Removal Method. For more precise output, the mammogram image is upgraded utilizing IEC (Image Enhancement Classifier) algorithm. Also alter the mammogram into gray scale if the contribution image is in RGB color [4, 12].

A small number of methodologies are there for noise reduction [8]. Speckle has negative impact on ultrasound imaging, Radical diminishment fascinatingly strength of mind may be in charge of the poor successful determination of ultrasound as contrasted with MRI. If there should arise an incidence of therapeutic written works, stain noise is otherwise called texture. Widespread model of the speckle [13] is represented by Eq. 1.

$$G(p, q) = f(p, q) * u(p, q) * \pi(p, q), \qquad (1)$$

where, G (p, q) is the observed image, f (p, q) is the multiplicative image part and u (p, q) is the added picture part section of the speckle noise. Here p and q means the axial and similar index of the image sample. For the ultrasound imaging, just multiplicative image part of the noise is to be careful and added image part of the noise is to be disregarded. Thus, mathematical Eqs. 1 and 2 could be altered as represented in Eq. 3.

$$G(p, q) = f(p, q) * u(p, q) + \pi(p, q) - \pi(p, q) \tag{2}$$

Therefore,

$$G(p, q) = f(p, q) * u(p, q) \tag{3}$$

In this research the speckle noise might be evacuated by then, once more select any sub-option of standard filter, wavelet filter and TV filter and so on. Before evacuating the noise, the image size is prolonged and cushioned as piece sort of sub-images, and filter connected, as a result the noise gets empty totally.

In this research, a determination development technique exploits IEC (Image Enhancement Classifier) process is future. By vigorously charitable a priori likelihood delivery right for an exact request surroundings at present carefully, the future method provides a general structure for depiction of good image excellence at the chosen firmness for a large class of image growth procedure [10, 15].

Algorithm 1.1: ARQA Let I be the input tumor image
  I = add_ noise (I, noisy type)
  I = remove _ noise (I, Filter type)
  [m,n] = image_segmentation (I, thresholdvalue)
  KL = Feature_Extraction (I, thresholdvalue)
  Train 1 = SVM – classifier (KL);
  Result = Train1
  Result is tumor or non-tumor segmented a particular boundary image.
  Give MRI image of breast as input and Convert it to gray scale image.
  Apply high-pass strain for image noise removal.
  Apply median filter to improve the quality of image.
  Compute threshold segmentation and Calculate divide segmentation.
  Finally output will be a multi-tumor region

Step 1.
Let x1... X M are N data points in the input image, let k be the number of Region which is given by the user.
Step 2.
Choose R_1.... R_k Region of centers.
Step 3.
Distance between each pixel and each cluster center is found.
Step 4.
The distance function is given by J = | Xi- Rj| for i = 1,..., N and for j = 1,...,k, where |Xk –Rk|, the absolute difference of the detachment among a information tip and the cluster center indicates the distance of the N data points from their respective cluster centers. Distribute the data points x among the k clusters using the relation X ∈ Ri if |x-Rj | < |X-Rk| for i = 1,2 .....R where denotes the set of data points whose Region cluster center.

Repeat the steps initial to Step IV until meeting is met. Following segmentation and discovery of the preferred area, there are chances for miss clustered regions to occur after the segmentation algorithm.

3 Simulation Results

To examine the efficiency of the proposed approach, the algorithm is implemented in MATLAB software and run on different datasets. Tumor detection was gathered from the consequence set. The results of the planned approach are shown in Fig. 3. Performance Evaluation to run a normal image in MATLAB software that takes a minimum of 14 s using PC with core i5 processor and 4 GB RAM. All the database images are individually called for training and testing the detection as well as categorization processes. The entire dataset is divided into 50% of training and 50% of testing categories. Finally ASVM classifier is trained with the training dataset. Then the each testing image is compared with the trained images and classified. This research mainly focuses in presenting the experimental results of the developed system in ASVM training results relative to the memorization and learning of the binary SVM classifier. In order to perform a comparative research, results are obtained after evaluating the developed framework using different machine learning algorithms other than the ARQA [11, 18, 19].

In mammogram segmentation Completeness (CM) and Correctness (CR), CM is the percentage of the GT region, which describes the segmented region, using the following Eq. 4.

$$CM = \frac{TP}{(TP + FN)} \tag{4}$$

CM range from 0 to 1, with 0 representing that none of the region is right partition, and 1 indicating that the entire region was segmented. For example,

(a)   **(b)**   **(c)**

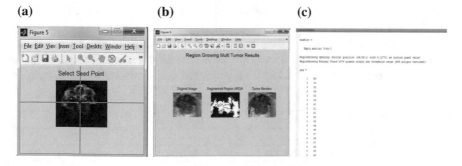

Fig. 3 **a** Seed point in multi-tumor detection **b** region growing result tumor detection **c** values in command window

a value of CM = 0.92 indicates a 92% partly cover with the GT image. Similarly, CR represents the percentage of properly segmented breast region (profile), using Eq. 5.

$$CR = \frac{TP}{(TP + FP)} \tag{5}$$

Similar to CM, the best value for CR is 1 and the smallest amount value is 0. A more universal measure of the mammogram segmentation presentation is achieved by combine CM and CR into a single measure known as excellence (Q), using Eq. 6.

$$Q = \frac{TP}{(TP + FN + FP)} \tag{6}$$

Likewise, the best value for Q is 1 and the least amount value is 0. When commerce with mammograms, it is known that pixels of tumor regions be inclined to have the utmost allowed digital value. Based on this in order, morphological operators such as dilation and erosion are used to notice likely clusters which contain amasses. Image features are then extracted to get rid of clusters that fit into backdrop or normal tissue as a first cut. Features used here comprise region area and peculiarity. The implementation process is shown in Fig. 2.

The Breast tumor location is found out by applying my proposed algorithm using MATLAB Simulator. A GUI (Graphical User Interface) is created to make the system user friendly. Gather the necessary input breast MR image from the database which is shown in Fig. 3 that shows the final clustering of Breast MR image after being process by my algorithm.

Table 1 shows that the principle of Skewness and kurtosis are fairly near in both the breasts for an asymptomatic volunteer, but they are far dissimilar in the contra lateral breasts of the indicative volunteer. Also the entropy principles are less in volunteer 1 as compared to unpaid helper 2 since volunteer 1 has asymmetric tumor patterns. The combined entropy is additional for the asymptomatic with symmetric patterns. The value of AHOG parameter is graphically represented for Volunteer 1 and 2 respectively. X axis indicate the HOG parameter and y axis indicates their values. For the asymptomatic Volunteer 2, the graphical lines for the left and the aright breast nearly agree [16–18].

The Proposed algorithm is implemented in this scenario; the image which ensures the overall multi-tumor cells presents Breast detection. Computer-based breast tumor discovery method has been proposed and analysed in the simulation

Table 1 AHOG (advance histogram gradient value)

Features	Symptomatic	Asymptomatic
Skewness	1.191	0.3409
Kurtosis	9.7398	2.167
Entropy	15.9552	21.1875
Joint entropy	4.3539	8.1141

environment. In Ref. [19], breast tumor cell is detected and secret using Region-based segmentation. Twenty six hundred sets of cell nuclei individuality obtain by applying image analysis techniques to microscopic slides. Method that has been extensively used is the watershed segmentation method [20] joint with other image processing methods. A mathematical set base technique called morphological operations is also used in detecting breast cancer [21]. This method is applied to a mammogram that produces a high contrast image, which can be further enhanced for segmentation steps which leads to an easy identification of cancerous portion. The result is obtained in command window successfully the values which belong to tumor affected cells dependent of mathematical calculation.

4　Summary/Conclusion

The present approach obtained a computational advance to detection of multi-tumor in breast thermo grams on medical MRI imaging. A proposed algorithm Advance Region Quantitative Approach is used for analysis MRI images. Asymmetry psychoanalysis is then performed using AHOG parameters, center calculation, and histogram generation. The investigational consequences show that HOG parameter namely Skewness, kurtosis, entropy, and joint entropy combined with center calculation and histogram psychoanalysis is used for detection of breast tumors. The GUI's created make the system real time and operating system is independent.

References

1. Ramani, R., Suthanthira Vanitha, N., Valarmathy, S.: A comparative study of algorithms for breast cancer detection in mammogram. Eur. J. Sci. Res. **91**(1), 100–111 (2012)
2. Siegel, R., Jemal, A.: Cancer Facts and Figures 2012. Americana Cancer Society Atlanta Georgia (2012)
3. Fassa, L.: Imaging and cancer: a review. Mol. Oncol. **2**, 115–152 (2005)
4. Baca, E., et al.: Mammogram Images, Descriptions and Details. Medical Review Board (2012)
5. Komen, S.G.: Breast Cancer Overview. American Cancer Society (2012)
6. Sheshadri, H.S., Kandaswamy, A.: Detection Breast Cancer based on Morphological Watershed Algorithm. Department of ECE, PSG College of Technology (2005)
7. Guliato, D., Rangayyan, R.M.: Carnielli, W.A., Zuffo, J.A., Desautels, J.E.L.: Segmentation of breast tumors in mammograms by fuzzy region growing. In Proceedings 20th Annul International Conference IEEE Engineering in Medicine and Biology Society, Vol II, pp. 002–1004, 29 Oct–1 Nov 1998
8. Sterns, E.E.: Relation between clinical and mammographic diagnosis of breast problems and the cancer/biopsy rate. Can. J. Surg. **39**(2), 128–132 (1996)
9. Li, H.D., Kallergi, M., Clarke, L.P., Jain, V.K., Clark, R.A.: Markov random afield for tumor detection in digital mammography. IEEE Trans. Med. Imag. **14**, 565–576 (1995)
10. Kopans, D.B.: Breast Imaging. Lippincott and Wilkons (2007)

11. Singh, S., Gupta, P.R.: Breast Cancer detection and Classification using Neural Network. Int. J. Adv. Eng. Sci. Technol. **6**(1), 004–009 (2011)
12. Guliato, D., Rangayyan, R.M., Zuffo, J.A., Desautels, J.E.L.: Detection of breast tumor boundaries using iso-intensity contours and dynamic thresholding. In Proceedings of the 4th International Workshop Digital Mammography, June 1998, pp. 253–260
13. Ponra, D.N., Jenifer, M.E., Poongod, P., Manoharan, J.S.: Morphologic al operations for the mammogram image to increase the contrast for the efficient detection of breast cancer. Eur. J. Sci. Res. **68**(1), 494–505 (2012)
14. Suckling, J., et al.: The mammographic images analysis society digital mammogram database. Exp. Med. Int. Congr. Ser. **1069**, 375–378 (1994)
15. Cerneaz, N.J.: Model-Based Analysis of Mammograms. PhD dissertation, Department of Engineering Science, University of Oxford, Oxford, U.K. (1994)
16. Wang, Lulu: Early diagnosis of breast cancer. Int. J. Sens. **17**(1572), 1–20 (2017)
17. Parmar, P., Thakur, V.: Multiple analysis of brain tumor detection based on FCM. Int. Res. J. Eng. Technol. (IRJET) **4**(7) (2017)
18. Head, J.F., Wang, F.L.: The important role of infrared imaging in breast cancer. IEEE Eng. Med. Biol. (2000)
19. Head, J.F., Elliott, R.L.: Infrared imaging: making progress in fulfilling its medical promise. IEEE Eng. Med. Biol. Mag. **21**, 80–85 (2002)
20. Schaefer, G., Zavisek, M., Nakashima, T.: Thermography based breast cancer analysis using statistical features and fuzzy classification. Elsev. Patt. Recogn. **47**, 1133–1137 (2009)
21. Michael, A., Lorin, M., Aleksey, N., Cheryl, B.: Detection of cancerous breasts by dynamic area thermometry. IEEE Eng. Med. Biol. (2001)

Categorizing Kidney Stones Using Region Properties and Pixel Intensity Matrix

Punal M. Arabi, Gayatri Joshi and Surekha Nigudgi

Abstract Kidney stones or renal calculi are crystals which are formed with in the kidney or in the urinary tract. When there is a decrease in urine volume or if there are more crystalline forming substances in urine, kidney stones are formed. The risk of getting more kidney stones is reduced by finding out the type of kidney stones that helps in identifying the cause for the formation of the stones. In most cases kidney stones larger than 5 mm in size are treated surgically and those lesser than 5 mm in diameter usually pass spontaneously in up to 98% of cases. Kidney stones form in the ureter, bladder, or in urethra. Based on the information obtained from patient history, physical examination, urine analysis, radiographic studies the kidney stones are diagnosed. Kidney stones if small in size are excreted out through urine and if they grow to be larger, they become lodged in the ureter and block the urine flow from that kidney and causes pain. We may need pain medication when there is discomfort. Treatment for the kidney stones is by medication, stone removal by surgery. This paper proposes a novel method for identification of three types of kidney stones namely stag horn, struvite, and calcium type based on Euler number using region properties and contrast which is calculated from pixel intensity matrix. The results obtained show that the Euler number is efficient in identifying calcium stones. Whereas the parameter contrast is calculated using the pixel intensity matrix of the kidney stone images is useful in segregating struvite/stag horn stones. The method proposed show 100% accuracy in the experimental case with limited number of samples; however the accuracy could be confirmed after experimenting the method with huge number of samples.

P. M. Arabi · G. Joshi (✉) · S. Nigudgi
Department of Biomedical Engineering, ACS College of Engineering,
Bangalore, India
e-mail: gayitrijoshi@gmail.com

P. M. Arabi
e-mail: Arabi.punal@gmail.com

S. Nigudgi
e-mail: sursanju@gmail.com

© Springer Nature Singapore Pte Ltd. 2019
P. K. Mallick et al. (eds.), *Cognitive Informatics and Soft Computing*,
Advances in Intelligent Systems and Computing 768,
https://doi.org/10.1007/978-981-13-0617-4_2

Keywords Kidney stones identification · Euler number · Contrast
Pixel intensity matrix · MAT LAB 2012a

1 Introduction

Kidneys remove waste materials from the blood; regulate electrolyte balance and
blood pressure. In addition to these functions kidney stimulate the cells which
produce red blood cells. Kidney stones are formed when urine contains more
crystalline forming substances namely calcium oxlate, uric acid than the fluid
content in it and also the substances that prevent the crystals sticking from together
may be of lesser quantity in urine. Current study shows that 30 million (1 in 11)
people suffer from kidney stone problem at one point or other in their life time and
about 50% of new cases would face a reoccurrence within 5 years of first diagnosis.
The steep rise in obesity, diabetes rate and the climate changes have a strong
influence over the increasing in the number of kidney stone cases. When dietary
minerals in the urine become supersaturated, crystals of urinary stones are formed.
Kidney stones are hard, solid particles that form in the urinary tract; these stones
form in ureter, bladder or in urethra which are known as urethral stones, usually the
kidney stones reached the bladder after passing down the urethra and then ejected
out with the urine but at times they get into the bladder where they grow larger. CT
scans, X-rays, or ultrasound images are used to diagnose kidney stones. Person
suffering from kidney stones experience pain in lower abdomen, in the side and
back, below the ribs and pain on urination; the pain often is accompanied by nausea
and vomiting. The patients may suffer from fever, chills and shivering. Different
types of kidney stones are identified by proper medication, lithotripsy, tunnel
surgery, ureteroscopy but methods like lithotripsy, tunnel surgery, ureteroscopy are
costly and for everyone it is not possible to go for those methods. By using imaging
processing one can easily identify three different types of kidney stones namely stag
horn, struvite, calcium type of stones.

This paper proposes a novel method for identification of three types of kidney
stones namely stag horn, struvite, and calcium type based on Euler number using
region properties and contrast calculated using pixel intensity matrix.

1.1 Literature Survey

Koizami et al. [1], proposed a non-invasive ultrasound the diagnostic system that
tracks movement in an affected area.

Kumar et al. [2], proposed and diagnose kidney stone disease by using three
different neural network algorithms which have different architecture and characteristics. They compared the performance of neural networks on the basis of
accuracy, and time taken to build model. Cunitz et al. [3] proposed a method called

"twinkling Artifact" (TA) under color Doppler Ultrasound and involves that the technique has better specificity than conventional B-mode imaging. Krishna et al. [4] proposed a computer aided automatic detection of abnormality in kidney on the ultrasound system and classified kidney normal and abnormal.

Mane et al. [5], proposed methodology for kidney stone diagnosis using neural network. Viswanath et al. [6], implemented level set segmentation algorithm for identifying location of kidney stones. Attia et al. [7], proposed computer-aided system for automatic classification of Ultrasound Kidney diseases. They considered images of five classes Normal, Cyst, Stone, Tumor and Failure. Devaraj et al. [8], developed a model of a 10 year study of kidney stone patients as herbal respondents and application of herbal therapy in management of the disease. Ebrahimi et al. [9], explored the development of a semi-automated program that used image processing techniques and geometry principles to define the boundary, and segmentation of the kidney area, and to enhance kidney stone detection. Hu et al. [10], discussed X-ray dark-field tomography for differentiation of different kidney stones. The important advantage of this method is its ability to image non-homogeneous kidney stones, i.e., to localize and identify the individual components of mixed-material kidney stones. They used weighted total-variation regularized reconstruction method to compute the ratio of dark-field over absorption signal (DA Ratio) from noisy projections. Wazir et al. [11], introduced filter method for suppression and detection of calculi in sonographic images of the kidney. The statistical features are extracted by decomposing the kidney stone images into different frequency sub-bands using wavelet transform. Asha et al. [12], presented the identification of ultrasound kidney stone images by using Reaction diffusion level set method to detect precise stone location. The proposed system is implemented with Virtex-2 Pro (FPGA) by writing the code in Verilog HDL.

2 Methodology

Figure 1 shows the block diagram of proposed method. Figures 2, 3 and 4 shows X-ray images of stag horn, struvite, calcium kidney stone images are acquired from the Rajarajeswari medical hospital. These RGB images are converted into gray, acquired RGB images are three-channel color images which takes three times as long as processing a grayscale image. Hence RGB to gray conversion is done to speed up the process. The gray images are then preprocessed as shown in Fig. 5. In the preprocessing stage the gray images are filtered by using median filter and enhanced by using contrast stretching.

The preprocessed images are converted into binary for which the Euler number is calculated using region properties; MATLAB version 12 is used for this work and Euler number is tabulated in the Table 1 as shown in results section. If the Euler number is found to be equal to 1 the stone is categorized as calculi stone, If not that might be either struvite or stag horn. Further to segregate stag horn and struvite stones the parameter contrast is found using pixel intensity matrix that is

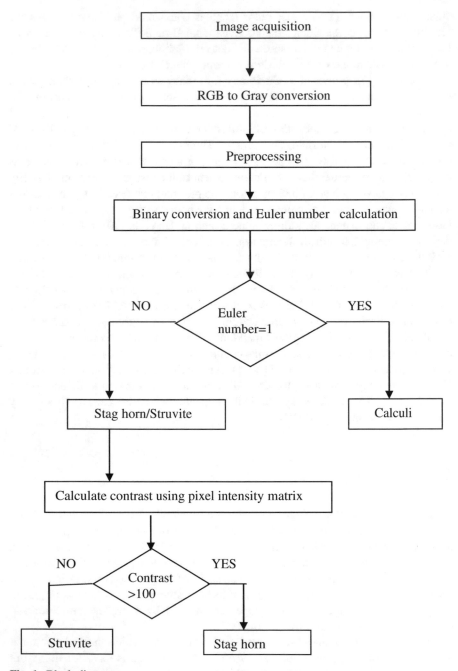

Fig. 1 Block diagram

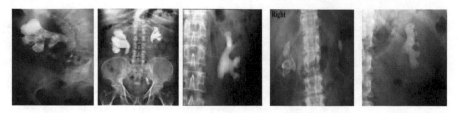

Fig. 2 Stag horn kidney stone images

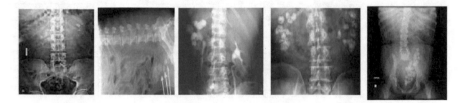

Fig. 3 Struvite kidney stone images

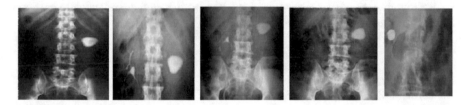

Fig. 4 Calcium kidney stone images

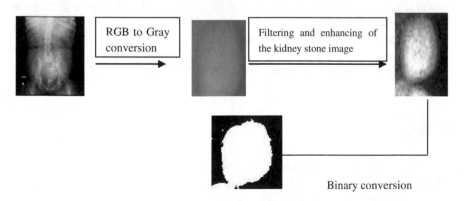

Fig. 5 Sample step of preprocessing of proposed method

Table 1 Euler number of kidney stone images

Sl. No	Stag horn	Struvite	Calcium
1	−5	−2	1
2	−4	−1	1
3	−3	−5	1
4	−6	−1	1
5	−7	−3	1

Table 2 Contrast value of stag horn and struvite kidney stones

Sl. No	Stag horn (contrast)	Struvite (contrast)
1	133	63
2	179	52
3	195	66
4	209	83
5	217	45

tabulated in Table 2. If the value of contrast is found to be greater than 100 the stone is categorized as stag horn; If not that is less than 100 it is labeled as struvite. For this work a set of five X-ray images of stag horn, struvite, and calcium each are taken for analysis.

2.1 Calculation

$$e^{ix} = \cos x + i \sin x, \tag{1}$$

Where e is the base of the natural logarithm, i is the imaginary unit and cos, sin are the trigonometric functions cosine and sine respectively with the argument x is given in radians.

Pixel intensity matrix parameter (contrast)

$$\text{(Maximum pixel intensity value − Minimum pixel intensity value)} \tag{2}$$

3 Results

Figures 2, 3 and 4 show the set of five images of stag horn, struvite, and calcium kidney stone images. Figure 5 shows the sample preprocessing step of struvite kidney stone image. Table 1 shows the Euler number of stag horn, struvite, and calcium kidney stones. Table 2 shows the contrast values of stag horn and struvite kidney stone images.

4 Discussion

A set of five images of each stag horn, a struvite and calcium kidney stone is taken for analysis. The images are resized and converted to gray; Gray images are then preprocessed by using filtering and image is enhanced by contrast stretching. The contrast stretched image is binarised for which the Euler number is found by using Eq. (1). If the Euler is not equal to 1 the image of kidney stone under test is identified as calcium stones. If not the image might be of either struvite or stag horn. To categorize the stones as struvite or stag horn the pixel intensity matrix is found using which the contrast value of image is calculated by using the Eq. (2). The Euler number and contrast values are calculated using mat lab software version R2012a.

Identification of these three types of kidney stones (stag horn, struvite, and calcium) is done based upon their Euler number and contrast. The Euler number is calculated using region properties; from the Table 1 it is seen that, the Euler number is found to be equal to 1 for calcium stones and more than one for struvite and staghorn stones. Hence a decision rule is framed so that if the Euler number for kidney stones is equal to 1 that may be categorized as calcium stone if not it may be of either struvite or staghorn type. If the stone falls into the category struvite/staghorn type the next step of categorization begins.

The contrast value of the stone of interest that may be belonging to struvite/staghorn is calculated using pixel intensity matrix of that stone. The contrast values found for struvite/staghorn stones are shown in Table 2; it is seen in Table 2 that the struvite stones are having contrast value lesser than 100 whereas these values are greater than 100 for staghorn stones. Hence the decision rule to categorize struvite/staghorn stones is framed as if the contrast value is greater than 100 the stone is of stag horn type and if it is lesser than 100 it is struvite.

To categorize the kidney stones as calcium, struvite, and stag horn is framed as if the Euler number of the image is equal to 1 then the stone is of calcium type, if not it is of either stag horn or struvite. Further categorization is done as the contrast value of image is greater than 100 the image is of stag horn, if contrast value of the image of interest is lesser than 100 then it is categorized as struvite. The accuracy of method is experimented with set of 15 images under test.

5 Conclusion

A set of 15 kidney stone images is taken for experimentation to see feasibility of proposed method. Using Euler number and contrast value the stones are categorized. A decision rule is framed as the Euler number of the kidney stone image is equals to 1 then it is categorized as calcium type of kidney stone, if not the kidney stone is categorized as struvite/stag horn. Further to segregate struvite/stag horn the parameter contrast value is calculated using pixel intensity matrix. If the contrast value is found greater than 100 the stone is categorized as stag horn, if the contrast value is lesser than 100 then it is categorized as struvite kidney stone. The results obtained show that the proposed method is suitable for identification of kidney stones (stag horn, struvite, and calcium) by using Euler number and contrast values and found to be 100% accurate in the experimental case with limited samples; However the accuracy of the method can be confirmed only after experimenting with many more images and after clinical studies.

Acknowledgements The authors thank the Management and Principal of ACS College of engineering, Mysore road, Bangalore for permitting and supporting to carry out the research work.

References

1. Koizumi, N., Seo, J., Lee, D., Funamoto, T., Nomiya, A., Yoshinaka, K., Sugita, N.: Robust kidney stone tracking for a non-invasive ultrasound theragnostic system servoing performance and safety enhancement. In: IEEE International Conference on Robotics and Automation Shanghai International Conference Center, pp. 2443–2450 (2011)
2. Kumar, K., Abhishek, B.: Artificial neural networks for diagnosis of kidney stones disease. I.J. Inf. Technol. Comput. Sci. **7**, 20–25 (2012)
3. Cunitz, B., Dunmire, B., Paun, M., Sapozhnikov, O., Kucewicz, J., Hsi, R., Lee, F., Sorensen, M., Harper, J.: Michael Bailey improved detection of kidney stones using an optimized doppler imaging sequence. In: IEEE International Ultrasonics Symposium Proceedings, pp. 452–455 (2014)
4. Krishna, D., Akkala, V., Bharath, R., Rajalakshmi, P., Mohammed, A.M.: FPGA based preliminary CAD for kidney on IoT enabled portable ultrasound imaging system. In: IEEE 16th International Conference on e-Health Networking, Applications and Services (Healthcom), pp. 257–261 (2014)
5. Mane, S.A., Chougule, S.R.: A review on neural network methodology for diagnosis of kidney stone. Int. J. Sci. Res. (IJSR) **4**(11), 300–302 (2015)
6. Viswanath, K., Gunasundari, R., Hussan, S.A.: International conference on intelligent computing, communication and convergence. Procedia Comput. Sci **48**, 612–622 (2015)
7. Attia, M.W., Abou-Chadi, F.E.Z., Moustafa, H.E-D., Mekky, N.: Classification of ultrasound kidney images using PCA and neural networks. Int. J. Adv. Comp. Sci. Appl. **6**(4), 53–57 (2015)
8. Devaraj, R.A., Devaraj, P., Dolly, D.R.J.: Magneto herbal treatment for clearing kidney stones. World J. Pharm. Res. **4**(05), 1272–1280 (2015)
9. Ebrahimi, S., Mariano, V.Y.: Image Quality improvement in Kidney stone detection on computed tomography images. J. Image Graphics **3**(1), 40–46 (2015)
10. Hu, S., Yang, F., Griffa, M., Kaufmann, R., Anton, G., Maier, A., Riess, C.: Towards quantification of kidney stones using X-ray dark-field tomography. In: IEEE 14th International Symposium on Biomedical Imaging (ISBI 2017), pp. 1112–1115 (2017)
11. Wazir: Detection of Kidney stone in MRI images using ideal filter. Int. J. Sci. Res. Manage. **5**(02), 5170–5175 (2017)
12. Asha, K.R., Swamy S.M.: Design and implementation of kidney stone detector using rdlss on FPGA. J. Emerg. Technol. Innovative Res. **4**(05) (2017)

Security Measures in Distributed Approach of Cloud Computing

K. Shirisha Reddy, M. Bala Raju and Ramana Naik

Abstract Cloud computing has given a new approach of data accessing in a distributed manner, where the users has the advantage of higher data accessing feasibility. This approach works on the approach of outsourcing the storage requirements in public provider. The distributed approach has the advantage of low cost data accessing, scalable, location independent and reliable data management. It is observed that Conventional approaches are focused to achieve the objective of reliable and secure data access in cloud computing. However the signaling overhead in these approaches was not explored: in the exchange of control signal in cloud computing, it is needed that lower effort should be made for authentication and data integrity, so as more accessing provision exists. A new monitoring scheme is proposed to minimize the signaling overhead by monitoring record systems. In this paper, security issues also focused. In this process, at each of the data exchange, security risk arises, which are evaluated by different security measure such as Mean failure cost (MFC), and multi-dimensional failure cost (M^2FC). To demonstrate our approach, To develop the suggested objectives, MATLAB interfacing with data feed toolbox was used, effectiveness of the proposed method has been shown by the experimental results.

Keywords Cloud computing · MFC · Distributed approach · MATLAB
Multi-dimensional failure cost

K. Shirisha Reddy
Department of Computer Science and Engineering, JNT University,
Kukatpally, Hyderabad, Telangana 500085, India
e-mail: Shirishakasireddy20@gmail.com

M. Bala Raju (✉)
Department of Computer Science and Engineering, KITE,
JNT University, Kukatpally, Hyderabad, Telangana 500085, India
e-mail: principla@kite.edu.in

R. Naik
Department of Computer Science and Engineering, Farah Engineering College,
Chevella, Hyderabad, Telangana 501503, India
e-mail: ramananaikb@gmail.com

© Springer Nature Singapore Pte Ltd. 2019
P. K. Mallick et al. (eds.), *Cognitive Informatics and Soft Computing*,
Advances in Intelligent Systems and Computing 768,
https://doi.org/10.1007/978-981-13-0617-4_3

1 Introduction

Cloud computing is an rising model of computing, which give computing an open software. It can be characterized as the conveyance of on-request registering belongings via the Internet on restitution for each utilization premise. Resources (as an example, processor computing time and information garage) are provisioned steadily finished the Internet and their subscribers are charged in mild of the uti lization of PC belongings. There are many analogues for Cloud Computing. For ideal, the National Institute of Standards and Technology defines Cloud Computing as "a version which grants handy, on-call for network get right of entry to a shared pool of configurable computing sources (e.g., networks, storage, server, packages, and offerings) that can be swiftly provisioned and released with minimum control attempt or service company interplay". Cloud computing affords its administrations as three layers of administrations that deliver infrastructure assets, application platform, and software program consisting of patron offerings. Infrastructure as a Service (IaaS) gives the essential framework to server registering, getting ready, garage, and structures administration.

The platform as a provider (PaaS) layer shows a layer where clients can send and introduce their packages. Software as a Service (SaaS) offers applications through a web application to a large wide variety of customers without installing on their PCs. Cloud computing provides every one of the blessings of an utility framework as a long way as frugality of scale, adaptability, and accommodation, but raises big problems like lack of manipulate and lack of protection. In any case, as extra information about people and businesses are placed in the cloud, troubles are beginning to expand mainly within the range of safety. Truth be told, the out-sourcing of records clients makes it tough to hold up the trustworthiness and safety of information, and accessibility, which causes proper results. Security is the extensive take a look at in allotted computing systems 3, 4, 5, 6, 7, 8, 9, 10, 18. Truth be informed, as in line with a look at directed by using International Data Group (IDG) [1], Cloud Computing.

2 Methodology

First, we developed a reliable link with minimum link overhand in terms of delay factor and data accuracy. The simultaneous signaling overhead to access a cloud server is optimize by allocating a higher reliable link with lower delay constraint and higher access data accuracy. The operational flow is shown in Fig. 1.

Next phase, over this link, security concern is focused. When a source sends a data it operates in two levels of accessing as shown in Fig. 2. In this process, client send an request for data to a registered cloud server and intern, the cloud server pass the data to host server to fetch the data. In this process, at each of the data exchange,

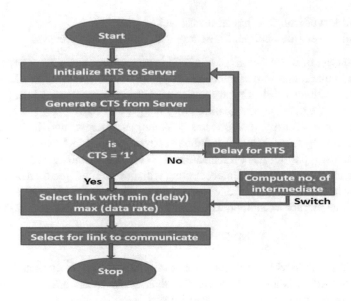

Fig. 1 Flow chart for operational flow

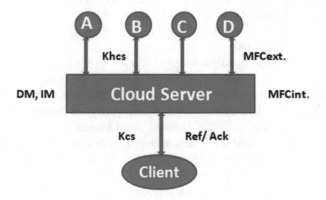

Fig. 2 Network modeling of monitoring system

security risk arises, which are evaluated by different security measure such as Mean failure cost (MFC), and multi dimension failure cost (M^2FC) [2].

During the data exchange, four security cases were observed, which could be an intentional (due to attacker) or unintentional (due to system or power failure) process.

(1) Cloud server giving true ack. but no data accessed.
(2) Cloud server giving false ack. and no data is accessed.

(3) Cloud server true ack. but host false ack.

(4) Cloud server true ack. And host true ack., but no data accessed.

The concept of mean failure cost (MFC) in general is a measure of dependability and in particular a measure of cyber security. The MFC represents to a hypothetical model that evaluates this arbitrary variable regarding commercial loss per unit of working time (e.g., $/h) because of security dangers. The MFC gives a safety effort that relies upon security prerequisites, stakeholder interest, and the architectural component of a system. Actually, it differs as per the partner and considers the change of wagers that a partner has on the satisfaction of every security necessity, the difference in the cost of disappointment starting with one prerequisite then onto the next, the failure from requirement of one segment to another and the change in the effect of failure of one partner to another. The mean failure cost is represented as [1]

$$MFC = ST * DP * IM * PT$$

MFC is a vector with the same number of contributions as there are framework partners and MFCi is an irregular variable that speaks to the cost to the Hi partner that can come about because of a security rupture.

- ST is the wagering grid: a framework where the lines represent the partners, the sections represent to the security necessities, and the ST cell (H, R) is the stake H has in fulfilling the R prerequisite. A wager is a budgetary intrigue that can be lost by an invested individual when R fails. The betting matrix is filled, push by push, by the relating partners.
- DP is the unwavering quality network: a cluster where lines represent to system security necessities, sections represent to framework segments, and DP (R, C) is the likelihood that the system does not meet prerequisite R if part C is submitted. The DP lattice is filled by the framework engineer who knows the part every segment plays in the accomplishment of every necessity.
- IM is the impact matrix: an exhibit where lines are framework segments, segments are security dangers, and IM (C, T) is the likelihood that the C part is bargained if a T danger appears. IM is filled by the check and approval group, who know how the different security dangers compromise segments.
- PT is the threat vector: a vector that has the same number of contributions as dangers in our risk model, and PT (T) is the likelihood that the T. PT danger is brimming with security gear, which knows the setup of the risk likelihood of event of every risk per unit of working time) inside which the framework works. The MFC demonstrate is utilized to evaluate security holes in some genuine word applications, for example, A web based business framework [1] and a Cloud Computing (CC) framework [3–5].

In the MFC coding, to govern the security measure Dependency matrix (DM) and Impact matrix (IM) is suggested. These two matrixes are used for data routing and security access. Where in DM is used for trafficking the data from a reliable Host to Source as per the entry of DM, the IM is used to record the data failure conditions arise.

Wherein DM and IM are used for link selectively and failure observation, there is a need to maintain a security metric for each link to define the trustiness property in cloud computing [6].

In this work to derive a trustiness factor, IM is used. IM consist of all data failure records observed during data exchange. Wherein in the conventional model it is used as a administrative record to define the link reliability, we define a security trust factor called 'multidimensional trust factor' (M^2TF) to improve the security concern in CC.

The M^2TF builds a reputation for each of the link from a source to a host at cloud server using two reputation factors α, and β ($= 1$). Where α defines the positive trust factor, and β is used for negative trust factor. A higher value of α is selected for data exchange on a request.

In this approach, an updation factor "k_i" is used as a reputation updation factor which is set as one for a successful data exchange or a zero for unsuccessful exchange.

$$\alpha := \alpha + k_i$$
$$\beta := \beta + (1 - k_i)$$

from the four cases as observed in the CC,
when,

Case 1   Occurs, K_{cs} security factor of cloud server to client
           (CS) is set 0., $K_{cs} = 0$

Case 2 $K_{cs} = 1$;

Case 3 $K_{ch} = 1$;

Case 4 $K_{ch} = 0$; $K_{cs} = 1$;

At each of the data access request, the two security metrics is observed and a link with ($\alpha > \beta$) is trusted for data exchange.

3   MATLAB Simulation Results and Observation

Figure 3 illustrates the link connections for a cluster network with deployed server and client nodes. The communication links are built on the possible communication range o each unit node. The client node communicates with the host server via registered cloud server to the host server. In this process, the client server request for a link to the host server via a cloud to exchange data between two nodes.

Figure 4 illustrates all possible paths from source to sink, which could be used for communication. The one hope links are used for the data communication based on communication range, a broadcasting of link request is generated, and all the

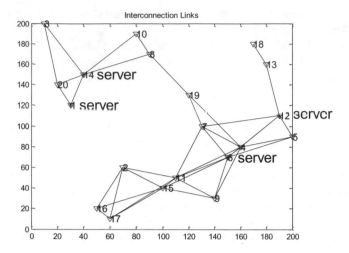

Fig. 3 Interconnection links for the network topology

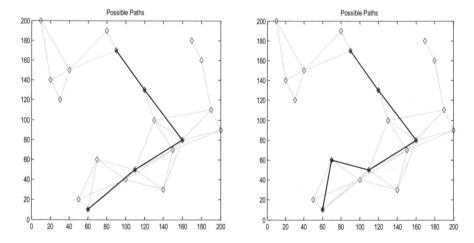

Fig. 4 Possible link paths from source to sink

possible links capable of data exchange acknowledge to offer a communication path for data exchange. In the process of communication, one of the best-fit paths is selected for communication.

The best-fit selected path is illustrated in Fig. 5. The possible paths a optimized based on the suggested weighted link optimization scheme, proposed in our earlier work. This selected path is used in data exchange, where the source node called "client" forward the data from source to sink called "Host" to fetch requested data. In this communication process, single or multiple cloud network servers were used in data exchange.

Fig. 5 Selected path from a source to sink

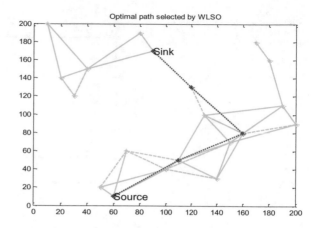

In the selection of this optimal link, the trustiness factor of the selected links is derived. A higher value of positive link forwarding factor (α) is selected, satisfying the condition of ($\alpha > \beta$). The optimization process allocates the server switching to a link with higher reliability which result in faster data exchange. The impact of measuring metric on the trustiness based link switch is as presented in (Fig. 6).

The network overhead is defined as the number of request generated over the number of request been successfully acknowledged. To validate the proposed M2TF approach, a conventional M2CF [2] approach is compared. The overhead is minimized by about 0.7% for the proposed approach, due to the faster data exchange as carried out on a reliable communication path. The initial overhead in this case is observed to be lower to a value of 2.5, however in the increase in communication iteration, the overhead gets increase due to repetitive link contention for the failure paths. This repetitive contention leads to higher overhead in the network, whereas a faster clearance due to reliable path leads to lower in overhead (Fig. 7).

The network throughput observed for the two methods is illustrated in above figure. It is observed that, the throughput of the proposed approach is increased by 9% as compared to the conventional M2CF coding. In the proposed approach, the

Fig. 6 Network overhead plot

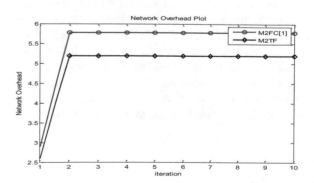

Fig. 7 Network throughput plot

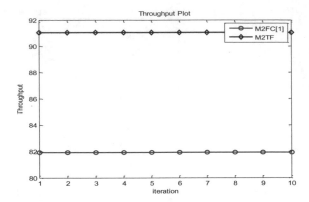

Fig. 8 End to End delay

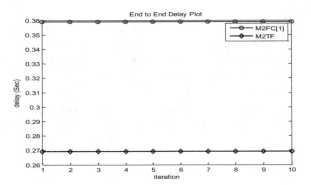

path with highest trust factor is chosen, which leads to lower in delivery failure and improves the network throughput. The two methods retains a steady throughput with course of communication iteration, as the link reliability are defined with observed successful data exchange, and links with M2TF approaches are observed to be more reliable compared to the conventional M2CF [2] approach (Fig. 8).

The link reliability directly impacts on the communication delay. The end to end delay is a measuring unit to define the delay factor in data exchange. This delay is defined as the total time taken for data exchange from a source to sink. The delay factor in the proposed approach is observed to be 0.09 s lower in comparison to the conventional M2FC approach. The delay minimization in due to a lower failure rate for the proposed approach due to higher link reliability. A similar case analysis is carried to observe the impact of higher client density in the network. The increase in number of client node increases the contention probability which effects the measuring metrics. The observed parameters are as illustrated below. The network density in this case is taken for 45 units (Figs. 9, 10, 11, 12, 13 and 14).

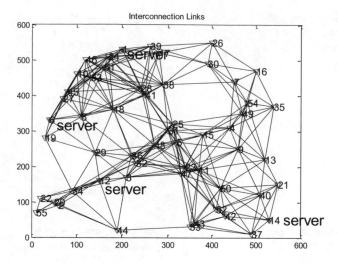

Fig. 9 Network layout with 45 communicating units

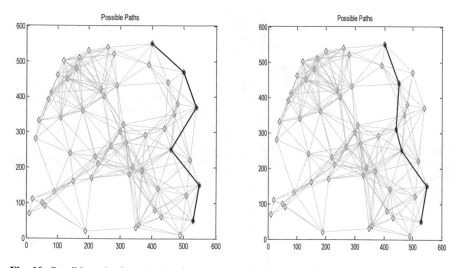

Fig. 10 Possible paths for selection from a source client to host sink

Fig. 11 Optimal path with
highest trustiness factor for
data exchange

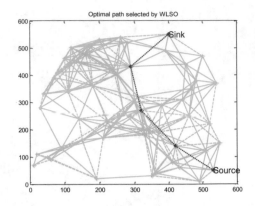

Fig. 12 Network overhead
plot for the developed
approaches at network
density = 45

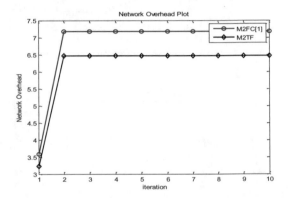

Fig. 13 Network throughput
plot for the developed
approaches at network
density = 45

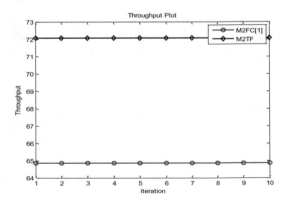

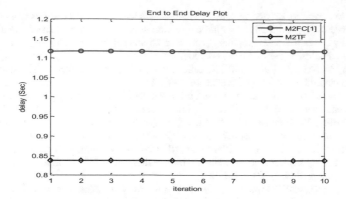

Fig. 14 End to end delay plot for the developed approaches at network density = 45

4 Conclusion

Sharing data in cloud when the cloud service provider is mistrusted is an issue. Risk assessment is an imperative system in Information Security Management. Enterprise has to adopt systematic and well-designed process for determining information security risks to its properties. However, we illustrated some approaches that protect data seen by the cloud service provider while getting shared among many users. The end to end delay is a measuring unit to define the delay factor in data exchange. This delay is defined as the total time taken for data exchange from a source to sink. The delay factor in the proposed approach is observed to be 0.09 s lower in comparison to the conventional M2FC approach. The delay minimization in due to a lower failure rate for the proposed approach due to higher link reliability.

Acknowledegements I thank for the facilities provided by VBIT-FIST R&D for my research and implementation.

References

1. Jouini, M., Ben Arfa Rabai, L., Ben Aissa, A., Mili, A.: Towards quantitative measures of Information Security: A Cloud Computing case study. Int. J. Cyber-Security Digital Forensics (IJCSDF) **1**(3), 265–279 (2012)
2. Jouini, M., Rabai, L.B.A.: Comparative study of information security risk assessment models for cloud computing systems, the 6th international symposium on frontiers in ambient and mobile systems (FAMS 2016). Procedia Comput Sci **83**, 1084–1089 (2016)
3. NIST. An introduction to computer security: The NIST handbook. Technical report, National Institute of Standards and Technology (NIST), Special Publication 800–12, Gaithersburg, MD: NIST (2012)

4. Hale, M., Gamble, R.: SecAgreement: Advancing security risk calculations in cloud services. In: Proceedings of 8th IEEE world congress on servicesmutum zico meetei and anita goel, 'security issues in cloud computing', 5th international conference on bio-medical engineering and informatics, 978-1-4673-1184-7 (2012)
5. Emam, A.H.M.: Additional authentication and authorization using registered email-id for cloud computing. Int J Soft Comput Eng **3**(2), 110–113 (2013)
6. Aldossary, S., Allen, W.: Data security, privacy, availability and integrity in cloud computing: issues and current solutions (IJACSA). Int. J. Adv. Comput. Sci. Appl. **7**(4) (2016)
7. Demchenko, Y., Gommans, L., De Laat, C.: Web Services and grid security vulnerabilities and threats analysis and model. In: Bas Oudenaarde, advanced internet research group, University of Amsterdam, Kruislaan 403, NL-1098 SJ Amsterdam, The Netherlands (2000)
8. Amazon.com, -Amazon s3 availability event: July 20, 2008, Online at http://status.aws.amazon.com/s3-20080720.html (2008)
9. ISO/IEC 27005:2007 Information technology-security techniques-information security risk management, Int'l Org. Standardization (2007)
10. Boehme, R., Nowey, T.: Economic security metrics. In: Irene, E., Felix, F., Ralf, R. (eds.) Dependability metrics, vol. 4909, pp. 176–187 (2008)
11. Wang, J.A., Xia, M., Zhang, F.: Metrics for information security vulnerabilities. Proc. Intellect. Base Int. Consortium **1**, 284–294 (2009)
12. Saripalli, P., Walters, B.: QUIRC: a quantitative impact and risk assessment framework for cloud security. In: Proceedings of the IEEE 3rd international conference on cloud computing, pp. 280–288 (2009)
13. Mell, P., Grance, T.: Effectively and securely using the cloud computing paradigm. ACM Cloud Comput. Secur. Workshop (2009)
14. Miller, E.H.: A multiple-replica remote data possession checking protocol with public verifiability-2010 second international symposium on data, privacy, and e-commerce., -a note on reflector arrays (Periodical style—Accepted for publication),‖ IEEE Trans. Antennas Propagat., to be published
15. Ben Aissa, A., Abercrombie, R.K., Sheldon, F.T., Mili, A.: Quantifying security threats and their potential impact: a case study. Innovations Syst. Softw. Eng **6**(1), 269–281 (2010)
16. Ben Arfa Rabai, L., Jouini, M., Ben Aissa, A., Mili, A.: A cybersecurity model in cloud computing environments. J King Saud Univ. Comput. Inform. Sci. (2013)
17. Jouini, M., Ben Arfa Rabai, L.: A security risk management metric for cloud computing systems. Int. J. Organ. Collect. Intell. (IJOCI) **4**(3), 1–21 (2014)
18. IDG cloud computing survey, cloud continues to transform business landscape as CIOs explore new areas for hosting, http://www.idgenterprise.com/news/press-release/cloud-continues-to-transform-business-landscape-as-cios-explore-new-areas-for-hosting/ (2014). Accessed 14 Jan 2016
19. Jouini, M., Ben Arfa Rabai, L.: Surveying and analyzing security problems in cloud computing environments. The 10th international conference on computational intelligence and security (CIS 2014), pp. 689–493 (2014)
20. Jouini, M., Ben Arfa Rabai, L.: Mean failure cost extension model towards a security threats assessment: a cloud computing case study. J. Comput. (JCP) **10**(3):184–194 (2015)
21. Jouini, M., Ben Arfa Rabai, L., Ben Aissa, A.: Classification of security threats in information systems. ANT/SEIT **32**:489–496 (2014)
22. Jouini, M., Ben Arfa Rabai, L., Khedri, R.A.: Multidimensional approach towards a quantitative assessment of security threats. ANT/SEIT, pp. 507–514 (2015)
23. Cloud security defence to protect cloud computing against HTTP-DoS and XML-DoS attacks, Ashley Chonka,YangXiang n, WanleiZhou, AlessioBonti, Elsevier

Performance of Time-Varying Particle Swarm Optimizer to Predict Cancers

T. R. Vijaya Lakshmi

Abstract Classification of tumors is a challenging task in the field of bioinformatics. The gene expression levels measured using microarray approach contains thousands of levels. Finding optimum number of genes expression levels to classify tumor samples is carried out in this paper using PSO. The conventional PSO algorithm works with constant social and cognitive coefficients. This paper proposes time-varying PSO in which the social and cognitive coefficients are allowed to vary with respect to time. The performance of the proposed particle swarm optimizer gives better results when compared to the conventional PSO in classifying the tumor samples.

Keywords Particle swarm optimization · Time-varying coefficients
Tumor classification · Gene expression levels

1 Introduction

Identification of tumors at the earlier stage is a crucial task [1, 2]. Most of the citizens face various types of cancers. Due to the abnormal growth of cells, tumors are formed. The different types of cancers/tumors considered in the current study are DLBCL outcome, Prostate, CNS, and Colon Tumor. The raw gene expression data is collected from Kent Ridge Biomedical Data Repository [3] for conducting experiments. These expression levels were measured using microarray method [1, 2]. Each dataset contains thousands of levels.

Identifying the gene levels from thousands of levels is a complex task. Reducing the number of dimensions/levels have advantages like improved predictive accuracy, computational cost is low, etc. [4, 5].

In conventional PSO the next generation population is generated by computing the fitness measure of all the particles in the swarm. The social and cognitive

T. R. Vijaya Lakshmi (✉)
MGIT, Gandipet, Hyderabad, India
e-mail: vijaya.chintala@gmail.com

© Springer Nature Singapore Pte Ltd. 2019
P. K. Mallick et al. (eds.), *Cognitive Informatics and Soft Computing*,
Advances in Intelligent Systems and Computing 768,
https://doi.org/10.1007/978-981-13-0617-4_4

coefficients are set to a constant. With this there is a tendency that the algorithm converges either faster or slower. To avoid this in the proposed time-varying PSO the social and cognitive coefficients are allowed to change according to time/ generation number.

2 Particle Swarm Optimization

The standard PSO is developed by Eberhart and Kennedy. Let the randomly generated initial population/swarm for PSO be POP_A. The initial swarm of the particles is uniformly distributed in the range $[L_b \; U_b]$, where the lower bound (L_b) and the upper bound (U_b) (gene expression levels).

The flowchart of Particle Swarm Optimization is shown in Fig. 1. Every particle in the swarm is initialized with random velocities and positions for the first generation. The classification fitness error is computed using k-NN classifier for every particle's position. Let the fitness of the particle $X_{j,g}$ be denoted by $f(X_j,g)$. The particle's previous best value, X_{best}, for the first generation is itself. Let the best value among the group/swarm be G_{best}.

The velocities of the particles for the next generation are updated by considering the particle's values. The velocities of the particles are updated using Eq. (1) with linearly varying weight.

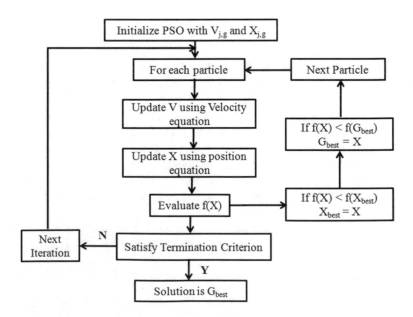

Fig. 1 PSO flowchart to find optimum subset of features

Table 1 Optimum values set for PSO parameters

Parameter	Optimum value
ω^1	0.4
ω^2	0.9
cog_1	2
soc_2	2

$$Vel_{i,g+1} = \omega \times Vel_{i,g} + \text{cog}_1 \times \Psi_1 \times (A_{best_i} - A_{i,g}) + \text{soc}_2 \times \Psi_2 (G_{best} - A_{i,g}),$$

$$(1)$$

where ψ_1 and ψ_2 are the two random numbers generated in the range [0 1]. ω is defined in terms of initial and final inertia weights ω^1 and ω^2, respectively as depicted in Eq. (2).

$$\omega = (\omega^1 - \omega^2) \times \frac{g^{max} - g}{g^{max}} \qquad (2)$$

The optimum values set for ω^1, ω^2, cog_1 and soc_2 are tabulated in Table 1.

The position of the particle is updated based on its current position and velocity, as depicted in Eq. (3).

$$A_{i,g+1} = A_{i,g} + Vel_{i,g+1} \qquad (3)$$

With new positions ($A_{i,g+1}$) and velocities ($Vel_{i,g+1}$), the particle compete in the search domain to achieve the best solution. For the new positions of the particle, the fitness errors are again computed. The previous best (A_{best}) and group best (G_{best}) values are updated based on the fitness error. These optimum features are taken from G_{best} positions of the particle.

3 Proposed Time-Varying PSO

In PSO, the cognitive (cog_1) and social (soc_2) acceleration coefficients are constant. However, in time-varying/self-organizing PSO (TVPSO), these coefficients are allowed to vary with respect to time/generations. The main idea behind this is to improve cognitive learning and to enhance global convergence during the later iterations of the search (to improve social learning). The time-varying acceleration coefficients are as follows:

$$\text{cog}_1(g) = 2.5 - 2 \times \frac{g}{g^{max}} \qquad (4)$$

$$soc_2(g) = 0.5 + 2 \times \frac{g}{g^{max}} \tag{5}$$

In the early iterations cog_1 is high and soc_2 is low to exploit the local search and during later iterations cog_1 is low and soc_2 is high to explore the global search.

4 Experimental Results

The proposed time-varying PSO is tested on various medical datasets. The gene expression (tumor) datasets used in the current work are described in Table 2. The four datasets used are DLBCL outcome, CNS, Prostrate and Colon Tumor.

The common control parameters set in the current study for the proposed time-varying PSO and conventional PSO techniques are tabulated in Table 3. The size of the initial population is set to 40 in the current work. The maximum generations in the algorithms is set to 200 (stopping condition). Each particle in the swarm/population is evaluated by computing classification error using k-NN classifier (fitness measure). The desired number of features/gene levels is allowed to vary in the simulations in steps of 10. Tests are conducted for a maximum subset size of 200 features/gene levels for each dataset.

The simulation results obtained with both proposed and conventional PSO are shown in Figs. 2, 3, 4 and 5 for the tumor datasets described in Table 2. It is observed from these results that the proposed Time-varying PSO performed well to classify the cancer samples when compared to the conventional PSO. Even with less number of gene levels the cancer samples are identified.

Table 2 Details of the cancer datasets

Dataset	No. of gene levels	Total samples	No. of normal samples	No. of abnormal samples
DLBCL	7129	58	32	26
Prostrate	12600	136	59	77
CNS	7129	60	21	39
Colon tumor	2000	62	22	40

Table 3 Classification rates obtained using various transformation techniques

Parameter	Value
Size of initial population (NP)	40
No. of generations (g^{max})	50
Lower bound (L_b)	1

Fig. 2 Optimum gene expression levels obtained for DLBCL (diffuse large B-cell Lymphoma)

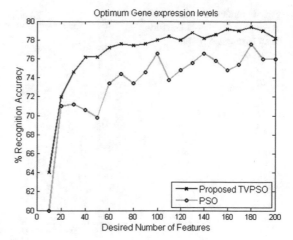

Fig. 3 Optimum gene expression levels obtained for prostate tumor

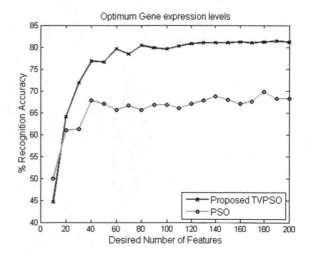

Fig. 4 Optimum gene
expression levels obtained for
CNS (Central Nervous
System)

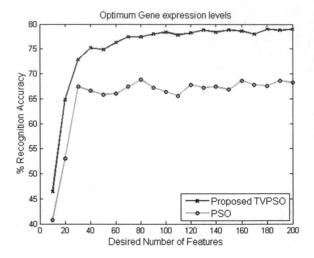

Fig. 5 Optimum gene
expression levels obtained for
colon tumor

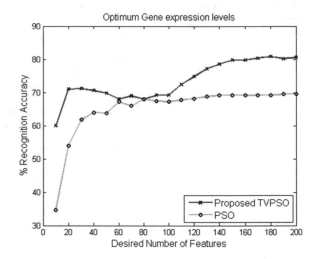

5 Conclusion and Future Scope

Detecting tumors/cancers with less number of gene expression levels is a crucial
task. In conventional PSO the social and cognitive coefficients are set to constant
values. This leads the algorithm to either converge faster or slower. To avoid this
time-varying PSO is presented in this paper. In time-varying PSO the social and
cognitive coefficients are allowed to change with respect to time. This improves the
searching capability. The performance of this algorithm is tested on various tumor

datasets to identify the tumor samples. The proposed time-varying PSO gave better results when compared to the conventional PSO for all the medical datasets considered. In future even more advanced algorithms can be used to classify the cancer samples.

References

1. Gunavathi, C., Premalatha, K.: A comparative analysis of swarm intelligence techniques for feature selection in cancer classification. Sci. World J. **2014**, Article ID 693831, http://dx.doi.org/10.1155/2014/693831 (2014)
2. Alonso, C.G.J., Moro-Sancho, I.Q., Simon-Hurtado, A., Varela-Arrabal, R.: Microarray gene expression classification with few genes: Criteria to combine attribute selection and classification methods. Expert Syst. Appl. **39**, 7270–7280 (2012)
3. http://leo.ugr.es/elvira/DBCRepository/index.html
4. Vijaya Lakshmi, T.R., Sastry, P.N., Rajinikanth, T.V.: Feature optimization to recognize Telugu handwritten characters by implementing DE and PSO techniques. In: International conference on FICTA, Springer, Odisha, India, pp. 397–405 (2016)
5. Vijaya Lakshmi, T.R., Sastry, P.N., Rajinikanth, T.V.: Feature selection to recognize text from palm leaf manuscripts. Signal, Image and Video Processing. Springer, Article in press, Berlin (2017). https://doi.org/10.1007/s11760-017-1149-9

An Analytical Review of Different Approaches for Detection and Analysis of Electrocardiographic ST Segment

Akash Kumar Bhoi, Karma Sonam Sherpa, Bidita Khandelwal and Pradeep Kumar Mallick

Abstract ST segment analysis is significantly a critical non-invasive pointer to identify ischemic unsettling influences. It has been considered from the start of the electrocardiography on the ground that critical cardiovascular conditions are one of the fundamental drivers of death on the planet and ischemic aggravations are a standout amongst the most vital heart conditions. Most basic Sudden Cardiac Deaths (SCDs) are triggered by Coronary heart disease (CHD). This study is for promising investigation for ST segment discovery and determination of Coronary heart disease. Different presented techniques and algorithms are concisely discussed on ST segment detection and analysis of this segment for possible identification of cardiovascular abnormalities.

Keywords Electrocardiography · ST segment · Ischemia · Arrhythmia
Coronary heart disease

A. K. Bhoi (✉) · K. S. Sherpa
Department of Electrical and Electronics Engineering, Sikkim Manipal Institute
of Technology (SMIT), Sikkim Manipal University, Gangtok, India
e-mail: akash.b@smit.smu.edu.in; akash730@gmail.com

K. S. Sherpa
e-mail: karma.sherpa@smit.smu.edu.in; karmasherpa23@gmail.com

B. Khandelwal
Department General Medicine, Central Referral Hospital and SMIMS,
Sikkim Manipal University, Gangtok, India
e-mail: drbidita@gmail.com

P. K. Mallick
Department of Computer Science and Engineering, Vignana Bharathi Institute
of Technology (VBIT), Hyderabad, India
e-mail: pradeepmallick84@gmail.com

© Springer Nature Singapore Pte Ltd. 2019   39
P. K. Mallick et al. (eds.), *Cognitive Informatics and Soft Computing*,
Advances in Intelligent Systems and Computing 768,
https://doi.org/10.1007/978-981-13-0617-4_5

1 Introduction

Ischemia is intricacy of the severity of heart and a foremost reason for the initiation of cardiac dysfunction and life threatening cardiovascular arrhythmias [1]. The depolarization of cellular resting membrane potential indicates the cellular level ionic imbalances due to ischemia. This leads to alteration in the potential of normal and ischemic tissue and initiated "injury current" [2]. Based on the anatomical position of heart and dipoles location by concerning the recording electrodes, the "injury current" is substantiated in the ECG either by ST elevation or depression [3].

ST segment carries low frequency and low amplitude signal content. ST segment measurement would be difficult while baseline wanders and muscle noise introduced. Reliable ST measurements (Fig. 1) could be possible with signal averaging technique as it does not changes rapidly [4].

Figure 2 shows the treatment strategy for a suspected patient with Acute Coronary Syndrome (ACS) by analyzing the morphological changes in ST segment [5].

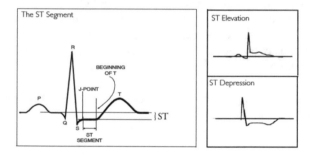

Fig. 1 ST segment measurement [4]

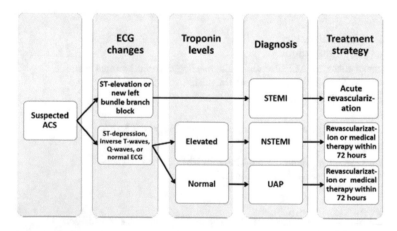

Fig. 2 Patient diagnosed and treatment strategy with acute chest pain (*ACS* Acute Coronary Syndrome, *STEMI* ST Elevation MI, *NSTEMI* Non ST Elevation MI, and *UAP* Unstable Angina Pectoris) [5]

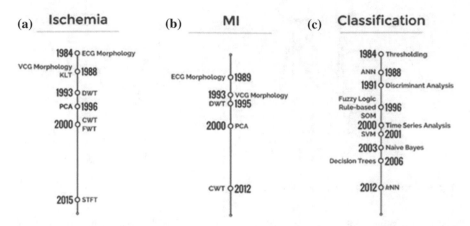

Fig. 3 Presented by Ansari et al. [5]

Figure 3 shows the timeline about the evolution of popular automatic detection approaches, feature extraction methods and classification techniques for ischemia and Myocardial Ischemic (MI) groups [6–17]. Ischemia then again, is an intense malady, which must be distinguished promptly, particularly in Critical Care Unit (CCU) situations. Various techniques have been proposed in the writing for ST detection in light of advanced sifting, time examination of the flag first subordinate, and syntactic strategies [18–20]. The ST segment examination is broadly utilized as a part of stress measurement as the investigation parameters can be identified as condition for Coronary Artery Disease (CSD) [21]. The ST segment investigation for the ambulatory ECG is a moderately new system that can generate more diagnostic parametric data for myocardial ischemia [22]. Steven K. White et al. have exhibited in their arbitrary investigation that Remote Ischemic Molding (RIC) started preceding essential Primary Percutaneous Coronary Intercession (PPCI) Myocardial Infarct (MI) estimate, decreased myocardial edema and expanded myocardial rescue in patients giving ST segment rise myocardial dead tissue [23]. Ramun Schmid et al. displayed a technique for remedying ST segment estimations based of AC coupled ECG which additionally helps in exact discovery of ST fragment for intense ischemia recognizable proof [24]. Template boundary algorithm was to determine the ST-T segment (repolarization) using quantitative measures [25].

Numerous calculations have been produced to analyze the ST segment [26, 27], however, a standard strategy still to be formulated and specialists are searching for alternative improvised arrangements consistently. The primary issues to be considered, such as: (i) Positioning and estimating the baseline, as ST deviation or elevation could be measured by taking baseline as reference, (ii) The QRS complex end point (offset) identification as QRS width is utilized to make the unpredictable grouping. Additionally, ST incline/slope could be computed with the help of J point (Fig. 4). (iii) ST estimations over the T wave can be seen frequently during increased heart rate. A strategy to examine the ST segment will help in avoiding

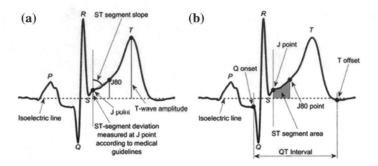

Fig. 4 Importance of J point and J80 point [5]

these challenges and lead to precise ST measurement [28]. Franc Jager et al. developed a robust Karhunen–Loeve Transform (KLT) algorithm which detects the ischemic ST episodes along with measures the duration of ischemia for ambulatory ECG [29]. ST segment is asymmetric in nature and Nonlinear Principal Component Analysis (NLPCA) approach is implemented for feature extractions, which classifies into normal, ST+ , ST− and artifact classes [30]. Multi-resolution wavelet approach for ST segment online analysis has been implemented along with the Digital Signal Processing (DSP) based processor (TMS320C25) [18]. A novel approach with filtered RMS difference series was implemented to detect the changes in repolarization phase (ST-T complex) [31]. Bulusu S.C et al. proposed beat classification method for transient ST Segment episode detection using the application of artificial intelligence and signal processing [32]. Gioia Turitto, Nabil El-Sherif have discussed in their study concluded that the ST alternans (Fig. 5) is an indication towards the severity of ischemia and antecedent for ventricular arrhythmias [33].

Several techniques have been developed and measurements of diagnostic parameters are taken care. One of the crucial approaches is by measuring the deviation or distance of a specific point (fiducial point) on the ST segment from the benchmark (baseline). Notwithstanding, the accompanying three techniques have

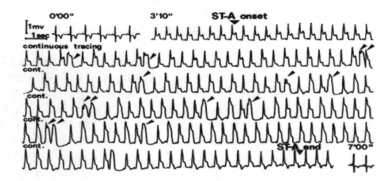

Fig. 5 Variant angina showing ST alternans and ventricular arrhythmias [33]

impacted: (1) J_{point} + x, 2) R ± x, and 3) Search windowed [34]. Few well-known popular techniques for ST segment detection and analysis being brought out and discussed in the following sections.

2 Presented Methodologies and Analysis

2.1 Approximation Method

E. Skordalakis has demonstrated how to recognize the shape of ST segment using approximation method. After applying this method on ST segment, its gives (1) the localization (i.e., onset and offset of ST segment) and (2) the suitable approximation equations for ST segment, i.e., either the straight line or parabola [35].

The initial two requirements are fulfilled if the auxiliary segment is chosen (Fig. 6a) which starts at the peak of the S wave. If the S wave is missing (Fig. 6b), then it will be chosen at the peak of the R wave. This ends at T $_{max}$, i.e., the peak of T wave [36].

2.2 Spectral Analysis

Repolarization Alternans (RPA) demonstrates substitute beat variances in the fleeting or spatial characteristics of the echocardiogram (ECG) STU segment which may speak to scattering in repolarization. Phantom decay has uncovered microvolt-level RPA which has been found to connect with Ventricular Tachycardia (VT) and fibrillation, and is progressively being utilized for clinical hazard stratification. In any case, while interferences in periodicity are known to influence phantom deterioration,

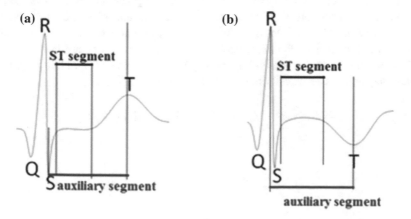

Fig. 6 Illustrations of the auxiliary segment [35]

their quantitative effect on RPA and its clinical utility have been ineffectively depicted. Sanjiv M. Narayan et al. along these lines studied the impact of variable arrangement, additional systoles, different beats and beat rejection on RPA greatness in reproductions and on the affectability and specificity of RPA for VT in a pilot clinical study. RPA size was dazzlingly delicate to QRS alignment with the end goal that ±1 ms irregular beat misalignment diminished it by 68% in recreations. Correspondingly, problematic QRS arrangement in clinical ECG's made the affectability of RPA for inducible VT tumble from 93% to as low as 63%; while JT arrangement was additionally less powerful for RPA recuperation. As an analysis in limiting morphometric inconsistencies in clinical ECG's, Sanjiv M. Narayan et al. discovered that RPA extent really fell when supplanting either quantifiably divergent or ectopic beats with more illustrative beats. Moreover, embeddings or erasing beats additionally diminished RPA extent in clinical groupings and recreations. These statistical investigations suggest that the accuracy of beat arrangement and interferences to ECG periodicity, which may happen physiologically, may incredibly decrease the clinical utility of RPA for VT. Dynamic changes in RPA because of grouping anomalies require additional study before RPA might be ideally connected to screen for ventricular arrhythmias [37].

2.3 Wavelet Transform

Automatic ST-T complex evaluation of the ECG signals was investigated. First, an adaptive filter of wavelet structure was utilized to expel the pattern meandering (baseline wander) of the ECG signals, which was fundamentally imperative for ST segment investigation. At that point, taking favorable circumstances of the numerous determination capacities of the wavelet transform, and strategy was created to recognize the ST segment fiducial purposes of the ECG signals at various wavelet decomposition scales and frequency bands. The proposed strategies were tested utilizing the standard MIT/BIH ECG ST segment database. The fiducial focuses recognizable proof outcomes were contrasted and those gotten physically by the accomplished cardiologists. This correlation demonstrated a decent coordinating, which suggested the dependability of the proposed strategy [38].

Wavelet transform is a straight operation that deteriorates a signal into segments at various scales. The wavelet transform can be characterized as

$$WT_a x(t) = \frac{1}{a} \int x(\tau) \psi\left(\frac{t-\tau}{a}\right) d\tau = x(t) * \psi_a(t), \tag{1}$$

where,

$$\psi_a(t) = \frac{1}{a} \psi\left(\frac{t}{a}\right) \tag{2}$$

$\psi_a(t)$ is a wavelet function, which has successfully restricted term and a normal estimation of zero and an is the wavelet transform scale [38].

2.4 Hidden Markov Model (HMM)

This model utilizes the Ambulatory ECG Monitoring (AECG) data which is derived from European ST-T database [39] where 48 ECG recording containing two channels each at 250 Hz sampling rate is considered for evaluation. The proposed methodology (Fig. 8) is comprises of two major parts, i.e., detection and segmentation of online beat (Fig. 7 showing all possible transitions) and ischemic episode detection (Fig. 8) [40, 41, 42].

Fig. 7 Beat segmentation model [42]

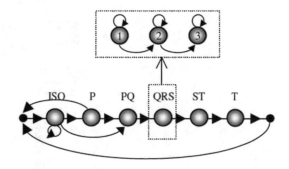

Fig. 8 Block diagram of Hidden Markov Model (HMM) [42]

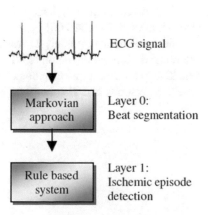

2.5   Time-Frequency Method

Here, time-frequency based approach has been implemented to analyze Heart Rate Variability (HRV) for the possible detection or identification of ischemic episodes and their classification [43]. The RR intervals were considered to be the feature for HRV analysis using Smoothed Pseudo Wigner Ville Distribution (SPWVD).

2.5.1   Smoothed Pseudo Wigner Ville Distribution (SPWVD)

The RR sequence, Instant Central Frequency (ICF) and Group Delay (GD) can be formulated [44] by Eqs. (3), (4) and (5) respectively for selected episodes

$$\text{SPWVD}_x(t,f) = \int_{-\infty}^{\infty} x\left(t - u + \frac{\tau}{2}\right) x^*\left(t - u - \frac{\tau}{2}\right) g(u) h(\tau) e^{-j2\Pi t f} d\tau \tag{3}$$

$$\text{ICF} = \frac{\int f \text{SPWVD}_x(t,f) df}{\int \text{SPWVD}_x(t,f) df} \tag{4}$$

$$D = \frac{\int t \text{SPWVD}_x(t,f) dt}{\int \text{SPWVD}_x(t,f) dt} \tag{5}$$

$$\frac{\text{LF}}{\text{HF}} = \frac{P_{\text{LF}}}{P_{\text{HF}}} \tag{6}$$

$$\text{LF}_a = \frac{P_{\text{LF}}}{P_{\text{LF}} + P_{\text{HF}}} \tag{7}$$

$$\text{HF}_a = \frac{P_{\text{HF}}}{P_{\text{LF}} + P_{\text{HF}}} \tag{8}$$

[Considered frequency bands: LF (0.04–0.15 Hz), AF (0–0.4 Hz), HF (0.15–0.40 Hz), VLF (0–0.04 Hz)]

The time-frequency analysis was performed at different instances, i.e., before, after, and during the deviation of ST segment with considered frequency bands. The sensitive parameters were extracted using Fisher Linear Discriminant from the detected episodes and also helps in ischemic classification [43].

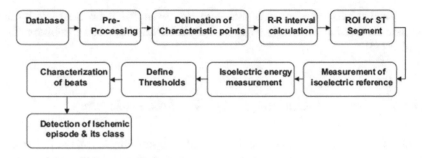

Fig. 9 Schematic diagram for proposed methodology [38]

2.6 Isoelectric Energy Function

Amit Kumar and Mandeep Singh have implemented a novel method for ischemia detection by analysis the deviations of ST segment using Isoelectric Energy Function (IEEF). The method comprises of five stages as presented in Fig. 9.

Fig. 10 STAT-MI Protocol
and Network [46]
(*EMS* = Emergency Medical
Service,
UMDNJ = University of
Medicine)

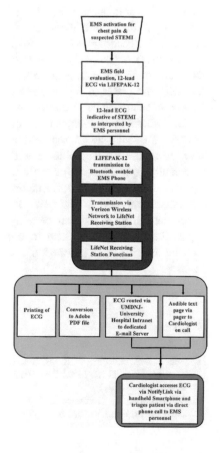

The methodologies were validated with European ST-T database (EDB) database and achieved average sensitivity (S_E) as 98.12% and average specificity (S_P) as 98.16% [45].

2.7 ST Segment Analysis Using Wireless Technology

Vivek N. Dhruva et al. have devised an automatic wireless network which is supported by preconfigured Bluetooth devices, dedicated e-mail servers, preprogrammed receiving/transmitting stations and smart phones for analysis of ST segment Elevation Myocardial Infarction (STEMI). Figure 10 shows the network protocol and functional of the system for STEMI [46].

3 Conclusion

Effective learning approaches have been discussed both in time and frequency domains. Different mathematical and analytical techniques have been evolved during these decades and significant way of understanding the morphophysiological changes in ST segment during certain cardiovascular diseases have been taken into consideration. Several detection methods has been discussed which precisely locate the onset and offset of ST segment having accuracy rate above 95%. The performances of many of these algorithms are comparable to the best of other reported systems which were evaluated using standard methods. Even though the involvement of ST events and ST segment drifts on ST segment, several proposed methods have overcome these problems with their preprocessing techniques. These preprocessing approaches further help in utilizing much wider contextual information while addressing critical noise induced ECG signals. This review, might certainly helps the researchers to find new solution in monitoring ST segment deviation and morphological changes over certain conditions and factors with improvised techniques. Further this will lead to overcoming certain challenges in the areas of decision making and automated investigation of life-threatening cardiovascular diseases.

References

1. Gettes, L.S., Cascio, W.E.: Effects of acute ischemia on cardiac electrophysiology. In: Fozzard, H.A., et al. (eds.) The Heart and Cardiovascular System, vol. 2, pp. 2021–2054. Raven Press (1991)
2. Goldschlager, N., Goldman, M.J.: Principles of Clinical Electrocardiography, East Norwalk, Connecticut, Prentice Hall Int (1989)

3. Gallino, A., Chierchia, S., Smith, G., Croom, M., Morgan, M., Marchesi, C., Maseri, A.: A computer system for analysis of ST segment changes on 24 hour Holter monitor tapes: comparison with other available systems. J. Arner. Coll. Cardiol. **4**, 245–252 (1984)
4. Philips Medical Systems Nederland, B.V.: ST Segment Monitoring, ST/AR Algorithm, IntelliVue Patient Monitor and Information Center, Application Note, Koninklijke Philips Electronics N.V. (2006)
5. Ansari, S., et al.: A review of automated methods for detection of myocardial ischemia and infarction using electrocardiogram and electronic health records. IEEE Rev. Biomed. Eng. **264–298**, 10 (2017)
6. Gallino, A., et al.: Computer system for analysis of ST segment changes on 24 hour Holter monitortapes: comparison with other available systems. J. Amer. College Cardiol. **4**(2), 245–252 (1984)
7. Oates, J., Cellar, B., Bernstein, L., Bailey, B.P., Freedman, S.B.: Real-time detection of ischemic ECG changes using quasi-orthogonal leads and artificial intelligence. Proc. IEEE Comput. Cardiol. 89–92 (1988)
8. Sun, G., et al.: Classification of normal and ischemia from BSPM by neural network approach. Proc. Annu. Int. Conf. Eng. Med. Biol. Soc. 1504–1505 (1988)
9. Passariello, G., Mora, F., De La Cruz, E., Gotoc, J., Cerreult, B.: Real time detection and quantification of ischemic ECG changes. In: Proceeding 12th Annual International Conference English Medical Biology Social, pp. 809–810 (1990)
10. Krucoff, M.W., et al.: Continuous computer-assisted electrocardiographic monitoring in patients with acute myocardial infarction: early experience. In: Proceedings IEEE Computers in Cardiology, pp. 197–200 (1989)
11. Jager, F., Mark, R., Moody, G., Divjak, S.: Analysis of transient ST segment changes during ambulatory monitoring using the Karhunen–Loeave transform. In: Proceedings IEEE Computers in Cardiology. pp. 691–694 (1992)
12. Thakor, N., Gramatikov, B., Mita, M.: Multiresolution wavelet analysis of ECG during ischemia and reperfusion. In: Proceedings IEEE Computers in Cardiology, pp. 895–898 (1993)
13. Brooks, D., On, H., MacLeod, R., Krim, H.: Analysis of changes in body surface potentials during PTCA-induced ischemia using the temporal wavelet transform. In: Proceedings IEEE Computers in Cardiology, pp. 329–332 (1994)
14. Sierra, G., et al.: Multiresolution decomposition of the signal averaged ECG in patients with myocardial infarction compared to a control group. In: Proceeding IEEE 17th Annual Conference English Medical Biology Social, vol. 2, pp. 1057–1058 (1995)
15. Lemire, D., Pharand, C., Rajaonah, J.-C., Dube, B., LeBlanc, A.-R.: Wavelet time entropy, T wave morphology and myocardial ischemia. IEEE Trans. Biomed. Eng. **47**(7), 967–970 (2000)
16. Gramatikov, B., Brinker, J., Yi-chun, S., Thakor, N.V.: Wavelet analysis and time-frequency distributions of the body surface ECG before and after angioplasty. Comput. Methods Programs Biomed. **62**(2), 87–98 (2000)
17. Banerjee, S., Mitra, M.: Cross wavelet transform based analysis of electrocardiogram signals. Int. J. Elect. Electron. Comput. Eng. **1**(2), 88–92 (2012)
18. Weisner, S.J., Tompkins, W.J., Tompkins, B.M.: A compact, microprocessor-based ECG ST-segment analyser for the operator room. IEEE Trans. Biomed. Eng. **29**, 642–649 (1982)
19. Skordalakis, E.: Recognition of the shape of the ST segment in ECG waveforms. IEEE Trans. Biomed. Eng. **33**, 972–974 (1986)
20. Hsia, P., Jenkins, J.M., Shitnoni, Y., Cage, K.P., Santigna, J., Pitt, B.: An automated system for ST segment and arrhythmia analysis in exercise radionuclide ventriculagraphy. IEEE Trans. Biomed. Eng. **33**, 585–593 (1986)
21. Anderson, C.M., Bragg-Remschel, D.A., Harrison, D.C.: An algorithm to analyse ST segment changes during ambulatory monitoring. In: Proceeding Computer in Cardiology, pp. 225–228 (1981)
22. ACCIAHA Special Report.: Guidelines for ambulatory electrocardiography. Circ. **79**(1), 206–215 (1989)

23. Steven K.W., et al.: Myocardial infarct size and edema in patients with ST-segment elevation myocardial infarction. J. Am. Coll. Cardiol. Intv. American Coll Cardiol. Found. **8**(1) (2015)
24. Schmid, R., et al.: A correction formula for the ST-segment measurements of AC-coupled electrocardiograms. IEEE Trans. Biomed. Eng. **64**(8) (2017)
25. Greenhut, S.E., Chadi, B.H., Lee, J.W., Jenkins, J.M., Nicklas, J.M.: An algorithm for the quantification of ST-T SEGMENT Variability, Academic Press, Inc. (1989)
26. Jane, R., Blasi, A., Garcia, J., Laguna, P.: Evaluation of an automatic threshold based detector of waveform limits in Holter ECG with the QT database. Comput. Cardiol. **24**, 295–298 (1997)
27. Jager, F., Moody, G.B., Tadder, A.: Performance measures for algorithms to detect transient ischemic ST segment changes. IEEE Comput. Cardiol. IEEE Comput. Soc. Press. 369–372 (1991)
28. Gonzalez', R., Caiiizares, M., Rodriguez, G., Meisijimilly, G.: A spatial study of the ST segment. In: Proceedings of the 25 annual international conference of the IEEE EMBS Cancun, Mexico September 17–21 (2003)
29. Jager, F., Mark, R.G., Moody, G.B., Divjak, S.: Analysis of transient ST segment changes during ambulatory monitoring using the karhunen-loeave transform. In: Computers in Cardiology, Proceedings of 1992 Oct 11, pp. 691–694. IEEE (1992)
30. Dimantaras, K.I., Stamkopoulos, T., Maglaveiras, N., Strintzis, M.: ST segment nonlinear principal component analysis for ischemia detection, Computers in Cardiology (1996)
31. García, J., Sornmo, L., Olmos, S., Laguna, P.: Automatic detection of ST-T complex changes on the ECG using filtered RMS difference series: application to ambulatory ischemia monitoring. IEEE Trans. Biomed. Eng. **47**(9) (2000)
32. Bulusu, S.C., Faezipour, M., Ng, V., Nourani, M., Tamil, L.S., Banerjee, S.: Transient ST-segment episode detection for ECG beat classification, life science systems and applications workshop. LiSSA), IEEE/NIH. (2011)
33. Turitto, Gioia, El-Sherif, Nabil: Alternans of the ST segment in Variant Angina Incidence, time course and relation to ventricular arrhythmias during ambulatory electrocardiographic recording. Chest **93**(3), 587–591 (1988)
34. Weisner, J.S., Tompkins, W.J., Tompkins, B.M.: A compact, microprocessor based ECG ST-segment analyzer for the operating room, IEEE Trans. Biomed. Eng. vol. BME-**29**, 624–648 (1983)
35. Skordalakis, E.: Recognition of the shape of the ST segment in ECG waveforms. IEEE Trans. Biomed. Eng. BME-**33**(10) (1986)
36. Pavlidis, D.: Waveform segmentation through functional approximation, IEEE Trans. Comput. C-**22**, 689–697 (1973)
37. Narayan, S.M., Lindsay, B.D., Smith, J.M.: Demonstration of the proarrhythmic preconditioning of single premature extrastimuli by use of the magnitude, phase, and distribution of repolarization alternans. Circ **100**(18), 1887–1893 (1999)
38. Li, X.Y., Wang, T., Zhou, P., Feng, H.Q.: ST-T complex automatic analysis of the electrocardiogram signals based on wavelet transform. In: Bioengineering Conference, 2003 IEEE 29th Annual, Proceedings, pp. 144–145. IEEE (2003)
39. Taddei, A., Biagini, A., et al.: The European ST-T database: development, distribution and use, Computers in Cardiology, Chicago, USA, 1990, pp. 177–80 (1991)
40. Andreão, R.V., Dorizzi, B. et al.: Transformée en ondelettes et modèles de Markov cachés pour la segmentation automatique du signal ECG. Proc. 19 ème colloque GRETSI, Paris, France (2003)
41. Andreão, R.V., Dorizzi, B., Boudy, J., Mota, J.C.M.: Online Beat Segmentation and Classification Through Hidden Markov Models. Proc. IIICLAEB, João Pessoa, Brazil (2004)
42. Andreao, R.V., Dorizzi, B., Boudy, J., Mota, J.C.M.: ST-segment analysis using hidden markov model beat segmentation: application to ischemia detection. Comput. Cardiol. **31**, 381–384© IEEE (2004)

43. Xing, W., Liang, X., Zhongwei, S., Zibin, Y., & Yi, P.: Heart rate variability analysis of ischemic and heart rate related ST-segment deviation episodes based on time-frequency method. In: Proceedings of NFSI & ICFBI 2007, Hangzhou, China, October 12–14 (2007)
44. Hu, G.S.: Course for modern signal processing. Tsinghua University Press, Beijing (2004)
45. Kumar, A., Singh, M.: Ischemia detection using isoelectric energy function. Comput. Biol. Med. **68**, 76–83 (2016)
46. Dhruva, V.N., et al.: ST-Segment Analysis Using Wireless Technology in Acute Myocardial Infarction (STAT-MI) Trial. J. Am. Coll. Cardiol. **50**(6), 509–513 (2007)

Survey on Sentiment Analysis Methods for Reputation Evaluation

P. Chiranjeevi⑩, D. Teja Santosh and B. Vishnuvardhan

Abstract Sentiment Analysis gathered huge attention in recent years. In this field, sentiments are analyzed and aggregated from the text. There are certain relevant sub-areas in research. This survey mainly concentrates on aspect-level (product feature) sentiment analysis. The aspects of the products are the noun phrases of the sentences. It is necessary to identify the goal and aggregate sentiments on entities in order to find the aspects of the entities. The detailed overview of study is given in such a way that the incredible evolution was already made in finding the target corresponding to the sentiment. The recent solutions are based on the aspect detection and extraction. In a detailed study, a performance report and evaluation related to the data sets are mentioned. In a variety of existing methods, an attempt is made to use the shared data values to standardize the evaluation methodology. The future research is in the direction of sentiment analysis which mainly concentrates on aspect centric reputation of online products.

Keywords Text mining · Linguistic processing · Machine learning
Aspect extraction · Sentiment analysis · Reputation

P. Chiranjeevi (✉)
Faculty of Computer Science and Engineering,
ACE Engineering College, Hyderabad, India
e-mail: chiruanurag@gmail.com

D. Teja Santosh
Department of Computer Science and Engineering,
GITAM University, Hyderabad, India
e-mail: tejasantoshd@gmail.com

B. Vishnuvardhan
Department of Computer Science and Engineering,
JNTUHCEJ, Nachupally Karimnagar, India
e-mail: mailvishnuvardhan@gmail.com

© Springer Nature Singapore Pte Ltd. 2019   53
P. K. Mallick et al. (eds.), *Cognitive Informatics and Soft Computing*,
Advances in Intelligent Systems and Computing 768,
https://doi.org/10.1007/978-981-13-0617-4_6

1 Introduction

The information age is also known as digital, in which technology provides the ability to transfer information quickly. Current generations are seamlessly using web connectivity to share their opinions with others. The traditional information sharing is to exchange data between a sender and receiver. Younger generations are showing lot of interest to share the content in web, so this trend is growing faster and may increase in future. To share their preferences and opinions with others, most of the people are depending on web and are using blogs and comments to rate the product. The comments in the web blogs represent certain opinion of the person. This kind of data is useful to rate the product.

From the past years, government organizations and commercial merchant organizations are struggling a lot to determine the customer or target opinion. But nowadays, by using the web, people voluntarily share their opinions. This is a social web which allows the producer or any organization to get immediate feedback of product, stocks, and policies. In recent times it is easy to get past data which is readily available on web. Participants can give their personalized opinion in the form of text or in the form of ratings or reviews [1]. Many people were influenced by the opinions which are shared on web. To purchase any product, many customers are depending on the reviews of old customers opinions [2]. Moreover, information provided by the individual is trust-worthy than the information given by the producer [2]. Every customer is important for vendors. So it is very important to know the customer likes and dislikes about product for its sale or to develop new products [3], as well as it is helpful to manage already existing ones [4]. Most of the companies or producers are depending on the reviews to increase their sales [5]. The opinion about the product on the web is just like a word-of-mouth of customer [6]. In financial markets sentiment analysis plays vital role [7].

Pang and Lee [4] specified terms related to the sentiment analysis and reputation. The area of sentiment analysis is driven by natural language processing, artificial intelligence and information retrieval system. There are different terms similar to this topic. Another most important term used in this survey is "*opinion mining*" the word which is coming from the area of information retrieval system and data mining. The opinion mining is to identify the group of people's opinion regarding one issue or product. "*Sentiment Analysis*" is also often used in this survey. The aim of sentiment analysis is to identify the orientation of the sentiment expressed in script. The term "*Subjectivity Analysis*" comes from Opinion Mining [8]. These are the important terms used in our research. This area of research labeled as Subjectivity Analysis studies the emotions, sentiment, opinion, manner and evaluation [9]. Throughout this survey these terms are considered as sentiment or opinion for easy understanding.

According to the definition opinion is either a decision of a belief and is not contained any conviction or proof [10]. In this context, this is contradictory to facts. Hence sentiments which express opinions are considered as subjective, whereas factual or real statements are considered as objective. Sentiment is nearly related to personal emotion [11] and this is used to evaluate the person's opinion. So this is clear that statements that convey a sentiment is both objective as well as subjective.

With the above context, an opinion is quadruple (e, o, s, t, h) [9], where "e" is the target entity, "o" is the opinion, "s" is the sentiment, "t" is the opinion expressed time, "h" is the possessor of particular opinion. Here the pair (o, s) is important to focus talk something about sentence and document level. The term "entity" denotes the target objects. For example, in product review system, a product is considered as an "entity" and the features of the product are related to aspects. Aspect-based sentiment analysis finds the sentiment as well as the sentiment for aspects of entity. The aspect-level sentiment analysis is product feature based as it attempts to identify aspects of target entity with in the sentence.

1.1 Overview of Aspect-Level Sentiment Analysis

There are mainly three phases in aspect-level sentiment analysis. These are namely aspect or entity identification, aspect opinion classification and aspect opinions aggregation [8]. However, survey specifies that all the methods are not implementing all these steps. In the first step, sentiment-target identification is carried out. The second step classifies the opinion-target pairs. This step determines whether the opinion values are positive, negative or neutral. At last, the opinion values for each and every aspect are aggregated towards the sentiment for better view.

1.2 Center of This Survey

There are certain sub-fields of sentiment analysis which need to focus. Sentiment Analysis [9] has three important levels those are document, sentence and aspect levels. The document level finds the sentiment polarity for entire document. In sentence level, it expresses sentiment polarity for each and every word and sentence. With the same sentence multiple entities are compared.

In this survey, the focus is laid down on document-level sentiment analysis, which is similar to the research by Pang and Lee, Tang et al. [4, 12]. It mainly focuses on document-level machine learning. There are certain other surveys [8], which are also, focuses on document-level machine learning approach. There are some other approaches for identifying the sentiment like dictionary-based approach and corpus-based approach. In dictionary-based approach, the Sentiment words are identified from the dictionary to find the sentiment. The corpus-based approach is applied to two scenarios. In the first scenario, from the given list of sentiment, (often general-purpose) sentiment words and their orientations are discovered. In the second scenario domain corpus is used to adapt sentiment lexicon to new sentiment.

2 Review-Based Opinion Mining

Opinion Mining is a kind of natural language processing which analyzes people's opinions, sentiments, emotions, attitudes, and evaluations towards entities such as individuals, services, organizations, goods, issues, subjects, attributes, and their events.

2.1 Feature Extraction

The aspects are mined from unstructured text and the opinion value is assigned for each feature. Feature extraction involves considering the large dataset and reducing to the required words. There are certain approaches for feature extraction in sentiment analysis and opinion mining.

The preprocessing considers the review words and removes the stop words and assigns the Parts of Speech class to the words. Then the nouns are analyzed for identification of product features. A product feature is morphological, frequent and implicit feature.

Opinion in a document is hidden in words or sentences. The goal of Opinion extraction is to find where the opinions are embedded in this document. From the document, the relations among the words and their contextual information are considered as clues to extract opinion. An opinion can be positive, negative or neutral appraisal, emotion or attitude about an aspect of the entity from an opinion holder. Opinion orientations are positive, negative, and neutral. The opinion is classified as positive or negative by the sentiment classifier. The sentiment classification uses supervised and unsupervised learning.

Sentiment Classification Using Supervised Learning:

The Sentiment classification is generally defined as classification with two classes namely positive and negative [13]. Ratings of online Reviews are pre defined as 1–5 stars; based on this predefined data set positive and negative reviews can be identified. For example the star rating with 4 and 5 star reviews are basically treated as positive review, and 1 and 2 star reviews are treated as a negative review. Neutral class is not used in most of the research.

Sentiment classification is a kind of text categorization. Text classification classifies the text based on the keywords, these keywords are basically topic related, e.g., sports, polity, and goods. In sentiment classification, opinion words are important which indicates positive or negative opinions, e.g., *good, better, poor, bad, ok, bad, best*, etc. All the existing supervised learning methods were applied to text classification problems. Different classification models such as Support Vector Machine (SVM), Naïve Bayes (NB) classification were used.

Sentiment Classification Using Unsupervised Learning:

In a document or text, sentiment words are important for sentiment classification. Identification of sentiment words is easy, sentiment words are very useful for sentiment classification in unsupervised learning. Here the classification takes place to identify some fixed syntactic patterns; finally these patterns are used to express particular context opinion. POS tags are used to compose syntactic patterns.

2.2 Feature Reputation

Reputation and opinion are two different terms. Each product has its own features. The popularity of product features is used to calculate the Reputation of the product. The feature reputation generated based on the strength of the opinion and sub-features of the features. For example, Feature is represented as fi, Aggregation of the positive and negative and opinion weights of its sub-feature fj are represented as Reputation of feature fi.

3 Various Explored Techniques

See Fig. 1.

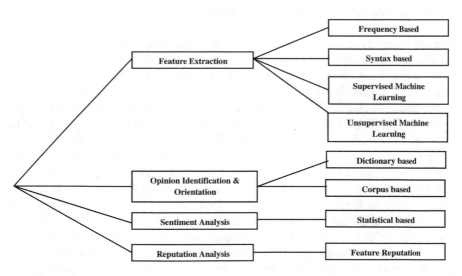

Fig. 1 Reputation analysis taxonomy

3.1 Aspect Extraction Approaches

3.1.1 Frequency-Based

Aspect extraction is one of the most important and extensively explored phases of sentiment analysis. It helps to carry out the classification of sentiment. Aspect Extraction is carried out with Topic-modeling, Frequency-based, supervised learning and Relation-based approaches. An aspect can be a noun, verb, adverb, or adjective. Aspects are the frequent nouns which are mostly used by the people in reviews and not all the aspects are frequent nouns. Hu et al. [14] work identifies the nouns as product aspects and extracts them. The close by adjectives to these nouns is treated as sentiment words. The infrequent aspects are also identified using the aspects polarity and using the extracted sentiment words. Popescu et al. [15] developed a system called OPINE. This is a domain independent, web-based information extraction system. This system identifies various types of product components as aspects. Frequency-based approaches have many advantages and these methods are very simple and effective. The limitations of these methods are it require manual tuning of various parameters and cannot detect low-frequency aspects. It also produces too many non-aspects.

3.1.2 Syntax-Based (Relation-Based Methods)

Syntax-based approach specifies that each sentiment expresses an opinion. Sentiments are commonly known or easy to find. The main goal of syntax-based approaches is to develop aspect-sentiment relationships from the underlying reviews collection in order to extract new aspects and sentiments. Liu et al. developed [16] Opinion Observer for differentiating consumer opinions of different products. The important task in the work is to replace aspects by a specified POS tag. e.g., "Resolution is great" "Resolution is tagged as NN is tagged as VB great is tagged as JJ" "[aspect] is tagged as NN is tagged as VB great is tagged as JJ" "[aspect] is tagged as NN space is tagged as VB and space is tagged as JJ". Then, the POS patterns learned are used to mine association rules among aspects and sentiments. Baccianella et al. [17] used polarity general inquirer, predefined POS patterns and Filtering variance of distribution across polarity. To improve the exacting matching and exploring the substructure Jiang et al. [18] used Tree kernel approach. The strength of syntax-based approaches is to find aspects of low-frequency and the limitation is to generate too many non-aspects matching with the patterns.

3.1.3   Sequence Modeling

Supervised Learning Techniques (sequence-based):
Sequence-based learning techniques applies a function from labeled data to unlabelled data and it identifies aspects, sentiments and their polarity. Hidden Markov Model (HMM) and Conditional Random Field (CRF) are the two popular sequence labeling approaches that are useful in sentiment analysis. Jin et al. developed [19] OpinionMiner, a tool for analyzing opinions from the reviews. It follows the underlying basic Hidden Markov model. The task in this research is to identify polarity, sentiments and their aspects by using POS information integrated with lexicalization technique. Li et al. worked [20] on Skip Tree CRF model. The task in this research is extracting polarity, sentiments and their aspects using the syntactic tree structure and conjunction structure. Lakkaraju et al. developed [21] CFACTS model whose underlying basic model adopted for the development is HMM. The task in this research is discovering polarity, sentiments, and their aspects. The method proposed in this research is integrated coherence in reviews. Wong et al. developed [22] a model using Hidden Markov model. The task in this research is extracting and grouping of aspects from various web pages using layout and content data. The other methods in this line of research are Sauper et al. [23], Yu et al. [24], Choi et al. [25], Jakob et al. [26] and Kobayashi et al. [27]. The strengths of these models are the frequency-based limitations are conquered and limitation is the requirement of a manually labeled data for training.

3.1.4   Unsupervised Learning Techniques (Topic-Modeling Based)

To model both sentiments and aspects in various contexts, the basic topic model is extended. At first, Latent Dirichlet Allocation (LDA), a statistical method for learning topics was devised. Blei et al. developed [28] LDA is the popular statistical model. It assumes that every document is a mixture of latent topics. These topics are related with vocabulary of the document collection on the basic of probability distribution. Mei et al. modified [29] LDA to produce Topic Sentiment Mixture (TSM). It has the underlying basic model as Hidden Markov model. The task in this research is to recognize aspects and their polarity using TSM where traditional LDA model cannot extract both aspects and polarities from the underlying document collection. Titov et al. produced [30] Multi-grain LDA (MG-LDA) model to extract aspects. The new method proposed in this research is considering local as well as global topics in order to extract aspects as topic words. The same researchers extended this MG-LDA in [31] to find the correspondence between topics and aspects. Lin et al. developed [32] Joint Sentiment Topic (JST) model to identify aspect and their orientations by using different aspect distributions for each polarity. The continuation to this work was carried out by He et al. in [33]. Li et al. proposed [34] Sentiment-LDA to identify aspects and their polarity by modeling different polarities for each aspect in the LDA graphical model. The polarity with respect to the local context was achieved using this work. Zhao et al. modified [35] the traditional LDA to propose MaxEnt-LDA. The task in this research is to identify aspects and sentiments with the POS tagged nouns as aspects, adjectives and adverbs as sentiments. Jo et al. proposed [36] Aspect-Sentiment Unification Model (ASUM) to identify aspects and their polarity by extending the work of JST. The extension is based on the premise that one sentence has one aspect and related polarity.

3.2   Opinion Identification and Orientation

3.2.1   Dictionary-Based Approaches

The works on dictionary-based approach for opinion word identification and orientation was presented in [37, 38]. Initially, the set of opinion words are gathered with well known orientations. Then, the size of the set is improved by adding the synonyms and antonyms by searching in the WordNet [39] or thesaurus [40]. The iteration is continued until no new words are found. After the completion of this entire cycle, to correct the errors or to remove the errors manual inspection is carried out. The major loop hole in dictionary-based approach [41] is it cannot able to find opinion words with context and domain specific orientations.

3.2.2   Corpus-Based Approaches

The corpus-based approaches identify the opinion words by using machine learning technique. Corpus-based method depends on syntactic patterns. The large corpus is used to find the opinion words. Mckeown and Hatzivassiloglou presented [42] one of the methods. They manually considered the opinion adjectives as list of seed, opinion words and their orientations. Following are the constraints like OR, AND, EITHER-OR, BUT, etc.; "AND" conjunction represents the same orientation. Some negating expressions are written in the reviews, such as "but", which changes the opinion. So, it is necessary to identify the orientation of adjective. To determine orientation of adjectives of the same or different orientations, learning is applied to a large corpus. Then, generate a graph by linking the adjectives and then perform clustering on the graph to produce positive and negative words. The Conditional Random Fields (CRF) method [43] is a statistical method used to extract the opinion phrases. This method was presented by Jiaoa and Zhoua [44] to distinguish sentiment polarities. Dictionary-based approach is efficient when compared to corpus-based approach. Using corpus-based approach alone is not efficient because it is not possible to define all English words in a corpus, but corpus-based approach has a great advantage to find orientations, and the context of certain opinion words using a domain corpus.

3.3 Sentiment Classification

3.3.1 Statistical-Based Approaches

The review in statistical-based approach is represented as a combination of latent aspects and their respective ratings. It is implicit that aspects and their ratings correspond to the multinomial distributions and clustering the head terms as aspects and sentiments as ratings.

Statistical approach:

Finding seed opinion words or co-occurrence patterns is carried out using statistical techniques. Fahrni and Klenner proposed [45] that finding seed words can be done by deriving posterior polarities using the co-occurrence of adjectives in a corpus. To construct the corpus, all the document words are included from the dictionary. So that word unavailability problem can be overcome with this approach [46]. The word polarity is identified by studying the word frequency in a large annotated corpus [47]. The polarity is considered to be positive when the word occurred more frequently.

Generating Recommendations:

After experiential cases generation by using the Eq. (1) for each product, the next step is to generate n recommendations for each and every product based on the product similarity and product feature sentiments. So, for overall score calculation we need product similarity and product sentiment [48].

$$\text{Case}(P) = \{[Fi, \text{Sent}(Fi, P), \text{Pop}(Fi, P)] : Fi \in F(P)\} \tag{1}$$

Product similarity computation:

To find the similarity between two products, classic content-based approach is relied upon. For this, cosine similarity between products based on the features popularity present in each product using (2) is computed.

$$\text{Sim}(Q, C) = \frac{\sum_{Fi \in F(Q) \cup F(C)} \text{Pop}(Fi, Q) \times \text{Pop}(Fi, C)}{\sqrt{\sum_{Fi \in F(Q)} \text{Pop}(Fi, Q)^2} \times \sqrt{\sum_{Fi \in F(C)} \text{Pop}(Fi, C)^2}} \tag{2}$$

The popularity of each word can be computed with the below Eq. (3)

$$\text{Pop}(Fi, P) = \frac{|\{RK \in \text{Reviews}(P) : Fi \in RK\}|}{|\text{Reviews}(P)|} \tag{3}$$

Product sentiment computation:

To calculate feature sentiment we use Eq. (4)

$$\text{Sent}(Fi, P) = \frac{\text{Pos}(Fi, P) - \text{Neg}(Fi, P)}{\text{Pos}(Fi, P) + \text{Neg}(Fi, P) + \text{Neut}(Fi, P)} \tag{4}$$

Finally, the score for each product pair with (5) has to be computed

$$\text{Score}(Q, C) = (1 - w) \times \text{Sim}(Q, C) + w \times \left(\frac{\text{Sent}(Q, C) + 1}{2} \right) \qquad (5)$$

3.4 Reputation Analysis

A Reputation analysis looks deep into the data of what people see when they search for in online and builds a strategy for improvement. Reputation is the opinion or belief of some one. Reputation analysis is a method that analyzes the information evaluation of products and services. The information includes their users' comments and rating information.

3.4.1 Feature Reputation

Generating the reputation of the product is based on product's reputation on its features. This is an important stage. Based on the strength of its sub-features and opinion orientation the reputation of each feature is computed. For each fi of the product, the reputation of the feature fi should be the aggregate of the positive and negative opinions weights for all of its sub-features fj, where (fi, fj)€L. For finding the feature reputation, the negative and positive weights are to be calculated first [49]. Elanor presented [50] a measure of online reputation of a company using social media content. To monitor the real-time evolution of company reputation, the researchers also presented an open platform that uses sentiment analysis algorithms on twitter traffic. For finding polarity, Cristina et al. presented [51] the task of reputation polarity detection of social media updates. They proposed a supervised and data driven approaches for extracting textual features, which can used to train a reputation polarity classifier. Three set of features are considered: (i) Surface, (ii) Sentiment and (iii) Textual.

4 Evaluation Techniques

4.1 Precision

Precision is defined as a fraction of relevant documents among the retrieved documents. It is also known as positive predictive value.

$$Precison = \frac{|\{Relevant documents\} \ \ddot{E}C\{Retrieved documents\} \ |}{|\{Retrieved documents\} \ |}$$

4.2 Recall

Recall is the fraction of relevant documents that have been retrieved over total relevant documents in the image. Recall is also known as sensitivity. Both precision and recall are based on measure called "relevance".

$$Recall = \frac{|\{Relevant documents\} \ \ddot{E}C\{Retrieved documents\} \ |}{|\{Relevant documents\} \ |}$$

4.3 F1-Score

The F1-score can be interpreted as a weighted average of the precision and recall.

$$F1 - score = 2 \times \frac{Precision \ \times \ Recall}{Precision \ \ + \ \ Recall}$$

5   Conclusion

The complete overview of aspect-level sentiment analysis is presented in this survey. In Sentiment Analysis most of the approaches are using machine learning which are derived in this survey. Supervised and Unsupervised learning models are well presented. The terms transparency and standardization evaluation methodology are highly stressed. Sentence level approach is discussed precisely. However the accuracy of systematic classification approaches is missing. Combining machine learning techniques with concept-centric approaches raise the ability of algorithm. The future direction of aspect level sentiment analysis is concept-centric aspect-level sentiment analysis.

References

1. Schouten, K., Frasincar, F.: Survey on aspect-level sentiment analysis. IEEE Trans. Knowl. Data Eng. **28**(3), 813–830 (2016)
2. Bickart, B., Schindler, R.M.: Internet forums as influential sources of consumer information. J. Interact. Mark. **15**(3), 31–40 (2001)

3. Van Kleef, E., Van Trijp, H.C., Luning, P.: Consumer research in the early stages of new product development: a critical review of methods and techniques. Food Qual. Prefer. **16**(3), 181–201 (2005)
4. Pang, B., Lee, L.: Opinion mining and sentiment analysis (Foundations and Trends (R) in Information Retrieval) (2008)
5. Chen, Y., Xie, J.: Online consumer review: word-of-mouth as a new element of marketing communication mix. Manage. Sci. **54**(3), 477–491 (2008)
6. Goldsmith, R.E., Horowitz, D.: Measuring motivations for online opinion seeking. J. Interact. Adv. **6**(2), 2–14 (2006)
7. Arnold, I.J., Vrugt, E.B.: Fundamental uncertainty and stock market volatility. Appl. Financ. Econ. **18**(17), 1425–1440 (2008)
8. Tsytsarau, M., Palpanas, T.: Survey on mining subjective data on the web. Data Min. Knowl. Disc. **24**(3), 478–514 (2012)
9. Liu, B.: Sentiment analysis and opinion mining. Synth. Lect. Hum. Lang. Technol. **5**(1), 1–167 (2012)
10. Collins, H.: Collins English Dictionary. Dictionary. com (2000)
11. Kim, S.M., Hovy, E. Determining the sentiment of opinions. In: Proceedings of the 20th international conference on Computational Linguistics (p. 1367). Association for Computational Linguistics (2004)
12. Tang, H., Tan, S., Cheng, X.: A survey on sentiment detection of reviews. Expert Syst. Appl. **36**(7), 10760–10773 (2009)
13. De Albornoz, J.C., Chugur, I., Amigó, E.: Using an emotion-based model and sentiment analysis techniques to classify polarity for reputation. In: CLEF (Online Working Notes/Labs/Workshop) (2012)
14. Hu, M., Liu, B.: Mining and summarizing customer reviews. In: Proceedings of the tenth ACM SIGKDD international conference on knowledge discovery and data mining (pp. 168–177). ACM (2004)
15. Popescu, A.M., Etzioni, O.: Extracting product features and opinions from reviews. In: Natural language processing and text mining (pp. 9–28). Springer, London (2007)
16. Liu, B., Hu, M., Cheng, J.: Opinion observer: analyzing and comparing opinions on the web. In: Proceedings of the 14th international conference on World Wide Web (pp. 342–351). ACM (2005)
17. Baccianella, S., Esuli, A., Sebastiani, F.: Multi-facet rating of product reviews. In: ECIR, vol. 9, pp. 461–472 (2009)
18. Jiang, P., Zhang, C., Fu, H., Niu, Z., Yang, Q.: An approach based on tree kernels for opinion mining of online product reviews. In: Data Mining (ICDM), 2010 IEEE 10th International Conference on, pp. 256–265. IEEE (2010)
19. Jin, W., Ho, H.H., Srihari, R.K.: OpinionMiner: a novel machine learning system for web opinion mining and extraction. In: Proceedings of the 15th ACM SIGKDD international conference on Knowledge discovery and data mining, pp. 1195–1204. ACM (2009)
20. Li, F., Huang, M., Zhu, X.: Sentiment analysis with global topics and local dependency. In: AAAI, vol. 10, pp. 1371–1376 (2010)
21. Lakkaraju, H., Bhattacharyya, C., Bhattacharya, I., Merugu, S.: Exploiting coherence for the simultaneous discovery of latent facets and associated sentiments. In: Proceedings of the 2011 SIAM international conference on data mining, pp. 498–509. Society for Industrial and Applied Mathematics (2011)
22. Wong, T.L., Lam, W., Wong, T.S.: An unsupervised framework for extracting and normalizing product attributes from multiple web sites. In: Proceedings of the 31st annual international ACM SIGIR conference on research and development in information retrieval, pp. 35–42. ACM (2008)
23. Sauper, C., Haghighi, A., Barzilay, R.: Content models with attitude. In: Proceedings of the 49th annual meeting of the association for computational linguistics: human language technologies, vol. 1, pp. 350–358. Association for Computational Linguistics (2011)

24. Yu, J., Zha, Z.J., Wang, M., Chua, T.S.: Aspect ranking: identifying important product aspects from online consumer reviews. In: Proceedings of the 49th annual meeting of the association for computational linguistics: human language technologies, vol. 1, pp. 1496–1505. Association for Computational Linguistics (2011)

25. Choi, Y., Cardie, C.: Hierarchical sequential learning for extracting opinions and their attributes. In: Proceedings of the ACL 2010 conference short papers, pp. 269–274. Association for Computational Linguistics (2010)

26. Jakob, N., Gurevych, I.: Extracting opinion targets in a single-and cross-domain setting with conditional random fields. In: Proceedings of the 2010 conference on empirical methods in natural language processing, pp. 1035–1045. Association for Computational Linguistics (2010)

27. Kobayashi, N., Inui, K., Matsumoto, Y.: Extracting aspect-evaluation and aspect-of relations in opinion mining. In: EMNLP-CoNLL, vol. 7, pp. 1065–1074 (2007)

28. Blei, D.M., Ng, A.Y., Jordan, M.I.: Latent dirichlet allocation. J. Mach. Learn. Res. **3**, 993–1022 (200)

29. Mei, Q., Ling, X., Wondra, M., Su, H., Zhai, C.: Topic sentiment mixture: modeling facets and opinions in weblogs. In: Proceedings of the 16th international conference on World Wide Web, pp. 171–180. ACM (2007)

30. Titov, I., McDonald, R.: Modeling online reviews with multi-grain topic models. In: Proceedings of the 17th international conference on World Wide Web, pp. 111–120. ACM (2008)

31. Titov, I., McDonald, R.T.: A joint model of text and aspect ratings for sentiment summarization. In: ACL, vol. 8, pp. 308–316 (2008)

32. Lin, C., He, Y.: Joint sentiment/topic model for sentiment analysis. In: Proceedings of the 18th ACM conference on Information and knowledge management, pp. 375–384. ACM (2009)

33. He, Y., Lin, C., Alani, H.: Automatically extracting polarity-bearing topics for cross-domain sentiment classification. In: Proceedings of the 49th annual meeting of the association for computational linguistics: human language technologies, vol. 1, pp. 123–131. Association for Computational Linguistics (2011)

34. Li, F., Huang, M., Zhu, X.: Sentiment analysis with global topics and local dependency. In: AAAI, vol 10, pp. 1371–1376 (2010)

35. Zhao, W.X., Jiang, J., Yan, H., Li, X. Jointly modeling aspects and opinions with a MaxEnt-LDA hybrid. In: Proceedings of the 2010 conference on empirical methods in natural language processing, pp. 56–65. Association for Computational Linguistics (2010)

36. Jo, Y., Oh, A.H.: Aspect and sentiment unification model for online review analysis. In: Proceedings of the fourth ACM international conference on Web search and data mining, pp. 815–824. ACM (2011)

37. Hu, M., Liu, B.: Mining and summarizing customer reviews. In: Proceedings of the tenth ACM SIGKDD international conference on Knowledge discovery and data mining, pp. 168–177. ACM (2004)

38. Kim, S.M., Hovy, E.: Determining the sentiment of opinions. In: Proceedings of the 20th international conference on computational linguistics, p. 1367. Association for Computational Linguistics (2004)

39. Miller, G.A., Beckwith, R., Fellbaum, C., Gross, D., Miller, K.J.: Introduction to wordnet: an on-line lexical database. Int. J. Lexicogr. **3**(4), 235–244 (1990)

40. Mohammad, S., Dunne, C., Dorr, B.: Generating high-coverage semantic orientation lexicons from overtly marked words and a thesaurus. In: Proceedings of the 2009 conference on empirical methods in natural language processing, pp. 599–608 (2009)

41. Qiu, G., He, X., Zhang, F., Shi, Y., Bu, J., Chen, C.: DASA: dissatisfaction-oriented advertising based on sentiment analysis. Expert Syst. Appl. **37**(9), 6182–6191 (2010)

42. Hatzivassiloglou, V., McKeown, K.R.: Predicting the semantic orientation of adjectives. In: Proceedings of the eighth conference on European chapter of the association for computational linguistics, pp. 174–181. Association for Computational Linguistics (1997)

43. Lafferty, J., McCallum, A., Pereira, F.C.: Conditional random fields: Probabilistic models for segmenting and labeling sequence data (2001)
44. Jiao, J., Zhou, Y.: Sentiment polarity analysis based multi-dictionary. Phy. Procedia **22**, 590–596 (2011)
45. Fahrni, A., Klenner, M.: Old wine or warm beer: target-specific sentiment analysis of adjectives. In: Proceeding of the symposium on affective language in human and machine, pp. 60–63. AISB (2008)
46. Turney, P.D.: Thumbs up or thumbs down?: semantic orientation applied to unsupervised classification of reviews. In: Proceedings of the 40th annual meeting on association for computational linguistics, pp. 417–424. Association for Computational Linguistics (2002)
47. Read, J., Carroll, J.: Weakly supervised techniques for domain-independent sentiment classification. In: Proceedings of the 1st international CIKM workshop on topic-sentiment analysis for mass opinion, pp. 45–52. ACM (2009)
48. Dong, R., Schaal, M., O'Mahony, M.P., McCarthy, K., Smyth, B.: Opinionated product recommendation. In: International conference on case-based reasoning, pp. 44–58. Springer, Berlin, Heidelberg (2013)
49. Abdel-Hafez, A., Xu, Y., Tjondronegoro, D.: Product reputation model: an opinion mining based approach. In: SDAD 2012 The 1st international workshop on sentiment discovery from affective data, p. 16 (2012)
50. Colleoni, E., Arvidsson, A., Hansen, L.K., Marchesini, A.: Measuring corporate reputation using sentiment analysis. In: Proceedings of the 15th international conference on corporate reputation: navigating the reputation economy (2011)
51. Gârbacea, C., Tsagkias, M., de Rijke, M.: Detecting the reputation polarity of microblog posts. In: Proceedings of the twenty-first european conference on artificial intelligence, pp. 339–344. IOS Press (2014)

A Study on the Turnout on the Use of E-Shopping Websites with Arabic Platform in Bahrain

Yousef A. Baker El-Ebiary and Najeeb Abbas Al-Sammarraie

Abstract As an e-commerce domain, online shopping is now in its golden era. It is now a potential contributor in e-commerce. E-commerce success has been documented by numerous countries and it contributed to these countries' economic growth. The BAHRAIN is witnessing growth of e-commerce at a yearly rate exceeding 20%. In this region, the majority of the population (>80%) use Internet. Within this portion, 15% do online shopping and approximately 10% use mobile devices for shopping online? 250 respondents in the Kingdom of Bahrain were collected in this study and gathered these through questionnaire has conducted online. In short, demonstrates a predictive model 60.1% of variation in behavioral intent (BI). This is directly clarified according to performance expectations, Expected effort, social impact, cost, awareness and the model represents 54.2% of variation in usage behavior. This is directly clarified by behavioral intention and Facilitating Conditions. As such, it is expected that the increase in online Arab websites will shift the online purchasing intentions of the customer to the local Arab websites. This will enhance the Arab countries' microeconomics. Online shopping or e-commerce adoption differs based on countries because of different affecting factors. Thus, factors that affect, inhibit, and encourage e-commerce that are unique to each country should be considered in order to increase online shopping.

Keyword E-Commerce · Technology acceptance model · UTAUT
Performance expectancy · Effort expectancy

Y. A. Baker El-Ebiary (✉) · N. A. Al-Sammarraie
Faculty of Computer Science and IT, MEDIU University, Selangor, Malaysia
e-mail: yousef.abubaker@mediu.edu.my

N. A. Al-Sammarraie
e-mail: dr.najeeb@mediu.edu.my

© Springer Nature Singapore Pte Ltd. 2019
P. K. Mallick et al. (eds.), *Cognitive Informatics and Soft Computing*,
Advances in Intelligent Systems and Computing 768,
https://doi.org/10.1007/978-981-13-0617-4_7

1 Introduction

Numerous countries as well as the developing countries including the Arab and Asian countries have experienced the rapid growth of electronic commerce (E-commerce). E-commerce comprises business transactions via electronic means where the delivery of services or goods is electronically done or in the tangible form [1]. E-commerce as the recent type of enterprise that manipulates the technology of Internet to assist the development of business [2]. The Middle East is experiencing rapid growth of online shopping, and over the last decade, it has grown by 1500%. The potential of online spending is speedily becoming one of the highest in the world and this is also factored by Dynamic young population who has one of the highest levels of Internet penetration worldwide. As the Internet grows rapidly globally, it offers high-speed, low-power, almost free infrastructure, global, instant, online everything, unrestricted by borders. All these impact both the society and the economies worldwide. Globally, there has been an increase in the utilization of advanced technology in learning, living, working, socializing, and entertainment via online services' devices including PCs, Tablets, Smartphones and Smart TVs.

2 Problem Statement

The Arab nations appear to be the slowest online shopping adopter [3], where almost half (48%) of the Internet users in this region have never engaged in online transaction. Risk issue appears to be the reason why Arab consumers were reluctant to engage in online shopping [4]. In support to this notion. The necessity of online retailer in increasing the trust perception and site quality so that users would be motivated to shop online. In short, the Arab consumers were majorly concerned with the issue of trust and websites quality when doing online shopping [5]. In fact, the lack of trust in the online transactions as a hindrance in online shopping acceptance [6].

Electronic Commerce allows the transactions of business to be electronically performed and such move has generated substantial operational and strategic benefits. Developed countries have aggressively implemented e-commerce and in fact, e-commerce has become an essential part of business activities in these countries [7]. E-commerce could reduce the economic and digital gap between developed and developing countries. However, the countries who under development still lagging in e-commerce adoption. Also, there remains a lack of research on e-commerce readiness research in the developing countries and so, the relevance of e-commerce in these countries is not yet fully understood. As such, the main objective of this study is to readiness of e-commerce in BAHRAIN. To achieve this objective, this study will assess aspects of technology, legality and the environment of these countries in relation to the use of e-commerce.

E-commerce is expected to become a megatrend among businesses in the BAHRAIN. In fact by 2018, e-commerce in this region was forecasted to worth \$10 billion (BHD 36.7 billion). Further remarked that albeit having the lowest percentage of online purchase in comparison to other countries worldwide, BAHRAIN will see e-commerce expanding as a major megatrend its businesses [8]. At the current time, it has been estimated that the overall e-commerce value in BAHRAIN is at 2.5 billion dollars (BHD 9.2 billions) and it is expected to grow and generate new business and employment opportunities. E-commerce seems to be the fastest growing business in Bahrain. The general trend is a global trend and transforms transformational change into companies, culture and society. The market will also be affected by this transformation by generating business models.

E-commerce requires Internet as its key medium. As such, countries that attempt to improve their economy through the use of e-commerce are obliged to adopt Internet and accept both its advantages and disadvantages. On the other hand, there are numerous factors that contribute to e-commerce's success and failure. As such, the factors of performance expectations, stress forecasts, social factors, cost and awareness that could potentially influence users' intention to engage in online transaction is examined in this study.

3 Research Methodology

A total of 250 people in the Kingdom of Bahrain were chosen as respondents while online data gathering. The questionnaire has only one part and there are seven (7) key constructs that are linked with the intention of user behavior in using e-commerce for assessing possibility of acceptance level. Each construct is covered by a number of items. The 5-point scale is used to gauge user's degree of acceptance. The regression analysis is used to gauge the linkage between the seven keys with the behavior of user intention for using e-commerce.

4 Theoretical Background and Current Research

In understanding what determines user acceptance of e-commerce, in the beginning should elaborates the Unified Theory of Acceptance and Use of Technology (UTAUT) Model as below:

4.1 UTAUT Model

UTAUT model consolidates the past research that is related to TAM model. UTAUT describes the user intentions to utilize the Information System (IS) and

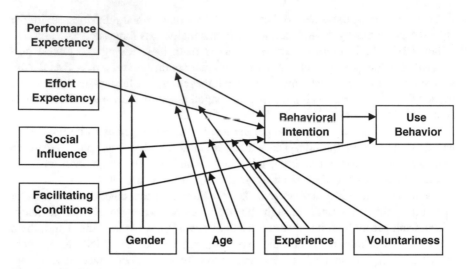

Fig. 1 UTAUT model [9]

ensuing user's behavior. UTAUT has four keys determine directly the user behavior and intention. As presented these determinants are: Performance Expectancy, Effect Expectancy, Social Influence, and Facilitating Conditions. Further, factors of Gender, Age, Experience, and Voluntariness of use age are theorized as the mediators to the impact of the aforementioned four key constructs [9]. The theoretical foundation was based on the review and consolidation of structures from eight models used by previous research on the behavior of the use of the information system, as appeared in Fig. 1.

4.2   Research Model Factors

Therefore this study propose a UTAUT with two more factors added. This model will be used to examine the two factors that impact in users' acceptance of e-commerce in Kingdom of Bahrain. The constructs of the proposed model are discussed as below:

4.2.1   Cost

As explained perceived price denotes the internalization or encoding of the objective selling price of a product or service. Then, when moving to m-commerce (MC) from wired e-commerce (EC), there will be added costs such as cost related to access, equipment and transaction [10]. In other words, MC is generally a more costly solution when compared to the wired EC. Further, financial and hardware/software resources appear to be crucial for users with respect to IS further added that compared to other factors that are linked to MC, cost is perceived as aspect for customers if they decide if they should or not purchase and utilize MC [11]. For instance, pointed out how important the factors of cost are, particularly with respect to innovation adoption. According to the author, the cost is a one of major factors that prevent Singapore users, also Australian users when they are using Internet Banking services [12].

In Malaysia, behavioral intention is found to considerably be affected by cost and also other factors [11]. Also Malaysian users, are respect perceived cost factor as one of the major obstacles that prevent m-commerce use among Malaysians. Also, in Finland, stated the factor of cost as significantly impacting users' implementation of 3G (Third Generation) services and in fact, in this country, the factor of cost is more important than the factors of privacy and security [13].

4.2.2 Awareness

Awareness is about familiarization of consumers by way of advertising and marketing of the brand, product, and services of the company, and also making the consumers aware of the special features and benefits of the products and services in a manner that is distinctive from the competitors particularly with respect to function or style. Stated that consumers who obtain information by way of mass media or word of mouth will have knowledge of the products and services [14]. Thus, promotion can generate awareness towards online services. In relation to this, reported that consumer purchase decisions are linked with e-promotions. Further, that site awareness has substantial impact on consumer's site commitment. As such, a positive relationship is assumed in this research as will be detailed in the section followed [15].

5　Research Framework and Hypothesis

Taking into account the factors that have the potential to impact users' acceptance of m-learning, three new constructs are added into TAM2 and IDT so that the factors that may affect the acceptance of university students towards m-learning can be scrutinized. These factors are: service quality, student readiness and trust. The condensed model can fully clarify m-Learning user (Fig. 2).

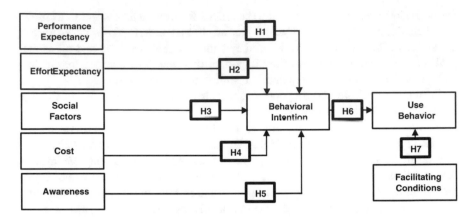

Fig. 2 The framework of research

5.1 Research Hypothesis

Hypothesis 1: Expected performance has a significant positive relationship with the behavioral intention to use e-commerce.

Hypothesis 2: Expected Effort has a great positive relationship with the behavioral intent to use e-commerce.

Hypothesis 3: Social factors have a great positive relationship with the behavioral intention to use e-commerce.

Hypothesis 4: Cost has a great positive relationship with the behavioral intention to use e-commerce.

Hypothesis 5: Awareness has a great positive relationship with the behavioral intention to use e-commerce.

Hypothesis 6: Behavioral intentionality has a great positive relationship with the use of behavior to use e-commerce.

Hypothesis 7: Facilitation of conditions has a great positive relationship with the use of behavior to use e-commerce.

6 Data Analysis and Results

Based on the results, each hypothesis shows importance based on the zero-order test. Thus, at this level, all hypotheses are supported. Moreover, the predictive model includes 60.1% of the variation in Behavioral Intent (BI) and is directly justified by expected performance, expected effort, social influence and cost as well as awareness. Also, the model involves 54.2% of the variation in usage behavior. This result is explained directly by behavioral intent and facilitation conditions. Figure 3 represents predictive models with R^2 as well as path parameters in the search model.

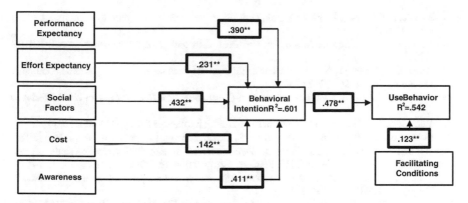

Fig. 3 Predictive model with R^2 and path coefficients

7 Conclusion

The expansion of e-commerce within the last decade in the developed nations has intrigued the developing counterparts particularly the Middle East; they understand how important it is to be a player in the e-commerce domain. It is important that the countries of the Middle East also benefit from Internet technology in trade. For most countries in the Middle East Internet users are still lagging in Internet's varied use when compared to their peers from the developed countries (i.e., Europe, America and Southeast Asia). In fact, there are very few e-commerce sites available in Arab countries. This study's findings fulfill its key main objective of increasing the intention to use of Arab users of Arabic e-commerce websites. Thus, it is expected that the increase in the number of Arab e-commerce websites would shift the online purchasing intentions of consumer to the local Arab websites. If such happens, the Arab countries will improve their microeconomics. The validated a framework model on the setting of Arabic Websites and online shopping industry. Has made a fundamental contribution to the range of knowledge and empirical results of the industry through a large model as the main guideline for academics and practitioners as well as in their business development practices.

References

1. Kotler, P., Keller, K.L., Ancarani, F., Costabile, M.: Marketing Management (2014)
2. Kalakota, R., Whinston, A.B.: Reading in Electronic Commerce. Addison Wesley Longman Publication Co., Inc, Boston (2015)
3. Rambo, K., Liu, K.: Culture sensitive virtual e-commerce design with reference to female consumers in Saudi Arabia. In: Virtual Worlds and e-Commerce (2010)
4. Ruba, F.M., Ahmed, H., Yousef, A.E.: Analysis the Uses of M-Shopping in Malaysian University, ASL (2017)

5. Megdadi, Y.A., Nusair, T.T.: Shopping consumer attitudes toward mobile marketing. IJMS **3** (2), 1 (2011)
6. Faqih, K.: Integration perceived risk and trust with technology acceptance model, ICRIIS (2011)
7. Ahmed, H., Ruba, M., Yousef, A.: The effect of trust based factors on using mobile commerce in Jordon. IJCCR **1**(2), 1–7 (2017)
8. Alhassan, K.A.: Consumer Confidence in e-Commerce: The Need for Reform in the Bahrain (2014)
9. Venkatesh, V.: User acceptance of information technology. MIS Q., 425 478 (2014)
10. Constantinides, E.: The 4P' s marketing mix model. Electro. Commer. Res. Appl. **1**(1) (2010)
11. Wu, J.H., Wang, S.C.: What drives mobile commerce? An empirical evaluation of the revised technology acceptance model. Inf. Manage. **42**(5), 719–729 (2015)
12. Anil, S., Ting, L.T.: Overcoming barriers of the successful adoption of mobile commerce in Singapore. IJMC **1**(1), 194–231 (2013)
13. Carlsson, C., Walden, P., Bouwman, H.: Adoption of 3G services in Finland. IJMC **4**(4), 369–385 (2016)
14. Martins, C., Oliveira, T., Popvoic, A.: Understanding the internet banking adoption: a unified theory of acceptance and use of technology and perceived risk application. IJIM **34**(1), 1–13 (2014)
15. Nuseir, M.T., Arora, N., Al-Masri, M.M.: Evidence of online shopping: a consumer perspective. Int. Rev. Bus. Res. Pap. **6**(5), 90–106 (2010)

A Novel Trimet Graph Optimization (TGO) Topology for Wireless Networks

Shanmuk Srinivas Amiripalli, Veeramallu Bobba
and Sai Prasad Potharaju

Abstract Topology plays an important role in modern wireless technologies like WSN, Adhoc networks, Cyber-physical system, and IOT. We have basic topologies like a bus, ring, star, mesh, tree, etc., which are not fulfilling all the requirements of the modern engineering problems. In this paper, we are making a new attempt to bridge the gap between existing topologies and network design problems. We are introducing a novel Trimet graph based optimized topology named as TGO (Trimet graph optimization) which is used to design a reliable, scalable, secure, simple, cost-effective topology for engineering problems. TGO topology was simulated extensively on various scenarios of wireless technologies using Cooja simulator. Final results are compared with the existing topologies.

Keywords Topology · Optimization · Wireless technologies · IOT
Trimet graph

1 Introduction

The basic partitions of network topology are physical and logical. In physical topologies devices are connected with cables called as wired networks, similarly, in logical topology devices are connected through signals called as wireless networks. Logical topology was mainly operated in two modes that are ad hoc and infrastructure. In ad hoc mode, we do not require central coordinator whereas in infrastructure mode we need a central coordinator. The usage of the topologies is

S. S. Amiripalli (✉) · V. Bobba · S. P. Potharaju
Department of Computer Science and Engineering, KL University,
Guntur 522502, Andra Pradesh, India
e-mail: shanmuk39@gmail.com

V. Bobba
e-mail: bvmallu@gmail.com

S. P. Potharaju
e-mail: psaiprasadcse@gmail.com

© Springer Nature Singapore Pte Ltd. 2019
P. K. Mallick et al. (eds.), *Cognitive Informatics and Soft Computing*,
Advances in Intelligent Systems and Computing 768,
https://doi.org/10.1007/978-981-13-0617-4_8

increasing nowadays and are applied in many engineering applications. The bases for these topologies are derived from graph theory a branch of mathematics. In graph theory, we have many specialized graphs; one among them is the complete graph or mesh fully connected.

Wireless Sensor Networks

WSN is a special type of wireless networks having a large number of independent units or devices. Each device is equipped with micro-controller, Transceiver, power source, memory, antenna and different type of sensors. The functionality of WSN is to collect data from the surroundings to the base station. Sensors are largely deployed and they are connected by star, tree, mesh topologies [1]. WSN are categorized into different types like terrestrial, mobile, underwater, multimedia wireless sensor networks and architecture consists of five layers namely physical, data link, network, transport, application layers and three cross layers namely task, mobility, power management.

Unequal Clustering

Unequal clustering is one of the solutions to the sinkhole problem with largely deployed sensors. In this process initially sensors are partitioned into different clusters and small clusters are formed near the base station and gradually increases the size of clusters at the next level [1]. These clusters will have cluster heads will transfer aggregated data to the sink node. Finally, from this process, we can increase the network lifetime.

Contiki OS

It is an open-source operating system for IOT and these devices are connected with optimal power, optimal process with low cost to the External word. The main design goal of this operating system is to communication devices with the optimal power to the Internet [2]. It can handle IPv6 IPv4, 6lowpan, RPL, CoAP, etc. Contiki OS is having a file structure consists of different files examples, app, CPU, regression-test, doc, tools, core, platform. The applications of Contiki OS are like health, industry, home monitoring systems.

2 Preliminaries

Trimet graph

G is simply connected and planar graph. One vertex called dominant vertex with the degree $(V - 1)$. Other $(V - 1)$ number of vertices with degree 2 and $(3V - 3)/2$ Number of edges if V is odd. One more vertex called Special Vertex with degree 3, other $(V - 2)$ number of hanging vertices with degree 2 and $(3V - 2)/2$ numbers of edges if V is even [3].

Properties of Trimet graph

- The minimum path distance of an Even GTEven Trimet $(n, (3n - 2)/2)$ and GTOdd $(n, (3n - 3)/2)$ is 1.

- The minimum path length between two vertices of inter sectors of Trimet GT (V, E) is 2.
- All Semi-Complete Graphs are not Trimet but all Trimet are Semi-Complete Graph and Trimet are not Bi-Partition Graphs.
- GTEven (n, (3n − 2)/2) is 3 vertices colorable, (n − 1) edge colorable and 3 regions colorable [4].
- GTOdd (n, (3n − 3)/2) is 3 vertices colorable, (n − 1) edge colorable and 2 regions colorable.
- GTEven (n, (3n − 2)/2), GTOdd (n, (3n − 3)/2) is a semi-Euler graph and Euler graph respectively.
- GTEven holds1 factor graphs where GTOdd does not.
- GTEven (n, (3n − 2)/2), GTOdd (n, (3n − 3)/2) holds Hand Shaking Lemma $\sum$degree (V_i) = (2*e).
- GTEven and GTOdd are even and odd Trimet respectively satisfies the Euler characteristic (χ) = ($R − E + V$) must equal to 2. Where R is the number of regions E is the number of edges and V is the number of vertices [4].

Semi-Complete Graph

A Graph G1 is said to be semi-complete if it holds the properties of simple and for any 02 vertices u1, v1 of G there is a vertex w of G1 such that w1 is adjacent to both u1 and v1 (in G) (i.e., Fu,w; VG is a path in G). K2 is complete, but not semi-complete. K1 is trivially semi-complete [5]. To avoid trivialities, throughout this paper we consider a nontrivial graph, with at least three vertices.

Properties of a Semi-complete graph:

- Any semi-complete graph is nontrivial (i.e., it has edges).
- Any Complete graph is semi-complete, but any semi-complete is not Complete graph [6].
- Any semi-complete graph G is connected and hence contains spanning tree. So, when G is a finite graph, "(G) , À (G) ¡1, where '"' and 'À' denote the number of edges and vertices respectively [6].
- Any edge of a semi-complete graph G lies in a triangle (cycle) and hence the graph has no cut edges, and so the graph is a block.
- Any vertex of a semi-complete graph G lies in a triangle and hence the degree of each vertex is at least 2. So the graph has no pendant vertices [7].

3 Proposed Approach

TGO Topology Mathematical Model:

Topology is a map of an internetwork that designates interconnection points, network divisions, and user communication paths. Designing a novel topology is the primary action in the mathematical model phase of the network design. So topology (interconnection points, network divisions, and user communication paths) could be transformed into a newly introduced TGO topology (vertices, sectors, edges) mathematical model.

Objectives

- To transform the concept of Trimet to defense segment for maintaining their own strategies at specific levels of the military hierarchy.
- Broadcasted messages to receive by authorized nodes in accordance with levels of hierarchy and distinctness over sectors.
- To invent a new network topology with own advantages over the network.
- To invent a new security, unlocking system by introducing sector key along with existing public and private keys for internal communication over the Trimet-based Network [8].
- To reduce the complexity or toughness over identifying or sectoring the node in the network.
- To provide tracking facilities for a military network to identify the victims by sectoring concept this results from the dynamic mesh.

Requirements

- Adjacency matrix formed by the node degrees of Trimet should possess the following properties.
- The determinant of even Trimet is Zero, a rank of an odd Trimet is the O (T) and Rank of an even Trimet is the O(T) − 1.
- Trace of the TRIMET shall always be Zero and If the sum of the elements in a row or column is referred as R then existence formula should satisfy $2 \leq R \leq O(T) - 1$.
- For $(2n + 2)$ number of vertices in even Trimet there exists $(3n + 2)$ number of edges where $n = 1, 2, 3...$
- For $(2n + 3)$ number of vertices in odd Trimet there exists $(3n + 3)$ number of edges where $n = 0, 1, 2, 3$.
- Highest degree among all the nodes of Trimet must be for dominant vertex and the value is O (T) − 1.

Adjacency matrices

Trimet is derived from the similar properties of semi-complete graph, possibly. Trimet T (V, E) can be represented mathematically by the concept of Adjacency Matrix. TRIMET can be designed with 3 minimum number of vertices. According to the formula derived for the required number of edges for the given vertices specified in the constraints section in the conceptual design phase. Adjacency Matrix for Tn, where n is even, can be anticipated with the following parameters If Vd is the Dominant vertex, then it holds $(n - 1)$ degree and V be the set of non dominant vertices, then $|V| = n - 1$ then one special vertex with constant degree 3 and the rest of the vertices will hold the degree 2 constantly.

Adjacency Matrix A of order $n \times n$ could be formed with the path lengths between all the vertices [9].For example, if $n = 8$ then the number of edges will be 11 in Fig. 1. We have a dominant vertex with the degree $(8 - 1)$ is 7, one special

$$T1=\begin{vmatrix} 0 & 0 & 0 & 0 & 0 & 0 & 1 & 1 \\ 0 & 0 & 1 & 0 & 0 & 0 & 0 & 1 \\ 0 & 1 & 0 & 0 & 0 & 0 & 0 & 1 \\ 0 & 0 & 0 & 0 & 1 & 0 & 0 & 1 \\ 0 & 0 & 0 & 1 & 0 & 0 & 0 & 1 \\ 0 & 0 & 0 & 0 & 0 & 1 & 0 & 1 \\ 1 & 0 & 0 & 0 & 0 & 1 & 0 & 1 \\ 1 & 1 & 1 & 1 & 1 & 1 & 1 & 0 \end{vmatrix} T2=\begin{vmatrix} 0 & 1 & 1 & 1 & 1 \\ 1 & 0 & 1 & 0 & 0 \\ 1 & 1 & 0 & 0 & 0 \\ 1 & 0 & 0 & 0 & 1 \\ 1 & 0 & 0 & 1 & 0 \end{vmatrix}$$

Fig. 1 Matrix representation of Trimet graph

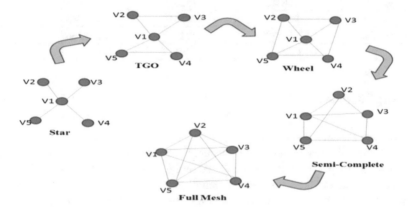

Fig. 2 Diagrammatic transformation of different hub topologies

vertex with degree 3 and leftover are hanging vertex with degree 2. If $n = 5$ then the number of edges will be 6, as per Fig. 2, we have a dominant vertex with the degree $(5 - 1)$ is 4, no special vertex for odd Trimet and leftover other are hanging vertex with degree 2.

4 Implementation and Testing

In this phase finally, we conducted the experiment on Cooja simulator using Contiki OS with different time's spans [10]. Throughout the experiment, we used only five sky motes for all topologies by taking Unit Disk Graph Medium and collect view shell. We have compared star, wheel and proposed TGO topology and we found TGO topology is performing well over star, but less effective than partial mesh, the same is given in Table 1 (Fig. 3).

Table 1 Comparison between different typologies

Parameter	Star topology	TGO topology	Partial Mesh (wheel)
Received	16.66667	16.58333	16.08333
Dups	0	0	0
Lost	0	0	0
Hops	1.75	1.5	1.5
Rtmetric	21.71383	28.05558	33.80392
ETX	1.789667	1.35975	1.37475
Churn	0.083333	0.25	0.333333
Beacon interval (s)	342.6667	340.6667	336
Reboots	0	0	0
CPU power	0.372833	0.364083	0.3585
LPM power	0.152167	0.1525	0.269083
Listen power	0.4275	0.436417	0.450833
Transmit power	0.225333	0.171667	0.172667
Power	1.224417	1.149167	1.112667

5 Applications

Defense and disaster management

In defense, disaster management systems we come across a situation like, initially a large group of people are connected and do rescue or search operation, latter these large groups will be divided into small groups and they become independent groups. For better communication among them dynamic mesh topology was introduced in IPV6. In these independent groups we required a edge survivable network to connect each other. TGO and wheel networks will have a better survivability are proposed in this paper.

Missiles and Plain Network Management

Inter node communication in missiles can increase the probability of hitting the target. This can be handle best by TGO topology because if four missiles were used on a specific target with time delay. Suppose if first target was missed then that target will give information to the next missile. So that second can hit the target by taking exact information from previous target because of its inter-communication in a sectors.

Cybe- physical system and IOT

In creating smart world, sensors, cyber-physical systems, IOT place an important role. The major challenges face by this industry is increase of sensors in larger numbers. Statistics says that in the year 2020, devices will reach more than 20 billion. To address these problems, we need a special topologies and network designs to fulfill these requirements. In this paper we proposed novel network models based on star, TGO, wheel graphs.

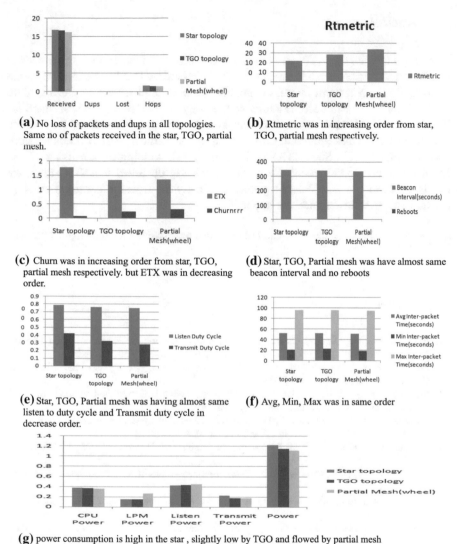

(a) No loss of packets and dups in all topologies. Same no of packets received in the star, TGO, partial mesh.

(b) Rtmetric was in increasing order from star, TGO, partial mesh respectively.

(c) Churn was in increasing order from star, TGO, partial mesh respectively. but ETX was in decreasing order.

(d) Star, TGO, Partial mesh was have almost same beacon interval and no reboots

(e) Star, TGO, Partial mesh was having almost same listen to duty cycle and Transmit duty cycle in decrease order.

(f) Avg, Min, Max was in same order

(g) power consumption is high in the star , slightly low by TGO and flowed by partial mesh

Fig. 3 Graphical representation of star, TGO, wheel topologies

6 Future Scope

In future, we will design and present a modeled heterogeneous complex networks based on TGO topology, which will address the problems facing in Internet networking from the broadcast of unauthorized messages to an entire connected network. We also look forward to simulating the performance analysis of TGO Topology with multiple topologies like the ring, mesh, and tree in heterogeneous complex wireless networks based on various routing techniques.

7 Conclusion

Trimet is a Graph GT (V, E) exists in two forms basing on the number of vertices in it. If we need to establish a simple and useful Network with a head control over the sectorized members with internal communication secured through a Sector, there the TGO topology will become a reliable topology. TGO topology is best suited for WSN, Adhoc networks, Cyber-physical system and IOT because of its Sectors, Dominant Vertex and Interconnection among other Vertices. The wireless network was compared with Star, TGO and partial mesh topologies for using Cooja Emulator. From experimental results, it was clearly observed that TGO topology lies in between star and Wheel topologies. Finally the research contribution is an optimal network for engineering applications.

Acknowledgements The authors would like to express their gratitude for the support offered by the GITAM University.

References

1. Rao, P.C.Srinivas, Banka, Haider: Novel chemical reaction optimization based unequal clustering and routing algorithms. Wirel. Netw. **23**(3), 759–778 (2017). Springer
2. Pietro Gonizzi, I., Duquennoy, S.: Hands on Contiki OS and Cooja simulator: Exercises (Part II), no. Part II, pp. 1–15 (2013)
3. Srinivas Amiripalli, S., Bobba, V.: Research on network design and analysis of TGO topology. Int. J. Netw. Virtual Organ. (in press 2017)
4. Srinivas Amripalli, S., Kumar, K., Tulasi, A.B.: TRIMET along with its properties and scope. Am. Inst. Phys. Conf. Proc. **1705**, 0200321–0200329 (2016)
5. Rao, I.H.N.R., Raju, S.V.S.R.: Semi-complete graphs. Int. J. Comput. Cogn. **7**(3), 50–54 (2009)
6. Rao, I., Rao, I.H.N.R., Raju, S.V.S.R.: Semi-complete graphs-II. Int. J. Comput. Cogn. **8**(3), 57–62 (2010)
7. Rao, I.H.N.R., Raju, S.V.S.R., Shanmuk Srinivas, A.: On path connector sets. IJMSC **2**(2), 55–65 (2012)
8. Kamalesh, V.N., Shanthala, K.V., Ravindra, V., Chandan, B.K., Pavan, M.P., Bomble, P.P.: On the design of fault tolerant k-connected network topologies. Int. J. Innov. Manag. Technol. **6**(5), 339–342 (2015)
9. Ali, Z.H., Ali, H.A., Badawy, M.M.: Internet of Things (IoT): definitions, challenges and recent research directions. Int. J. Comput. Appl. **128**(6), 37–47 (2015)
10. Wikipedia. [online] https://en.wikipedia.org/ (Accessed 07 February 2016)

Performance Analysis of Reliability and Throughput Using an Improved Receiver—Centric MAC Protocol and Itree-MAC Protocol in Wireless Sensor Networks

K. S. Ananda Kumar and R. Balakrishna

Abstract Wireless sensor networks have been identified as one of the key areas in the field of wireless communication. When an event has occurred, the number of data packets will be generated. A MAC protocol considered for this category of wireless sensor networks, here MAC protocols able to adapt to both light and heavy traffic load situations of the network. Accessible MAC protocols are used for light traffic for the energy efficiency concern. Our proposed systems An Improved Receiver-Centric MAC protocol and Itree-MAC protocol implemented in NS2 Simulator, the significance of this paper is highlight the idle listening, infrequent, and light weight synchronization in wireless sensor networks. Here evaluating the performance analysis of Improved Receiver-Centric MAC protocol and Itree-MAC protocol; considering comparing parameters like throughput, reliability and end-to-end delay in different sensor nodes in wireless sensor networks. Comparing MAC, receiver-centric MAC, Improved Receiver-Centric MAC with Itree-MAC protocol, the results show that IRC-MAC, Itree-MAC protocols are more efficient than the RC-MAC and IEEE 802.11 MAC protocol.

Keywords Improved receiver-centric MAC · Itree-MAC · Receiver-centric MAC · Reliability · Throughput

K. S. Ananda Kumar (✉)
Dept of CSE, Department of Information Science and Engineering,
RajaRajeswari College of Engineering, Bangalore-74, Bangalore, India
e-mail: anandgdk@gmail.com

R. Balakrishna
Department of Computer Science & Engineering,
RajaRajeswari College of Engineering, Bangalore-74, Bangalore, India
e-mail: rayankibala@yahoo.com

© Springer Nature Singapore Pte Ltd. 2019
P. K. Mallick et al. (eds.), *Cognitive Informatics and Soft Computing*,
Advances in Intelligent Systems and Computing 768,
https://doi.org/10.1007/978-981-13-0617-4_9

1 Introduction

Wireless Senor Networks are playing a major role in environmental monitoring, agriculture, military, etc. WSN contains nodes, which collect data from different nodes and sharing that data with other nodes. A sensor node is a tiny device containing a communication unit, processing unit, sensor unit and battery. Energy, reliability, and delay are important parameters for wireless Senor Networks. MAC protocols have an important role in life of wireless Senor Networks. An efficient design of MAC protocols will decrease the consumption of battery, increase the network lifetime [1].

Wireless MAC protocols are divided into two types, distributed and centralized. In distributed, random access is one type, again centralized MAC protocols have divided into three categories, (i) random access (ii) guaranteed access (iii) hybrid access. Wireless medium makes the MAC layer design have important issues are (a) half duplex operation (b) errors in burst channel (iv) time varying channels [2].

The organization of the paper is as follows. The MAC Related work is explained in Sect. 2, the methodology of the work is explained in Sect. 3, Interpretation and obtained results are explained in Sect. 4, the conclusion of the paper explained in Sect. 5.

2 Related Work

Verma et al. proposes work on MAC protocols for WSN. Here they discussed the different MAC layer protocols, issues related to MAC layer protocols like collisions, protocol overhead, idle listening and complexity [3].

Vikram et al. worked on multichannel MAC protocol for WSN, here they discussed the new protocol used for cross-layer multi-channel in MAC protocol. For network performance using hidden Markov model, ensures that minimum interference and enhance network performance [4].

Kakria et al. worked on synchronous MAC protocols for WSN, here they discussed about various synchronous MAC protocols [5].

Singh et al. proposes a MAC protocol for transmission delay reduction in WSN, here they discussed about joint routing and MAC protocol, uses a novel cycle structure, reduces the idle listening duration, showing less delay in many to one communications [6].

Agarwal et al. working on energy consumption and life time of WSN, here they are using contention based MAC protocol, they present the stochastic energy model is used for determining the energy of sensor nodes [7].

Anwar et al. worked on DORMS, here they investigate the different MAC variants, power usage and latency, it can lead to both energy savings and performance enhancements [8].

Dou et al. proposes a reliable protocol used in hybrid WSN, they design, reliable hybrid MAC, it proposes two mechanisms (i) hybrid polling (ii) hybrid duty cycle scheduling, 21.5% of EHSN's energy level in each duty cycle [9].

Kuo et al. worked with sensor MAC, here they are using the NS2 simulator, in this work results showed that throughput of S-MAC is more than the bidirectional traffic, here proposed work saves 25 and 47% energy bidirectional and unidirectional respectively, idle listening problem is solved in this approach [10].

Rodríguez-Pérez et al. worked on receiver initiated MAC for WSN. Here results shows that collisions are reduced and delivery time is maintained in duty cycle, it shows better results in terms of performance of EH-MAC compared to PW-MAC [1].

3 Methodology

Figure 1 explains the methodology for Improved Receiver-Centric MAC protocol. In this methodology start with tree construction of topology and then attribute based selection. Then it moves to position deployed stage, if yes, it goes to route selection, data gathering and Transmission of packets. If No at position deployed it returns to tree construction. Then repeats the same process for completion of all nodes.

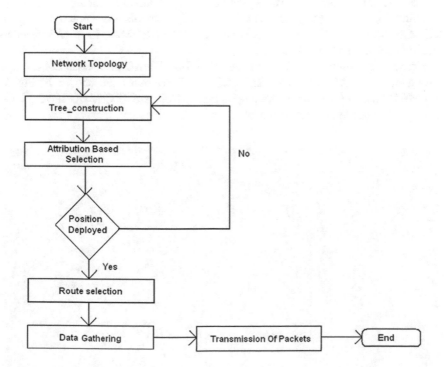

Fig. 1 Methodology for IRC-MAC protocol

4   Results and Discussion

In simulation, the work is done by using the NS2 simulator. In this paper, we are comparing the performance of the RC-MAC, IRC-MAC and Itree-MAC protocol with various parametric measures like delay, reliability and throughput. In the experimental setup, Analysis of RC-MAC, IRC-MAC and Itree-MAC protocols for WSN are performed using NS2 simulator. In experimental setup we are using AODV routing protocol, wireless channel, packet size of 780 bytes and no of nodes used are 100.

4.1   Performance Analysis of MAC, RC-MAC and IRC-MAC Protocols

In this research paper, we considered the parameters like end-to-end delay, reliability and throughput.

Here we are comparing MAC, RC-MAC and IRC-MAC protocols. The average end-to-end delay of MAC, RC-MAC and IRC-MAC protocols are shown in Fig. 2. RC-MAC, IRC-MAC protocols have less delayed compared to the MAC protocol.

Here we are comparing MAC, RC-MAC, and IRC-MAC protocols. The average reliability of MAC, RC-MAC and IRC-MAC protocols are shown in Fig. 3. RC-MAC have 22.37% more reliable than MAC and IRC-MAC have 12.82% more reliable than the RC-MAC protocol.

Here we are comparing MAC, RC-MAC and IRC-MAC protocols. The average throughput of MAC, RC-MAC and IRC-MAC protocols are shown in Fig. 4. RC-MAC have 4.03% more throughput than MAC and IRC-MAC have 3.84% more throughput than MAC protocol.

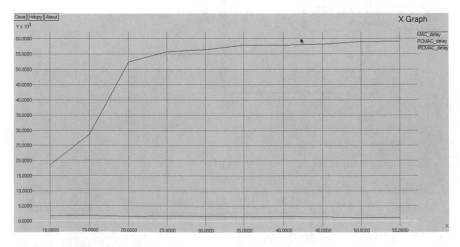

Fig. 2 Graph for end-to-end delay versus simulation time

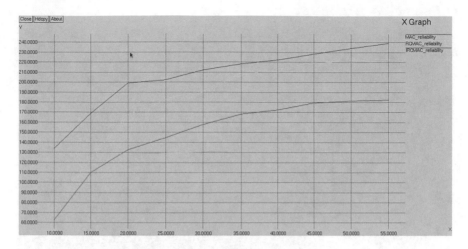

Fig. 3 Graph for reliability versus simulation time

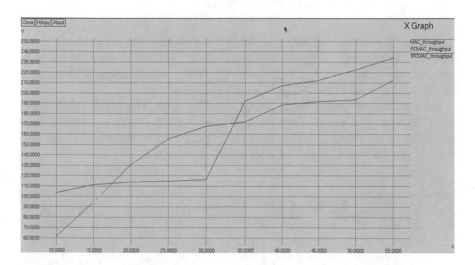

Fig. 4 Throughput versus simulation time

4.2 Performance Analysis of MAC, RC-MAC, IRC-MAC and ITREE-MAC Protocols

Here we are comparing MAC, RC-MAC, IRC-MAC, and ITREE-MAC protocols. The average end-to-end delay of MAC, RC-MAC, IRC-MAC and ITREE-MAC protocols are shown in Fig. 5. IRC-MAC, ITREE-MAC protocols have less delayed compared to the RC-MAC and MAC protocols.

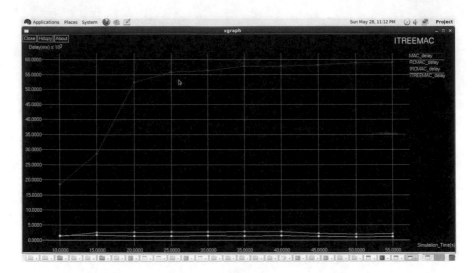

Fig. 5 Graph for end-to-end delay versus simulation time

Here we are comparing MAC, RC-MAC, IRC-MAC, and ITREE-MAC proto-cols. The average reliability of MAC, RC-MAC, IRC-MAC and ITREE-MAC protocols are shown in Fig. 6. IRC-MAC, ITREE-MAC protocols have more reliability compared to the RC-MAC and MAC protocols.

Here we are comparing MAC, RC-MAC, IRC-MAC, and ITREE-MAC proto-cols. The average Throughput of MAC, RC-MAC, IRC-MAC and ITREE-MAC protocols are shown in Fig. 7. ITREE-MAC protocol has more throughput com-pared to the IRC-MAC, RC-MAC, and MAC protocols.

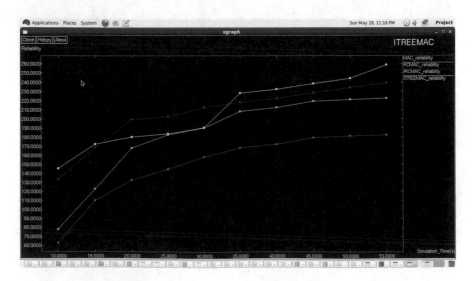

Fig. 6 Graph for reliability versus simulation time

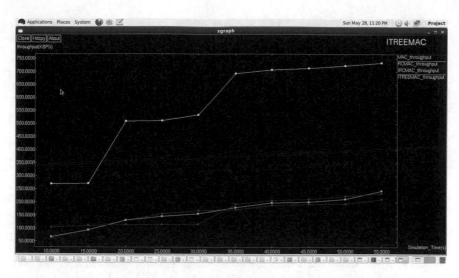

Fig. 7 Graph for throughput versus simulation time

5 Conclusion

Currently many researchers working on different MAC layer protocols of wireless sensor networks. In this research paper, we have implemented MAC protocols for various parameters like throughput, reliability and delay. Here compared the MAC, RC-MAC, IRC-MAC, and Itree-MAC protocols for various parameters like throughput, reliability, and end-to-end delay. Results show that RC-MAC, IRC-MAC protocols have less delayed compared to the MAC protocol. RC-MAC have 4.03% more throughput than MAC and IRC-MAC have 3.84% more throughput than MAC protocol. RC-MAC have 22.37% more reliable than MAC and IRC-MAC have 12.82% more reliable than the RC-MAC protocol. Itree-MAC have more throughput and Reliability compared to IRC-MAC, ITree-MAC have less Delay compared to IRC-MAC protocol.

Acknowledgements The authors would like to express sincere thanks for the encouragement and constant support provided by the Management RRGI, and Principal RajaRajeswari College of Engineering, Bangalore-74, India during this research work.

References

1. Rodríguez-Pérez, M., Sergio, H.-A., Manuel, F.-V., Cándido, L.-G.: A self-tuning receiver-initiated MAC protocol for wireless sensor networks. IEEE Wirel. Commun. Lett. **4**(6), 601–604 (2015)

2. Kumar, K.S.A., Balakrishna, R.: An improved receiver centric mac protocol for effective power management in wireless sensor networks. Int. J. Eng. Sci. Technol. (IJEST™), **9**(09), ISSN: 0975–5462 (2017, Sep)
3. Verma, A., Singh, M.P., Singh, J.P., Kumar, P.: Survey of MAC protocol for wireless sensor networks. In: 2015 IEEE Second International Conference on Advances in Computing and Communication Engineering (ICACCE), pp. 92–97 (2015, May)
4. Vikram, K., Narayana, K.V.L.: Cross-layer, multi channel MAC protocol for wireless sensor networks in 2.4-GHz ISM band. In: IEEE International Conference on Computing, Analytics and Security Trends (CAST), pp. 312–317 (2016, Dec)
5. Kakria, A., Aseri, T.C.: Survey of synchronous MAC protocols for wireless sensor networks. In: 2014 IEEE Recent Advances in Engineering and Computational Sciences (RAECS), pp. 1–4 (2014)
6. Singh, R., Rai, B.K., Bose, S.K.: A joint routing and MAC protocol for transmission delay reduction in many-to-one communication paradigm for wireless sensor networks. IEEE Internet Things J. (2017)
7. Agarwal, V., DeCarlo, R.A., Tsoukalas, L.H.: Modeling energy consumption and lifetime of a wireless sensor node operating on a contention-based MAC protocol. IEEE Sens. J. **17**(16), 5153–5168 (2017)
8. Anwar, A.-K., Kim, J.-K.: DORMS: design of multi-objective optimized RPL and MAC protocols for wireless sensor network applications. In: 2017 IEEE Ninth International Conference on Ubiquitous and Future Networks (ICUFN), pp. 147–152 (2017)
9. Dou, S., Liu, D.: A reliable MAC protocol for hybrid wireless sensor networks. In: 2016 IEEE International Conference on Internet of Things (iThings) and IEEE Green Computing and Communications (GreenCom) and IEEE Cyber, Physical, and Social Computing (CPSCom) and IEEE Smart Data (SmartData), pp. 211–216 (2016)
10. Kuo, Y.-W., Liu, K.-J.: Enhanced sensor medium access control protocol for wireless sensor networks in the NS-2 simulator. IEEE Syst. J. **9**(4), 1311–1321 (2015)

A Secure and Computational-Efficient Multicast Key Distribution for Wireless Networks

B. Srinivasa Rao⑩ and P. Premchand

Abstract In the present research work a simulation mechanism has been designed and implemented for a secure, computation efficient and dynamic multicast key distribution system that can be applicable to wireless networks. The present model uses the concept of Maximum Distance Separable (MDS) codes to enhance the computation efficiency of the multicast key distribution system. Also an attempt has been made to improve the security of the multicast key distribution system against the previous multicast group members that had turned into adversaries. By encrypting the newly generated keys with the older one may provide improved security to the system. The DES algorithm has been used for encryption of the new session keys. The implemented results have presented and discussed.

Keywords Multicast key distribution · Security · Session keys
Computational efficiency · Wireless network · Key security · Dynamical distribution

1 Introduction

It is well understood that the multicast key distribution system plays a significant role in providing security and computation efficiency in various wireless network applications. In multicast key distribution system the security issue is attended by some key distribution systems using dynamic or time variant session keys. At the

B. Srinivasa Rao (✉)
Department of Computer Science and Engineering,
Gokaraju Rangaraju Institute of Engineering and Technology
(Affiliated to Jawaharlal Nehru Technological University Hyderabad),
Bachupally, Hyderabad 500090, Telangana, India
e-mail: bsrgriet2015@gmail.com

P. Premchand
Department of Computer Science Engineering, University College
of Engineering, Osmania University, Hyderabad 500007, Telangana, India
e-mail: p.premchand@uceou.edu

© Springer Nature Singapore Pte Ltd. 2019
P. K. Mallick et al. (eds.), *Cognitive Informatics and Soft Computing*,
Advances in Intelligent Systems and Computing 768,
https://doi.org/10.1007/978-981-13-0617-4_10

same time the efficiency of the communication system is decided by network features like low computation complexity, communication cost, storage cost, etc. The balance between these two mechanisms yields a secure and efficient multicast key distribution system. One of the multicast key distribution methods is the distributed key management system in which the security design is based on either a cryptographic or non-cryptographic approaches. Over a time period several secure and efficient multicast models have been developed and implemented using any one of these two approaches. In this section, a brief review over the progress and development of multicast key distribution models has been presented. It can be seen that the recent research in this direction has paid much attention to cryptographic approaches [1–10] in comparison with non-cryptographic methods. This is may be due to the possibility of achieving ambient computation-efficient features in their respective models by using different cryptographic tools conveniently. Dutta et al. [1] proposed a key distribution model that recovers a session key that might have been lost during broadcast communication in WSN. For achieving less power consumption in networks a cluster based key distribution mechanism was proposed by Suganyadevi et al. [2]. A time-invariant key structure model was introduced by Zhow [3] that manifests low computational complexity at server level. The SGCMKDS model reported by Palanisami and Annadurai [4] was designed for confidentiality and integrity using group key security. A highly secured model was designed for key distribution mechanism to provide key authentication and filter the malicious adversaries by Vijaykumar et al. [5]. Elliptic curve cryptography based multicast key distribution mechanisms have been reported to provide reduced computational and communication complexity and improved security [6, 7]. A standard protocol was designed incorporating cryptographic methods for an efficient key management and distribution mechanism by Hanatani et al. [8]. Recently dynamically variant cluster concept has been incorporated into multicast key distribution model to solve the 'one affects many' problem in wireless network systems [9]. In order to overcome the intruder problem in the wireless networks Zhu [10] has presented a new protocol that adopts private key system and cryptographic random number generation. Many other models in this direction can be referred to the vast literature. On the other hand scant attention has been paid on non-cryptographic approaches. The necessity of development of diversified approaches and difficulty of achieving justified efficiency may be reasons for scant attention on non-cryptographic approaches. However, a computational efficient multicast key distribution scheme with reduced computational complexity was proposed implementing MDS codes by Xu et al. [11]. Some other models have been reported considering MDS codes for secure and multicast key distribution [12, 13]. Recently, mobility-based key distribution mechanism has been developed for providing security in mobile and ad hoc networks [14]. A comparative study of the above-discussed models indicates that in either of the above-discussed approaches, it may be difficult to have a standard and sophisticated model for secure and computational-efficient multicast key distribution with out any compromise [15–17]. This may be due to various inherent weaknesses and vulnerabilities of the wireless network systems. Further continuous improvement of these models to achieve high level security and efficiency may be a possible solution. Hence, in the present research work, an improved secure and computation-efficient multicast key distribution system using MDS codes has been proposed and implemented. But some of these models may have a security threat from former group members who could become adversaries. In the present model the security is improved by encrypting new session keys using the older one. As it would be difficult for a group member to have access to the group after leaving the group, the adverse group members cannot obtain the old keys [5]. Thus encryption of new keys with previous keys enhances security of the system further. In this paper, the existing multicast key distribution mechanism has been discussed in Sect. 2. In Sect. 3 the improved secure multicast key distribution system has been described. The detailed simulation mechanism of the present model has been described in Sect. 4. The implementation part is presented in Sect. 5. Finally the results have been discussed and concluded in Sect. 6.

2 Computation-Efficient Multicast Key Distribution System (CEMKDS)

Xu and Huang [11] proposed a Computing-Efficient Multicast Key Distribution scheme for wireless sensor networks to provide reduced computational complexity. The model used Maximum Distance Separable (MDS) codes to achieve the security and reduced computational complexity [11]. The complete details of the model and its implementation are available in Ref. [11]. Here the mechanism of the model is briefly discussed. The interesting and complex feature of the multicast key distribution is the change of the session keys of the key distribution system dynamically due to frequent time variant inclusion and exclusion of group members of the system [11]. In the multicast groups it is essential to provide the keys only to authorized users through central system designated as Group Controller (GC) [11]. The GC requires communication storage and computation resources to provide session keys to the group members [11, 16, 18]. While joining or leaving the group by a multicast group member, the GC selects a fresh value to compute the new session keys and old one are discarded [11]. But it is also essential to consider a security threat from members who leave the group and becomes adversaries later [5].

3 Present Secure and Computing-Efficient Multicast Key Distribution (SCEMKD) System

The present secure and computation-efficient multicast key distribution system has been designed based upon the CEMKDS proposed by Xu and Huang [11] by enhancing the security of the system by considering the case of a former group member becomes a potential adversary [5]. The present model is designed in four modules as: 1. Group Controller. 2. Client. 3. Key_generation. 4. Rekeying [7, 13]. 1. Group Controller: It is a server that provides keys for chat members for secured chatting and communication. 2. Client: A new member authorized by the GC intended to join the chatting group. 3. Group Key Generation: It generates keys and supplies to the group/chat members to communicate information in secured manner. 4. Rekeying: Generation of new keys to replace the old keys when member joins or leaves a chat group [7, 11, 13]. During the rekeying process the security of the system can be further improved by considering the case of a former group member may become a potential adversary [5]. The newly generated keys may be encrypted by using old session key and any symmetric algorithm and communicated to the group members securely. It is very difficult to the adversary to recover the old key after leaving the group [5]. Thus the present model has an improved security. The simulation mechanism of presently proposed model is detailed in the next section.

4 Simulation Design

The present section describes the design and development of the simulation mechanism intended for implementation of the secure and computation-efficient multicast key distribution system. The simulation mechanism is designed in the framework of database design model to describe all the required logical and behavioural features of the multicast key distribution system effectively. As mentioned earlier, the important components of the computation-efficient multicast key distribution are: group controller, client, group key generation and rekeying [7, 11, 13]. The functionality of these components is discussed elsewhere [7, 11, 13]. The overall architecture of the SCEMKD system can be visualized as shown in Fig. 1. Initially, the functional mechanisms of the individual component are designed. Later these mechanisms are combined into an overall single unit. In this process the database model concepts are used to design the mechanisms at different contexts.

The behavioural diagrams of the client and group controller are shown in Fig. 2. The client is authorized to login to have secure communication with fellow group mates by getting the session key and secret key from the group controller during registration.

The roles, functionality, behaviour of individual objects and the overall operation of the system in real time are shown the collaboration diagram as shown Fig. 3.

As mentioned earlier, the client logins with secret and session key to the group controller to communicate with a group in the system. The group controller verifies the existence of the particular multicast group in the database and forwards the request to group key generation module for activation of chat module. The client logins with a secret key and session key which is sent as a request to the group controller. The group controller (GC) checks the database thoroughly which

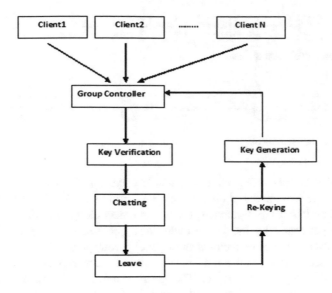

Fig. 1 Architecture of the computing-efficient multicast key distribution design

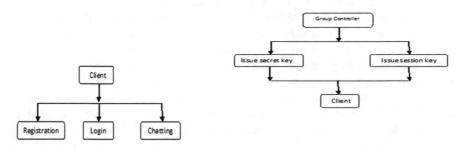

Fig. 2 Behavioural diagrams of the client and group controller

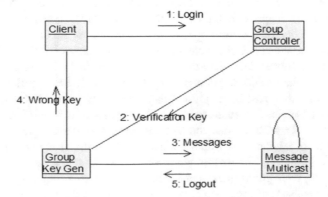

Fig. 3 Design of simulation technique for computing-efficient multicast key distribution

Fig. 4 Client model class diagram

Fig. 5 Server model class diagram

consists of multicast group table, session key table and user details table. The multi-groups table consists of different kinds of groups involved in an organization like the administrator, programmer, tester and others along with their session keys which are generated by the group controller. The session table consists of the registered user's names or member name, their group name and the session keys that they possess. The user detail's table consists of users name, their secret key and the password they use for login. The group controller checks all these tables thoroughly and then the client is allocated to the chat room. The relations and source code dependencies can be depicted by class diagrams for both client and server model as shown in Figs. 4 and 5. The client class diagram consists of four classes namely a login class, verification class, key verify class and SGSclient class. Each class consists of three fields namely name of the class; the fields present in that class; and the methods associated with them. The above classes posses a direct association relationship between each other since each class is dependent on other class for functioning. The client has to login using username and password which is then verified in the verification class. Hence, the secret key and session key which are entered are verified in the key verification class. If all the verifications are true then the client enters the chat room and can communicate with the server belonging to the same group.

The server class diagram as shown in Fig. 5 consists of three classes namely Select Group class, session key class and SGSserver class. The server is initialized by selecting a group. In the next step the session key of that particular group acquired from the database. In this model the session key is generated by group key generation module. Then a form opens which is called as the Server form. All the information from the client who have joined the chat room or left the chat room will be stored in the server along with the communication that takes place.

The entity–relationship diagram (in Fig. 6) describes three entities which represent the tables that exist in the present database. They are Multi-Groups, Session Key and User Details. The multi groups table consists of different kinds of groups involved in an organization like the administrator, programmer, tester and others along with their session keys which are generated by the group controller. It also consists of a group ID. The session table consists of the registered user's names or member name, their group name and the session keys that they possess. The user detail's table consists of users name, their secret key and the password they use for

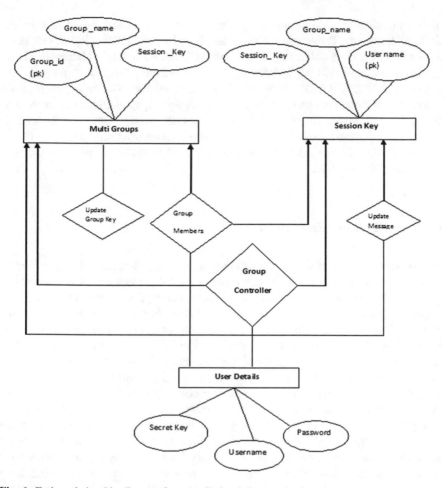

Fig. 6 Entity relationship diagram for overall simulation mechanism

login. The group controller is in one-to-many relationships between with each and every table in the database. The multi-groups table relates with the session key during the rekeying process. The user details table makes track of the members that have registered, i.e. new and old members. Session table is also related to the user members and it also updates messages.

5 Implementation of the Present Model

The implementation phase comprises of several activities. 1. The required hardware/software acquisition is carried out. 2. The required software (code) is developed, verified and tested. For encryption of the new keys with old keys DES algorithm is used. 3. The tested system is used for simulation of the present SCEMKD system. The required code is developed for the scheme in C#.NET using Mrosoft.NET Technologies, Microsoft-SQL-Server-2005, Microsoft-Visual-Studio-2008, and MS_ Windows_XP. The Hardware Requirements are Intel Pentium, RAM: 512 MB (*Minimum*), Hard disk: 40 GB. The deployment of the resources for implementation of the presently designed simulation mechanism is in the following steps. Step1: Installing Microsoft visual Studio 2008. Step 2: Installing Microsoft SQL Server 2005. Step3: Attaching the database file to Microsoft SQL Server 2005. Step4: Execution of Software Step5: Collection of results into screen shots.

6   Results and Conclusion

The implementation results of the simulation process are presented in the form of screen shots as shown in Fig. 7. Each screen shot describes a location and action taken at that location. The sequence of these descriptions constitutes the entire implementation of the present designed scheme. For better understanding and clarity of the outcome of the present simulation technique the descriptions of the screen shots have been tabulated and presented in Table 1. From Fig. 7 and Table 1 it is very clear that the present simulation technique is working properly in simulating the present secure and computation-efficient multicast key distribution scheme. The Group controller, Client, Key generation and Rekeying modules have been implemented successfully. Both the entry and exit secrecy of the multicast group system has been ensured along with improved security by changing the session keys dynamically during the simulation. However, the improved security with use of old keys for encrypting the new keys is not directly visible. The internal security measures like validation of user id, password and password length have been taken properly. Thus a secure and computation-efficient multicast distribution system has been designed and implemented successfully. Scope of the present work: Using of cryptographic algorithm for further improvement of the security may project the present system as a hybrid model. Also it is necessary to check the influence of this additional computation on computation complexity and other features of the entire model. In addition, a systematic comparison with presently existing models would give a better picture about the proposed model. Conclusion: Thus the present scheme may be a prototype for further extension of the scheme for various multicast key distribution systems and other efficient key distribution systems.

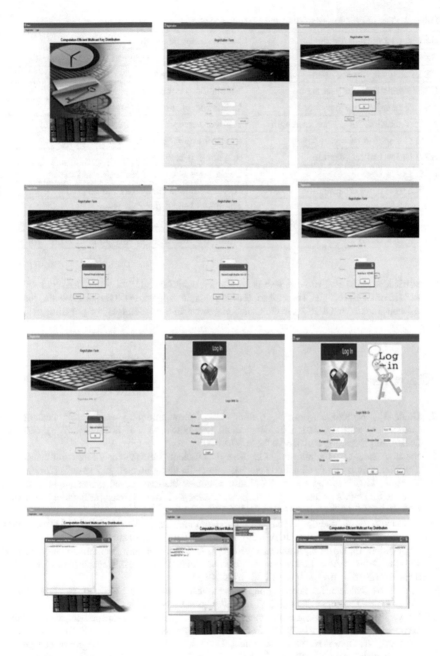

Fig. 7 Screen shots during the implementation of the simulation process

Table 1 The simulation results of the present scheme

Screen shot during simulation	Result during simulation
SS 1: Client start	Starting page for the client for login/registration
SS 2: Registration for client	Secret key generation
SS3: Validations for user name	Validation for the client side security
SS4: Validations for password	Validation of the password-client security
SS5: Validation for password length	Password length should be >6 & <10: validation
SS6: Generation of secret key	A secret key is generated for the client
SS7: Sucessfull registration	Registration is successful
SS8: Login page	Login page is opened for registered users
SS9: Login page continuation	Server IP and security key are entered
SS10: Chat room	The client joined the chat room
SS11: Client-server communication	Communication between the client and server
SS12: Chat room	Multiple numbers of clients joined the chat room

Acknowledgements B. Srinivasa Rao is very much thankful to Dr. L. Pratap Reddy, Professor, Department of ECE, JNTUH, Hyderabad, for his valuable suggestions. The author is also thankful to the Management of GRIET for their encouragement and cooperation for pursuing his Ph.D. work.

References

1. Datta, R., Mukhopadhyay, S., Collier, M.: Computationally secure self healing key distribution with revocation in wireless ad hoc networks. Ad hoc Netw. **8**(6), 597–613 (2010). http://doi.org/10.1016/j.adhc.2009.11.005
2. Suganyadevi, D., Padmavathi, G.: Energy efficient CBMT for secure multicast key distribution in mobile and adhoc networks. Procedia Comput. Sci. **2**, 248–255 (2010)
3. Zhou, J., Ou, Y.: Key tree and Chinese reminder theorem based group key distribution scheme. In: Hun, A., Chang, S.L. (eds.): Algorithms and Architectures for Parallel Processing, ICA3PP2009(LNCS), pp. 254–265 (2009)
4. Planisamy, V., Annadurai, P.: Secure group communication using multicast key distribution scheme in adhoc networks. IJCA **1**(25), 86–91 (2010) (www.ijcaonline.com)
5. Vijaykumar, P., Bose, S., Kannan., Deborah, L.J.: Computation and communication efficient key distribution protocol for secure multicast communication. KSII Trans. Internet Inf. syst. **7**, 878–894 (2013). https://doi.org/10.3837/tiis.2013.04.016
6. Manjul, M., Mishra, R.: Secure group communication based on elliptic curve cryptography. Trans. Netw. Commun. **2**(1) (2014). https://doi.org/10.14738/tnc.21.7
7. Vijay, A., Kumar, D.S.: Elliptic curve for secure group key management in distributed networks. IJCER **6**(6), 21–28 (2016) (www.ijceronline.com)
8. Hanatani, Y., Ogura, N., Ohba, Y., Chen, L., Das, S.: Secure multicast group management and key distribution in IEEE 802.21. In: Chen, L., et al. (eds.): Security Standardisation Research, SSR2016, LNCS10074, pp. 227–243 (2016). https://doi.org/10.1007/978-3-319-49100-4_10
9. Yadav, A.K., Soni, S.: Secure multicast key distribution in mobile and adhoc networks. Adv. Wirel. Commun. **10**(4), 781–782 (2017) (www.ripublication.com)

10. Zhu, H.: An efficient protocol for secure multicast key distribution in the presence of adaptive adversaries. Sci. China Inf. Sci. **60**, 52109 (2017). https://doi.org/10.1007/s11432-014-0911-8
11. Xu, L., Huang, C.: Computing efficient multicast key distribution. IEEE Trans. Parallel Distrib. Syst. **19**(5), 577–586 (2008)
12. Sivani, Y.V., Sudha, T.: A novel approach for secured symmetric key distribution in dynamic multicast network. IJERA **1**(4), 1441–1447 (2011) (www.ijera.com)
13. Rani, D., Babu, K.G.P.: Computationally efficient group keying for time sensitive applications. IJCER **2**, 589–595 (2012) (www.ijceronline.com)
14. Madhusudhan, B., Chitra, S., Rajan, C.: Mobility based key management technique for multicast security in mobile and adhoc networks. Sci. World J. **2015** (2015) (Article id: 801632). https://dx.doi.org/10.1155/2015/801632
15. Amara, S.O., Beghdad, R., Oussalah, M.: Securing wireless sensor networks: a survey. J. EDPACS **47**(2), 6–29 (2013)
16. Seetha, R., Saravanan, R.: A survey on group key management schemes. Cybern. Inf. Technol. **15**(3), 3–25 (2015). https://doi.org/10.1515/cait-2015-0038
17. El-Bashary, M., Abdelhafez, A., Anis, W.: A comparative study of group key management in MANET. IJERA **5**(8), 85–94 (2015) (www.ijera.com)
18. Fu, K., Kamara, S., Kohno, T.: Key regression: enabling efficient key distribution for secure distributed storage. In: Proceedings of the NDSS'06 (2006) (www.microsoft.com)

Data Mining Approaches
for Correlation and Cache Replacement
in Wireless Ad Hoc Networks

Y. J. Sudha Rani and M. Seetha

Abstract Wireless ad hoc network is a collection of wireless nodes that are self-configured without fixed infrastructure. In such nodes caching is the mechanism that makes use of high speed memory known as cache memory. This memory holds frequently used data so as to reduce data transmission. It also leads to better data management. It is very useful especially in ad hoc networks that do not have fixed infrastructure. Many researchers contributed towards better data management through caching and cache replacement techniques. Base line algorithms like LRU and NRU need to be improved for better performance. Data mining technique approaches like Association Rule Mining (ARM) can be used to have better performance in cache coordination and cache replacement. The FP-Growth algorithm is used for this purpose by some researchers. However, this algorithm consumes more resources. Therefore, a light weight algorithm for discovering association rules is needed. To achieve better performance in data management and cache replacement, in this paper FIN ARM used as a part of methodology. The simulation result shows the improved performance when compare with LRU and FP-Growth algorithms.

Keywords Wireless ad hoc network · Caching · Cache replacement
Association rule mining

Y. J. Sudha Rani (✉)
Research Scholar, JNTUH, SMICH, Hyderabad, Telangana, India
e-mail: su.joyfull@gmail.com

M. Seetha
Department of Computer Science & Engineering, GNITS,
Hyderabad, Telangana, India
e-mail: smaddala2000@yahoo.com

© Springer Nature Singapore Pte Ltd. 2019
P. K. Mallick et al. (eds.), *Cognitive Informatics and Soft Computing*,
Advances in Intelligent Systems and Computing 768,
https://doi.org/10.1007/978-981-13-0617-4_11

1   Introduction

Wireless ad hoc network is the network with collection of nodes without having fixed infrastructure. On the other hand some wireless networks are with fixed infrastructure. The difference between them is shown in Fig. 1. Ad hoc networks have important utility in the real-world applications. Ad hoc networks have the feature known as caching. Caching helps nodes to store data that might be needed frequently in order to provide better data management. There are schemes like Least Recently Used (LRU) and Most Recently Used (MRU) techniques for cache replacement. However, these are baseline algorithms that need improvement. Therefore, in this paper, described data mining techniques that are widely used for different applications. Here especially mainly focus on use of association rule mining in order to have better data management in ad hoc networks.

As shown in Fig. 1, it is evident that ad hoc network is a network of wireless devices. The devices are automatically configured to form a network without any fixed infrastructure. The network is formed on demand without having pre-defined infrastructure. On the other hand the other kind of wireless network (shown left in Fig. 1), there is need for infrastructure for forming a network.

In the ad hoc networks caching helps in faster query processing and better data management. There is provision for reusing data which is frequently used instead of transmitting it every time. To process a query, it is possible to reuse data in cache memory. Then it is known as cache hit. When cache hit ratio is more, it is said to have better performance. Many researchers contributed towards caching and cache replacement techniques. Section 2 throws light on this. The methodology work of this paper makes use of FIN ARM for generating knowledge that can help in making cache replacement decisions. The available paper is structured as follows. Section 2 provides review of literature and implementation details. Section 3 presents the proposed system in detail. Section 4 shows experimental results while Sect. 5 concludes the paper.

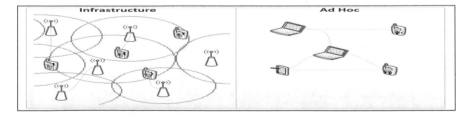

Fig. 1 A typical difference between wireless ad hoc network and other network

2 Related Works

This section provides review of literature on different aspects of caching and cache replacement. Information density based cache replacement techniques are explored in [1, 2]. In [3] caching techniques are used to improve information retrieval performance by using neighbour group data caching. Counter based cache replacement technique is used explored in [4]. Information solidity evaluation based approach for caching strategy is used in [5] for resource efficient access of information. On-demand cache-based routing was used in [6] for improving performance of wireless ad hoc networks. Cooperative caching strategies with different case studies are presented in [7] while data caching schemes with coherence are proposed in [8]. Cache discovery over a network with multiple hops is presented in [9] with power control and better cooperation.

Cooperative caching and its implementation is done in [10] for social wireless networks that are formed by wireless devices. Different caching techniques are explored in [11–14] for efficient data management and information retrieval. FP-tree [15] is used to have association rule mining. This algorithm is used for generating association rules based on given support and confidence. The association rules are used in the ad hoc nodes to have better data management. However, the FP-tree-based algorithm consumes more resources. It needs further improvement as the ad hoc networks are resource constrained. To overcome the drawbacks of existing algorithms that are based on FP-tree [15], proposed a novel methodology based on FIN ARM which provides more light weight processing for improving performance. The cache hit ratio is improved with the proposed system.

3 Proposed Methodology

In ad hoc networks for cache replacement and correlation proposed a novel methodology, that framework can be shown in Fig. 2. There is query manager to take care of queries of nodes and processing them. The query processing is done by hitting cache in order to perform data transmission faster. This is the case in general.

Fig. 2 Proposed methodology based on FIN ARM

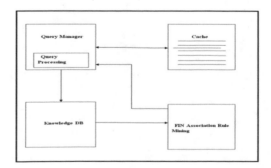

Sometimes, when data is not found the cache, it needs to be transmitted explicitly causing more overhead. Improving cache hit ratio is the main aim of the proposed methodology. Cache is the high-speed memory that holds frequently used data which is reused by nodes in ad hoc networks. Knowledge DB is the database containing data that can be updated by using query processing results. The FIN association rule mining algorithm has light-weight processing which can help query manager to perform well and the overall system will get benefits of correlation among data items and cache replacement in an effective manner.

The proposed system is based on FIN algorithm which is more efficient as it reduces complexity in producing frequent item sets. FIN was taken from [16]. FIN makes use of POC tree for storing data that is used for association rule mining. This tree structure also make discovery of frequent items faster. This will reflect in the accurate decision-making with respect to cache replacement. In order to illustrate the tree and its structure, Table 1 is used to represent data in the form of POC tree.

As shown in Table 1, it is evident that the POC tree is generated from the data presented and the tree appears as follows.

As shown in Fig. 3, the POC tree is represented with given data. This tree helps faster processing of data when compared with traditional tree-based data structures. The framework in this paper makes use of it with respect to FIN association rule mining.

Table 1 Sample data

ID	Items	Ordered frequent items
1	a, c, g, f	c, f, a
2	e, a, c, b	b, c, e, a
3	e, c, b, i	b, c, e
4	b, f, h	b, f
5	b, f, e, c, d	b, c, e, f

Fig. 3 Generated POC tree-based on the data of Table 1

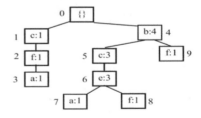

Algorithm: FIN ARM
Inputs: Dataset D, support sup, confidence $conf$
Output: Causal rules
01 Initialize vector **POC** to hold POC tree
02 Initialize vector **AR** to hold association rules
03 Initialize vector **CR** to hold causal rules
04 Constant **POC** from **D**
05 Find frequent1-itemsets
06 Scan POC tree for finding frequent2-itemsets
07 Mine all frequent (>2)itemsets
08 Mine **AR** rules that satisfy **sup** and **conf**

This algorithm takes data, support and confidence as inputs and generates association rules that are used to make cache replacement decisions. The support and confidence are the statistical measures that are used to have good quality association rules.

4 Experimental Results

Experiments are made with simulations done using a custom simulator. The results are observed in terms of cache hit ratio with different number of nodes in ad Hoc network. LRU, FP-tree mining and proposed methodology are compared for performance evaluation.

As shown in Table 2, the results are presented. The results contain the hit ratio of LRU, FP-tree-based AR- and FIN-based ARM (proposed).

As shown in Fig. 4, it is evident that there is hit ratio on different algorithms including the proposed algorithm.

As shown in Table 3, it is evident that there are results related to hit ratio with different number of nodes.

As shown in Fig. 5, it is evident that the hit ratio of the proposed system is better than other algorithms.

Table 2 Cache hit ratio comparison

Proposed	Fp-TreeARM	LRU
0.15	0.13	0.11
0.25	0.2	0.15
0.34	0.3	0.21
0.29	0.2	0.21
0.34	0.23	0.25
0.25	0.15	0.2
0.2	0.17	0.15
0.19	0.23	0.17
0.28	0.25	0.2
0.29	0.23	0.23
0.27	0.23	0.25
0.32	0.26	0.2
0.18	0.24	0.1
0.29	0.13	0.21
0.26	0.21	0.23
0.17	0.21	0.11
0.18	0.13	0.14
0.16	0.16	0.12

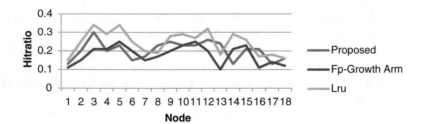

Fig. 4 Cache hit ratio comparison

Table 3 Cache hit ratio comparison

Proposed	FP-Growth ARM	LRU
0.18	0.14	0.11
0.24	0.2	0.14
0.28	0.27	0.14
0.29	0.25	0.2
0.2	0.26	0.18
0.19	0.2	0.17
0.23	0.17	0.18
0.16	0.15	0.1
0.29	0.17	0.27
0.12	0.13	0.08
0.25	0.25	0.19
0.29	0.27	0.23
0.28	0.26	0.22
0.18	0.15	0.11
0.19	0.21	0.11
0.28	0.14	0.21
0.29	0.24	0.19
0.25	0.18	0.19

Fig. 5 Hit ratio for different algorithms with different number of nodes

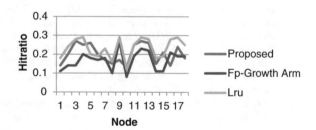

5 Conclusion and Future Work

In this paper, described the concept of caching and cache replacement in wireless ad hoc networks. It is understood that cache replacement needs improved algorithms. The traditional algorithms like LRU and NRU are not sufficient to have better data management. The FP-tree-based solutions came into existence are suffering from the problem of resource consumption. Therefore a novel methodology with an underlying algorithm was proposed. The algorithm is based on light-weight FIN ARM which is efficient. Thus the proposed algorithm is able to improve cache replacement with more accurate decisions. The cache hit ratio is improved further. The results of the proposed system are compared with FP-tree and base line algorithm such as LRU. The results revealed that the proposed system performs better. This research can be extended further by implementing caching mechanisms in real ad hoc networks.

References

1. Vimala, S., Anbarasu, R.: Efficient cache replacement techniques based on information density over wireless ad hoc networks. ISSN **2**(3), 209–214 (2012)
2. Nagaraju, D., Srinivasa Rao, L., Nageswara Rao, K.: Caching strategies based on information density estimation in wireless ad hoc networks. IJCER **6**, 139–146 (2012)
3. Shanmugavadivu, K., Madheswaran, M.: Caching technique for improving data retrieval performance in mobile ad hoc networks. IJCSIT **1**(4), 249–255 (2010)
4. RadhaVasan, A., Vinoth, N., AbdulRahuman, M., Silambarasan, S.: Wireless ad hoc network using counter based cache replacement algorithm. JARCCE **3**(1), 5039–5042 (2014)
5. Praksah, A., Durga, K.B.K.S., Divya, P., Jayanth, B.: Caching strategies based on information solidity evaluation in wireless ad hoc networks. IJARCSSE **4**(10), 123–129 (2014)
6. Dhobale, Dhanashri D., Ghorpade, V.R.: Wireless ad-hoc networks with on-demand cache routing. IJCTE **3**(6), 785–789 (2011)
7. Rao, A., Kumar, P., Chauhan, N.: Cooperative caching strategies for MANETs and IMANETs. ICACCT, 89–95 (2011)
8. Mandhare, V.V., Thool, R.C.: Improving performance in mobile ad-hoc network using data caching schemes: a review. IJCTA **5**(5), 1780–1788 (2014)
9. Joy, P.T., Jacob, K.P.: Cache discovery over a multihop wireless ad hoc network. IJIP **8**(2), 71–80 (2014)
10. Suganya, S.L., Dr, R., Gandhi, Indra: Implementation of cooperative caching in social wireless networks. IJECS **3**(5), 6147–6152 (2014)
11. Zam,A., Movahedinia, N.: Performance improvement of cache management in cluster based MANET. IJCNIS **10**, 24–29 (2013)
12. Yao, L., Deng, J., Wang, J., Wua, G.: A-CACHE: Ananchor-based public key caching scheme in large wireless networks. ELSEVIER **87**(0), 78–88 (2015)
13. Badru, D., Deepthi, P., Sankaraiah, B.: Analysis on intrusion & detection of sybil attacks in mobile adhoc networks using classification. IJARF **4**(3), 1–5 (2017)
14. Hallousha, R., Liub, H., Dongc, L., Wu, M., Radhae, H.: Hop-by-hop content distribution with network coding in multihop wireless networks. ELSEVIER **3**(0), 47–54 (2017)
15. Islam, N., Shaikh, Z.A.: Exploring correlation among data items for cache replacement in ad-hoc networks. IEEE, pp 1–4 (2010)
16. Deng, Z., Lv, S.: Fast mining frequent item sets using nodesets. Elsevier Exp. Syst. Appl. **41**, 4505–4512 (2014)

An Approach to Detect an Image as a Selfie Using Object Recognition Methods

Madhuri A. Bhalekar, Mangesh V. Bedekar and Saba Aslam

Abstract Selfie is the act of taking self portrait through the front camera of the mobile. By visualizing the captured image one can identify the details regarding the image such as number of objects, location and much more. By the method of object recognition in the image we can identify whether the image taken is a selfie or not. For this we should first segregate both the foreground and background details from an image. From the details of foreground one can identify the object (i.e., the person taking the selfie) and from the background we can tell about the location. Using various object recognition methods such as exhaustive search, segmentation, selective search, Gaussian mixture model the information regarding objects, foreground and background can be detected. And further the value of foreground and background will be compared with a certain threshold value and according to the obtained result can recognized whether an image is a selfie or not. In this paper we are presenting an approach which can be used to detect an image is a selfie.

Keywords Object recognition · Selfie · Exhaustive search · Segmentation
Selective search · Gaussian mixture model

1 Introduction

Internet has become a very major part of our lives. Earlier the Internet was used for other purposes such as research work, official work, military, communication, etc. As the times have changed, the entertainment section has taken its space. People use

M. A. Bhalekar (✉) · M. V. Bedekar · S. Aslam
Department of Computer Engineering, MAEER'S Maharashtra Institute of Technology,
Savitribai Phule Pune University, Pune, Maharashtra, India
e-mail: madhuri.bhalekar@mitpune.edu.in

M. V. Bedekar
e-mail: mangesh.bedekar@mitpune.edu.in

S. Aslam
e-mail: saba.cs14@gmail.com

© Springer Nature Singapore Pte Ltd. 2019
P. K. Mallick et al. (eds.), *Cognitive Informatics and Soft Computing*,
Advances in Intelligent Systems and Computing 768,
https://doi.org/10.1007/978-981-13-0617-4_12

the Internet to watch movies, share movies, upload picture, connect through social media and much more. As the social media has become a very famous, it generates a lot of data. The data generated is mainly photographs and very less of textual data. The photographs these days have become more and more of selfies and groupfies.

Photographs have been a very essential and important part of our lives. The traditional method was storing the photographs for one's memories but in due course of time, photographs are now used for the self representation over the Internet. These are now used to create an image, to make them visible to the world. They are used to define the geographical mobility, status of living, culture, and social circles. The term 'Selfie' was coined in 2013. It is defined as the self-photography using the front camera of a mobile. The first selfie taken was back in 1839 by Robert Cornelius.

1.1 Selfie and Social Media

The social media platforms which are popularly used like Instagram, Flickr, Twitter, Facebook, Tumblr, and many more are the major sources of generating the selfies. These websites act as a platform for commercializing the pictures taken by people. As and when anyone (like general public, actors, singers, dancers, you tubers, influential speakers, politicians, etc.) posts his/her picture, the social media generates a big amount of data and hence these can tell a lot about a person's life, interests, mobility, social circles, etc.

1.2 Selfie Detection

The detection of selfie is done by identifying the foreground and the background in the picture as shown in Fig. 1. This segregation helps in identifying various objects, locations, persons, animals and many more. Selfie detection can further helps in

Fig. 1 Portrait which represent a selfie

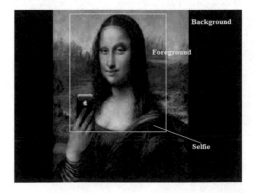

finding the behavior and the lifestyle of the person in the selfie. It could also highlight a person's hobby, like and dislikes, interests.

2 Overview

As the social media has become a very famous and widely used, it generates a lot of data in the form of images. To manage this large amount of data and help in classifying whether the photograph that has been taken by the user is a selfie or not is the new challenge. Figure 2 Shows the proposed approach system flow for detecting an image is a selfie or not.

An image consists of many elements. It can contain object or a group of objects, people, nature or just simple single colored images. To identify the objects, people or nature, we can apply various object recognition algorithms. Some of the object recognition algorithms are exhaustive searching, segmentation, selective search, Gaussian mixture model and many more. The paper is organized as in the following sequence. In Sect. 3 related works is surveyed, mathematical model given in Sect. 4 and the proposed approach is elaborated in Sect. 5.

Fig. 2 System flow for the proposed approach to detect an image is a selfie

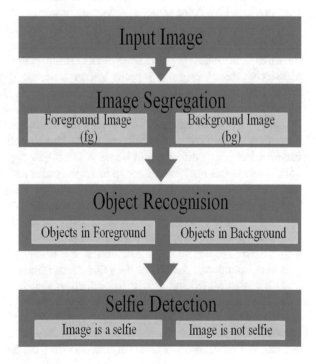

3   Related Work

The related work is segregated in two parts. In first part we focus on selfie culture in today's world and in second part elaborates different methods that are used for object recognition.

Orekh et al. [1] discuss about the selfie culture in today's world. It explains how people have taken this culture as a way of self acceptance and self representation. It helps to build and destroy many people's self-esteem as well. It also discusses about the information sharing. It can be used to obtain information about the standard of living, geographical mobility, friends circle, work place, availability of exclusive places and things, hobbies and many more. It also discusses how important people take their representation over the Internet.

Du Preez [2] presents another aspect regarding selfie culture. It presents a study on how sometimes taking a selfie can be dangerous. The emphasis falls on the analysis of selfies of death that overlaps with the sublime experience almost entirely, and it becomes nearly impossible to distinguish between selfie and sublimity.

Dalal and Triggs [3] describe how different objects like various people's can be detected in a single photograph. First, the faces in the photograph are recognized. Further the face with the best expression is selected and finally synthesizing photo gets produced by color transfer and image completion.

Harzallah et al. [4] have reviewed and analyzed that Histogram of Oriented Gradient (HOG) approach performs best for human object detection.

Viola and Jones [5] mentions about finding the localized objects and then classifying them using the Support Vector Machine (SVM). The images are classified into two categories. A positive training set is the set of images that were identified correctly and then classifier SVM was able to classify them in the right category. The other category is the negative examples. These are not used for the classifiers as they have been rejected by it. The classifier either cannot classify the identified objects correctly and hence it is tagged negative example or the object identification has not taken place in the right order and as it has failed to recognize and identify the object, it is termed as a negative example.

Felzenszwalb et al. [6] describe a robust step by step machine learning approach for visual object detection. This method provides more accurate and faster results.

Alexe et al. [7] proposed segmentation based greedy decision approach with graph based representation. Mentioned approach give much clear and distinguishing boundaries for object detection in the image.

Uijlings et al. [8] introduced the selective search which is the combination of exhaustive search and segmentation. It helps in taking all objects into consideration and diversifies the searching. Mean Average Best Overlap (MABO) value is used to evaluate the quality of the object hypothesis that is found. After the average best overlap is found, mean is taken in order to find the MABO value.

Anghelescu et al. [9] describes iterative clustering approach for efficient edge detection. Sun—Jung Kim et al. [10] discusses about the differentiation of the foreground and the background of the image. This segregation helps in finding the location of the person.

4 Methods Surveyed for Object Recognition

4.1 Exhaustive Search

The exhaustive search is a brute force search method. This means that it takes a very large area into consideration and goes through every necessary and unnecessary detail in the picture. This makes the exhaustive search computationally very expensive. It uses classifiers such as Support Vector Machines (SVM) and Histogram of Oriented Gradient (HOG) descriptors [6].

4.2 Segmentation

Segmentation helps in generating a class independent object hypothesis. The contour lines are drawn around the objects present in the photograph. This is done by convoluting the edges using the gray scale imaging along with local derivatives. Contouring is done around the objects and hence the background is separated from the picture and the entire focus is on the objects and person present in the photograph.

The input image is given and then hierarchical segmentation is done on the image. The proposed regions are the ones which can be used as the images in the classifiers. The ranked regions are the images which are fed into the classifiers. It helps in localizing the objects [6].

4.3 Bag of Words Approach

The Bag of Words model helps in predicting the object location. This is done using jumping windows and then the relation between individual visual words and the object location is learned. To represent an image is treated as a document in the Bag of Words method. To define the words, we need to follow certain steps:

1. Feature detection: The different objects present in the image are detected using the different points of the objects. For example, if a door is to be detected, the four corners would be the points of recognition of the door.

2. Feature description: The description is made for every possible angle of the object. The door when opened might have two points in the object for the recognition of the door whereas a closed door might have four points.
3. Codebook generation: These entries are then enclosed in a single codebook. The entry helps for the future use in identifying objects.

4.4 Selective Search

The selective search strategy is an improved method of object recognition. It follows certain design conditions:

Fast to compute: This method of object recognition is faster than the methods discussed above.

Diversification: The method of searching objects within the photographs must be diversified. It should not follow just one method or path to identify the objects. The strategy for choosing the parameters of selection of objects should be different and efficient.

Capture all scales: The object size can vary in a picture. It can have a full boundary of partial boundary. The algorithm should take every single detail into account and identify the objects present in the image [8].

The object recognition method in selective search is better than the above methods. The selective search algorithm helps to find objects present in an image of different textures. This is done using the hierarchical algorithm. The hierarchical algorithm helps in detecting object over an object or any object that is contained in it. For example, if there is a bowl of salad on a table, both made out of wood, the hierarchical algorithm helps in detecting both the wooden material different from each other.

The comparative analysis of exhaustive search, segmentation and selective search methods is shown in Table 1.

Table 1 Analysis of the different methods used in object recognition

	Technique	Characteristic
Exhaustive search	Image is divided into regular grids	– Effective algorithm for recognizing objects in an image – It aim to captures all possible object locations [6]
Segmentation	Image is recognized using contour lines	– Scan the image by dividing in grid patterns – It tries to find a single partitioning of the image into its unique objects before any recognition [6]
Selective search	Combination of exhaustive search and segmentation	– Supports hierarchy of the image – As compared to an exhaustive search, it provides reduced number of locations by uniting the advantages of segmentation exhaustive search methods [8]

4.5 Gaussian Mixture Model

In this, two burst shots are taken and these are then compared to find the foreground, (i.e., the person taking the selfie) and the background. This is then using tri-map blurs the background of the image that is being taken and gives the result. A tri-map is constructed. The tri-map consists of four regions which are foreground (FG), probably foreground (PRFG), probably background (PRBG), and background (BG). The estimated shoulder points are defined by learning RGB information from initial foreground (FG) and background (BG) region using Gaussian Mixture Model (GMM). The PRFG and PRBG region are located within a certain margin from the FG region [10, 11].

To gather different information from the selfie, it is needed to separate the foreground from the background. In order to do so, the motion vectors are used. The motion vectors define the objects that have moved in the two burst shots taken. The burst shots are the images which are taken one after the other in a very minimum interval of time. This time is in some milliseconds. The photographs taken are very closely timed. This is done by comparing the motion vectors of both the burst shot images. When pictures are normally compared, not much of a difference can be found. This difference is pointed out by the motion vector as it can detect even the slightest change in the foreground as the background of image stays the same due to it being stationary.

5 An Approach to Identify Image as a Selfie

A selfie itself is an image, so for detecting whether an image is an selfie or not we need to perform some preprocessing on an image which mostly includes following steps:

- Image Preprocessing
- Feature Extraction
- Segmentation
- Object Recognition
- Detecting a Selfie

The main objective of image preprocessing is to improve the image data that suppresses unwanted distortions as well as to enhance some image features which are required for further processing.

An image consists of many elements in it which can be detected as objects. To identify the objects we can apply various object recognition algorithms. Here we are considering following methods for object recognition: exhaustive searching, segmentation, selective search, and Gaussian mixture model.

5.1 Mathematical Modeling

For the proposed approach we considered the selfie image taken by the human hands and not by the selfie stick. The flow of proposed mathematical model is shown in Fig. 3.

Let us consider an image with attributes/characteristics such as width, height, color, texture, aspect ratio, orientation of the image, resolution, etc. Further to detect that the image can be classified as a selfie, let a system considered be S.

It is represented as S = {Input (I), Function (F), Output (O)}

The function F can be defined as

$$F = \{f_1, \ f_2, \ f_3, \ f_4, \ f_5\}$$

The different functions considered are as follows:

$$f_1 = \{\text{Gaussian Mixture Model}\}$$
$$f_2 = \{\text{Exhaustive method}\}$$
$$f_3 = \{\text{Segmentation}\}$$
$$f_4 = \{\text{Selective method}\}$$
$$f_5 = \{\text{Bag of Words}\}$$

These functions are discussed in Sect. 4.

The input goes through the function f_1 and finds the following outputs:

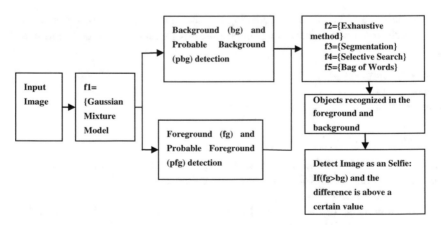

Fig. 3 Flow of mathematical modeling

$$O_1 = \{Foreground\}$$
$$O_2 = \{Background\}$$
$$O_3 = \{Probable\ Foreground\}$$
$$O_4 = \{Probable\ Background\}$$

The output O_1 and O_2 help in finding the foreground and the background that is present in the input image. The outputs O_3 and O_4 help in finding the undefined region, that is, the region which could not be classified as the foreground or background definitively, to be classified as a the probable foreground or the probable background.

Objects in the foreground can be recognized using the f_2, f_3, f_4, f_5 methods. As the image is separated into two regions, the object like human figure in the foreground area is identified using the other desirable functions, namely f_2, f_3, f_4, f_5. The background area is then checked for the locations recognition by the objects present. A value is assigned to them which in turn is used in determining if the given input image is a selfie or not. This is done by the following comparison:

If value of foreground is greater than the value of background and the difference is above a certain limit, we can predict the input image can be a selfie. Using the above-mentioned approach we can identify different objects from an image and by analyzing its foreground and background details can further detect whether an image is an selfie or not.

6 Conclusion

In this paper we present an approach to detect whether a given image is a selfie or not. For object recognition the different methods such as exhaustive search, segmentation, and selective search were analyzed. The exhaustive search takes a very large area into consideration and goes through every necessary detail in the image. Segmentation is the searching process in which it forms a contour line. The contour lines help in defining the object boundary. According to the different methodology analyzed, the algorithms used for the object recognition can be used for selfie detection. The Gaussian Mixture Model helps in differentiating the foreground and the background through which the selfies can be detected.

References

1. Orekh, E., Sergeyeva, O., Bogomiagkova, E.: Selfie phenomenon in the visual content of social media. IEEE Conference (2016)
2. Du Preez, A.: Sublime selfies: to witness death. Eur. J. Cult. Stud. Published 10 August (2017)

3. Dalal, N., Triggs, B.: Histograms of oriented gradients for human detection. In: CVPR (2005)
4. Harzallah, H., Jurise, F., Schmid, C.: Combining efficient object localization and image classification. In: ICCV (2009)
5. Viola, P., Jones, M.: Rapid object detection using a boosted cascade of simple features. CVPR 1, 511–518 (2001)
6. Felzenszwalb, P.F., Huttenlocher, D.P.: Efficient graph-based image segmentation. Int. J. Comput. Vision 59, 167–181 (2004)
7. Alexe, B., Deselaers, T., Ferrari, V.: What is an object? In: CVPR (2010)
8. Uijlings, J.R., Sande, K.F., Gevers, T., Smeulders, A.W.: Selective search for object recognition. Int. J. Comput. Vision 104(2), 154–171 (2013)
9. Anghelescu, P., Iliescu, V.G., Mara, C., Gavriloaia, M.B.: Automatic thresholding method for edge detection algorithms. In: ECAI—International Conference, 8th edn (2016)
10. Kim, S.J., Kim, B.S., Kim, H.I., Hong, T.H., Son, J.Y.: The method for defocusing selfie taken by mobile frontal camera using burst shot. IEEE Conference (2016)
11. Kamate, S., Yilmazer, N.: Application of object detection and tracking techniques forunmanned aerial vehicles. Published by Elsevier, Science Direct (2015)

T Wave Analysis: Potential Marker of Arrhythmia and Ischemia Detection-A Review

Akash Kumar Bhoi, Karma Sonam Sherpa, Bidita Khandelwal
and Pradeep Kumar Mallick

Abstract T wave is the end potential waveform or segment of cardiac cycle. It is basically originated by the different layers of ventricular myocardium and their differences in repolarization time. This study analyzes the conventional techniques and automatic popular methods associated with T Wave Alternans (TWA) and their approaches towards ischemic and arrhythmic interventions. This review work is divided into two major parts i.e. (i) Analysis of T wave: association of T wave with ischemia and arrhythmia (ii) T wave detection techniques. This analytical literature survey also leads to the conclusion, where the importance of T wave analysis significantly inculcates ideological researcher and clinical mindset for approaching critical cardiac diseases in multimodal approaches.

Keywords T wave · T wave alternans (TWA) · Arrhythmia · Ischemia

A. K. Bhoi (✉) · K. S. Sherpa
Department of Electrical and Electronics Engineering,
Sikkim Manipal Institute of Technology (SMIT),
Sikkim Manipal University, Gangtok, India
e-mail: akash.b@smit.smu.edu.in; akash730@gmail.com

K. S. Sherpa
e-mail: karma.sherpa@smit.smu.edu.in; karmasherpa23@gmail.com

B. Khandelwal
Department General Medicine, Central Referral Hospital
and SMIMS, Sikkim Manipal University, Gangtok, India
e-mail: drbidita@gmail.com

P. K. Mallick
Department of Computer Science and Engineering, Vignana Bharathi
Institute of Technology (VBIT), Hyderabad, India
e-mail: pradeepmallick84@gmail.com

© Springer Nature Singapore Pte Ltd. 2019
P. K. Mallick et al. (eds.), *Cognitive Informatics and Soft Computing*,
Advances in Intelligent Systems and Computing 768,
https://doi.org/10.1007/978-981-13-0617-4_13

1 Introduction (Analysis of T Wave)

Martínez and Olmos [1] have proposed a unified framework, which holds the comparison of existing T Wave Alternans (TWA) ("TWA, also called repolarization alternans, is a phenomenon appearing in the electrocardiogram (ECG) as a consistent fluctuation in the repolarization morphology on an every-other-beat basis") analysis methods. This study also focuses on the cardiac instability and increased arrhythmogenicity [1]. Hanninen et al. have examined in their study that transient myocardial ischemia could be detected by analyzing ST-T area and T wave amplitude, which is also proved to be a sensitive and specific marker for ischemia. This analysis adds information complementary to the conventional ST depression [2]. TWA magnitude could be fluctuated (evidentially increased) due to myocardial ischemia and was observed during coronary artery occlusion of animals [3–5] and in human during angioplasty [3, 6]. Salah et al. have concluded in their investigation that Complexity Ratio (CR) of Holter electrocardiogram (T wave with middle-aged groups) was averaged over 24 h (CR_{24h}), which could predict ischemia. Their study also suggested the distorted T wave morphology was due to the prolonged subclinical myocardial ischemia, which alters the ventricular repolarization and might be initiated by early coronary artery disease [7].

TWA is associated with several Cardiovascular Diseases (CAD) and electrical disorder which leads to Sudden Cardiac Death (SCD). Numerous literature studies appearances TWA during stress tests [8], Long QT Syndrome (LQTS) [9], coronary angioplasty [10], pacing [11], ambulatory recordings [12], printzmetal angina [13], infarction [14], ischemia [12] and dilated cardiomyopathy [15]. TWA detection by analyzing ST-T complex is very difficulty in stress testing or Holter recordings as it comprises with non-invasive variation and subtle. Blanco-Velasco et al. have developed Empirical Mode Decomposition (EMD) technique for addressing such issues [16]. Narayan has described that the ventricular arrhythmia could be predicted from the T Wave Alterans (TWA) analysis [17] (Fig. 1).

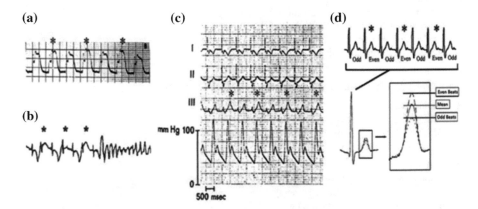

Fig. 1 Narayan [17] depicted the work of Kleinfeld MJ and Rozanski JJ (Fig. 1a) [18], Raeder et al. (Fig. 1b) [19], Wellens (Fig. 1c) [20], Smith et al. (Fig. 1d) [21]

2 Popular T Wave Detection Approaches

Gritzali et al. have proposed "length" transformation method for localizing onset and offset boundaries of P and T waves [22]. Boix et al. have applied Wavelet Transform (WT) on electrocardiogram to detect TWA and it is measured by the amplitude difference between the augmented T waves and normal T waves [23]. Costas et al. has developed new rule based computerized system with fast and robust performance to detect the T wave alterations [24]. Elgendi et al. have proposed T wave detection algorithm using two moving averages and dynamic threshold, which is possibly implemented on battery-driven mobile devices [25]. Pan and Tompkins algorithm also detect the T wave, which can be summarized as presented in Fig. 2 [26].

Experimental and clinical literature studies already establishes fundamental link between susceptibility to malignant arrhythmias and T Wave Alternans (TWA) [27–42]. Richard and Bruce [43] have devised "Modified Moving Average Beat Analysis" (MMA), (Fig. 3) a time domain method which is accurately measure T Wave Alternans (TWA) in freely moving subjects.

Tomas et al. have explored correlation-to-template-based algorithm for automatic T wave localization in low signal to noise ECG [44]. Krimi et al. have described a wavelet based approach called as "Wavelet Transform Modulus Maxima Method" (WTMM), which deals with low amplitude and alternating morphological segment (i.e., T wave) and sensibly detect T wave. Figure 4 shows the adjustments which were made with its scale and threshold for WTMM method to improvise detection of T wave [45].

Khaustov et al. [46] have described an open source algorithm package for TWA detection and quantification. This software package developed using MATLAB,

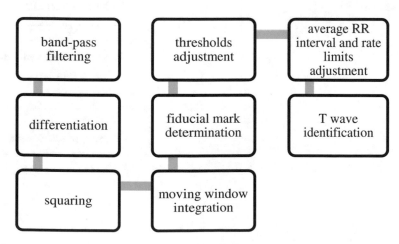

Fig. 2 T wave detection using Pan and Tompkins algorithm

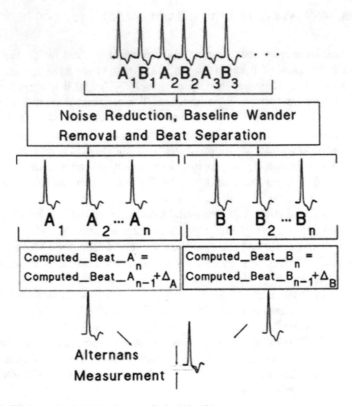

Fig. 3 Modified moving average beat analysis [42, 43]

Fig. 4 Adaptation in
WTMM [45]

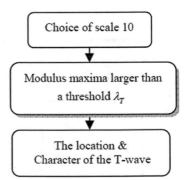

which performs the Spectral Method and Modified Moving Average (MMA) techniques. The procedure can be summarizes as below:

(1) Pre-filtering: simple baseline wander filtering
(2) QRS and T detection: Discrete Wavelet Transform (DWT) method proposed by Martinez et al. [47]
(3) Spectral methods: (window uses 128 beats)

    (i) Beat alignment and rejection: implemented in repolarization cycles to avoids false TWA
    (ii) Alternans series: evaluated with every lead and every offset in the ST-T segment
    (iii) Periodogram and alternans values: implemented over every offset in the ST-T segment and then averaged.

(4) Modified Moving Average

    (i) Initial beat selection
    (ii) Beat averaging: individually calculated for Q-S and ST-T segments.
    (iii) Alternans value: calculated every 15 s

Mehta et al. have presented a Support Vector Machine (SVM) method for detection and delineation of P and T wave [48] (Fig. 5).

Goya-Esteban et al. [49] have proposed TWA detection and estimation method using signal processing stages, as depicted below

(1) ECG Preprocessing Stage

    (i) Coarse Low Pass Filtering (CLPF)
    (ii) Baseline Cancelation (BLC)

(2) R-peak detection stage

    (i) Band Pass Filtering
    (ii) QRS detection
    (iii) R-peak detection

(3) Discard Non-Valid Beats (DNVB)
(4) T wave segmentation and synchronization stage

    (i) Fine Low Pass Filtering
    (ii) ST-T segmentation
    (iii) Windowing
    (iv) Synchronization

(5) TWA detection and estimation stage.

Goldwasser et al. have discussed about the possibility of atrial activity (i.e., P wave) over-shadowed in T wave and the importance of proper identification of T wave. Moreover, the T wave filtering was performed using wavelet transformation, which improvises the detection possibility of atrial activity through surface ECG [50].

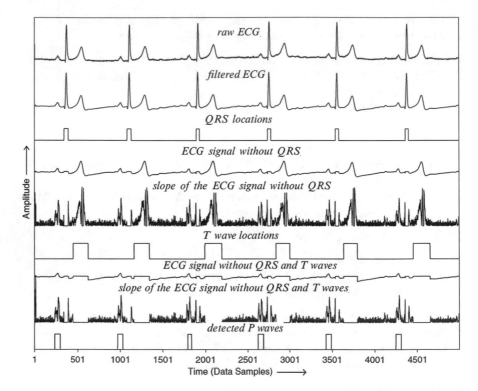

Fig. 5 Obtained results with SVM [48]

TWA and its associated risk stratification were examined with more than 12,000 patients using frequency domain Spectral Method and a modified time domain Moving Average Method (MMA) method [51] (Fig. 6).

Wana et al. have presented Least Squares Curve Fitting Techniques (LSCF) based method for TWA assessment and the performance is compared with MMA algorithm [54]. Latif et al. have implemented Empirical Mode Decomposition (EMD) as denoising tool on ST-T complexes for improvising the TWA estimation and detection accuracy. After that the state of art approaches (i.e., Correlation Method (CM), Modified Moving Average Method (MMAM), Spectral Method (SM) and Median Matched Filter (MMF) method) were carried out for TWA analysis [55]. Cesari et al. have implemented wavelet-based algorithm for automatic calculation of TWA, by attaining sensitivity of 99.50% for T peak detection on QT database. It was also suggested that TWA interpretation proven to be a significant marker for the automatic ventricular intervention [56]. Ramanan et al. later found that manual adjustment of T wave window had not improvised the Microvolt T Wave Alternans (MTWA) in 98.2% of the patients, where spectral MTWA remains effective in structural heart diseases [57].

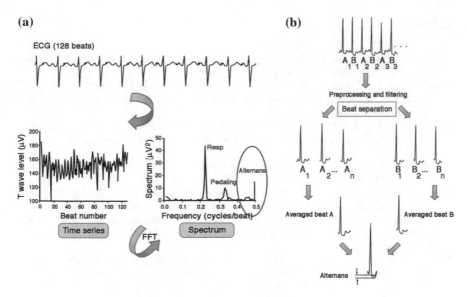

Fig. 6 a Spectral TWA method [52] derived from Cohen [3], **b** MMA TWA method [51] derived from nearing BD [53]

3 Conclusion

As T wave associated with the ventricular repolarization of the cardiac cycle, most critical cardiovascular diseases which are incubated at ventricles reflect the morphological changes in T wave geometry. Different frequency-time domain approaches and automatic techniques are discussed in compliance with T wave analysis and detection. T wave analysis could be a potential marker for cardiac ischemia and arrhythmia detection and can lead to reduce the Sudden Cardiac Death (SCD) ratio. This review also suggests the alternative approaches associated with T wave analysis as well as overcoming the conventional analysis approaches *such as* arrhythmia detection with QRS complex morphology and ischemia detection by analyzing ST segment. The present study redefines the need of T wave analysis along with accurate detection and brings out new research possibility in the areas of ventricular repolarization. The further scope lies in the collective analysis of the complexes (i.e., ST-T, QT), feature extraction, classification and so on.

References

1. Martínez, J.B., Olmos, S.: Methodological principles of t wave alternans analysis: a unified framework. IEEE Trans. Biomed. Eng. **52**(4) (2005)
2. Hänninen, H., Takala, P., Rantonen, J., et al.: ST-T integral and T-wave amplitude in detection of exercise-induced myocardial ischemia evaluated with body surface potential mapping. J. Electrocardiol. **36**(2) (2003)
3. Nearing, B.D., Oesterle, S.N., Verrier, R.L.: Quantification of ischaemia induced vulnerability by precordial T wave alternans analysis in dog and human. Cardiovasc. Res. **28**, 1440 1449 (1994)
4. Nearing, B.D., Huang, A.H., Verrier, R.L.: Dynamic tracking of cardiac vulnerability by complex demodulation of the T wave. Sci. **252**, 437–440 (1991)
5. Nearing, B.D., Verrier, R.L.: Modified moving average analysis of T-wave alternans to predict ventricular fibrillation with high accuracy. J. Appl. Physiol. **92**, 541–549 (2002)
6. Martinez, J.P., Olmos, S., Wagner, G., Laguna, P.: Characterization of repolarization alternans during ischemia: time-course and spatial analysis. IEEE Trans. Biomed. Eng. **53**, 701–711 (2006)
7. Salah, S.A., Kristen, N.R., Mary, G.C.: Increased T wave complexity can indicate subclinical myocardial ischemia in asymptomatic adults. J. Electrocardiol. **44**, 684–688 (2011)
8. Estes III, N.A., Michaud, G., Zipes, D.P., El-Sherif, N., Venditti, F.J., Rosenbaum, D.S., Albrecht, P., Wang, P.J., Cohen, R.J.: Electrical alternans during rest and exercise as predictors of vulnerability to ventricular arrhythmias. Am. J. Cardiol. **80**, 1314–1318 (1997)
9. Zareba, W., Moss, A.J., Le Cessie, S., Hall, W.: T wave alternans in idiopathic long QT syndrome. J. Am. Coll. Cardiol. **23**, 1541–1546 (1994)
10. Kwan, T., Feit, A., Alam, M., Afflu, E., Clark, L.T.: ST-T alternans and myocardial ischemia. Angiol. **50**(3), 217–222 (1999)
11. Rosenbaum, D.S., Jackson, L.E., Smith, J.M., Garan, H., Ruskin, J.N., Cohen, R.J.: Electrical alternans and vulnerability to ventricular arrhytimias. N. Engl. J. Med. **330**(4), 235–241 (1994)
12. Verrier, R.L., Nearing, B.D., MacCallum, G., Stone, P.H.,: T-wave alternans during ambulatory ischemia in patients with stable coronary disease. Ann. Noninvasive Electrocardiol. pt. 1 **1**(2), 113–120 (1996)
13. Turitto, G., El-Sherif, N.: Alternans of the ST segment in variant angina. incidence, time course and relation to ventricular arrhythmias during ambulatory electrocardiographic recording. Chest **93**, 587–591 (1988)
14. Verrier, R.L., Nearing, B.D., LaRovere, M.T., Pinna, G.D., Mittleman, M.A., Bigger, J.T., Schwartz, P.J.: Ambulatory electrocardiogram- based tracking of T wave alternans in postmyocardial infarction patients to assess risk of cardiac arrest or arrhythmic death. J. Cardiovasc. Electrophysiol. **14**(7), 705–711 (2003)
15. Adachi, K., Ohnisch, Y., Shima, T., Yamashiro, K., Takei, A., Tamura, N., Yokoyama, M.: Determinant of microvolt-level T-wave alternans in patients with dilated cardiomyopathy. J. Am. Coll. Cardiol. **34**(2), 374–380 (1999)
16. Blanco-Velasco, M., et al.: Nonlinear trend estimation of the ventricular repolarization segment for T-Wave alternans detection. IEEE Trans. Biomed. Eng. **57**(10) October (2010)
17. Narayan SM.: T-Wave alternans testing for ventricular arrhythmias. progress in cardiovascular diseases. **51**(2), 118–127 (September/October) (2008)
18. Kleinfeld, M.J., Rozanski, J.J.: Alternans of the ST segment in Prinzmetal's angina. Circ. **55**, 574–577 (1977)
19. Raeder, E.A., Rosenbaum, D.S., Bhasin, R., et al.: Alternans of electrocardiographic T-wave may predict lifethreatening ventricular arrhythmias. N. Engl. J. Med. 271–272 (1992)
20. Wellens, H.J.: Isolated electrical alternans of the T wave. Chest **62**, 319–321 (1972)
21. Smith, J.M., Clancy, E., Valeri, C., et al.: Electrical alternans and cardiac electrical instability. Circ. **77**, 110–121 (1988)

22. Gritzali, F., Frangakis, G., Papakonstantinou, G.: Detection of the P and T Waves in an ECG. Comput. Biomed. Res. **22**, 83–91 (1989)
23. Boix, M., Cantó, B., Cuesta, D., Micó, P.: Using the wavelet transform for T-wave alternans detection. Math. Comput. Model. **50**, 738–742 (2009)
24. Costas, P., Dimitrios, I.F., Aristidis, L., Christos, S.S., Lampros, K.M.: Use of a novel rule-based expert system in the detection of changes in the ST segment and the T wave in long duration ECGs. J. Electrocardiol. **35**(1) (2002)
25. Elgendi, M., Eskofier, B., Abbott, D.: Fast T wave detection calibrated by clinical knowledge with annotation of P and T waves. Sens. **15**, 17693–17714 (2015)
26. Pan, J., Tompkins, W.J.: A realtime QRS detection algorithm. IEEE Trans. BME **32**, 230–236 (1985)
27. Surawicz, B., Fisch, C.: Cardiac alternans: diverse mechanisms and clinical manifestations. J. Am. Coll. Cardiol. **20**, 483 (1992)
28. Verrier, R.L., Nearing, B.D.: Electrophysiologic basis for T-wave alternans as an index of vulnerability to ventricular fibrillation. J. Cardiovasc. Electrophysiol. **5**, 445 (1994)
29. Verrier, R.L., Cohen, R.J.: Risk identification and markers of susceptibility. In: Spooner P, Rosen MR. (eds.) Foundations of cardiac arrhythmias. Marcel Dekker, New York, NY, p. 745 (2000)
30. El-Sherif, N., Turitto, G., Pedalino, R.P., et al.: T-wave alternans and arrhythmia risk stratification. Ann. Noninvasiv. Electrocardiol. **6**, 323 (2001)
31. Smith, J.M., Clancy, E.A., Valeri, C.R., et al.: Electrical alternans and cardiac electrical instability. Circ. **77**, 110 (1988)
32. Nearing, B.D., Huang, A.H., Verrier, R.L.: Dynamic tracking of cardiac vulnerability by complex demodulation of the T wave. Sci. **252**, 437 (1991)
33. Rosenbaum, D.S., Jackson, L.E., Smith, J.M., et al.: Electrical alternans and vulnerability to ventricular arrhythmia. N. Engl. J. Med. **330**, 235 (1994)
34. Nearing, B.D., Oesterle, S.N., Verrier, R.L.: Quantification of ischaemia-induced vulnerability by precordial T wave alternans analysis in dog and human. Cardiovasc. Res. **28**, 1440 (1994)
35. Hohnloser, S.H., Klingenheben, T., Yi-Gang, L., et al.: T-wave alternans as a predictor of recurrent ventricular tachyarrhythmias in ICD recipients: Prospective comparison with conventional risk markers. J. Cardiovasc. Electrophysiol. **9**, 1258 (1998)
36. Adachi, K., Ohnishi, Y., Shima, T., et al.: Determinant of microvolt-level T-wave alternans in patients with dilated cardiomyopathy. J. Am. Coll. Cardiol. **34**, 374 (1999)
37. Klingenheben, T., Zabel, M., D'Agostino, R.B., et al.: Predictive value of T-wave alternans for arrhythmic events in patients with congestive heart failure. Lancet **356**, 651 (2000)
38. Gold, M.R., Bloomfield, D.M., Anderson, K.P., et al.: A comparison of T-wave alternans, signal averaged electrocardiography and programmed ventricular stimulation for arrhythmia risk stratification. J. Am. Coll. Cardiol. **36**, 2247 (2000)
39. Hennersdorf, M.G., Neibch, V., Perings, C., et al.: T-wave alternans and ventricular arrhythmias in arterial hypertension. Hypertens. **37**, 199 (2001)
40. Nearing, B.D., Verrier, R.L.: Modified moving average method for T-wave alternans analysis with high accuracy to predict ventricular fibrillation. J. Appl. Physiol. **92**, 541 (2002)
41. Ikeda, T., Saito, H., Tanno, K., et al.: T-wave alternans as a predictor for sudden cardiac death after myocardial infarction. Am. J. Cardiol. **89**, 79 (2002)
42. Verrier, R.L., Nearing, B.D., LaRovere, M.T., et al.: Ambulatory ECG-based tracking of T-wave alternans in post-MI patients to assess risk of cardiac arrest or arrhythmic death. J. Cardiovasc. Electrophysiol. **14**, 70S (2003)
43. Richard, L.V., Bruce, D.N.: Ambulatory ECG monitoring of T-Wave alternans for arrhythmia risk assessment. J. Electrocardiol. **36**(Supplement) (2003)
44. Kulvicius, T., Tamošiunaite, M., Vaišnys, R.: T wave alternans features for automated detection. Informatica **16**(4), 587–602 (2005)

45. Krimi, S., Ouni, K., Ellouze, N.: T-Wave detection based on an adjusted wavelet transform modulus maxima. world academy of science, engineering and technology,. Int. J. Med. Health Sci. **1**(3) (2007)
46. Khaustov, A., Nemati, S., Clifford, G.D.: An open-source Standard T-Wave alternans detector for benchmarking. Comput. Cardiol. **35**, 509–512 (2008)
47. Martinez, J., Olmos, S., Laguna, P.: Evaluation of a wavelet based ECG waveform detector on the QT database. Comput. Cardiol. **2000**(27), 81–84 (2000)
48. Mehta, S., Lingayat, N., Sanghvi, S.: Detection and delineation of P and T waves in 12-lead electrocardiograms. Expert Syst. **26**(1) February (2009)
49. Goya-Esteban, R., et al.: Nonparametric signal processing validation in T-Wave alternans detection and estimation. IEEE Trans. Biomed. Eng. **61**(4) April (2014)
50. Goldwasser, D., et al.: A new method of filtering T waves to detect hidden P waves in electrocardiogram signals. Europace **13**, 1028–1033 (2011)
51. Verrier, R.L., et al.: Microvolt T-Wave alternans, physiological basis, methods of measurement, and clinical utility—consensus guideline by international society for holter and noninvasive electrocardiology. J. Am. Coll. Cardiol. **58**(13) (2011)
52. Cohen, R.J.: TWA and Laplacian imaging. In: Zipes D.P., Jalife J, (eds.) Cardiac electrophysiology: from cell to bedside, vol. 889, 5th edn. Saunders, Philadelphia (2009)
53. Nearing, B.D., Verrier, R.L.: Modified moving average analysis of T-wave alternans to predict ventricular fibrillation with high accuracy. J. Appl. Physiol. **92**, 541–549 (2002)
54. Wana, X., Li, Y., Xia, C., Wu, M., Liang, J., Wang, N.: A T-wave alternans assessment method based on least squares curve fitting technique. Meas. **86**, 93–100 (2016)
55. Latif, M., Bakhshi, A.D., Ali, U., Siddiqui, RA.: Empirical mode decomposition on T-Wave alternans detection. J. Image Graph. **4**(2) December (2016)
56. Cesari, M., Mehlsen, J., Mehlsen, A.B., Sorensen, H.B.D.: A new wavelet-based ECG delineator for the evaluation of the ventricular innervation. IEEE J. Transl. Eng. Health Med. **4**(5), 2000215 July (2017)
57. Ramanan, T., et al.: Does manual T-wave window adjustment affect microvolt T-wave alternans results in patients with structural heart disease? J. Electrocardiol. **49**(6), 967–972, Nov—Dec (2016)

Functional Link Artificial Neural Network (FLANN) Based Design of a Conditional Branch Predictor

Abhaya Kumar Samal, Pradeep Kumar Mallick, Jitendra Pramanik, Subhendu Kumar Pani and Rajasekhar Jelli

Abstract Conditional branch predictor (CBP) is an essential component in the design of any modern deeply pipelined superscalar microprocessor architecture. In the recent past, many researchers have proposed varieties schemes for the design of the CBPs that claim to offer desired levels of accuracy and speed needed to meet the demand for the architectural design of multicore processors. Among various schemes in practice to realize the CBPs, the ones based on neural computing—i.e., artificial neural network (ANN) has been found to outperform other CBPs in terms of accuracy. Functional link artificial neural network (FLANN) is a single layer ANN with low computational complexity which has been used in versatile fields of application, such as system identification, pattern recognition, prediction and classification, etc. In all these areas of application, FALNN has been tested and found to provide superior performance compared to their multilayer perceptron (MLP) counterpart. This paper proposes design of a novel FLANN-based dynamic

A. K. Samal (✉)
Department of Computer Science and Engineering, Trident Academy
of Technology, Bhubaneswar, India
e-mail: kabhaya1@gmail.com

P. K. Mallick · R. Jelli
Department of Computer Science and Engineering, Vignana Bharathi
Institute of Technology (VBIT), Hyderabad, India
e-mail: pradeepmallick84@gmail.com

R. Jelli
e-mail: Sekharcse504@gmail.com

J. Pramanik
Biju Patnaik University of Technology, Rourkela, Odisha, India
e-mail: bidun35@gmail.com

S. K. Pani
Department of Computer Science and Engineering, Orissa Engineering
College, Bhubaneswar, India
e-mail: skpani.india@gmail.com

© Springer Nature Singapore Pte Ltd. 2019
P. K. Mallick et al. (eds.), *Cognitive Informatics and Soft Computing*,
Advances in Intelligent Systems and Computing 768,
https://doi.org/10.1007/978-981-13-0617-4_14

131

branch predictor and compares the performance against a perceptron-based CBP. The proposed FALNN-based CBP has been implemented in C++. The performance of the proposed CBP has been evaluated using standard benchmark trace data files and is found to have performance comparable to the existing perceptron-based predictions in terms of speed and accuracy.

Keywords Conditional branch predictor (CBP) · Functional link artificial neural network (FLANN) · Multilayer perceptron (MLP) · Multicore technology

1 Introduction

Growing demand for designing computers having high performance forced the processor architects to come up with parallel processing architecture with increased degrees of instruction-level parallelism (ILP). Instruction-level parallelism (ILP) has been used extensively as one of the most preferred approach to come up with high-performance computing architecture; and the outcome is super-pipelining and superscalar designs [1]. The goal of pipelining is to maximize the utilization of all the independent functional units of a microprocessor at the same time, in parallel fashion. The major difficulty encountered due to extensive use of parallelism is the existence of branch instructions in the set of instructions presented to the processor for execution: both conditional and unconditional branch instructions in the pipeline. If the instructions under execution in the pipeline do not bring any change in the control flow of the program, then there is no problem at all. However, when the branch instruction puts the program under execution to undergo a change in the flow of control, the situation becomes a topic of concern as the branch instruction breaks the sequential flow of control, leading to a situation what is called pipeline stall and levying heavy penalties on processing in the form of execution delays, breaks in the program flow and overall performance drop. Changes in the control flow affects the processor performance because many processor cycles must be wasted in using the pipeline already loaded with instructions from wrong locations and again reading-in the new set of instructions from right address. It is well understood that, branch instructions can break the smooth flow of instruction fetch-decode-execute cycle in highly parallel computer system. This results in delay, since the instruction issuing must often wait until the actual branch outcome is known. Therefore, making the pipeline deeper makes the situation worse by incurring more delay, and thus greater is the performance loss.

Branch predictors attempt to improve the performance of a pipelined microprocessor by accurately predicting with maximum possible accuracy, irrespective of whether or not a change in the control flow will occur. Therefore, to bring down delay, predicting the direction of a branch instruction that it will take and start fetch, decode, and issue instructions before the branch decision is made appears productive. Branch prediction is a technique to guess the outcome of a conditional branch instruction before they are actually executed. However, a wrong prediction may produce more delay as the instructions fetched incorrectly would occupy useful functional units, like the ALU, reservation stations, and memory bus. This builds up a need for highly accurate branch prediction strategies in place. Accurate prediction of the behavior and the direction of a branch instruction will help in improving the processor performance [2]. The need for predicting the behavior of branch instruction arises due to extensive use of pipelining technique in the design of modern microprocessors in order to meet the growing performance need. With the growing demand for designing systems with deeper pipelines, the need for branch prediction mechanisms with high degree of accuracy becomes crucial since a huge amount of speculative perfecting may need to be used out of the pipeline in case any misprediction takes place [3]. Due to advent of multicore architecture and use of deeper pipelines where the misprediction penalties becomes prohibitively large, throttling the performance of the microprocessor, the need for integrating more accurate and fast branch prediction mechanism as a part of instruction execution framework has become a highly demanding requirement [4].

To tackle with the situation, varieties of branch prediction schemes have been proposed by the architect and are in active deployment over past two decades [5]. It is observed that a small improvement in the quality of branch prediction, in terms of accuracy and speed, could result in a significant boost in overall processor performance [5]—a demanding need for modern computing platforms required in diverse fields of application. Therefore, continual progress in branch prediction research still is in active state [5].

The key objective of this paper is to present the design and implementation of a two dynamic branch predictor simulators: one based on functional link artificial neural network and the other one based on piecewise linear perceptron network, coded in C++ programming language, modeled in accordance with the idea presented by Jimenez [5] along with other assorted ideas collected from different Internet resources.

The rest of the paper is structured as follows. In the following Sect. 2, a short survey of literature has been presented, followed by discussions on key background concepts presented in Sect. 3. Details of different dynamic branch prediction schemes are discussed in Sect. 4. In Sect. 5, the implementation approach of the proposed FLANN-based DBP scheme is presented. Discussions on simulation approach and results of simulation are presented in Sect. 6 and finally, the concluding remarks are presented in Sect. 7.

2 Survey of Literature

Milenkovic et al. [6] have presented experimental flow for the benchmark tests that determine the organization and size of a branch predictor using on-chip performance monitoring registers. Technical note by McFarling [7] presents discussion on how the implementation leading to degree of instruction-level parallelism plays an advantageous role in boosting computing performance and suggested a method for combining different types of branch predictors for maximizing prediction accuracy for a given predictor size. Yeh and Patt [3] have introduced the idea of dynamically collecting branch history information at two different levels, namely, branch execution history and pattern history; the scheme being called Two-level Adaptive Branch Prediction, where the prediction is adjusted according to the runtime behavior of the branch instruction encountered. Major portion of this paper is influenced by the work of Jimenez [8], the idea of implementation of dynamic branch predictor using coding in C++ was taken from the predictor header file: http://faculty.cse.tamu.edu/djimenez/ snp_my_predictor.h [9], information regarding the benchmark trace file format was taken from the source code file "trace.cc" supplied along with the Branch Prediction Competition infrastructure resource http://faculty.cse.tamu.edu/djimenez/ 614/competition/cbp2-infrastructure-v2.tar [10], and the documentation on Infrastructure for Branch Prediction Competition: http://faculty.cse.tamu.edu/ djimenez/614/competition/cbp2-infrastructure-v2/doc/ [11]. Implementation of functional link artificial neural network (FLANN) based dynamic branch prediction is influenced by the core idea taken from the papers [12, 13]. Fundamental idea regarding different types of branch prediction schemes, such as static branch prediction, two-level adaptive branch prediction, local and global branch prediction, hybrid branch prediction, neural branch prediction, etc., was available from the Wikipedia—the free encyclopedia [14]. Concept regarding modularization and organization of the code for the design of the trace-driven branch predictor was taken from the work of Wilamowski [15]. The motivation behind the whole work rests on various fundamental concepts taken from the work of Jimenez and Lin [2, 4] and Jimenez [5, 8, 16]. The branch predictor binary was successfully built using different compilers and tested under different platforms: (1): GNU C++ under Linux distribution CentOS Release 4.8, i386, 32 bit; (2): Cygwin 32 bit environment under Windows 7 and (3): Bloodshed Developer C++ (Dev C++ 5.0.2). Literature survey has revealed that the proposed approach is unique and novel.

3   Background Concepts

Instructions executed by a processor, in general, are of the types: *load*, *store*, *move*, *add*, *compare* or *jump*; a collection of which forms the Instruction Set Architecture (ISA) of any microprocessor. A jump is a control flow instruction, which can be broadly divided into two categories: (a) *Conditional Branch* (CB) and (b) *Unconditional Branch* (UB). Based on a runtime condition, CBs can be further classified as Forward CBs (FCBs), also called a *forward jump*, where the Program Counter (PC) is changed so as to point to an address ahead of the current position in the instruction stream; and Backward CBs (BCBs), a *backward jump*, where the PC is changed to point backward in the instruction stream. This is pictorially shown in Figs. 1 and 2, respectively.

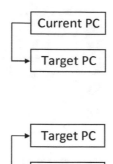

Fig. 1 Forward jump

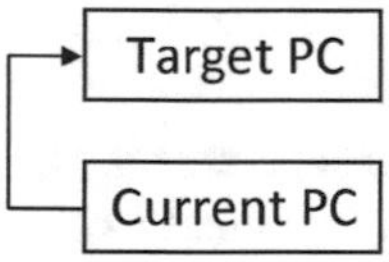

$$\boxed{\text{Target PC}}$$

$$\boxed{\text{Current PC}}$$

Fig. 2 Backward jump

An UB instruction always causes transfer of control to the specified target instruction and are usually called "*jump*". On the other hand, a CB instruction causes transfer of control to different (more than one) target instruction based on the evaluated outcome of some conditional statement. Such CB constructs are usually called "*branch*". In general, UBs includes *jumps, procedure calls,* and *returns* that have no specific condition. Again, based on specific processor architecture, the addressing scheme followed adds extra level of complexity, as in case of Intel $\times 86$ architecture, where with the segmented addressing scheme, the jumps can be either "*near*" (within a segment) or "*far*" (outside the segment). Each type has different effect on the branch prediction schemes implemented.

On the other hand, a *conditional branch* instruction upon execution transfers the control to the target address to execute the instruction based upon the outcome of evaluation of some conditional construct. The behavior of the branch instruction is such that the branch is either *Taken* (T) or *Not-Taken* (NT). Such instructions are further categorized as *Conditional Jump*, often called as *branch* (such as *if-then-else* and *switch*) and *Loop* (such as *do-while, repeat-until,* and *for* loops).

The phases an instruction passes through during the course of its execution in CPU are instruction fetch (IF), decode (D), data fetch (DF), execute (EX) and write back (WB) in sequence, as shown in Fig. 3.

Pipelining concept for instruction overlaying has been considered to be one of the most trusted and effective way to improve execution performance of a processor [17]. Pipelining is a mechanism to increase instruction throughput by performing multiple operations during the same time slot, thereby improving CPU performance without reducing instruction latency, i.e., the time needed to complete a single instruction. But, for this, called Instruction-Level Parallelism (ILP), the CPU has to pay a price rather by increasing the latency due to additional overhead due to (a) pipelining overhead,

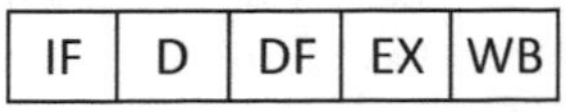

IF	D	DF	EX	WB

Fig. 3 Pipeline stages

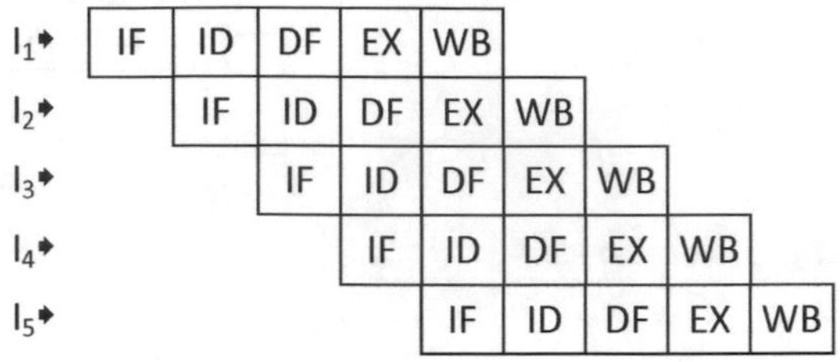

Fig. 4 Pipelined instruction
execution

I_1◆	IF	ID	DF	EX	WB				
I_2◆		IF	ID	DF	EX	WB			
I_3◆			IF	ID	DF	EX	WB		
I_4◆				IF	ID	DF	EX	WB	
I_5◆					IF	ID	DF	EX	WB

i.e., breaking the computation into separate steps, and (b) pipeline flushing overhead to handle possible pipeline stalls. Pipelined execution of instruction carried out in a number of stages is shown below in Fig. 4. In a pipelined processor, a branch instruction may give rise to pipeline stalls if the processor does not know which instruction to be fetched next into the pipeline. Sequential flow of control is interrupted by every branch taken. To get to the right location for the next instruction to fetch from, a huge amount of work already done would have been wasted by the time the current instruction reaches the end of the execute (EX) stage in the pipeline. Until steps are taken to prevent this, such wasteful way of instruction execution process will cause the pipeline to flush a part of the work already done resulting in a lot of CPU time wastage and processor inefficiency. Thus, there is a growing need to solve this problem and this is a vital issue in the design of modern microprocessor which is equipped with longer and deeper pipelines.

Suppose the control flow in a program arrives at a point where it encounter a branch instruction of the form "if $x > y$ then ...". In fact, we assume here that a branch instruction has moved through the pipeline and now is at the EX stage where the processor knows whether or not the branch will take place. The result of this decision is based on the compare between the data items "x" and "y". The problem arises because, until the comparison is done, the processor has no scope of knowing the location of the next instruction where to fetch from and execute. However, the stages prior to EX have already been speculatively fetched with respect to the in-process instruction following sequentially the branch instruction. If the branch does not take place, there is no problem, as the instructions already in the pipeline are in correct sequential flow of control. But, if the branch happens to take place, instructions already occupying different functional components of the pipeline are definitely not the right ones. Hence, they must be flushed and IF must start from the branch target location. Flushing pipeline wastes many cycles of execution time resulting in poor processor performance.

In order to prevent wastage of processor cycles, some mechanisms need to predict the direction of each branch instruction before the next instruction is fetched. If the prediction is correct, the next instruction to execute after the branch executes is already in the pipeline, avoiding the chance of pipeline stall. If the prediction comes out to be wrong, however, the pipeline must be flushed and next instruction from the correct address location must be read into the pipeline. Hence, correctness in the prediction of branch outcome in advance is a key to successful implementation of pipelining technique.

4 Branch Prediction Strategies

In general, branch prediction strategies can be broadly divided into two basic categories: A: *Static Branch Prediction Schemes* and B: *Dynamic Branch Prediction Schemes* depending upon the way they make use of available information to make branch predictions. These are outlined below

A. **Software-Based Static Branch Prediction**:

Static branch prediction (SBP) schemes, frequently implemented at the software level, are usually carried out by the optimizing compiler to decide the direction of a branch instruction (T—Taken or N—Not-Taken). As a matter of convention, all backward pointing branches are always predicted to be taken (T): loops checking exit condition at the end of the loop. Likewise, all forward going branches are always predicted to be not-taken (N): this leads to miss-prediction at the last iteration of the loop construct. Widely used software-based static branch prediction strategies are:

1. Taken (T): Predict that all the branches will be taken.
2. Not-taken (N): Predict that all the branches will not be taken.
3. Back-taken: Predict that all the backward branches will be taken; and predict that all the forward branches will not be taken.
4. Predict that all the branches with certain operation codes will be taken; Predict that the others will not be taken.

Static predictions are advantageous for loops but not for if-then-else constructs.

B. **Dynamic Branch Prediction**:

Dynamic branch prediction (DBP) schemes, frequently implemented at the hardware level, utilize runtime behavior of branch instructions to make predictions. Usually information about the outcomes of previous branch instruction occurrences, i.e., branch history information, is used to predict the outcome of the current branch. There are varieties of DBP schemes as listed below:

1. Two-bit Counter Branch Prediction Scheme
2. Bimodal Branch Prediction Scheme
3. Correlated Branch Prediction Scheme
4. Local Branch Prediction Scheme
5. Global Branch Prediction Scheme
6. Global Selection Branch Prediction Scheme
7. Global Sharing Branch Prediction Scheme
8. Selective Branch Predictors Scheme

Dynamic prediction technique involves processor hardware to decide the direction (i.e., **T**-*Taken* or **N**-*Not-Taken*) based on the availability of dynamic information, such as history of certain branch behavior. Basic idea is to forecast the future action of a branch based on the record of its past behavior. Dynamic Branch

Prediction (DBP) predicts the branch outcome depending on the branch behavior at the runtime and will change if the branch changes its behavior during the execution.

Since different data sets (set of instruction constituting the program to be executed by the processor) will let the programs have different dynamic branch behaviors, therefore usually hardware-based dynamic branch prediction strategies have better prediction accuracy compared to the software-based branch prediction schemes.

Broadly, branch prediction, or simply, the prediction algorithms are categorized on the basis of their prediction approach: Static Prediction and Dynamic Prediction.

(a) **Static Prediction**

Implemented in software, usually carried by the optimizing compiler to decide the direction (T—Taken or N—Not-Taken). As a matter of convention, all backward pointing branches are always predicted to be taken (T): loops checking exit condition at the end of the loop. Likewise, all forward going branches are always predicted to be not-taken (N): this leads to miss-prediction at the last iteration of the loop construct. Static predictions are advantageous for loops but not for if-then-else constructs.

(b) **Dynamic Prediction**

Dynamic prediction technique involves processor hardware to decide the direction (i.e., T—Taken or N—Not-Taken) based on the availability of dynamic information, such as history of certain branch behavior. Basic idea is to forecast the future action of a branch based on the record of its past behavior. Dynamic Branch Prediction (DBP) predicts the branch outcome depending on the branch behavior at the runtime and will change if the branch changes its behavior during the execution.

DBP is based on two key techniques:

1. Branch Outcome Predictor (BOP): to predict the direction or behavior of branch, i.e., whether T—Taken or NT—Not-Taken.
2. Branch Target Predictor (BTP): to predict the branch target address in case of taken branches.

Functional units implementing these techniques are used by the IF unit to predict the next instruction to read into instruction cache (IC).

• If the branch is NT	→ PC is incremented to maintain sequential flow
• If the branch is T	→ BTP supplies the target address, i.e., next address

DBP relies on the information available in the Branch History Table (BHT), also known as Branch Prediction Buffer (BPB), Branch Prediction Table (BPT), and Branch History Register (BHR).

- BHT records previous outcomes (actual) of branch instructions.
- Prediction using BHT is attempted at the IF stage of the current instruction (say I_1) in the pipeline.

- The prediction is acted upon in the ID stage of the current branch instruction (I_1) to fetch the next instruction (I_2), otherwise, without prediction; the next instruction (I_2) fetch would have taken place after the EX stage of the current branch instruction (I_1).

If the prediction is not made, the default assumption about the branch instruction is "Not-Taken". On the other hand, if the prediction is made, and then if the prediction comes out to be true, it triggers no further action, while if the prediction comes out to be wrong, it causes delay to carryout different housekeeping works necessary for the purpose.

There are many approaches to implement BHT. As shown above (Fig. 5), in the simplest form, BHT is implemented in a small piece of memory indexed by k lower order bits of the address (PC content) of the current branch instruction. Entries in the table contain one-bit data (0 or 1) to represent branch instruction outcomes: "1" if the branch was Taken (T), "0" if the branch was Not-Taken (NT or N). Based on the prediction, the action triggered is that the next instruction fetch takes place in the predicted direction. BHT has no tag, i.e., every access to BHT is a hit and the prediction bit could have been placed there by another branch instruction with the identical low-order address bits. But this does not matter as the prediction is just a hint.

One-bit BHT and 2-bit BHT are discussed further in detail below. All DBP schemes involve a mechanism to update information to support improve the accuracy of the prediction process. When a branch instruction is executed in the pipeline, the DBP unit is asked for a guess in the IF stage. After the execution stage, the current address and the actual outcome are supplied to the update process to store the information.

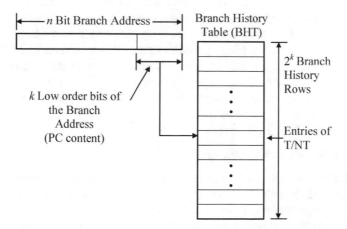

Fig. 5 Branch history table (BHT) implementation

Fig. 6 State machine
representation of the one-bit
BHT

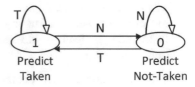

Predict   T   Predict
Taken   Not-Taken

(c) One-Bit Branch History Table:

One-bit BHT remembers only the last branch outcome (T/N). As part of the PC (low-order bits) is used to index into the table, multiple branch history may share the same bit. Update process may invert the bit if the prediction comes out to be wrong. Backward branches for operations in a loop will be miss-predicted twice: one upon loop entry and then again on next loop entry, that is, even if the branch is almost always taken, and not-taken once, the One-bit BHT will miss-predict twice.

- At the last iteration of a loop, prediction is *taken*, but actually it does not take place, i.e., *not-taken*. Branch history is flipped.
- At reentry, the prediction is exit out of loop, i.e., *not-taken* (as per history), but actually it is *taken*.

State machine representation as well as the structure of the one-bit BHT is shown in Fig. 6:

Implementation of the one-bit BHT is shown in Fig. 7. Though one-bit BHT helps in saving hardware space, it is inefficient because of its poor performance. Better solution is 2-bit BHT.

(d) 2-Bit Branch History Table:

2-bit BHT is based on the concept of 2-bit Saturating Counter Model as shown in Fig. 8. In 2-bit BHT, the prediction changes only if miss-prediction happens twice. In a loop branch, at the last loop iteration, the prediction need not be changed. For each index in the BHT, two bits are used to represent the four states of a finite-state machine, i.e., the 2-bit saturating counter. MSB of the 2-bit state value represents prediction status. 00 and 01 both represents *not-taken*, 00 as strongly *not-taken* and 01 being weakly *not-taken*. Similarly, 10 and 11 both represents *taken*, 11 being strongly *taken* and 10 being weakly *taken*.

Structure of a 2-bit BHT is shown Fig. 9.

(e) Correlating Branch Predictor:

The 2-bit BHT uses only the recent behavior of a single branch to predict the future behavior of that branch. BHT contains prediction regarding the next branch—whether it will be a taken (T) or not-taken (N), but does not supply the target PC value to initiate fetch of next instruction into the pipeline. Branch predictors that use knowledge about the behavior of other branches to make prediction are called Correlating Predictors or Two-Level Predictors, where:

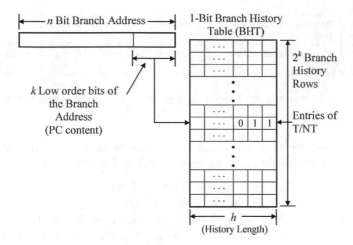

Fig. 7 1 bit branch history table (BHT) implementation

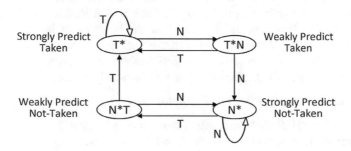

Fig. 8 2-bit saturating counter state machine implementing the 2-bit BHT

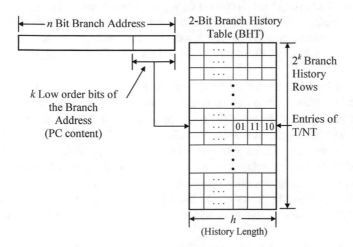

Fig. 9 2-bit branch history table (BHT) implementation

1st level	Stores the branch history in branch history table (BHT) or branch prediction buffer (BPB)
2nd level	Stores prediction pattern in the pattern history table (PHT) or branch target buffer (BTB)

Traditionally, Correlating Predictors specified as (m,n) are classified as

- (m,n) Predictors

    - $m \rightarrow$ History bits, i.e., number of branches in the history (History Length).
    - $n \rightarrow$ Prediction Bits (one-bit Prediction or 2-Bit Prediction)

- $(0,1) \rightarrow$ one-bit Branch Prediction Buffer or BHT
- $(0,2) \rightarrow$ 2-bit Branch Prediction Buffer or BHT
- $(2,2) \rightarrow$ two branches in the history, 2-bit Branch Prediction Buffer or BHT
- $(5,2) \rightarrow$ five branches in the history, 2-bit Branch Prediction Buffer or BHT

For example, a $(1,1)$ Correlating Predictor means one-bit predictor with one-bit correlation: the behavior of last branch is used to choose among a pair of one-bit branch predictors. Correlating or 2-level Predictors also look at, i.e., correlate the behavior of other branches for clues to guess the behavior of the current branch. The basic idea is the behavior of the recent branches is correlated, i.e., the recent behavior of other branches rather than just the current branch we are trying to predict can influence the prediction of current branch behavior.

(f) **Two-Level Adaptive Branch Predictor**

To improve branch prediction, many authors have suggested to use two levels of branch history information. However, to achieve a higher degree of accuracy in branch prediction, Yeh and Patt [16] introduced the idea of *Two-Level Adaptive Branch Prediction* where the branch history information is collected dynamically at two different levels: (1) the first level being the branch execution history—history of last k branch instructions encountered and (2) the second level is the pattern history—a record of branch behavior for the last j occurrences of the specific pattern of these k branches; all these data being collected at runtime and maintained in tables.

Differing by the way the 1st level branch history information is maintained in BHT, i.e., global (G) or on per-address (P) basis, and the way the second level PHTs are associated with the BHT, i.e., global (g) or on per-address basis (p), Yeh and Patt [18] have presented three variations of the Two-Level Adaptive Branch Prediction schemes. These schemes are identified as GAg, PAg, and PAp, embedded A being signifying "Adaptive" and GAp being the *Correlating Branch Predictor*. Considering the addresses that contains branch instructions partitioned into sets (represented by S in the first level and by s in the second level), the *Two-Level Adaptive Branch Prediction* scheme yields nine possible variations, as listed in Table 1.

The most widely used hardware-based dynamic branch prediction strategies (implemented as a functional component of the CPU) are

Table 1 Varieties of two-level adaptive branch predictor

Scheme	Description
GAg	Global adaptive branch prediction using one global pattern history table
GAs	Global adaptive branch prediction using per-set pattern history tables
GAp	Global adaptive branch prediction using per-address pattern history tables
PAg	Per-address adaptive branch prediction using one global pattern history table
PAs	Per-address adaptive branch prediction using per-set pattern history tables
PAp	Per-address adaptive branch prediction using per-address pattern history tables
SAg	Per-set adaptive branch prediction using one global pattern history table
SAs	Per-set adaptive branch prediction using per-set pattern history tables
SAp	Per-set adaptive branch prediction using per-address pattern history tables

1. One-bit: One-bit branch prediction buffer.
2. Two-bit: Two-bit branch prediction-counter.
3. GAg.
4. PAg.
5. PAp.

5 Implementation of FLANN-Based DBP

Branch prediction, in general, can be considered as an optimization problem where the objective demands attainment of lowest possible miss rate, low power consumption and low complexity with minimal resource. Branch prediction is an old trick to boost CPU performance which still finds its place in the design of modern processor architectures. While the simple prediction techniques provide fast lookup of branch address, they suffer from high misprediction rate. On the other hand, complex branch prediction schemes—either ANN based or variants of two-level branch prediction mechanisms—provide better prediction accuracy but increase the complexity exponentially. In addition, the rise in complexity in the prediction techniques involved reflects an increase in the time taken to predict the branches; a moderately very high overhead—ranging from two to cycles compared to the execution time of actual branches.

In this paper, the proposed dynamic branch predictor is designed using Functional Link Artificial Neuron (FLAN) model, based on the idea of perceptron-based branch predictor proposed by Jimenez and Lin [3]. Since functional link networks are capable of solving linearly non-separable problem using appropriate input representation, in this paper, this is achieved by increasing the dimension of the input space. Now, this expanded input data along with the corresponding link weight is used for training the FLAN instead of the actual data. There are many possible schemes available for achieving the function expansion of the input data: use of trigonometric function (T-FLAN), Chebyshev polynomials (C-FLAN), Legender polynomials (L-FLAN) and Power Series expansion

Fig. 10 3 point functional expansion of an input data item

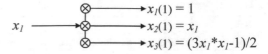

Fig. 11 Modular representation of FEB of above diagram

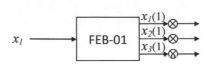

(P-FLAN) are common. In this paper, a three point expansion of input data using Legender polynomial has been adopted, as shown Fig. 10. For convenience, each functionally expanded input data is represented as a "functional expansion block" or FEB, shown Fig. 11.

Figure 12 shoes the proposed L-FLAN model designed based on the idea of perceptron-based branch predictor proposed by Jimenez and Lin [2, 4].

As per the proposed scheme, each input signal x_i undergoes a function expansion into $x_1(i)$, $x_2(i)$, $x_3(i)$, ..., $x_z(i)$, where z represents the number of lines to which each incoming signal is expanded. For example, an input data x_5 by undergoing a three-point functional expansion becomes $x_1(5)$, $x_2(5)$, and $x_3(5)$.

In this paper, a dynamic branch predictor (DBP) has been proposed based on L-FLAN which is a correlating DBP that makes a prediction of the current branch outcome based on the history pattern obtained from the past branch behavior.

A L-FLAN-based DBP uses a weight matrix W of signed short integers having dimension $N \times (z * h + 1)$ where N represents the number of L-FLANs taking part in the DBP, z is the number of lines to which each incoming signal is expanded and h is the history length. In simple term, weight matrix W constituted of N weight vectors of the form $w_i[0 \ldots N]$, where the first weight $w_i[0]$ corresponds to the bias weight w_0 of the ith L-FLAN, as shown below. Thus, the first column of W contains

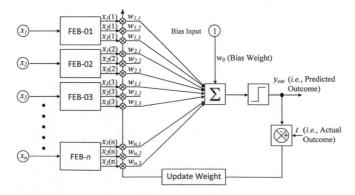

Fig. 12 Model of the proposed FLANN-based DBP

the bias vector for N number of L-FLANs. Each row of the matrix represents one L-FLAN, characterized by a weight vector of size $(z * h + 1) = (3 * h + 1) = (k + 1)$ where $k = 3 * h$. The Boolean vector G represents the global history shift register which contains the {taken (T), not-taken (N)} values to store the outcome of the h previous branches.

In order to make a prediction of the outcome of a branch instruction located at the memory address M loaded in the PC, a (PC mod N) operation helps index onto a particular row in the W matrix. In fact, M is the lower order byte of the full PC content. This row represents an L-FLAN that is responsible for making the prediction for that particular branch instruction. Note that a single L-FLAN could be responsible for the prediction of multiple branch instructions. The prediction made is based on the weight values in that row and on the outcome of the most recent h branches, which is stored in the global history shift register (GHSR) G. This most recent branch outcome is pushed onto the GHSR, dropping the oldest branch outcome information (Fig. 13).

Once the actual outcome is known after completion of EX stage in the pipeline, the L-FLAN is trained (i.e., weights are updated) depending on whether the prediction was correct and also based on the pattern history stored, that is stored in the global history register G. This is accomplished by updating the weights in the corresponding row of the W matrix. If the prediction was correct, nothing is done. However, if the prediction is found to be wrong, then the most recently pushed branch outcome onto the GHSR is now flipped.

A. **Aliasing**: Aliasing problem refers to the case of overlapping of more than one branch addresses (low-order byte of the PC content) to the same entry in the branch history table. Obviously, aliasing takes place due to limited table space,

Fig. 13 Design of global history shift register (GHSR)

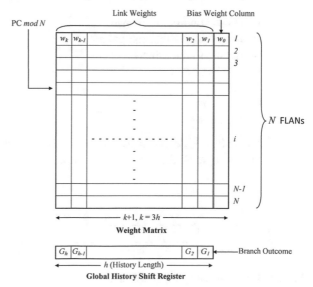

as a result of which, more than one branch address get a chance to index into the same BHT entry. However, since indexing into BHT is a hint, aliasing is not considered to be a serious problem.

B. **Simulation Environment**: The DBP algorithms were written in C++ language and was build using GNU g++ compiler Release 3.4 under CentOS Release 4.8 linux in a Pentium Core i3-based machine environment, with Flex and Bison installed. The same code has also been used to build binary under Windows 7 environment, using Bloodshed Developer C++ (Dev C++) and Cygwin environment. Further tuning will yield superior performance.

C. **The Trace file**: The trace file contains information regarding outcome of following types of instructions in addition to a four-byte little-endian branch address—the address in memory where the first byte of the branch instruction is available, along with a four-byte little-endian branch target—the address in memory to where the branch takes place. The branch types could be one of the following types:

- Taken conditional branch
- Not-taken conditional branch
- Unconditional branch
- Indirect branch
- Call
- Indirect call
- Return

D. **The Trace Data Format**: A trace is a piece of information about a branch [10]. The external representation of a trace is 9 bytes, as shown Fig. 14.

- A one byte "Code". The lower 4 bits are the ×86 Op-code for a conditional branch (modulo 16). The upper four bits are one of the following:

  1: taken conditional branch
  2: not-taken conditional branch
  3: unconditional branch
  4: indirect branch
  5: call
  6: indirect call
  7: return

- A four-byte little-endian Branch Address. This is the address in memory of the first byte of the branch instruction.

Fig. 14 Record format in the trace data file

- A four-byte little-endian Branch Target. This is the address in memory where the branch jumped.
- If the upper four bits of the first byte read are either 0 or 8 then the byte indicates that the trace has been compressed from the 9 byte representation to a 1 or 2 byte representation. Since the branch target is an additional piece of information included in the trace record which is not at all useful for making branch prediction, during actual prediction simulation run, a trace preprocessor program is used to filter out the redundant information and produce a trace data file (in both binary as well as ascii format), so that each trace record contains only *<branch address>:<outcome >* .

E. Computing Output:

$$y = w_0 + \sum_{i=1}^{n} x_i w_i$$

where y is the computed output, which is the predicted branch outcome; w_0 is the bias weight; x_i, i.e., $x_1, ..., x_n$, are the input values, which is represented by the branch history, n is the dimension of the input vector, i.e., branch history width; and w_i, i.e., $w_1, ..., w_n$, are the link weight values.

F. **Algorithm to Control Extent of Training Done**

Based on the value of the computed output y_{out}, decision is taken whether to train the network further or not. In a simple language, training the network is always triggered whenever there is a disagreement between the computed output y_{out} and the actual branch outcome t, irrespective of the threshold value θ. However, if the magnitude of the network output is less than or equal to the threshold value θ, then network training is automatically triggered. Training is performed as per the algorithm shown Fig. 15.

G. *The Predict Algorithm* (Fig. 16).

H. *The Training Algorithm* (Fig. 17).

Fig. 15 Training control algorithm

```
if sign (yout ≠ t or |yout| ≤ θ) then
        for i := 0 to n do
                wi := wi + txi
        end for
end if
```

```
function predict (address: integer): boolean
begin
    (* output is initialized to bias weight *)
    output := W[address,0,0]
    (* sum weights (or their negations) chosen using
    the addresses of the last GHL branches *)
    for i in 1::GHL do
        if GHR[i] = true then
            (* if the ith branch in *)
            output := output + W[address,GA[i], i]
        else
            (* otherwise subtract it *)
            output := output - W[address,GA[i], i]
        end if
    end for
    (* predict the branch taken if the output is at least 0 *)
    predict := output ≥ 0
end
```

Fig. 16 Predict algorithm

```
procedure train (address: integer; taken: boolean)
begin
    if |output| < θ or output ≥ 0 ≠ taken then
        if taken = true then
            W[address,0,0] := W[address,0,0] + 1
        else
            W[address,0,0] := W[address,0,0] □ 1
        end if
        for i in 1::GHL
            if GHR[i] = taken then
                W[address, GA[i],i] := W[address,GA[i],i] + 1
            else
                W[address, GA[i],i] := W[address,GA[i],i] - 1
            end if
        end for
    end if
    GA[2..GHL] := GA[1..GHL - 1]
    GA[1] := address
    GHR[2..GHL] := GHR[1..GHL - 1]
    GHR[1] := taken
end
```

Fig. 17 Training algorithm

6 Simulation and Results

This infrastructure used in the simulation process has been tested on $\times 86$ hardware-based Intel Pentium P4 Core i3 Desktop PC running Fedora Core 4 operating systems as well as the Cygwin environment on Microsoft Windows XP operating system.

The trace files used represents the branches encountered during the execution of 100 million instructions from the corresponding standard benchmark. The traces include branches executed during the execution of benchmark code, library code, and system activity such as page faults. The traces sent to the CBPs under study are bit-for-bit identical to the traces collected from the running benchmarks.

Outputs from typical simulation run for both the FLANN-based and Perceptron Network-based branch prediction schemes are discussed below. With the help of a comparative study of the predictor performance we can analyze their performances explicitly. The output dump is adequate enough to assess the performance figure of implemented branch prediction programs. Following figures presents a comparison of result of simulation of FLANN-based dynamic prediction scheme against the perceptron-based dynamic prediction scheme. Results of simulation clearly establishes the claim made by the author in the paper [8] regarding the superiority of the perceptron-based branch prediction scheme as compared with the other dynamic branch prediction scheme. Comparative performance plots are also provided to study the performance. Implementation of the FLANN-based branch predictor algorithm needs further parameter tuning to achieve superior performance as literature review has already established [12, 13] the superiority of the FLANN-based approach over many multilayer perceptron networks.

Given below are the outputs of conditional branch prediction simulation of the FLANN-based scheme (immediate below) and Perceptron-based scheme (the next one) on the same trace data file. Figures 18 and 19 presents the execution run status of the FLANN-based trace-driven Dynamic Branch Prediction Simulator (DBPSim) on standard trace data file of different lengths for predicting branch outcomes records (small trace data file and large trace data file, respectively). These traces were generated from running real programs and tracking if branches were taken or not-taken. Figures 20 and 21 are plots that present influence of warm-up value on prediction rate and prediction accuracy, respectively on small trace data file. Figures 22 and 23 are similar plots (like Figs. 20 and 21) attributed to large trace data files.

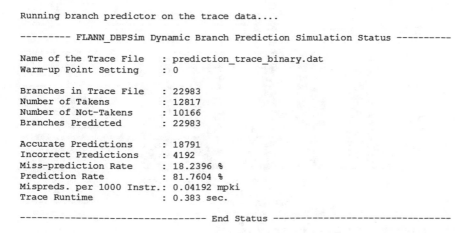

```
Running branch predictor on the trace data....

--------- FLANN_DBPSim Dynamic Branch Prediction Simulation Status ----------

Name of the Trace File    : prediction_trace_binary.dat
Warm-up Point Setting     : 0

Branches in Trace File    : 22983
Number of Takens          : 12817
Number of Not-Takens      : 10166
Branches Predicted        : 22983

Accurate Predictions      : 18791
Incorrect Predictions     : 4192
Miss-prediction Rate      : 18.2396 %
Prediction Rate           : 81.7604 %
Mispreds. per 1000 Instr.: 0.04192 mpki
Trace Runtime             : 0.383 sec.

------------------------------ End Status ----------------------------
```

Fig. 18 Small trace data file

```
Running branch predictor on the trace data....

--------- FLANN_DBPSim Dynamic Branch Prediction Simulation Status ----------

Name of the Trace File    : prediction_trace_binary_2.dat
Warm-up Point Setting     : 0

Branches in Trace File    : 210971
Number of Takens          : 109865
Number of Not-Takens      : 101106
Branches Predicted        : 210971

Accurate Predictions      : 169775
Incorrect Predictions     : 41196
Miss-prediction Rate      : 19.5269 %
Prediction Rate           : 80.4731 %
Mispreds. per 1000 Instr.: 0.41196 mpki
Trace Runtime             : 3.257 sec.

------------------------------ End Status -------------------------------
```

Fig. 19 Large trace data file

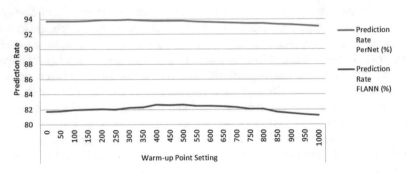

Fig. 20 Prediction rate on small trace data file

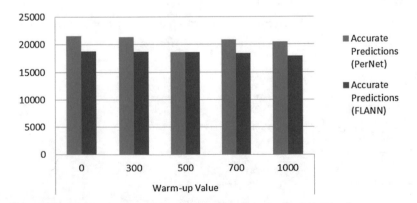

Fig. 21 Prediction accuracy on small trace data file

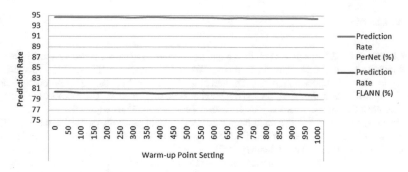

Fig. 22 Prediction rate on large trace data file

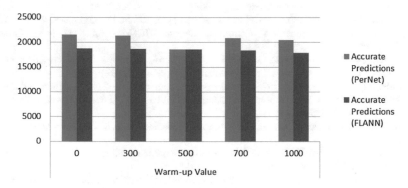

Fig. 23 Prediction accuracy on large trace data file

7 Conclusion and Future Work

From the simulation-based performance comparative study of the implementation of FLANN-based CBP with respect to using standard benchmark trace data files revels that a further fine tuning of the parameters of the implementation will enable the FLANN-based CBP to achieve performance at par with and better than that of the perceptron-based CBP implementation. Combining other CBP schemes has also not been studied during this work which allows for a good scope for further expansion in the study.

Acknowledgements The authors would like to thank the anonymous reviewers for their valuable comments and suggestions.

References

1. Burguiere, C., Rochange, C.: On the complexity of modeling dynamic branch predictors when computing worst-case execution times. Technical report, IRIT—CNRS (Institute de Recherche en Informatique de Toulouse â Toulouse Institute of Computer Science Research; Centre National de la Recherche Scientific que â The National Center for Scientific Research), University of Toulouse, Jean Jaures, France (2007)
2. Jimenez, D.A., Lin, C.: Dynamic branch prediction with perceptrons. In: Proceedings of the 7th International Symposium on High-Performance Computer Architecture (HPCA-2010), pp. 197–206, Busan, 19–24 January 2001
3. Yeh, T.-Y., Patt, Y.N.: A comparison of dynamic branch predictors that use two levels of branch history. In: Proceedings of the 20th Annual International Symposium on Computer Architecture, pp. 257–266, San Diego, CA, USA, 16–19 May 1993. IEEE
4. Daniel A. Jimenez and Clavin Lin. Neural method for dynamic branch prediction. ACM Transactions on Computer Systems, 20(4):369 397, November 2002
5. Jimenez, D.A.: An optimized scaled neural branch predictor. In: Proceedings of the 29th IEEE International Conference on Computer Design (ICCD), pp. 113–118, Amherst, Massachusetts, 9–12 October 2011
6. Milenkovic, M., Milenkovic, A., Kulick, J.: Microbenchmarks for determining branch predictor organization. Softw. Prac. Exper. **34**(5), 465–487 (2004)
7. McFarling, S.: Combining branch predictors. Technical Note TN-36, Digital Equipment Corporation (DEC), Western Research Laboratory (WRL) 250 University Avenue Palo Alto, California 94301 USA, June 1993
8. Jimenez, D.: Idealized piecewise linear branch prediction. J. Instr. Lev. Parallelism **7**, 1–11 (2005)
9. C++ code idea. Available Online as http://faculty.cse.tamu.edu/djimenez/snp_my_predictor.h. Accessed on 24 February 2016
10. Branch prediction competition. Available Online as http://faculty.cse.tamu.edu/djimenez/614/competition/cbp2-infrastructure-v2.tar. Accessed on 24 February 2016
11. Infrastructure for branch prediction competition. Available Online as http://faculty.cse.tamu.edu/djimenez/614/competition/cbp2-infrastructure-v2/doc/. Accessed on 24 February 2016
12. Patra, J.C., Pal, R.N., Chatterji, B.N., Panda, G.: Identification of nonlinear dynamic systems using functional link artificial neural networks. IEEE Trans. Syst. Man Cybern. Part B: Cybern. **29**(2), 254–262 (1999)
13. Patra, J.C., Panda, G., Baliarsingh, R.: Artificial neural network-based nonlinearity estimation of pressure sensors. IEEE Trans. Instrum. Meas. **43**(6), 874–881 (1994)
14. MultiMedia LLC. MS Windows NT kernel description. Available Online as http://en.wikipedia.org/wiki/Branch_predictor. Accessed on 14 February 2016
15. Falk Wilamowski. Branch Prediction simulation with archc. http://falk.muellerbande.net/blog/user/admin/. Accessed on 5 February 2016
16. Jimenez, D.A.: Fast path-based neural branch prediction. In: Proceedings of the 36th Annual IEEE/ACM International Symposium on Microarchitecture (MICRO-36, 2003), pp. 243–252, San Diego, CA, USA, 3–5 December 2003. IEEE
17. Yeh, T.-Y., Patt, Y.N.: Two-level adaptive training branch prediction. In: Proceedings of the 24th Annual International Symposium on Microarchitecture, Two-Level Adaptive Training Branch Prediction, pp. 51–61, 11–13 November 1991. ACM New York, NY, USA
18. Yeh, T.-Y., Patt, Y.N.: Alternative implementations of two level adaptive branch prediction. In: Proceedings of the 19th Annual International Symposium on Computer Architecture, pp. 124–134, Gold Coast, Australia, 19–21 May 1992. IEEE Computer Society, Washington

Development of a Model Recommender System for Agriculture Using Apriori Algorithm

M. B. Santosh Kumar and Kannan Balakrishnan

Abstract Recommender System (RS) has become very popular recently and being used in variety of areas including movies, music, books and various products. This study focused on the development of a model RS for agriculture (ARS) using Apriori algorithm. Prediction of the Agri-items (vegetables/fruits) can be made and the RS can provide the recommendations of the products which the customers can order. The data obtained for a period of 8 months about the consumption of the various items ordered through the website were used for designing and implementing the RS model. Preprocessing of the data is done followed by dimensionality reduction to make the data more refined. A hybrid web-based RS was modeled using Apriori algorithm with associated rule mining to recommend the various items, that will help the farmers to produce optimally and thus increasing their profit.

Keywords Data pre-processing · Dimensionality reduction · Association rule mining · Apriori algorithm

1 Introduction

The state of Kerala is tagged as one of the agricultural hubs for many food crops like rice, coconut, banana etc., due to the advantages in climatic conditions. But improper planning of the production of items ultimately will lead to a huge loss to the farmers. So a proper prediction and recommendation of items is required. The modeled RS act as a tool that provides suggestions to the customers and farmers

M. B. Santosh Kumar (✉)
Division of Information Technology, Cochin University of Science
and Technology, Kalamassery, India
e-mail: santo_mb@cusat.ac.in

K. Balakrishnan
Department of Computer Applications, Cochin University of Science
and Technology, Kalamassery, India
e-mail: mullayilkannan@gmail.com

© Springer Nature Singapore Pte Ltd. 2019
P. K. Mallick et al. (eds.), *Cognitive Informatics and Soft Computing*,
Advances in Intelligent Systems and Computing 768,
https://doi.org/10.1007/978-981-13-0617-4_15

153

[1–3]. In this study focus is on vegetables and fruits and its allied products such as chili-powder, rice-powder, banana-chips, jackfruit-halwa, etc.

By analyzing the buying behavior of the items purchased by the customer, the RS was designed. Using the system, farmers will get information about what all items to be cultivated for the next season by analyzing the interaction with the customers/users through the web interface. For example Amazon employs a RS to personalize each customer trading through the online store [4]. Most of the RS are personalized, since the buying behavior of one person differs from another, also there exists non-personalized RS [5]. To complete such tasks, customer's preferences, which are explicitly expressed, in the form of implicit ratings of the items are important for the design of RS [6].

The decisions made by other customers play a prominent role in making the recommendation by the individuals for the purchase of the items [1, 7]. For example people rely on the reviews of the movie that film critics have made before watching the movie, similar condition were checked in the purchase of Agri-items.

2 Review of Related Work

RS has evolved as a prominent and independent research area by mid-1990s [1, 7–9]. The terminology, RS indicates systems that normally filter information. Various recommendations based on the end customer needs and preferences were obtained. One of the important task is to give proper prediction of the items which can be viewed or purchased and thus generate the recommendation [10]. Different types of problems arise to make the correct prediction of the items. A proper ranking of the items was done using apriori algorithm [11].

The architecture like semantic web can be utilized in agricultural database for generating agricultural recommendations [12]. "GO-ORGANIC" is the name of the RS which can be accessed by the customers or farmers by logging in with a username and password. The result of the query related to any items will be displayed on the website and also in the Android App [13]. Customers can also have access to the various items online [14]. Based on the interest of the customers say for example their long-term and short-term preference for various items can be found out by rating, tagging and using some other metrics. Thus combining all these together a set of personalized record can be generated [15].

In content-based systems, prediction for a new item depends on the similarity of the items which were rated with a high percentage [16]. Consider Collaborative filtering, here the recommendations were done by considering the preferences of previous users ordering items of same interests [17]. Different approaches and algorithms of data filtering and recommendations exist [18]. Increasing the production of various crops considering the climatic condition, soil fertility etc., is an important task. Genetic algorithm based simulation software are available to maximize crop yields [19]. The predictive accuracy of various methods can be compared using a set of representative problem domains [20].

3 Methodology

The Recommender System works mainly on two techniques namely collaborative- and content-based filtering. The simple and efficient algorithm for association rule discovery is Apriori [21], which was considered while designing the model. The active customers have some agreement with the customers in the past, then the recommendation will have more relevance and of more interest to the active users [22].

A need for filtering the whole range of available alternatives to make the recommendations is necessary. It was found that choices of the items were good, but too much choice is not good [23]. RS generate recommendation with the help of customer data, various knowledge, the available item and historical data about the transactions with reference to the database. The customer can browse through the RS and accept or reject items; hence a feedback is received. All the transactions are stored in the database and a new recommendation is generated during the successive ordering of the items [24].

The RS developed utilizes item and customer-based approaches. In item based approach many of the products ordered by the customers remain more or less the same. Similar items help in building neighborhoods depending on the appreciations made by the customers [25]. Later the system will create recommendation based on the items, customer would prefer. In customer-based approach customers plays a prominent role. Customers ordering the same items are grouped together. The modeled system is a hybrid which combines content and collaborative filtering methods. Figure 1 depicts the model architecture of the Agriculture Recommender System (ARS).

Fig. 1 Model architecture of ARS

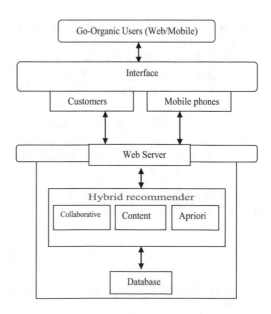

The system was developed as a web-based graphical user interface (GUI) and as a mobile app. The entire requests made by the customers were running on a web server. All the inputs were stored in the database and hybrid systems which can make approximate recommendations in the form of prediction of purchase of various items were designed.

Considering all the items ordered by each customer, the cumulative order of the same item by all the customers helps to predict the usage of a particular item at a particular season of the year. Hence such a system will help to make prediction of the consumption of various products by the customers. Thus the farmers can make the production of vegetables and fruits based on the RS.

The model utilizes association rule mining algorithm for designing and implementing the RS. India touted to be an agricultural country still using traditional ways of recommendations in agricultural sector. Presently the knowledge and recommendations are passed through interaction between farmers and experts. The recommendations made will be different for different experts. Our system helps the customers to get the information and recommendations based on their need and preferences. This helps to reduce the gap between technology, customers, and farmers. The customers and farmers can access products through the website "http://nss.cusat.ac.in/goorganic/login.php". An android based mobile application is also developed and the App can be downloaded from "http://nss.cusat.ac.in/goorganic/downloads/Green-RS.apk".

4   Design and Implementation of an Agricultural Recommender System (ARS)

Recommender System was modeled based on Association Rule Mining which is depicted in Fig. 2.

Dimensionality reduction was done by removing unwanted columns like serial-number, Order Id, customer email-ID etc. Column sets used from the available table were: User-ID (mobile number), Item ID, Purchased quantity of items, Order date.

Apriori Algorithm was used to analyze the data based on Frequent Item Sets generated. Then association rules were generated, and items included in rules with minimum confidence 75%, were stored as recommended products by using the formula given below.

Association rules $X \rightarrow Y$ & $Y \rightarrow X$:
If item set is X, Y then confidence is

$$\text{conf}(X \rightarrow Y) = \frac{\text{Supp}(X \cup Y)}{\text{Supp}(X)}$$

It means probability of buying Y if user bought X.

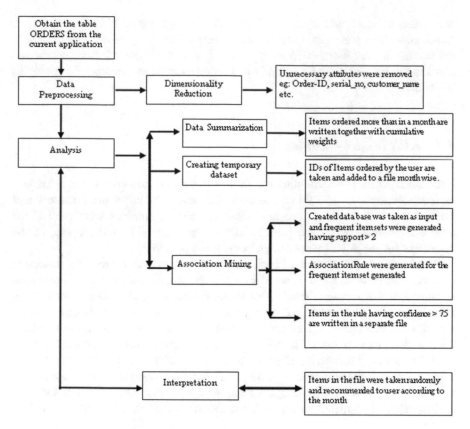

Fig. 2 Flow chart for design and implementation for ARS

If item set is X, Y, Z.
Then, association rules formed would be

$$XY \rightarrow Z, XZ \rightarrow Y$$

$$Z \rightarrow XY, Y \rightarrow XZ \text{ etc.}$$

And confidence would be calculated as,

$$\text{conf}(XY \rightarrow Z) = \frac{\text{Supp}(XY \cup Z)}{\text{Supp}(XY)}$$

Analysis phase consists of (a) Data summarization—here, the items ordered more than once in a month were considered along with cumulative weights. (b) Creating temporary data set—IDs of item ordered by the customers were taken and added to file, month wise. (c) Association rule mining—the created data base was taken as input and item set (frequent) were generated having support >2. Association rules were generated for the frequent item set generated and items in the rule having confidence >75 were written in a separate file. (d) Interpretation—items in the file were taken randomly and recommended to customer according to the month.

4.1 ARS Model—Tables

Four tables were designed initially. Table 1 indicates customer details; Table 2 indicates various items, Table 3 indicates the order placed by the customer and Table 4 indicates the list of ordered items. The mobile numbers were used as the unique ID which acts as the primary key for the table. The sample size of the customers for the ARS model design constituted around 350.

Transactional data related to each customer was extracted from the available customer database and record was made for the amount of each item purchased month wise. Transactional data refers to the columns which includes order date, item id, purchased quantity etc. present in "Order" table of the database which contains all the orders made for a period of eight months (May–December). Separate Graphs were made for each item, based on its sale during the same period.

After forming required number of tables and eliminating the item sets having support less than the minimum support i.e., 2, the association rules were applied on the final set. For example, taking the dataset of items purchased for 8 months.

(Repeated items mean that many times the item is purchased in that month).

Table 1 Customer

Customer_id (primary key)	Mobile (unique)	First_name	Last_name	Email_id	Password

Table 2 Item

Item_id (primary)	Item_name	Item_price

Table 3 Order

Order_id (primary key)	Mobile	Order_time	Order_status	Payment_status	Total

Table 4 Ordered items

Id (primary key)	Order_id	Item_id	Item_name	Item_quantity	Item_price

| May: 3, 7, 9, 10, 1, 3, 8, 1, 3, 6, 7, 10 |
| June: 2, 3, 6, 11, 12, 2, 12, 14, 3, 4, 6, 9, 10, 11, 2, 4, 6, 11, 12, 14 |
| July: 2, 3, 6, 12, 14, 1, 2, 3, 6, 12, 4, 6, 10, 1, 2, 3, 4, 6, 8, 10, 12, 14 |
| August: 1, 2, 6, 12, 3, 5, 14, 15, 1, 2, 3, 4, 5, 6, 12, 14 |
| September: 1, 2, 12, 14, 15 |
| October: 1, 2, 4, 6, 7, 12, 14, 15 |
| November: 2, 4, 6, 11, 12, 14, 15, 1, 2, 6, 7, 12, 15, 1, 8, 1, 2, 6, 7,12 |

So support of each item was stored in Table 5.

Taking minimum support = 2, and eliminating the rows with support value less than 2, a new Table 6 was obtained.

Now, two items present in the updated Table 6 were paired and the support of both of them appearing at same time was obtained and stored in a new Table 7.

A similar approach as above was applied and the item sets having support less than 2 were removed from the table to form another new table. The same process of grouping the items was applied till there was no group present with support less than the minimum support. After forming the final item set table, association rules were applied and confidence was calculated.

Now items with confidence more than 75% were noted in a separate file and recommended to the customers. Recommendations are shown to the customers based on their previous purchases they have made. User Interface was designed for the customers where they can view the recommended items along with the items they want to purchase. Amount of items to be produced by farmers will be based on the sales graph made for each month and recommendations made for each customer which is depicted below.

Table 5 Item versus support

Item_no	1	2	3	4	5	6	7	8	9	10	11	12	13	14	15
Support	11	13	10	7	2	14	5	3	2	5	4	13	0	9	5

Table 6 Item versus support with minimum support = 2

Item_no	1	2	3	4	5	6	7	8	9	10	11	12	14	15
Support	11	13	10	7	2	14	5	3	2	5	4	13	9	5

Item no. 13 was eliminated

Table 7 Item versus support with minimum support >2

Item_no	1, 2	1, 3	–	–	1, 11	–	–	–	3, 10	–	–	–	–	14, 15
Support	8	6	–	–	1	–	–	–	5	–	–	–	–	4

4.2 Graphs

The graphs below indicate the consumption of various types of agricultural products. Figure 3 indicates the consumption of Yam, with *x*-axis indicating the quantity of consumption in kilogram (kg) and y-axis indicating the month. Figure 4 indicates the consumption of water melon. Similarly graphs can be plotted for all the items.

Fig. 3 Consumption of Yam

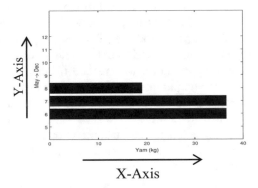

Fig. 4 Consumption of water melon

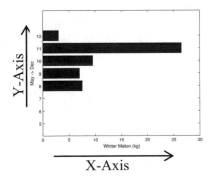

5 Results and Discussion

The recommender system used in Go-Organic recommends products to the users based on two aspects:

5.1 Purchases Made by the Customers (Previous)

RS is based on the transactions made in the previous year. The database contains information about every transaction. Each transaction has several products related to it and every product in the database is distinguished by a unique id. Now to give the recommendations following steps were followed: All the transactions of the specified month of a single user were taken and association rules were generated according to Association Rule Mining which also contains product ids. Whenever customer makes a purchase in that month, the product ids were matched with the generated association rules. Whenever the purchasing pattern of any of the association rules matches, the corresponding products present in the association rule were recommended. In case, the purchasing patterns did not match a single association rules then recommendations were made by matching them with multiple association rules. These recommendations were different for every user i.e.; according to the taste of the user. This is significant for the existing users who have some transaction history to make their purchases more interactive.

5.2 The Best-Selling Products

Most of the products sold from the previous year were recommended to the customers. The amount for each product in the previous year was calculated and the top five products ranked on the basis of the amount were taken as the most sold products and recommended to the user. These recommendations remained same for every user. This is significant for the new users who do not have a purchasing history.

6 Conclusion

A recommender system model for predicting and recommending the consumption of various agricultural items was designed and developed using Apriori algorithm. The system developed was able to make the predictions and recommendation on the basis of the customer's previous buying behavior and their peer's recommendation of the items. The model ARS can make prediction of items consumed by all the customers, so that the farmers can produce the items according to their choice. Thus the cumulative prediction can help the farmers to plan and make the cultivation of the Agri-products (Items) for any season, so that there will not be any wastage of the items produced by the farmers. Thus the system developed, helped the customers to recommend more items than the predicted one.

References

1. Tariq, M., Ricci, F.: Improving recommender systems with adaptive conversational strategies. In: Proceedings of the 20th ACM conference on Hypertext and hypermedia. ACM (2009)
2. Paul, R., Varian, H.R.: Recommender systems. Commun. ACM **40**(3), 56–58 (1997)
3. Robin, B.: The adaptive web. Chap. Hybrid Web Recommender Syst. 377 (2007)
4. Francesco, R., Rokach, L., Shapira, B.: Introduction to recommender systems handbook. In: Recommender Systems Handbook. Springer US, pp. 1–35 (2011)
5. Dias, M.B., et al.: The value of personalised recommender systems to e-business: a case study. In: Proceedings of the 2008 ACM conference on Recommender systems. ACM (2008)
6. Glance, N.S.: Recommender system and method for generating implicit ratings based on user interactions with handheld devices. U.S. Patent No. 6,947,922. Accessed on 20 September 2005
7. Frank, M., Mironov, I.: Differentially private recommender systems: building privacy into the net. In: Proceedings of the 15th ACM SIGKDD international conference on Knowledge discovery and data mining. ACM (2009)
8. Goldberg, D., et al.: Using collaborative filtering to weave an information tapestry. Commun. ACM **35**(12), 61–70 (1992)
9. Sarabjot Singh, A., Mobasher, B.: Intelligent techniques for web personalization. In: Proceedings of the 2003 International Conference on Intelligent Techniques for Web Personalization. Springer (2003)
10. Francesco, R., Rokach, L., Shapira, B.: Introduction to recommender systems handbook. In: Recommender Systems Handbook. Springer, US, pp. 1–35 (2011)
11. Hegland, M.: The apriori algorithm—a tutorial. Math. Comput. Imaging Sci. Inform. Process. **11**, 209–262 (2005)
12. Berger, T.: Agent-based spatial models applied to agriculture: a simulation tool for technology diffusion, resource use changes and policy analysis. Agric. Econ. **25**(2-3), 245–260 (2001)
13. Vikas, K., et al.: Krishimantra: agricultural recommendation system. In: Proceedings of the 3rd ACM Symposium on Computing for Development. ACM (2013)
14. Iorshase, A., Charles, O.I.: A well-built hybrid recommender system for agricultural products in Benue State of Nigeria. J. Softw. Eng. Appl. **8**(11), 581 (2015)
15. Mehdi, E., et al.: Interaction design in a mobile food recommender system. In: CEUR Workshop Proceedings. CEUR-WS (2015)
16. Van Meteren, R., Van Someren, M.: Using content-based filtering for recommendation. In: Proceedings of the Machine Learning in the New Information Age: MLnet/ECML2000 Workshop (2000)
17. Anvitha, H., Shetty, S.K.: Collaborative filtering recommender system. Int. J. Emerg. Trends Sci. Technol. **2**(7) (2015)
18. Badrul, S., et al.: Item-based collaborative filtering recommendation algorithms. In: Proceedings of the 10th International Conference on World Wide Web. ACM (2001)
19. Olakulehin, O.J., Omidiora, E.O.: A genetic algorithm approach to maximize crop yields and sustain soil fertility. Net J. Agric. Sci. **2**(3), 94–103 (2014)

20. Breese, J.S., Heckerman, D., Kadie, C.: Empirical analysis of predictive algorithms for collaborative filtering. In: Proceedings of the Fourteenth conference on Uncertainty in artificial intelligence. Morgan Kaufmann Publishers Inc. (1998)
21. Hamid Reza, Q., Nasiri, M., Minaei-Bidgoli, B.: Multi objective association rule mining with genetic algorithm without specifying minimum support and minimum confidence. Expert Syst. Appl. **38**(1), 288–298 (2011)
22. Dietmar, J., et al.: Recommender systems: an introduction. Cambridge University Press (2010)
23. Barry, S.: The paradox of choice: why less is more. Ecco, New York (2004)
24. Herlocker, J.L., et al.: Evaluating collaborative filtering recommender systems. ACM Trans. Inform. Syst. (TOIS) **22**(1), 5–53 (2004)
25. Pampın, H.J.C., Jerbi, H., O'Mahony, M.P.: Evaluating the relative performance of neighbourhood-based recommender systems. In: Proceedings of the 3rd Spanish Conference on Information Retrieval (2014)

Emotion Speech Recognition Based on Adaptive Fractional Deep Belief Network and Reinforcement Learning

J. Sangeetha and T. Jayasankar

Abstract The identification of emotion is a challenging task due to the rapid development of human–computer interaction framework. Speech Emotion Recognition (SER) can be characterized as the extraction of the emotional condition of the narrator from their spoken utterances. The detection of emotion is troublesome to the computer since it differs according to the speaker. To solve this setback, the system is implemented based on Adaptive Fractional Deep Belief Network (AFDBN) and Reinforcement Learning (RL). Pitch chroma, spectral flux, tonal power ratio and MFCC features are extracted from the speech signal to achieve the desired task. The extracted feature is then given into the classification task. Finally, the performance is analyzed by the evaluation metrics which is compared with the existing systems.

Keywords Emotion recognition · Adaptive fraction deep belief network
Reinforcement learning

1 Introduction

Emotion recognition systems purpose at distinctive emotions of human subjects from underlying knowledge with desirable accuracy. Audio signals are the number one modalities of human emotion belief have earned the foremost attention in developing smart structures for natural interaction. An emotion recognition machine should robotically perceive the human emotional states besides his or her voice. It is

J. Sangeetha (✉)
Department of Information Technology/SOC,
SASTRA Deemed University, Thanjavur, Tamil Nadu, India
e-mail: sangita.sudhakar@gmail.com

T. Jayasankar
Department of ECE, University College of Engineering,
BIT Campus, Anna University, Tiruchirappalli, Tamil Nadu, India
e-mail: jayasankar27681@gmail.com

© Springer Nature Singapore Pte Ltd. 2019
P. K. Mallick et al. (eds.), *Cognitive Informatics and Soft Computing*,
Advances in Intelligent Systems and Computing 768,
https://doi.org/10.1007/978-981-13-0617-4_16

believed that speech emotion recognition can enhance the performance of speech recognition systems and it is consequently very beneficial for sinful investigation, smart help surveillance and detection of potentially unsafe occasions and human–computer interaction. To efficaciously apprehend emotions from speech, the real features need to be extracted from raw speech information and converted into appropriate codecs which are suitable for in addition processing.

It is a significant challenge in speech-based emotion recognition to extract efficient speech features and overcoming the variability of the speaker. The continuous dialog change emotional-based recognition system is laid low with the individual characteristics of the speaker. The major issue is to recognize the emotion by different feelings such as anger, joy, neutral, boredom, and sadness. Hence, the Human–Computer Interaction (HCI) is an emerging task where the user exploits smart phones and PCs to show their feelings by emotions. Furthermore, the HCI device is used to govern the movement of the computer in regards to the human emotion.

HCI system is used to control the action of the computer with regard to the human emotion. Therefore, speech plays an important role in the human communication interaction system which is used to exhibit their emotions, cognitive states, and intentions to each other. To attain Speaker Emotion Recognition (SER), the classification is then performed by the extracted features.

Support vector machine (SVM) based emotion recognition [1] has been evolved to pick out various feelings of the speaker. The emotional features are extracted in continuous scale by means of Valence, Activation, and Dominance. Only some discriminative features, elected via by PSO methods are in addition applied for processing and recognition. Final, Emotional state labels are received through the SVM regression. Fisher vector [2] feature representation based speaker verification is produced out of Fisher Kernel then represents every speech signal into a large-dimensional vector with the aid of encoding the derivatives of the log likelihood of the UBM model with recognize to its mean and variances.

Related works based on emotional speech have addressed the problem using different approaches. In [3] speaker-independent recognition is carried out using the hierarchy approach and also gender is detected before the emotion recognition system is used. In [1] the sequence factor analysis model with shorter segments method to produce of smaller intersecting speech sounds. The sequence factor model debts for the dependency of the adjoining segments because of the overlap among every adjacent segments. In addition, [4–6] proposed the use of convolutional neural based model approach for stable and robust recognition.

E. Hinton and Simon focus on Deep Belief Network [7]. In this work, the "complementary priors" were used to reduce the effects that construct inference tricky in densely connected belief nets which have various hidden layers. This proposed model offers higher digit classification than the best discriminate learning algorithms. Omid Ghahabi and Javier Hernando describe deep learning based speech recognition [8]. The authors have suggested an impostor selection algorithm and a conventional model adaptation procedure in a hybrid machine primarily based on Deep Belief Networks (DBN) and Deep Neural Networks (DNN) to discriminatively version each goal speaker.

Franices Cruz and Cornelius Weber proposed interactive reinforcement learning (IRL) algorithm [9] to speed up the learning process such as cleaning a table and compare three different learning methods. In [10–13], proposed the use of the Reinforcement learning (RL) model, for emotion-based speech system by initiating psychological emotional aspects as an essential motivation. In terms of the quantification of emotions, they notably were targeting the thanks to improve the performance of present RL algorithms, and planned three basic emotional dimensions; in phrases of the mixing with existing RL frameworks.

Tamil emotion database has been developed from different Kollywood movies. At first all files are slice in .MP3 format and then transformed into (.wav) and Berlin emotion corpus of German language was utilized for feature extraction. DWT and MFCC are used for feature extraction. ART for BERLIN database and Tamil database were used for performance evaluation respectively.

In this paper, we developed an emotion speech recognition system using hybrid model of AFDBN and Reinforcement Learning. Initially, the speech signal is given as input to SER. The signal features are extracted for the speaker emotion recognition system such a spectral flux, MFCC, pitch Chroma and tonal power ratio. Based on the feature vectors, a hybrid technique such as an Adaptive Fractional Deep Belief Network (AFDBN) and Reinforcement Learning is utilized to find out the weight vectors optimally.

2 Proposed Emotion Speech Recognition System

The proposed emotion speech recognition system is shown in Fig. 1. The speaker emotional recognition is performed in two vital stages namely training phase and testing phase. The four spectral features are extracted from the input speech signals initially. The feature vectors are given to the hybrid classifier. The training process is mathematically modeled by the AFDBN and Reinforcement Learning (RL). The proposed hybrid classifier is used to find out the weights iteratively. On the other hand in the testing phase, the learned weights are applied to understand the emotion of the speech through the extracted feature parts.

2.1 Tonal Power Ratio

The tonal power ratio (T) is used according to measurement the tonalness of the speech signal. Tonal power ratio (T) is expressed in Eq. (1) and the ratio ranges from 0 to 1 where the zero-noise pattern; high value—tonal spectrum.

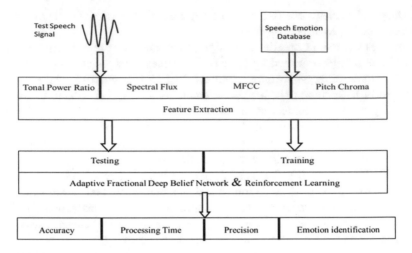

Fig. 1 Block diagram for proposed methodology

$$T = GI/ \sum_{i=0}^{n/2-1} |X(k+c)|^2, \tag{1}$$

where *GI*—Tonal power; $X(k, c)$—spectrum of the input speech signal s_n

2.2 Spectral Flux

The spectral components of the signal are extracted using the spectral flux. The spectral components are the important aspect because if the spectral content of the signal is changed over the time, then the recognition performance gets degraded. The, spectral flux *Pr* of the input speech signals (*n*) is given in Eq. (2)

$$Pr = \sum_{k=1} N/2[|x(k)| - |x_{r-1}|[k]]^2 \tag{2}$$

2.3 Pitch Chroma

The pitch chroma extraction is the powerful tool to categorize the pitches significantly and approximates to the equal tempered pitch scale. In order to extract the features using pitch chroma, the frequency of the Fourier transform is mapped with the 12 semi-tones pitch classes *C*. Its miles help to reduce the noise in the signals. Thus, the Pitch chroma is given in Eq. (3) [14].

$$C(I) = \sum_{n';c(n'=l)} D(n'); l \in [0, 12] \tag{3}$$

2.4 MFCC

Mel Frequency Cepstral Coefficient [15] is some over of the popular techniques because extracting the considerable feature component from the speech signal is shown in Fig. 2. It has the advantage of using perceptual frequency bands for the cepstral analysis. In order to compute the DFT, the window perform is employed to smooth the signal which leads to attenuate both ends of the signal towards zero. The input signal s_j is divided into a set of frames where the (DFT) is applied which is denoted by $S_l(i)$. And also, the period gram is computed based on the estimated power spectral for every speech frame $S_l(i)$.

Thus, the power spectrum of the speech signal is obtained. In order to convert the power spectrum into Melcepstrum, the Q point of FFT is utilized. Subsequently, the m triangular filters are required to design the Mel-spaced filter bank which poses $q(m)$ vectors of length Q. After that, the energy of the filter bank $g(m)$ is calculated by way of multiplying every filter bank $H_m(k)$ with the power spectrum when that, takes the log when together with those values.

In the end, the DCT of the all log filter bank energies is estimated to obtain the "m"cepstral coefficients is given by means of Eq. (4).

$$M(m) = \begin{cases} \frac{1}{2}\left(G_0 + (-1)^k G_{m-1} + \sum_{q=1}^{M-2} G_q \cos\left[\frac{\pi}{m-1}qm\right]m\right), \\ = 0, \ldots, M-1 \end{cases} \tag{4}$$

where $M(m)$ is the desired m cepstral coefficient.

2.5 AFDBN

The learning algorithm of AFDBN is finished the usage of the subsequent steps.

Fig. 2 MFCC basic blocks

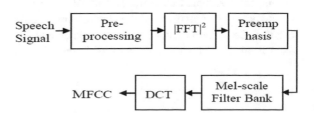

Step 1   Initialize weights of hidden units y and bias terms.

Step 2   Updating of weight vectors

$$y^{t+1} = y^t + y_{inc}^{t+1},$$

where y^t—weights of the current iteration; y_{inc}^{t+1}—Increment in weight for the next iteration. Fractional calculus (FC), to enhance the solution searching in the predefined search space.

$$D^\beta[y^{t+1}] = y_{inc}^{t+1}$$

Step 3   Weight vector of every iteration

$$y^{t+1} = y_{inc}^{t+1} + \beta * y^t + 1/2 * \beta * y^{t-1}$$

Increment of weights can be found out using Eq.

$$y_{inc}^{t+1} = \alpha * y_{inc}^t + \eta * [dy^t/w - \gamma * y^t]$$

Step 4: Updating of bias term b and bias weight of next iteration

$$b^{t+1} = b^t + b_{inc}^{t+1}$$

$$b_{inc}^{t+1} = \alpha * b_{inc}^t + \eta * [db^t/w]$$

3   Results and Discussion

3.1   The Database

In this work, databases were created with five different kinds of emotions from the different kind of speaker. Two different languages were utilized to extract the emotions, i.e., Tamil and Telugu. The emotional speech of the speaker was recorded by using the voice recorder in mobile phone. The noises introduces due to the environmental conditions like echoes were canceled by using the software called GOLDWAVE.

The Tamil databases were created from 12 different kinds of speakers with various environmental conditions. The 12 speakers including both genders (Male and Female) were involved. All the emotional voices have taken the duration with less than 15 s. While creating Telugu databases, some emotional speeches were taken from the Telugu movies with different emotions.

3.2 Performance Measures

The performance evaluation of the proposed algorithm is discussed in this section.

(a) **Accuracy**

The accuracy of a measurement system is how close it gets to a quantity's actual (true) value. It increases the performance of the system.

$$Accuracy = \frac{True\ positive\ +\ True\ negative}{True\ positive\ +\ True\ negative\ +\ Forecast\ positive\ +\ Forecast\ negative}$$

(5)

(b) **Precision**

$$Precision = \frac{True\ positive}{True\ positive\ +\ Forecast\ negative}$$

(6)

(c) **Recall**

$$Recall = \frac{True\ positive}{True\ positive\ +\ Forecast\ negative}$$

(7)

Figure 3a–d shows our proposed method Accuracy, Precision, Recall, and processing time analysis are better when compared to DBN, FDBN, and AFDBN.

3.3 Comparison Between Different Algorithms

From Table 1, Accuracy, Precision, Recall, and Average Processing Time for RL is more when compared with Deep Belief Network, Fractional Deep Belief Network, Adaptive Fractional Deep Belief Network Based Algorithm. In RL the Accuracy increased by 17%, Precision increased by 17%, Average processing time reduced by 360 ms and Recall increase by 17%.

From Fig. 4, it is found that the RL has higher Accuracy, Precision, and Recall compare with all other Classifiers.

From Fig. 5, it is found that the RL has low processing time compare to other algorithms.

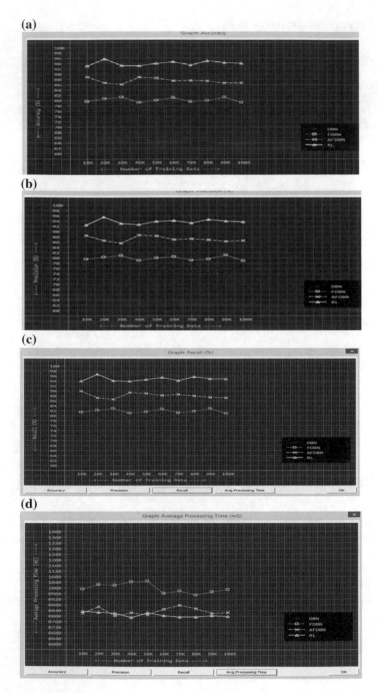

Fig. 3 **a** Accuracy versus no. of training data. **b** Precision versus no. of training data. **c** Recall versus no. of training data. **d** Average processing time versus no. of training data

Table 1 Comparison between different algorithms

	DBN	FDBN	AFDBN	RL
Accuracy	76%	80%	89%	93%
Precision	75%	80%	89%	92%
Recall	79%	81.9%	90%	96%
Avg. processing time	1160 ms	1000 ms	880 ms	800 ms

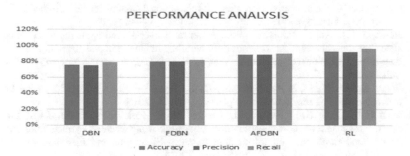

Fig. 4 Performance analysis of various classifiers

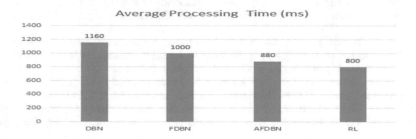

Fig. 5 Average processing time

4 Conclusion

Emotion speech recognition system has been implemented using AFBDN and RL. The experiment was conducted by using five types of emotional speeches from the different kind of speakers including male and female to recognize emotional state of the speaker. To analyze the performance of AFDBN and RL the feature vectors namely tonal power ratio, MFCC, pitch chroma, and spectral flux were extracted from the acoustic signal to measure the performance metrics like Accuracy, Recall, Average processing time, and Precision. From the analysis the accuracy of the RL is higher when compared to other classifiers. Similarly the precision and recall metrics also gives the higher performance compare to others. Comparatively the average processing time also gets reduced in the Reinforcement Learning. Therefore the proposed work is an efficient one to recognize the Emotion state of the speaker to enhance the Human–Computer interaction.

References

1. Mencattini, A., Martinelli, E., Costantini, G., Todisco, M., Basile, B., Bozzali, M., Di Natale, C.: Speech emotion recognition using amplitude modulation parameters. Knowl.-Based Syst. **63**, 68–81 (2014)
2. Omar, M.K.: A factor analysis model of sequences for language recognition. In: Spoken Language Technology Workshop (SLT), pp. 341–347. IEEE, California (2016)
3. Lu, C.-X., Sun, Z.-Y., Shi, Z.-Z., Cao, B.-X.: Using emotions as intrinsic motivation to accelerate classic reinforcement learning. In: International Conference on Information System and Artificial Intelligence (ISAI), pp. 332–337. IEEE, China (2016)
4. Newland, E.J., Xu, S., Miranker, W.L.: A neural network-based approach to modeling the allocation of behaviors in concurrent schedule, variable interval learning. In: Fourth International Conference on Natural Computation, ICNC'08, vol. 2, pp. 245–249. IEEE, China (2008)
5. Wang, K., An, N., Li, B.N., Zhang, Y., Li, L.: Speech emotion recognition using Fourier parameters. IEEE Trans. Affect. Comput. **6**(1), 69–75 (2015)
6. Jang, E.-H., Park, B.-J., Kim, S.-H., Chung, M.-A., Park, M.-S., Sohn, J.-H.: Emotion classification based on bio-signals emotion recognition using machine learning algorithms. In: International Conference on Information Science, Electronics and Electrical Engineering (ISEEE), vol. 3, pp. 1373–1376. IEEE, Japan (2014)
7. Hinton, G.E., Osindero, S., Teh, Y.-W.: A fast learning algorithm for deep belief nets. Neural Comput. **18**(7), 1527–1554 (2006)
8. Ghahabi, O., Hernando, J.: Deep learning backend for single and multisession i-vector speaker recognition. J. IEEE/ACM Trans. Audio Speech Lang. Process. **25**(4), 807–817 (2017)
9. Cruz, F., Twiefel, J., Magg, S., Weber, C., Wermter, S.: Interactive reinforcement learning through speech guidance in a domestic scenario. In: IEEE International Joint Conference on Neural Networks (IJCNN), pp. 1341–1348, Killarney, Ireland (2015)
10. Kim, E.H., Hyun, K.H., Kim, S.H., Kwak, Y.K.: Improved emotion recognition with a novel speaker-independent feature. IEEE/ASME Trans. Mechatron. **14**(3), 317–325 (2009)
11. Mao, Q., Dong, M., Huang, Z., Zhan, Y.: Learning salient features for speech emotion recognition using convolutional neural networks. IEEE Trans. Multimedia **16**(8), 2203–2213 (2014)
12. Hoque, S., Salauddin, F., Rahman, A.: Neighbour cell list optimization based on cooperative q-learning and reinforced back-propagation technique. In: Radio Science Meeting (Joint with AP-S Symposium), 2015 USNC-URSI, pp. 215–215. IEEE, Canada (2015)
13. Gharsellaoui, S., Selouani, S.-A., Dahmane, A.O.: Automatic emotion recognition using auditory and prosodic indicative features. In: 2015 IEEE 28th Canadian Conference on, Electrical and Computer Engineering (CCECE), pp. 1265–1270. IEEE, Canada (2015)
14. Lerch, A.: An Introduction to Audio Content Analysis: Applications in Signal Processing and Music Informatics, p. 272. Wiley IEEE Press, July 2012
15. Peeters, G.: Chroma-based estimation of musical key from audio-signal analysis. In: Proceedings of the 7th International Conference on Music Information Retrieval. Victoria (BC), Canada (2006)

Concept-Based Electronic Health Record Retrieval System in Healthcare IOT

P. Sony and N. Sureshkumar

Abstract The primary objective of Healthcare Internet Of Things (H-IOT) is to provide an early diagnosis and better treatment mechanism by establishing the connections with every active object in the healthcare world so that each individual objects can communicate with each other and they can take decisions up to some extent. Objects in H-IOT can be a wide variety of smart objects, from a wearable device attached to a patient body in one end of the spectrum to a data server stored in the healthcare provider at the other end. The data produced by these wearable's, implanted devices or any other medical devices, are collected and stored with the previous patient records and it is analyzed. As different healthcare providers are following various standard formats for Electronic Health records (EHR), the huge amount of heterogeneous data gathered from multiple healthcare enterprises needs additional processing for the efficient retrieval of the medical documents as well as medical data. This paper presents how an intelligent data server can be designed to retrieve medical data and medical documents.

Keywords Electronic health records · Healthcare IOT · UMLS
SVD · LSI

1 Introduction

CAN you imagine a world without Internet? In a very near future, we will have a world with completely Internet of objects rather than the Internet of Computers. The concept of the Internet of things was mentioned by Peter T. Lewis as he presented a talk at the Legislative weekend conference organized by U.S Federal Communications Commission (FCC). Internet of things or Internet of Everything is

P. Sony (✉) · N. Sureshkumar
Vellore Institute of Technology, Vellore, India
e-mail: sony.p2016@vitstudent.ac.in

N. Sureshkumar
e-mail: sureshkumar.n@vit.ac.in

© Springer Nature Singapore Pte Ltd. 2019
P. K. Mallick et al. (eds.), *Cognitive Informatics and Soft Computing*,
Advances in Intelligent Systems and Computing 768,
https://doi.org/10.1007/978-981-13-0617-4_17

meant for network of any objects in any place at any time and it is later modified for several applications such as Internet of Vehicles [1], Internet of Agriculture [2], Internet of Education [3], Internet of Smart cities [4] and Internet of Healthcare and so many.

HIOT (Healthcare IOT) or Internet of Health (IoH) is a network of Intelligent objects connected in the e-Healthcare world. These objects can be wearable's attached to a patient body, implanted devices, sensor connected to the medical equipment, a smart computer, smart medicine box or a smart data server etc. The concept of Healthcare IOT is an extension of the concept e-health (Electronic health) [5]. As an initial step, India has introduced the concept of telemedicine [6] and telehealth in most of the states with the objective of telemedicine is to provide a highly effective treatment mechanism for the people who are living in remote places. Local hospitals transfer the details of the patient records to a well established distant hospital electronically for a better treatment so that patient receives the service of a specialized doctor without moving physically to the specialized hospital (Fig. 1).

Even though the deployment of IPV6 [7] and standardization of protocols are being the major barriers in the IOT world, the RFID, sensor or other types of devices connected to the Internet is becoming enormously large [8]. Companies viz Philips and QUALCOMM announced a healthcare ecosystem which can avoid readmission cost for senior citizen suffering chronic diseases like diabetes [9]. With the effective use of IOT-Medical Devices [9], people can check their basic medical parameters like Blood Pressure, Glucose, Albumin, HB, TSH, etc., in their home and remedies are taken through online and thus the frequency of the patients visiting the hospital can be reduced. Wearable's attached to their body will

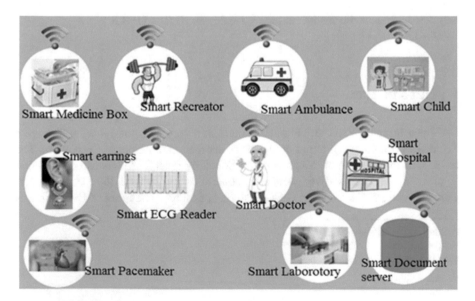

Fig. 1 Healthcare IOT

represent the health condition and caregivers can monitor the events and respond accordingly. As an example: Treatment of dialysis patient can be administered at their home and details of the dialysis input parameter and its effects captured by the IOT-Medical devices can be communicated with a well-defined healthcare enterprise so that patient does not need to move from one healthcare care settings to another to receive his treatment. Thus a cost-effective patient-centric rather than a hospital-centric treatment mechanism can be visualized by the wide adoption of Healthcare Internet of Things.

1.1 Preliminaries

Unified medical language system (UMLS) [10] is a medical ontology which makes an open door for the computer system to understand the linguistics of healthcare and it is developed by National Library of Medicine (NLM) after adopting the medical terminologies from various sources like Snomed CT, ICD 10, etc. Metathesaurus, Specialist lexicon, and semantic network are the three main knowledge sources of UMLS.

UMLS Metathesaurus is designed for representing vocabulary of various medical terms and its related information such as relationships, concepts, etc., in several languages. Medical terms and its associated concepts are represented in Mrconso table. The semantic network is used to provide the classifications of the medical concepts represented in the metathesurus. It also depicts the relationships among the represented concepts. UMLS contains more than 130 semantic types and more than 50 relationships in the 2017 release. Special lexicon furnishes several lexical tools which aid in natural language processing. Computational similarities and dissimilarities are represented by special lexicon and lexical programs.

Singular value decomposition Let A be a P * Q medical document by Medical term matrix. There is singular value decomposition (SVD) of A of the form

$$A = A_1 A_2 A_3 \tag{1}$$

Here A_1 be a $P \times P$ square matrix whose columns are the orthogonal eigenvectors of AA^{T} and A_2 be the $Q \times Q$ matrix whose columns are the orthogonal eigenvectors of $A^{\mathrm{T}} A$. Transpose of A is denoted by A^{T}. The eigen values $\lambda_1, \lambda_2, \lambda_3 \ldots, \lambda_n$ of AA^{T} are same as eigen values of $A^{\mathrm{T}} A$. Square root of the eigen values are represented as diagonal elements in the second matrix.

2   Different Applications in Health Care IOT

2.1   Real-Time Continuous Monitoring

Effective Device–Device communication can be established and thus cost-effective treatment mechanism can be employed especially in case of treatment on chronic diseases. Connected devices can collect vital information at periodic intervals or when any abnormal conditions occurred in the patient body and the collected information are then aggregated and communicated to other side smart device so that the device can trigger an event (e.g., Take additional medications or can inform present condition of the patient to the concerned doctor) immediately. This kind of monitoring is beneficial for the patient living in the remote areas.

2.2   Special Care for Elderly People

Efficient and effective treatment mechanism can be employed for senior citizens using the intelligent wearable's attached to their body and thus readings on vitals such as BP, sugar, pulse, body temperature, etc., can be captured and sent to the healthcare providers at a constant time.

2.3   Early Detection of Diseases

Prevention is better than cure. Some diseases are simultaneously occurring and some others are occurring as a side effect or after effects of other diseases. By the continuous communication of the devices, diseases either can be detected at the very initial stages or it can be prevented.

2.4   Overdose can be Prevented

Some medical practitioners are knowingly or unknowingly prescribe overdose medications and this medication may cause some ill-effects in the patient body or can even cause death of the patient. The presence of overdose can be detected by two dimensions. First, the overdose can be in the form of over amount of the intended medicine and second, the existence of other unwanted pharmacological substances in the prescribed medications. Such kinds of overdose can be easily detected by the device to device communication between the intelligent medicine box and data server.

2.5 Monitoring of the Physical Exercise

Wearable's attached to a human body can gather information like frequency, duration, type, etc., of the physical exercises. Collected data can be analyzed and advice on physical fitness of the body can be informed to the person.

2.6 Insurance Company can Offer Insurance to Deserving Patient

Most of the insurance companies refuse to give insurance to the patient in several situations such as if the patient has already under treatment, or if he is alcoholic, etc. But nowadays these situations mainly rely on laboratory result and doctors report. By the adequate communication with the wearable's attached to human, intelligent server and sensor node attached to the insurance company, they can easily find out the legitimate applicant.

2.7 Service of a Specialized Doctor

Most of the rural hospitals lack advanced facilities. Many hospitals situated in remote areas are not equipped with the advanced facilities or specialized doctors even in this twenty-first century. As Tele-medicine suggests, patients situated in rural hospitals can acquire the service of a specialized doctor without moving physically to a multi specialty care settings.

2.8 Better and Efficient Referral Program

People may be referred to various health care providers such as hospitals, specialists, labs for emergency treatment, and care. The traditional referral program can be less efficient, less effective and more dangerous unless the process is well coordinated, well shared and well communicated. To yield a better treatment, patient may be referred from one healthcare setting to another automatically in H-IOT.

2.9   Sleeping Disorders/Psychiatric Disorders can be Detected

Behavioral changes in humans are one of the consequences of disorders like sleep. Several kinds of sleeping and psychiatric disorders in humans are one of the common problems in the advanced medical world. Using smart bed or device implanted inside the human body or a sensor placed in the bedroom or even smart earrings can detect the sleeping problems [11] or mental disorders.

3   Problem Definition

It is impossible for a human being to visit a single hospital for his treatment in the entire lifespan. As a result, heterogeneous medical records are stored by different healthcare providers. The high availability of several biomedical sensors (e.g., Temperature sensor—max30205, Pulse sensor SEN-11574, Respiration sensor-X2 M 200) in the market makes a platform for the connected e-healthcare world. Ubiquitous computing and wireless sensor network play a great role in the development of Healthcare Internet of things. Designing an intelligent medical server which can integrate all the conventional paper-based medical records, all structured and semi-structured medical records with the streaming data from the IOT-Medical devices and be able to respond very quickly to any query submitted to them without leaking the privacy of the patient as well as the doctor.

4   General Architecture

The proposed architecture is based on concept-based vector space model and has got has four different modules namely HIOT data aggregator, EHR preprocessor, concept generator, and similarity calculator.

4.1   HIOT Data Aggregation

DN CXX library [12] is used to implement Named Data Network services for gathered data from IOT-MD. Smart Document server in the Healthcare IOT has to deal with two kinds of data. (1) Data from health records and (2) data originating from Medical Devices. There are no de facto standards for storage data format or communication data format for IOT-MD since it is still depending on the manufactures. So in this work, we designed a data format for IOT—Medical device as (Mid, Pid, Tm, Tm, Message). Mid is the machine identifier, Pid is the patient identifier, Td represents the type of the device whether it is an implanted device or a wearable device, message represents the message sent or received by the H-IOT device Mid.

4.2 EHR Preprocessor

Document preprocessing technique is an essential part of medical document processing as in the case of web documents. Data processing in HIOT may fall into two different categorics; Text processing in Electronic Health Records and Stream processing in real-time sensor data. Unlike normal text processing [13], Medical data processing is slightly different and more complicated. Collection of EHRs is given to a sentence extractor module, where the splitting of the sentence is done. Separated sentences are given to the parser module, where various parse of speech tagging is done. This is beneficial for feature extraction module, where various features such as noun tagged words; adjective tagged words, etc., are extracted.

4.3 Concept Generator Module

These extracted phrases from the previous step are checked in UMLS Metamorphosis for getting concept unique identifier (CUI). If CUI is present, the phrase is given to the Indexer module. Otherwise, it is given to stemmer and stop word remover module and finally to Indexer. All types of phrase variants can be generated using unique identifiers of phrases returned from the medical ontology UMLS. The detailed concept processing algorithm is given below.

Concept Processing Algorithm
Input: phrases
Output: ns[],un[], nw[],id(identifier for each phrase)
Terms: CUI(unique identifier for concept), SUI(Unique Identifier for String), LUI(unique Identifier for Term), nw[]-normalized word array, ns[]-normalized string, un[]-un normalized string, STR-String
Tables used: MRCONSO.RRF (concept table), MRXNS ENG.RRF (Normalized string table), MRXNW ENG.RRF (normalized word table)

$\forall PhraseP$, *do*

$P\varepsilon NormalizedString$, *do*

$\exists(CUI, SUI, LUI)$*of PinMRXNS_ENGG*
return CUI, SUI, LUI as C1, S1, L1
if (CUI, LUI) of P in concept table = (C1, L1)
Add STR to ns [] and set id as (C1, L1)

End

PεNormalizedword, do

> $\exists(CUI, SUI, LUI)of\ PinMRXNW\_ENGG$
> return CUI, SUI, LUI as C1, S1, L1
> if (CUI, SUI) of P in concept table = (C1, S1)
> Add STR to nw [] and set id as (C1, S1)

End

PεRegularword, do

> $\exists(CUI, SUI, LUI)of\ PinMRXW\_ENGG$
> return CUI, SUI, LUI as C1, S1, L1
> if (CUI, SUI) of P in concept table = (C1, S1)
> Add STR to nw [] and set id as (C1, S1)

End

End

A phrase is called polysemous or ambiguous if that phrase has different meaning or senses in different situations. The terms like cold, hyperthermia, etc., are polysemous, as it returns more than one concept. Hyperthermia gives two different concepts one as fever sense and another as temperature sensing.

Two words are called *synonymous* if both return same concept or meaning. The term fever and hyperthermia return same CUI from UMLS Metathesaurus. Different strings of same concept word return different CUI values. Table 1 represents the CUI, SUI values of four different terms. The first three terms represent the same concept (CUI), but their SUI is different (Table 2).

Variant generation for hypernym/hyponym kind of words If a word A is a *Hyponym* of another word B, then there exists an\is-a "or\Parent" relation. Then the word B is called hypernym of word A. For example liver cancer is a hyponym of cancer and cancer is a hypernym of liver cancer. This type of relationships is solved by using two tables, MRREL.RRF and MRCONSO.RRF. Relation representing table MRREL.RRF table includes\PAR "or\is-a" (representing Parent), "CHD" or "Inverse Is–a" (representing Child), RQ(representing classified as) and similar relationships.

Variant generation for Meronymy/holonymy relation If a word A is a *Meronymy* of another word B, then there exist a\has a" or\part of\relation. Then the word B is called holonymy of word B. As the finger is part of the arm, it is meronym of arm and arm is holonym of the finger. This type of relations is also solved by using two tables-relationship table MRREL.RRF and concept table MRCONSO.RRF.

Table 1 Data files in UMLS

Table name sample record	Purpose	Some attributes
MRCONSO	Concepts	C0001175kL0001175kA0019182kAIS
MRSTY	Attributes	C0001175kDiseaseorsyndromekjAT 17683839j3840
MRREL	Relationships	C0002372kA0021548kC0002372kA16796726kRO
MRXNW ENGG normalized word index		anemiakC0002871kL0002871kS0013742
MRXNW ENGG normalized word index		disorderkC0002871kL2818006kS3448137

Table 2 Identifers in UMLS

String	SUI	CUI
Carcinoma	C0007097	S0022073
CARCINOMA	C0007097	S0360186
Cancer	C0007097	S0021809
Liver cancer	C2239176	S0129926
Carcinoma	C0007097	S0022073

4.4　Document Vector Creation Module

Generated indices from EHR along with the IOT-MD indices are used to create concept document matrix. Generated Cue-Phrases are first used to create a term-document matrix. This term-document matrix is compressed to form a phrase document matrix using the two techniques namely Latent Semantic Indexing [13] and Singular Value Decomposition. Phrase document matrix is then converted to form concept document matrix. As an example, the terms cold, high temperature, hyperthermia are different terms in a term-document matrix. In phrase document matrix, the cold and high temperature is collapsed into a single term. But in concept document matrix all synonyms words are merged to form a single concept.

4.5　Similarity Calculation Module

Similarity of a query "q" with a document "d" can be calculated using the equation

$$S_{q,d} = \frac{\sum_p W_{d,p} W_{q,p}}{W_d W_q} + \frac{\sum_p W_{d,hm} W_{q,hm}}{W_d W_q} + \frac{\sum_p W_{d,hh} W_{q,hh}}{W_d W_q} \tag{2}$$

$$W_{d,p} = f_{d,p} \log\left(\frac{N}{f_p}\right) \tag{3}$$

$$W_{q,p} = f_{q,p} \log\left(\frac{N}{f_p}\right) \tag{4}$$

$$W_{d,hm} = f_{d,hm} \log\left(\frac{N}{f_{hm}}\right) \tag{5}$$

$$W_{q,hm} = f_{q,hm} \log\left(\frac{N}{f_{hm}}\right) \tag{6}$$

$$W_{d,hh} = f_{d,hh} \log\left(\frac{N}{f_{hh}}\right) \tag{7}$$

$$W_{q,hh} = f_{q,hh} \log\left(\frac{N}{f_{hh}}\right) \tag{8}$$

5   Implementation

The proposed system is implemented in Java Netbeans IDE using the tools like Standford parser, MySql, and UMLS database. Accuracy is verified by creating three different kinds of matrices namely, term-document matrix creation, Phrase document matrix creation, Concept document matrix creation. Two types of data are emerging in HIOT (1) Streaming data from the IOT-MD and (2) Data from the EHR. Streaming Data from IOT is also considered as normal health record.

5.1   Term-Document Matrix Creation

Each electronic health records are given to a sentence extractor module. The extracted sentence is given to a stop word removal module. Stopword removed sentences are given to a parser module and then to a stemmer module. All derivational and inflectional morphologies are removed from the word and thus form the word stem. The generated stem is used as index terms in the term-document matrix. In term-document matrix, columns consist of the set of EHRs and terms are represented in the rows.

5.2   Phrase Document Matrix Creation

Standford NLP Parser is used to parse the sentences. Noun phrases, Adjective phrases, adverbial phrases, verb phrases and prepositional phrases are generated.

Usually nouns and noun phrases are used as index terms in normal processing systems. But in phrase document matrix, CUI (concept unique Identifier) of each relevant phrases are considered.

$$NP \rightarrow DTJJNN \qquad NP \rightarrow DTJJNNS$$
$$NP \rightarrow DTJJNNNN \qquad NP \rightarrow DTJJJJNN$$
$$NP \rightarrow DTJJCDNNS \qquad NP \rightarrow RBDTJJNNNN$$
$$NP \rightarrow RBCTJJJJNNS$$

5.3 Concept Document Matrix Creation

Extracted phrases in the previous step are undergone phrase expansion. Phrase expansion can be done in three cases (1) Expansion using synonyms words (2) Expansion using meronymy and holonymy words (3) Expansion using hypernymy and hyponymy words. Expanded phrases are termed as concepts as they are concepts in lower dimensional space than the phrases.

6 Performance Evaluations

To evaluate the performance of the system, a dataset consisting of electronic medical documents of more than 1000 patients were collected along with the 100 IOT-MD data. The scenario of IOT-MD data is modeled by creating an application that automatically generates the streaming data. The experiment is conducted in three different kinds of retrieval system namely HIOT Retrieval System Based on Term-Document Matrix, HIOT Retrieval System Based on Phrase Document Matrix and HIOT Retrieval System Based on Concept Document Matrix. Following metrics are calculated on to estimate the goodness of the retrieval system provided by the proposed algorithm. Precision is the fraction of the retrieved medical documents that are relevant to the medical query.

$$\text{Precision} = \frac{|\text{relevant } EH\,Rs \cap \text{retrieved } EH\,Rs|}{|\text{retrieved } EH\,Rs|}$$
$$\text{Recall} = \frac{|\text{relevant } EH\,Rs \cap \text{retrieved } EH\,Rs|}{|\text{relevant } EH\,Rs|} \tag{9}$$

Precision is used to measure the correctness of the proposed system, while Recall measures the completeness. Neither recall nor precision alone can judge the goodness of a retrieval system. Therefore Accuracy, or F1 score which is the harmonic mean of precision and recall, is used to find the performances system. Retrieval performances are evaluated using the measures precision and recall.

Out of the three graph, concept-based retrieval shows better recall and precision values in H-IOT. Precision is used to measure the correctness of the proposed system. Recall measures the completeness. Neither recall nor precision alone can judge the goodness of a retrieval system. Therefore Accuracy, or F1 score which is the harmonic mean of precision and recall, is used to find the performances of the system (Figs. 2, 3 and 4).

$$\text{Accuracy} = \frac{2 * \text{Precision} * \text{Recall}}{\text{Precision} + \text{Recall}} \qquad (10)$$

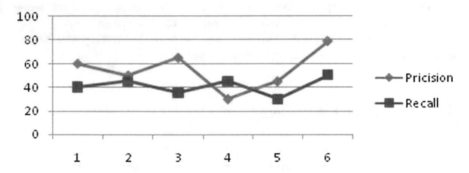

Fig. 2 HIOT retrieval system based on term-document matrix

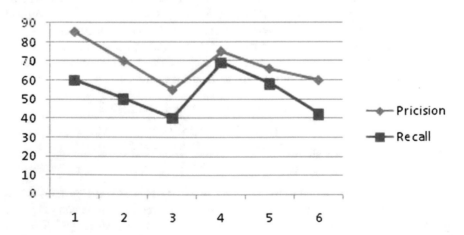

Fig. 3 HIOT retrieval system based on phrase document matrix

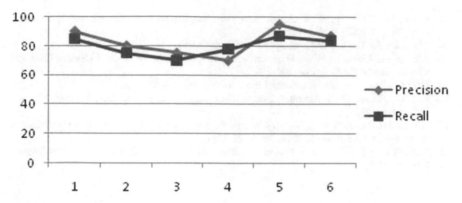

Fig. 4 HIOT retrieval system based on concept document matrix

7 Conclusion

There has been a global shift in the e-healthcare world using the ubiquitous computing paradigm. This has been necessitated by the invention of new medical sensors and advances in information communication technology. This research paper helps to assess how the Internet of things technology adoption should be organized and managed to enhance the retrieval performance of an intelligent data server in the e-healthcare world. By integrating the UMLS ontology, all lexical, morphological, and semantic variants of a medical term can be retrieved and these all variants form as medical concepts by the implementation of latent semantic indexing. Thus the proposed system achieves high precision and recall value since it is mainly focusing on the concepts rather than traditional term as the index term. In future work, the proposed system can be implemented in multiple document servers by parallelizing the operations of these servers.

References

1. Eskandarian, A.: Handbook of Intelligent Vehicles. Springer, Berlin (2012)
2. Dlodlo, N., Kalezhi, J.: The internet of things in agriculture for sustainable rural development. In: International Conference on Emerging Trends in Networks and Computer Communications, at Windhoek, Namibia (2015)
3. Selinger, M., Sepulveda, A., Bucha, J.: White Paper on Education and the Internet of Everything CISCO (2013)
4. Zanella, A., Vangelista, L.: Internet of things for smart cities IEEE. Internet Things J. **1**, 22–32 (2014)
5. Sebestyen, G., Hangan, A.: eHealth solutions in the context of internet of things. In: IEEE International Conference on Automation, Quality and Testing, Robotics (2014)
6. Bhowmik, D., Duraivel, S., Singh, R.K., Kumar, K.P.S.: Telemedicine—an innovating healthcare system in India. Pharma Innov. J. **2**, 1 (2014)

7. Evans, D.: The Internet of Things How the Next Evolution of the Internet Is Changing Everything. CISCO White Paper (2011)
8. Markman, J.: The IoT Is Coming to Healthcare (2016)
9. Khanna, A., Misra, P.: White Paper on the Internet of Things for Medical Devices Prospects, Challenges and the Way Forward
10. UMLS reference manual: National Library of Medicine (2016)
11. Kayyali, H.: Remotely Attended Home Monitoring of Sleep Disorders, Telemedicine and E-Health (2008)
12. Saxena, D., Raychoudhury, V., Sri Mahath, N.: SmartHealth-NDNoT: Named Data Network of Things for Healthcare Services ACM (2015)
13. Baeza-Yates, R., Neto, B.R.: Modern Information Retrieval, 3rd edn. Addison Wesley, Boston (1991)

Hybrid CAT Using Bayes Classification and Two-Parameter Model

**Nikita Singhal, Amitoz S. Sidhu, Ajit Kumar Pandit,
Shailendra Pratap Singh Sengar and Tutu Kumari**

Abstract Much research and implementation has been done in the field of adaptive learning, while many such platforms exist almost none of them have tackled the problem of maintainability of such high demand systems. This paper proposes a new system using naive Bayes classifier and two-parameter model of IRT to develop a low cost, easy to maintain, self-evolving test platform. The proposed model harnesses the knowledge of the community while implementing powerful test theory. The paper discusses in detail the major modules of the system along with the related theory. The proposed model incorporates machine learning and IRT to provide a state of the art system while still being a community powered platform. The scope of the proposed model is visited. This paper provides a direction and precedent for the development of a new breed of low maintenance high capability test platforms.

Keywords Item response model · Naive Bayes model · CAT
(Computer adaptive Test) · Two-parameter model · Recommendation system

N. Singhal · A. S. Sidhu (✉) · A. K. Pandit · S. P. S. Sengar · T. Kumari
Army Institute of Technology, Pune, India
e-mail: itsamitoz4u@gmail.com

N. Singhal
e-mail: ngupta@aitpune.edu.in

A. K. Pandit
e-mail: ajitk9304@gmail.com

S. P. S. Sengar
e-mail: spssaj@gmail.com

T. Kumari
e-mail: tutu.singh001@gmail.com

© Springer Nature Singapore Pte Ltd. 2019
P. K. Mallick et al. (eds.), *Cognitive Informatics and Soft Computing*,
Advances in Intelligent Systems and Computing 768,
https://doi.org/10.1007/978-981-13-0617-4_18

1 Introduction

Growing popularity of the long-distance courses and the highly competitive nature of the education industry have led to many developments and innovations. There has been a definite shift from traditional methodologies to more advanced systems of assessment and content delivery. This has led to research and development of new psychometric paradigms like the IRT (Item Response Theory) and adaptive testing [1, 2]. Adaptive testing is a new strategy for computer-assisted learning and e-learning. The objective is to increase the accuracy of estimating the learner's true ability.

While the implementation of such a system is not unprecedented, the existing systems have drawbacks. This paper aims to address these and propose implementation of innovative features to develop a more comprehensive system.

2 Present Versus Proposed System

While high-end system like GRE and AMCAT are available and highly relied upon, there are no easily maintainable low-cost systems. The main problem with these systems is the difficulty of maintaining the databases. With questions becoming more popular and their answers well known they lose their ability to effectively test the examinee. Thus, the same questions cannot be used in the examinations over a period.

Some of the most popular and well-developed question banks and preparation systems/platforms are community run. It is this power of a community run platform that we wish to harness for developing our system. The problem of repetition of questions will be tackled by making the scoring of the questions dynamic, where the trends in answers are used to decide the value of the question in terms of difficulty and discrimination.

We will also be employing two-parameter model of IRT [3] to increase the magnification in the scoring scale to better grade the examinee thus providing the capabilities of a high-end test system with the flexibility of a community run platform.

3 Proposed System

The project sets itself apart by harnessing the powers of both machine learning and IRT. The paper proposes a system, with an evolving database, which implements powerful testing and scaling theory with a comprehensive report generation module. The project aims to implement a community run forum that is used to populate the database where the questions are dynamically scored on difficulty and discretion. The proposed model not only provides a percentile score but also an in-depth performance comparison with the other examinees including the time taken to answer the question, allowing for a better understanding of the strengths and weaknesses of examinee and his current standing in the crowd.

Implementing the aforementioned model meets the goal of developing a system with increased accuracy of estimating the learner's true ability while addressing the drawbacks of the existing system

The proposed system can be broadly divided into three main modules.

(a) Database manipulator.
(b) Adaptive test and report generator.
(c) Forum.

First, the forum module that employs the Bayes Model of classification. This module will be used to populate the database after correct classification of the questions picked from the forum [4].

Second, we have the database manipulation module; this module is responsible for the dynamic scaling of the questions in the database based on their discretion and difficulty. This scaling is based on the data about the performance of different examinees on that question. This data contains the number of correct responses recorded and the average time to achieve the said responses.

The final module is the actual test that the examinee takes, this employs Item Response Theory Model and adaptive test principles to rate tested trait of the examinee and generate a comprehensive performance report (Table 1).

Table 1 Use case table

Use case ID	01
Use case name	Hybrid Computer Adaptive Test
Description	
Actors	Examinee/Contributor/System
Precondition	Initial dataset classified and scaled
Post condition	Successful emailing of report
Priority	High
Frequency of use	High
Normal steps	1. Register and login 2. Select company 3. Test 4. Track behavior 5. Evaluate the result 6. Generate report

3.1 Database Manipulation Module

This is an important innovative module that aims at making a truly adaptive system which also accounts for the prior knowledge about the solution of a question. The aim behind a question is to judge the problem-solving skill of the examinee but a question that is already common knowledge cannot do so. That is why this module will learn through the examinees answer patterns and judge weather a question should maintain its position on scale of difficulty and discrimination or should it be increased or decreased.

Each question will have two parameters "difficulty" and "discrimination". The difficulty will be inductive of the frequency by which the examinees get a correct solution, this can be represented in the form of percentage of correct attempts. The discrimination on the other hand is inductive of the time taken by the examinee to achieve the answer, this is helpful in differentiating between two examinees who get the correct solution, i.e., two examinees of similar standing (Fig. 1).

A record will be maintained for each question and after a period the system will decide whether to increase or decrease the score of the question or to even remove the question completely from the database. If a significant percentage of the examinees are not able to correctly answer a question then that question will be moved into a bracket of higher scoring questions, on the other hand if most them are able to answer it correctly or answer it uncharacteristically fast then the question will be demoted or even removed.

In this way the system will maintain a flexible and aptly scored dataset that will truly allow it to judge the standing of an examinee against the rest.

3.2 Test and Report Generator Module

Two-parameter model—we are proposing to use the two-parameter model as it provides an excellent scale to differentiate between examinees.

$$P(X = 1|\theta, a, b) = \frac{e^{a(\theta-b)}}{1 + e^{a(\theta-b)}}$$

Two-parameter model assumes that the data has no guessing. It uses both item difficulty and item discrimination to calculate the latent ability of a person. In the above expression "a" stands for "discrimination", "b" is the difficulty and "θ" is the trait measure. We chose 2two-parameter model because the changing difficulty of the question provides us with an effective scale for differentiating examinee of different capabilities while discrimination provides us with means of magnifying the scale and differentiating between two people of similar capabilities effectively. Discrimination is representative of the time taken to get solution while difficulty is the representative of the correctness.

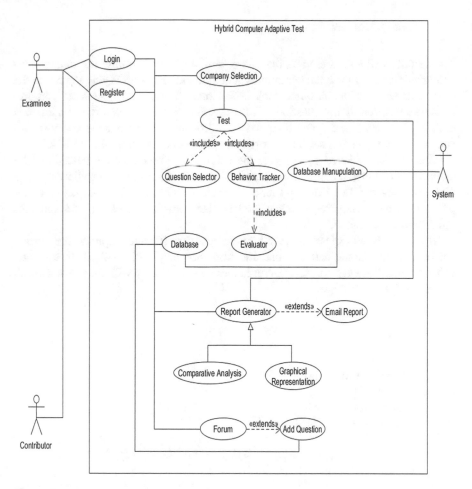

Fig. 1 Use case diagram

The proposed system will produce the score sheets which will inform the users how much time they spent on one question, difference between the user and the topper's performance and overall performance among the users.

The report generator sub-module will make the final assessment report which will include a time-based review, i.e., the examinee will get a comparative study of how long it took him to solve a question against the other examinees in the past. Thus, allowing for a more comprehensive report.

3.3 Classification Module

The module will aim at tapping into the potential of a community run database by taking submissions from the community and automatically classifying them into the correct database table. A quick look at the most popular and the most comprehensive collection of any question data set reveals that it is community run, it is this high level of involvement that if tapped can allow the system to achieve a powerful and up to date question set that can rival the high-end systems like AMCAT.

This will be achieved by applying the naive Bayes model of classification. The assumption of independence among predictors holds in this case thus allowing for a high accuracy and fast learning model. In fact, in this case Naive Bayes classifier performs better compare to other models like logistic regression [5] and less training data is needed.

This module will classify the submitted question [6] and populate the correct database table. Along with the classification the entry will also get an assumed value for the difficulty and the discrimination. These values will change according to the manipulation module

$$P(c|x) = \frac{P(x|c)P(c)}{P(x)}$$

$P(c|x)$ Posterior Probability
$P(x|c)$ Likelihood
$P(c)$ Class Prior Probability
$P(x)$ Predictor Prior Probability

$$P(c|X) = P(x_1|c) * P(x_2|c) * \cdots * P(x_n|c) * P(c)$$

4 Summary/Conclusion

In this paper we looked at the rapid changes taking place in the field of adaptive test systems. We looked at a major drawback of the currently implemented systems and discussed possible solution. We also took a dive into the proposed system and discussed the major modules that overcome the aforementioned drawbacks.

The first contribution of this system is the database manipulation module. We use the frequency of correct response and the time of getting the said response as factors to keep a challenging and flexible database.

The second contribution is the forum which is an attempt at making a self-populating database where questions are automatically classified into the correct table. In the end, this paper discusses a possible system of testing that is reliable, flexible, evolving, and self-sufficient. This provides for development of a new class of testing systems that can change the way people prepare for competitive exams.

References

1. Green, B.F., Bock, R.D., Humphreys, L.G., Linn, R.L., Reckase, M.D.: Technical guidelines for assessing computerized adaptive test. J. Educ. Measur. **21**(4), 347–360 (1984)
2. Hambleton, R.K., Swaminathan, H.: IRT Principles and Applications. Kluwe-Nijhoff, Boston (1985)
3. Harris, D.: Comparison of 1-, 2- and 3- Parameter IRT Models (1984). https://doi.org/10.1111/j.1745-3992.1989.tb00313.x
4. Mathew, J., Das, S.N.: Question classification using Naive Bayes. IJETT **23**, 155–160 (2015)
5. Ng, A.Y., Jordan, M.I.: On Discriminative vs Generative Classifiers: A Comparison of Logistic Regression and Naïve Bayes (2002)
6. McCallum, A., Nigam, K.: A Comparison of Event Models for Naive Bayes Text Classification (1998)

A Change Detection Technique Using Rough C-Means on Medical Images

Amiya Halder, Kasturi Saha, Apurba Sarkar and Arnab Sen

Abstract Change Detection plays an important role in detecting various kinds of dissimilarities between two images of the same object over a period of time. This has extensive use in medical imaging and remote sensing. We propose a method of change detection where we first apply a change vector analysis (CVA), then apply clustering on the image by Rough C-means (RCM) and finally threshold it to obtain the Difference image (DI). RCM provides the concepts of upper approximation n and lower approximation which lead to better clustering and decrease the error rate. Experimental results of the proposed method are compared with existing change detection algorithms and it has been found to perform evidently better than the others.

Keywords Rough set · Image segmentation · Change vector analysis
Thresholding

1 Introduction

Change detection is a powerful technique for detecting various kinds of abnormalities in different medical images. The scope of change detection is huge as it assists people to identify the affected areas. A lot of effective change detection

---

A. Halder (✉) · K. Saha · A. Sen
Department of CSE, STCET, Kolkata, West Bengal, India
e-mail: amiya.halder77@gmail.com

K. Saha
e-mail: kasturisaha14@gmail.com

A. Sen
e-mail: arnb.sen@hotmail.com

A. Sarkar
Department of CSE, IIEST, Howrah, India
e-mail: as.besu@gmail.com

© Springer Nature Singapore Pte Ltd. 2019
P. K. Mallick et al. (eds.), *Cognitive Informatics and Soft Computing*,
Advances in Intelligent Systems and Computing 768,
https://doi.org/10.1007/978-981-13-0617-4_19

algorithms using clustering and segmentation methods are present such as K-means
[1], FCM [2], GKC [3] and thresholding and also state-of-art change detection
algorithms such as techniques proposed by Soumya Dutta and Madhurima
Chattopadhyay [4], Huang et al. [5], Kapur et al. [6], Ridler and Calvard [7], Tsai
[8], Otsu [9], Rosin [10], Shading Model (SM) [11], Linear Dependence and Vector
Model (LDD) [12], Statistical Change Detection (SCD) [13] and Wronskian
Change Detection Model (WM) [12]. Algorithms like LDD, SM and SCD detect
undesirable object boundaries leading to erroneous results. On the other hand,
Huang and Wang and Rosin give extremely low threshold values resulting in
improper detection of the affected area.

2   Ground Work

2.1   Change Vector Analysis

CVA [14] plays a very important role in any change detection technique. This
method is used to get the net change between the images over a span of time. Since
we are comparing two images it is very essential to use this technique. Let there be
two images, one unchanged referred to as original image, and the other, the change
image, referring to the next image taken after a duration of time. We refer to them
as OI and CI respectively from now on. The general form of the equation (for three
components) is as follows:

$$CM = \sqrt{(CI_1 - OI_1)^2 - (CI_2 - OI_2)^2 - (CI_3 - OI_3)^2}, \tag{1}$$

where CM is the magnitude of change and the subscripts 1, 2, 3 refer to the
individual components.

2.2   Rough Sets

Rough Sets, proposed by Pawlak [15], introduced the concept of lower approximation and upper approximation in contrast with hard-clustering algorithms. An
information system $\gamma = (\Phi, \Psi)$, where Φ and Ψ are finite, nonempty sets known as
universe and set of attributes, respectively. The predominant concept in rough sets
is indiscernibility relation. For an object having a particular set of attributes we
cannot clearly say that it truly defines a particular characteristic but has some
features of that characteristic. An object may satisfy all of the attributes for a
characteristic (lower approximation) and may satisfy some of the attributes of a
characteristic upper Approximation). Thus, for an information system, $\gamma = (\Phi, \Psi)$,

$\Lambda \subseteq \Phi$, $\chi \subseteq \Psi$. Therefore, the sets $\underline{\chi}$ and $\overline{\chi}$ are said to be lower and upper approximations respectively defined as

$$\underline{\chi} = \cap_{\alpha \in \varphi} \{\chi(\alpha) : \chi(a) \subseteq \Lambda\} \tag{2}$$

and

$$\overline{\chi} = \cap_{\alpha \in \phi} \{\chi(\alpha) : \chi(\alpha) \cap \Lambda\} \tag{3}$$

Lingras and Georg [16] applied the concept as proposed by Pawlak to clustering.

2.3 Thresholding and Median Filtering

Thresholding is a very common technique that is used for converting an image to its binary form. A particular value or a "threshold" is chosen such that all the pixel values above the threshold is converted to one value and the remaining pixels are set to another value. In this way, construct the binary image. We use this technique to convert the clustered image to the initial difference image which is referred to as DI from now on. The initial DI is further processed to remove unwanted noise with the help of a median filter [17]. The final DI is obtained after the filtering.

3 Proposed Algorithm

In the proposed method, explain the various parts of algorithm in the section above, we will give a stepwise detailing in this section

Step 1: Perform Change Vector Analysis (CVA) on the OI and CI and obtain an image $\int$:

$$\int = \sqrt{(CI_k - OI_k)^2} \tag{4}$$

Step 2: Determine the number of clusters 'n' and initialize the initial centres.
Step 3: Determine the nearest centre:

$$\beta_{\min} = \min(|\beta - \alpha_i|) \tag{5}$$

for $i = 1$ to n and $\alpha \in \int$

Step 4:

for $j = 1$ to n and $i \neq j$ **do**

if $\left| \left| (\alpha - \beta_j) \right| - |\alpha - \beta_{min}| \right| \leq \varepsilon$ **then**

$\alpha \in \overline{\chi}(\beta_j)$ and

$\alpha \in \overline{\chi}(\beta_{min})$ and $\alpha \in \underline{\chi}(\beta_{min})$

else

$\alpha \in \overline{\chi}(\beta_{min})$ and $\alpha \in \underline{\chi}(\beta_{min})$,

end if
where $\in$ is any suitable constant
end for

Step 5: Calculation of new cluster centres

if $\left(\underline{\chi}(\beta_j) \neq \varphi \right)$ and $\left(\overline{\chi}(\beta_j) - \underline{\chi}(\beta_j) \right) = \varphi)$ **then**

$$\beta_j = \frac{\sum_{\alpha \in \underline{\chi}(\beta_j)} \alpha}{\left| \underline{\chi}(\beta_j) \right|}$$

else

if $\left(\underline{\chi}(\beta_j) = \varphi \right)$ and $\left(\overline{\chi}(\beta_j) - \underline{\chi}(\beta_j) \neq \varphi \right)$ **then**

$$\beta_j = \frac{\sum_{\alpha \in \overline{\chi}(\beta_j) - \underline{\chi}(\beta_j)} \alpha}{\left| \overline{\chi}(\beta_j) - \underline{\chi}(\beta_j) \right|}$$

else

$$\beta_j = \varpi_{lower} \times \frac{\sum_{\alpha \in \underline{\chi}(\beta_j)} \alpha}{\left| \underline{\chi}(\beta_j) \right|} + \varpi_{upper} \times \frac{\sum_{\alpha \in \overline{\chi}(\beta_j) - \underline{\chi}(\beta_j)} \alpha}{\left| \overline{\chi}(\beta_j) - \underline{\chi}(\beta_j) \right|}$$

end if

end if
for $j = 1$ to n, where $\varpi_{lower} + \varpi_{upper}$

Step 6: Repeat Steps 3 to 5 until the cluster centres change no more.
Step 7: Replace the pixel values in f with the nearest cluster centre.
Step 8: Choose an appropriate threshold and obtain a binary image, the initial DI.
Step 9: Apply the median filter to remove the unwanted noise and generate final DI.

4 Result and Analysis

The performance of our change detection method and other existing algorithms are compared in this section. Many different medical images are used for testing. We have employed the standard analysis method and calculated the PCC, JC, YC, TP, TN, FP and FN [18] between resultant difference image (DI) and groundtruth image. The comparative results are produced in Tables 1 and 2 for two medical images. The simulation results are compared the PCC, JC, YC and error using the different algorithms such as Huang and Wang algorithm, Yager method, LDD method, SM, FCM, GKC, Kapur, K-means, Otsu, Ridler Calvard, Rosin, Standard Deviation, Statistical change detector, Tsai, Wrokskian, Yager, and proposed algorithm. The output of the medical images is shown in Figs. 1 and 2. From the results of medical image1 and image2, the algorithms like Huang Wang, Yager, LDD, SM give highly erroneous results. For Image 2 (Table 2), the algorithms of Huang Wang and Yager give erroneous results. It is noticed that the proposed algorithm gives more accurate results, for both the images.

Table 1 The comparison of error and accuracy values of the different techniques with proposed method for medical Image1

Methods	TP	FP	TN	FN	PCC	JC	YC	Error
Proposed method	1220	243	15086	351	96.49	0.6725	0.8112	3.5148
Standard deviation	988	17	15312	583	96.45	0.6222	0.9464	3.5503
K-means	935	151	15178	636	95.34	0.5430	0.8207	4.6568
FCM	826	103	15226	745	94.98	0.4934	0.8425	5.0178
Rosin	1188	2879	12450	383	80.70	0.2670	0.2623	19.3018
Ridler Calvard	714	197	15132	857	93.76	0.4038	0.7302	6.2367
Tsai	681	177	15152	890	93.69	0.3896	0.7382	6.3136
Otsu	694	188	15141	877	93.70	0.3945	0.7321	6.3018
Kapur	601	122	15207	970	93.54	0.3550	0.7713	6.4615
Huang Wang	1134	2160	13169	437	84.63	0.3039	0.3121	15.3669
Yager	1134	2160	13169	437	84.63	0.3039	0.3121	15.3669
Gustafson Kessel	1340	507	14882	231	95.63	0.6449	0.7102	4.3669
LDD	1292	6762	8567	279	58.34	0.1550	0.1289	41.6627
SCD	1123	239	15090	448	95.93	0.6204	0.7957	4.0651
SM	1172	5910	5910	399	62.67	0.1567	0.1249	37.3314

Table 2 The comparison of error and accuracy values of the different techniques with proposed method for medical Image2

Methods	TP	FP	TN	FN	PCC	JC	YC	Error
Proposed method	1904	2	23367	327	98.71	0.8527	0.9851	1.2852
Standard deviation	1866	105	23264	365	98.16	0.7988	0.9313	1.8359
K-means	1312	0	23369	919	96.41	0.5881	0.9622	3.5898
FCM	1123	242	23127	1108	94.73	0.4541	0.7770	5.2734
Rosin	1294	7	23362	937	96.31	0.5782	0.9561	3.6875
Ridler Calvard	1398	25	23344	833	96.65	0.6197	0.9480	3.3516
Tsai	1275	5	23364	956	96.25	0.5702	0.9568	3.7539
Otsu	1436	35	23334	795	96.76	0.6337	0.9433	3.2422
Kapur	2126	3829	19540	105	84.63	0.3508	0.3517	15.3672
Huang Wang	2180	8008	15361	51	68.52	0.2129	0.2107	31.4805
Yager	2161	6028	17341	70	76.18	0.2617	0.2599	23.8203
Gustafson Kessel	1868	2	23367	163	98.57	0.8365	0.9836	1.4258
LDD	1558	104	23265	673	96.96	0.6672	0.9099	3.0352
WM	1851	122	23247	380	98.04	0.7867	0.9221	1.9609
SCD	1905	57	23312	326	98.50	0.8326	0.9572	1.4961
SM	1603	150	23219	628	96.96	0.6732	0.8881	3.3091

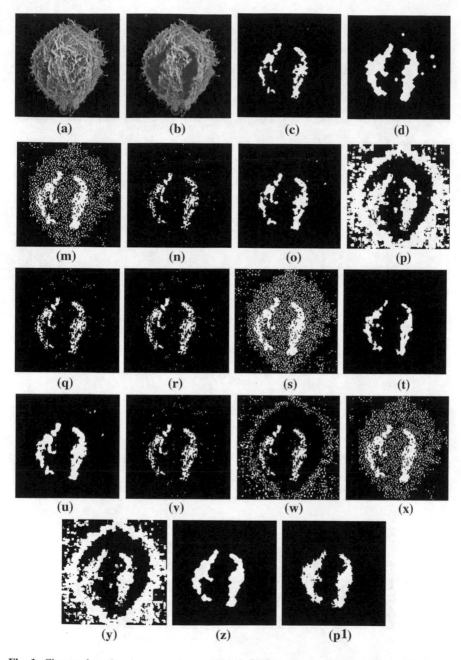

Fig. 1 Change detection images using **c** FCM **d** GKC **m** Huang Wang **n** Kapur **o** K-means **p** LDD **q** Otsu **r** Ridler Calvard **s** Rosin **t** standard deviation **u** statistical change detector **v** Tsai **w** Wrokskian **x** Yager **y** shading model **z** proposed algorithm and **p1** is the groundtruth image, **a**, **b** are two original images (Medical Image 1)

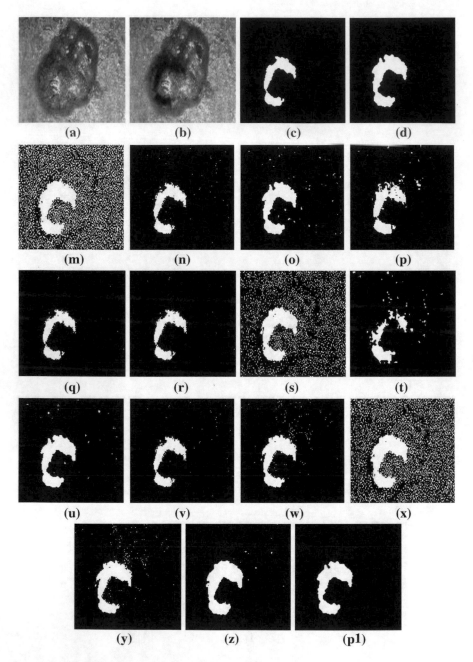

Fig. 2 Change detection images using **c** FCM **d** GKC **m** Huang Wang **n** Kapur **o** K-means **p** LDD **q** Otsu **r** Ridler Calvard **s** Rosin **t** standard deviation **u** statistical change detector **v** Tsai **w** Wrokskian **x** Yager **y** shading model **z** proposed algorithm and **p1** is the groundtruth image, **a**, **b** are two original images (Medical Image 2)

5 Conclusion

This proposed algorithm presents a change detect technique for medical images to check the disease or disease-free images using CVA, Rough sets and thresholding techniques. It gives better results as compared to other algorithms that have already been proposed. The Shading Model, LDD detect unwanted edges in the images and the Huang and Yager and Rosin give low thresholds resulting in a lot of unwanted noise. Therefore, comparing the PCC, Jaccard and Yule coefficients of the different algorithms, the proposed algorithm perform better than the existing ones. Hence, it is evident that Rough Set Theory has a promising future ahead.

References

1. Jain, A.K.: Data clustering: 50 years beyond K-means. Pattern Recogn. Lett. **31**(8), 651–666 (2010)
2. Bezdek, J.C.: Pattern Recognition with Fuzzy Objective Function Algorithms. Plenum Press, New York, London (1981)
3. Gustafson, D.E., Kessel, W.C.: Fuzzy clustering with a fuzzy covariance matrix. In: IEEE Conference on Decision and Control (1979)
4. Dutta, S., Chattopadhyay, M.: A Study of Change Detection Algorithm for Medical Cell Images, IJITKM (2011)
5. Huang, L.K., Wang, M.J.: Image thresholding by minimizing the measures of fuzziness. Pattern Recogn. **28**(1), 41–51 (1995)
6. Kapur, J., Sahoo, P.K., Wong, K.C.: A new method for grey-level picture thresholding using the entropy of the histogram. CVGIP **29**(3), 273–285 (1985)
7. Ridler, T., Calvard, S.: Picture thresholding using an iterative selection method. IEEE Trans. Syst. Man Cybernet. **8**, 630–632 (1978)
8. Tsai, W.: Moment-preserving thresholding. Comput. Vision Graph. Image Process. **29**, 377–393 (1985)
9. Otsu, N.: A threshold selection method from gray-level histograms. IEEE Trans. Systems Man Cybernet. **9**, 62–66 (1979)
10. Rosin, P.: Unimodal thresholding. Pattern Recogn. **34**, 2083–2096 (2001)
11. Skifstad, K., Jain, R.: Illumination independent change detection for real world image sequences. CVIP **46**, 387–399 (1989)
12. Durcan, E., Ebrahimi, T.: Improved Linear Dependence and Vector Model for Illumination Invariant Change Detection. SPIE, Bellingham (2001)
13. Aach, T.: Statistical model-based change detection in moving video. Signal Process. **31**, 165–180 (1993)
14. Tewkesbury, A.P., Comber, A.J., Tate, N.J., Lamb, A., Fisher, P.F.: A critical synthesis of remotely sensed optical image change detection techniques. Remote Sens. Environ. **160**, 1–14 (2015)
15. Pawlak, Z.: Rough Sets: Theoretical Aspects of Reasoning About Data. Kluwer Academic Publishers, USA (1992)
16. Lingras, P., Georg, P.: Applying Rough Set Concepts to Clustering: Rough Sets: Selected Methods and Applications in Management and Engineering. Springer, London (2012)
17. Gonzalez, R.C., Woods, R.E.: Digital Image Processing. Pearson Education, London (2002)
18. Sehairi, K., Chouireb, F., Meunier, J.: Comparison study between different automatic threshold algorithms for motion detection. In: 4th International Conference on ICEE (2015)

Malware Architectural View with Performance Analysis in Network at Its Activation State

Sisira Kumar Kapat and **Satya Narayan Tripathy**

Abstract Malware is an international issue which results in the loss of integrity, security, and authenticity. Some malwares work autonomously whereas some malware use a host and some malware change their identity each time. Researchers are constantly working for the defense mechanism. This paper focuses to analyze malware architecture to give a detailed study of malware which can be helpful to design a strong defense mechanism. Although many authors analyzed and classified the malwares in several categories, this paper classifies malware into four categories as per their architecture at the time of infection and discusses the mechanism behind the malware architectures. This paper also observed the performance of network at the time of infection.

Keywords Malware architecture · Independent malware · Appending malware
Prepending malware · Position-independent malware · Categories of malware

1 Introduction

The malware is a piece of code statement or a program which is designed by script kiddies for malicious purpose. Malware infects any malicious or benign file and turns them into malicious file. Malware may affect/infect other programs directly or indirectly [1]. Malicious purpose refers to replicate the self-program, copy other programs, erase data or modify other programs resulting unknown or unwanted state of the computing system behavior including DOS (denial of service), DDoS (distributed denial of service) and DeOS (destruction of service) [2] attacks. Sometimes malware does not affect other programs but pop-up some sort of

S. K. Kapat (✉) · S. N. Tripathy
Department of Computer Science, Berhampur University,
Berhampur, Odisha, India
e-mail: skk.rs.cs@buodisha.edu.in

S. N. Tripathy
e-mail: snt.cs@buodisha.edu.in

© Springer Nature Singapore Pte Ltd. 2019
P. K. Mallick et al. (eds.), *Cognitive Informatics and Soft Computing*,
Advances in Intelligent Systems and Computing 768,
https://doi.org/10.1007/978-981-13-0617-4_20

advertisements or it may steal user confidential data and send to third party. Some malware infect other programs and changes its structure. The infection may carry out in a standalone system or a networked environment. Considering the file structure after infection, this paper divides malware into four categories which are discussed in Sect. 3.

Malware also propagates in a networked environment to infect computing systems connected in the same network or different connected networks. This paper also analyzes the behavior of network with the propagation of malware from one computing system to another computing system. Some virtual network is designed using NetLogo (existing) tool and considered for analyzing the infection. The performance is analyzed in Sect. 4.

Malware coders normally use PE [3] (Portable Executable) files as their vector. Because PE files has the ability to load into windows environment in random memory locations using Relative Virtual Addressing (RVA) technology. So it is easier to inject one PE file into memory as the base address of the PE file is unknown at compile time, resulting difficulty to detect it.

The Script-Kiddies already know that they are being watched, in other words the malware program designed by malware coders are being studied and analyzed. So they are conscious to write sinister malwares to make them hard to be detected.

According to the major findings of Cisco 2017 Midyear Cyber security report [2], BEC (Business email compromise) is a major vector for the attackers. There is an increase in spam volume since mid-2016. Malicious hackers are working to breach the corporate cloud environment. Many organizations underestimate Spyware, which is present in 20% of the business organizations. Ransomware attack is an instance of this kind which is being distributed by some downloader malware like Nemucod.

The number of malware is increasing rapidly with different infection techniques. Frederick Cohen [4] proved that a single mechanism or detection technique cannot detect all the future malware in a finite time. But anti-malware engines are still updating their techniques with the change of malware techniques. So there is a requirement to analyze malware to detect malwares easily. This paper analyzes some of the architectures of malwares.

2 Related Works

Marco and Matt [5], in their paper proposed a propagation model which is similar to divide-and-conquer type propagation model. They proposed to divide the whole malware program into several parts. These subprograms of malware then are attached to different files for propagation or entry into the system. The files may be executable or non-executable file. The subprograms have neither complete code signature nor complete behavior signature so that it can be detected by any anti-malware programs. The sub-parts act like modules of an application. So, after injection and before execution, it must be assembled into a single malicious program.

In and Ulrich [6] discussed the different file formats of virus files. Three malware files formats were discussed in their paper; they are parasitic virus, macro virus, and polymorphic virus. According to them, parasitic virus appends the virus code at the end and the virus changes the header in such a way that, the virus code runs first. Macro virus can be injected after the macro-section of the Microsoft Word and Excel file, which is presented by a figure in their paper. According to them, polymorphic viruses can be represented by parasitic virus or macro virus. Their paper is to detect malware by using Self-Organizing Maps (SOM). Here we concentrate only on the architectures.

Aman [7] discusses the subtypes of malwares (viruses). According to him, malware can be classified in four classes as memory-based, target-based, obfuscation technique-based and payload-based. Again memory-based classification is discussed in [7, 8] and it is subdivided into various classes.

So it is required to analyze the malware and hence to design a detection technology which can detect the malware in standalone as well as networked environment.

3 Malware Analysis Based on Code Position While Infection

Malware infection is not same in all cases. Sometimes malware simply append itself to the host program and sometimes it simply prepends. So malware can be broadly categorized in four categories according to their infection strategy which can be called as the architecture of malware. The four categories of malware are, Independent malware, appending malware, prepending malware, and position-independent malware.

3.1 Independent Malware

Independent malware are designed not to affect other programs but to perform individually. The malware run (execute) directly on the operating system, so it does not need any host program. It can simply pop-up the ads, gather user confidential data and send to third party, etc. The program may replicate itself so that the performance of the system slows down or sometimes the system seizes to work.

The malware can be represented mathematically as,

$$M \cup \emptyset = \emptyset \cup M = M \tag{1}$$

Equation (1), states that Independent malware, work independently without the hosting of other programs and not interrupt other running programs of the system.

3.2 Appending Malware

According to webopedia and techopedia [9, 10], an appending malware inserts a copy of malicious code at the end of the host program. It does not harm the source program, rather it modifies the host program to hold the malicious code and run itself.

Let,

E: Any executable or non-executable program exists in the user computing system. It may be user file or system file.

M: Any Malware program which appends a portion of malicious program (set of instructions) or copy of whole program to the end of "E".

EM: The resultant malicious program which is executable or non-executable.

Then Appending operation can be represented mathematically as,

$$EM = \{Em : m \in M\} \tag{2}$$

Equation (2), states that, the executable or non-executable file "E" is constant and do not change its code whereas the malicious file "M" appends a part ("m") or copy of itself ("M") for each infection. This may behave as polymorphic from time to time.

In executable programs, the malicious code executes directly, each time the program runs. In non-executable programs, the malicious code executes by the help of another program. Some programs may call the malicious program as a subroutine which is at the server end (e.g., tongji.js).

Tongji.js is a malicious javascript program whose URL is shown in Fig. 1. Although this program has a variant of versions, this paper presents only one variant as shown in Fig. 2. This code statement appends itself to an "html" file and

```
<script type="text/javascript" src="http://web.nba1001.net:8888/tj/tongji.js">
</script>
```

Fig. 1 Complete URL of "tongji" malware

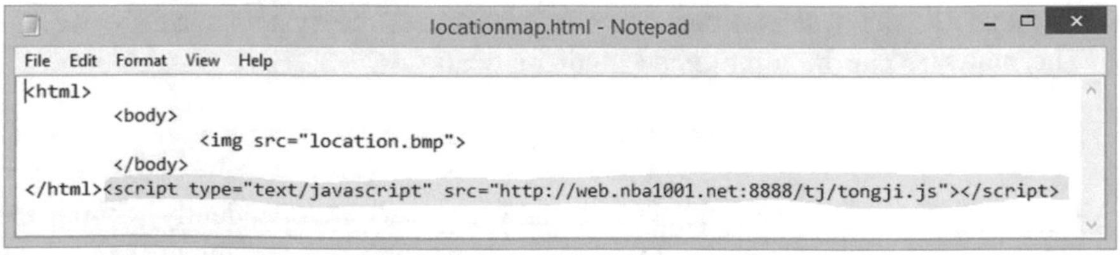

Fig. 2 An example of appending malware

Fig. 3 Detection ratio of tongji malware in virustotal

executes itself each time the program is executed. This malware resides at the server end and a link is appended to the source program at the client side. According to BitDefender anti-malware of virustotal (www.virustotal.com), this malware belongs to Trojan .JS .Agent .JBZ family. Besides BitDefender some other anti-malware of virustotal also detects this file as malicious. The detection ratio of virustotal is 43/58 as shown in Fig. 3.

3.3 Prepending Malware

This type of malware, adds some lines of malicious code to the host programs at the beginning of the program. The general architecture of prepending malware is shown in Fig. 4.

Let,

E: Any executable or non-executable program exists in the user computing system. It may be user file or system file.

M: Any Malware program which prepends a portion of malicious program (set of instructions) or copy of whole program to the beginning of "E".

ME: The resultant malicious program which is executable or non-executable.

Then prepend operation can be represented mathematically as,

$$ME = \{mE : m \in M\} \tag{3}$$

Fig. 4 Architecture of prepending malware

Malicious Code (Prepend)
Original Program

Equation (3), states that, the executable or non-executable file "E" is constant and do not change its code whereas the malicious file "M" prepends a part ("m") or copy of itself ("M") for each infection. For polymorphic malware, "m" may change for, each iteration.

3.4 Position-Independent Malware

According to the property of malware, the union of malicious code or a malware program with a malicious or benign program results a malicious program.

This can be represented mathematically as,

$$M \cup P = M, \tag{4}$$

where "M" is any malicious code or malicious program, and "P" is any malicious or benign program. The structure and mechanism of position-independent malware is shown in Fig. 5.

As the name suggest, the "position-independent malware" has no specific position. Sometimes it may show variants in case of polymorphic and metamorphic malware.

In some cases, the malware is divided into several parts. Say, "M" is subdivided into "i" parts as, $[i] = \{M_1, M_2, M_3, \ldots, M_i\}$. In the infection phase, the "i" parts will be injected into the host program and makes space to be stored. Let the source program "P" is split into "j" parts as $p[j] = \{P_1, P_2, P_3, \ldots, P_j\}$.

According to the property of position-independent malware,

$$M \cup P = M$$

And hence,

$$\{M_1, M_2, M_3, \ldots, M_i\} \cup \{P_1, P_2, P_3, \ldots, P_j\} = M \tag{5}$$

Example: malware can be inserted into an array of matrix (pixel) of an image as shown in Fig. 6. This type of malware code is known as position-independent code.

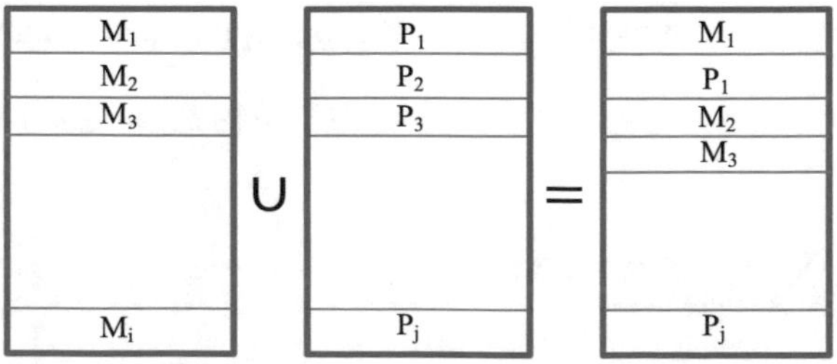

Fig. 5 Mechanism of position-independent malware

Fig. 6 Example of
position-independent code

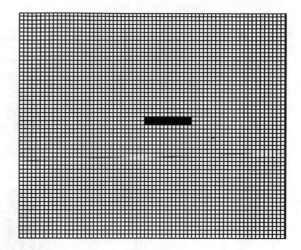

Another type of position-independent malware is web-bug which can acquire a very small area in the image irrespective of the position.

4 Malware Analysis Based on Propagation in a Network

To analyze the properties of propagation, some virtual networks have been designed using NetLogo tool [11] with an existing model known as virus on a network [12]. The observations are noted in different conditions as shown in Table-1 and analyzed the inference. NetLogo uses SIR model [13, 14], where the user has to set the number of nodes, average node degree, virus spread chance, initial outbreak size (initial number of malwares in a network), the recovery chance of each nodes and the gain resistance chance. Though the network in NetLogo is virtual in nature, but it is the simulation of a network, which can be used for analysis purpose. The gain-resistance chance is the resistance acquired after the recovery of a node from infection. In real time the user can assume that, after infection, a system can recover from the infection using some anti-malware tools, which can resist the same malware to infect for the next time. The observations for different network structure are presented in Table 1.

A malware continues to move in a network until it is detected and eliminated by an anti-malware tool. After removing a malware, the node may gain some resistance power which is capable to resist the future infection.

Let,

SN: Serial Number   VSC: Virus Spread Chance
NON: Number of Nodes   RC: Recovery Chance
AND: Average node degree   GRC: Gain resistance chance

Table 1 Observations for different network structure

SN	NON	AND	VSC (%)	RC (%)	GRC (%)	Network structure	Network status
1	150	6	2.5	5	5		
2	180	7	2	7	5		
3	200	10	3	7.5	5		
4	220	6	2.6	7	6		
5	225	7	3	7.5	6		

Fig. 7 Network status layout

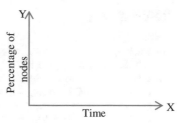

In the observations the network status shows the percentage of nodes susceptible, infected or resistant with respect to time. In the above figure the X-axis shows the time and Y-axis shows the percentage of nodes which is presented in Fig. 7.

Here five observations are considered and each time, the values of different attributes mentioned above are changed and observed the network status. The observations are mentioned below.

- Initially, the infection is increasing rapidly up to a certain time and then decreases slowly to zero.
- The number of susceptible node is increased or decreased inversely with respect to the number of infection, i.e., if the infection increases, then susceptible decreases and vice versa.
- Each time the number of nodes with resistance after infection is increasing.
- This is also observed that a single malware can perform different each time for different iteration.

In the above observation, the number of initial outbreak size is fixed and the same epidemic multiples each time. In that case, no external infection interrupts the network. But in real-time network, if new malware infects each time to the network, then the infection and curing process may continue forever.

5   Conclusion and Future Work

This paper presents four different architectures of malware and the mechanism associated with those architectures. This paper observed that, a single malware can perform different in different networking environment. This activity is observed in the simulation environment of netlogo. In real-time network, the network is not fixed and the number of nodes increases or decreases with respect to time. The architectures and observations can be used together in malware detection purpose which will detect a malware in standalone as well as networked environment.

References

1. Kramer, S., Bradfield, J.C.: A general definition of malware. J. Comput. Virol. **6**, 105–114 (2010). https://doi.org/10.1007/s11416-009-0137-1
2. Cisco 2017 Midyear Cybersecurity Report, https://www.automation.com/pdf_articles/cisco/Cisco_2017_MCR_Embargoed_til_072017_5_AM_PT_8_AM_ET.pdf
3. https://msdn.microsoft.com/en-IN/library/ms809762.aspx
4. Cohen, F.: Computer Viruses Theory and Experiments. Elsevier Science Publishers B.V., North Holand (1987)
5. Ramilli, M., Bishop M.: Multi-Stage Delivery of Malware, pp 91–99, IEEE (2010)
6. Yoo, I.S., Ultes-Nitsche, U.: Non-signature based virus detection-Towards establishing a unknown virus detection technique using SOM. J. Comput. Virol. **2**, 163–186 (2006). https://doi.org/10.1007/s11416-006-0013-1
7. Hardikar M.A.: MALWARE 101—VIRUSES. SANS Institute InfoSec Reading Room (2008) https://www.sans.org/reading-room/whitepapers/incident/malware-101-viruses-32848
8. Szor, P.: The Art of Computer Virus Research and Defense. Addison Wesley, Boston (2005)
9. http://www.webopedia.com/TERM/A/Appending_Virus.html
10. https://www.techopedia.com/definition/34/appending-virus
11. Wilensky, U.: NetLogo. Center for Connected Learning and Computer-Based Modeling, Northwestern University, Evanston, IL (1999) http://ccl.northwestern.edu/netlogo/
12. Stonedahl, F., Wilensky, U.: NetLogo Virus on a Network model. Center for Connected Learning and Computer-Based Modeling, Northwestern University, Evanston, IL (2008) http://ccl.northwestern.edu/netlogo/models/VirusonaNetwork
13. Martcheva M.: Introduction to Epidemic Modeling. An Introduction to Mathematical Epidemiology, ISBN: 978-1-4899-7611-6 (2015)
14. Sneha, S., Malathi, L., Saranya, R.: A survey on malware propagation analysis and prevention model. Int. J. Adv. Technol. (2015). https://doi.org/10.4172/0976-4860.1000148

Non-linear Analysis of Time Series Generated from the Freeman K-Set Model

F. Anitta, R. Sunitha, N. Pradhan and A. Sreedevi

Abstract Brain signals such as EEG and MEG are the only available dynamical measures of functional status of the brain. Over past several years EEG has been found to have nonlinear and chaotic properties. The nonlinear dynamical measures have been linked to brain functioning including the most complex cognitive behavior of man. Our study focuses on showing evidence of nonlinear chaotic behavior of simulated EEG. We have simulated the EEG at the mesoscopic level by using the biologically realistic Freeman K-sets. Here the behavior of the time series at every level of the olfactory system as modeled in the Freeman-KIII set is obtained by solving a set of second-order differential equations using Euler method in MATLAB. The generated low-dimensional- and high-dimensional time series is subjected to a nonlinear analysis using Higuchi fractal dimension, Lyapunov exponent, and Detrended Fluctuation analysis to validate the chaotic behavior. The study indirectly points to suitability of Freeman model for large-scale brain simulation.

Keywords Freeman K-set model · Higuchi fractal dimension · Lyapunov exponent · Detrended fluctuation analysis

F. Anitta · R. Sunitha (✉)
Department of Electronics and Communication Engineering,
Amrita School of Engineering, Bengaluru Amrita Vishwa Vidyapeetham,
Bengaluru, India
e-mail: r_sunitha@blr.amrita.edu

F. Anitta
e-mail: anittamercy@gmail.com

N. Pradhan
Department of Psychopharmacology, National Institute of Mental Health
and Neurosciences (NIMHANS), Bengaluru, India
e-mail: nprnimhans@gmail.com

A. Sreedevi
Department of Electrical and Electronics Engineering,
R. V. College of Engineering, Bengaluru, India
e-mail: sreedevia@rvce.edu.in

© Springer Nature Singapore Pte Ltd. 2019
P. K. Mallick et al. (eds.), *Cognitive Informatics and Soft Computing*,
Advances in Intelligent Systems and Computing 768,
https://doi.org/10.1007/978-981-13-0617-4_21

217

1 Introduction

Brain and its chaotic behavior has been a challenging area in neuroscience for a long period. The modeling of brain dynamics and mimicking the brain activity is still an interesting area for scientists [1]. Many models have been developed to mimic the brain dynamics and among them, the model developed by Walter J Freeman has been found to be more realistic [2]. By taking into consideration the thermodynamic behavior of any living system suggested by Aharon Katzir-Katchalsky, Freeman had modeled the olfactory system. He came up with an ensemble of neurons which interacted with each other either in excitatory or in inhibitory fashion. Skarda and Walter J Freeman came up with a set of mathematical expressions that could mimic the chaotic brain activity of the olfactory system [2, 3].

The chaotic nature of EEG and the neural activities come under the context of nonlinear dynamics and theory of deterministic chaos [4]. These nonlinear systems have a property to self-organize and evolve towards a strange attractor in state phase domain [5]. Thus nonlinear techniques such as Higuchi Fractal Dimension, Detrended Fluctuation analysis, and Lyapunov Exponent could analyze an EEG signal in a better way. These nonlinear techniques have applications in the analysis of sleep, epilepsy and cognitive tasks of brain. The abnormal functions of brain due to seizures, dementia, depression, autism, Alzheimer's disease were also studied using the nonlinear techniques in [6, 7].

In this paper, the chaotic EEG for low-dimensional- and high-dimensional activities has been simulated using the Freeman K-III model. The simulated EEG was analyzed using the nonlinear techniques such as Higuchi Fractal Dimension, Detrended Fluctuation Analysis, and Lyapunov Exponent.

2 Methodology

2.1 The Freeman K-III Model

The Freeman K-III model includes an ensemble of neurons that are arranged in a nested hierarchical level. The ensemble of neurons that are either excitatory or inhibitory in nature can interact with each other creating a positive or negative feedback loop [8]. These resulted in oscillation of the systems. K0 set is the basic node of Freeman K-III model. The K0 set consists of a collection of neurons that can be either excitatory (K0e) or inhibitory (K0i) in nature [9]. The activities of those neuron ensemble are analyzed using second order nonlinear differential equations which is bifurcated into, $F(t)$ and $G(v)$, a linear time-dependent and a nonlinear time-invariant parts respectively [10]. The equation solves for voltage v, which depends on time t given by

$$abF(v_n) = \ddot{v}_n + (a+b)\dot{v}_n + abv_n, \tag{1}$$

where $a = 220/s$ and $b = 720/s$ represent the fixed rate biological constants obtained using physiological experiments [11]. The equation for the nonlinear part contains the input variable v called as dendritic current density in A/sq cm at the cortical surface area and output variable p, called as axonal pulse density in pulse/s/cm^2 [10, 11].

When two K0 set of same nature come together with a positive feedback give rise to KI sets which are either excitatory (KI$_e$) or inhibitory (KI$_i$). When these KI sets interact with a negative feedback give rise to KII set [12]. Negative feedback connection between KIe and KIi forms the KIIei. An array of KII subsets, complete with both feedforward and detained feedback connections between the layers, forms the K-III network [11]. The KII set itself is enough to simulate the response of all three layers of the KIII model [13].

The Freeman K-III model has made its relevance as a better classifier in many real-time applications. The application of Freeman K-sets as a classifier in handwriting recognition [14], face recognition [15], tea classification [16], speech recognition [17], iris data analysis [18] has been discussed. The Freeman K-III set model was found to require less data set and was more immune to noisy environment. This made it as a better classifier for many of the real-time applications.

2.2 Nonlinear Techniques

The nonlinear behavior of the EEG signal is analyzed using the Higuchi Fractal Dimension, Detrended Fluctuation Analysis, and Lyapunov exponent.

Higuchi Fractal Dimension (HFD). Among the available fractal dimension algorithm Higuchi fractal dimension is found to be simpler, fast, and reliable method with accurate results [19]. The HFD is calculated from the mean length $L(k)$, for a given time series having N samples over the k values. If the mean length $L(k)$ and the fractal dimension FD satisfies the Eq. (2), then the curve is said to be a fractal with dimension FD.

$$L(k) \propto k^{-FD} \tag{2}$$

Then in the plot of $\ln(L(k))$ versus $\ln(k)$, the dimension is expected to fall on a straight line with slope equivalent to FD. Therefore, FD can be enumerated by means of a least squares linear best-fitting technique [20].

Lyapunov Exponent. Lyapunov exponent calculates the sensitivity of a system to the initial conditions with reference to the Largest Lyapunov Exponent (LLE) value, [21]. As the system tries to deform its shape with respect to the change in initial conditions from being an n-sphere to an n-ellipsoid the Lyapunov exponent of ith one dimension is defined in terms of the length of the ellipsoid principal axis $p_i(t)$ as in Eq. (3),

$$\lambda_i = \lim_{t \to \infty} \frac{1}{t} \log_2 \frac{p_i(t)}{p_i(0)}, \tag{3}$$

where λ_i are ordered from largest to smallest. A positive Largest Lyapunov Exponent (LLE) is typically taken as a sign to indicate that the system is chaotic in nature. A negative value implies that the trajectories tend to a common fixed point; and a zero exponent value implies that the directions keep up their positions, meaning that they are on a steady attractor [22]. The biggest Lyapunov type acts as a marker of long-term behavior of EEG and can be scarcely be anticipated [23].

Detrended Fluctuation Analysis (DFA). Detrended Fluctuation Analysis (DFA) method is used for the study of multi-fractal characterization of non-stationary biological signals [24]. A signal in time series is converted to an unbounded process and is fractionated into time windows of length samples each. Later, a local least square straight line fit (the local trend) is computed by minimizing the squared errors within the individual time window. Then, the root mean square deviation from the trend, the fluctuation, is deliberated as

$$F(n) = \sqrt{\frac{1}{N} \sum_{t=1}^{N} (X_t - Y_t)^2} \tag{4}$$

In the end, this procedure of detruding taken after by the fluctuation measurement is iterated over a scope of various window sizes and a log-log graph of $F(n)$ against n is formulated [25, 26].

3   Results and Analysis

Here we discuss the steps involved in the simulation of chaotic EEG represented using second-order differential equations implemented in MATLAB. The second-order differential equations were resolved using Euler method with a time step of 0.01 and with an initial external stimulus given as random signal following a uniform distribution.

3.1   Simulation of Chaotic EEG Using Freeman K-Sets

The basic building unit K0 sets were coupled together to form KI sets. The KI sets were coupled together to produce KII sets. The KII sets with feedbacks made the entire KIII sets which mimic the olfactory system. The external stimulus had been

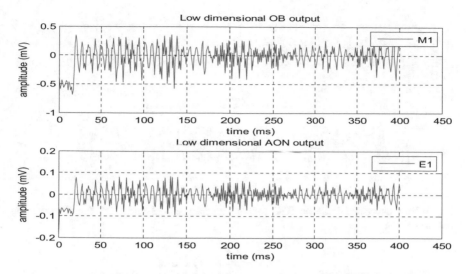

Fig. 1 Low-dimensional chaotic EEG pattern generated at OB and AON for 400 ms using Freeman K-sets

given to the periglomerular cells and also to the mitral cells in olfactory bulb. The high dimensional- and low-dimensional chaotic EEG is produced by establishing and removing the connection from PC to AON and OB. The low-dimensional EEG yields a low-level chaotic output and high-dimensional EEG yields the hyper chaos state of brain. Figures 1, 2, 3 and 4 give the chaotic EEG pattern simulated at the OB, AON, and PC for low dimensions and high dimensions respectively.

Figure 1 gives the low-dimensional chaotic EEG taken from the mitral (M1) of the OB and from the excitatory node (E1) of the AON. Low-dimensional EEG is the EEG that is recorded directly from the brain itself, or in this case, directly from

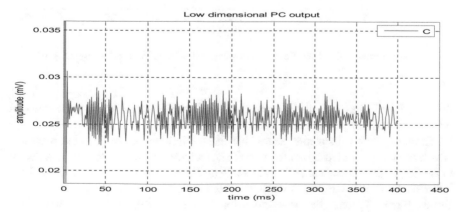

Fig. 2 Low-dimensional chaotic EEG pattern generated at PC for 400 ms using Freeman K-sets

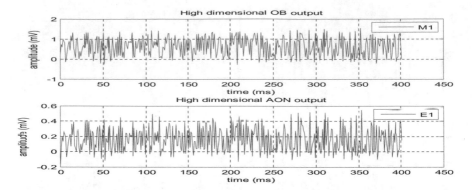

Fig. 3 High-dimensional chaotic EEG pattern generated at OB and AON for 400 ms using Freeman K-sets

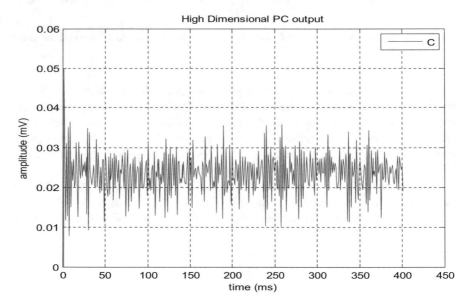

Fig. 4 High-dimensional chaotic EEG pattern generated at node PC for 400 ms using Freeman K-sets

the Olfactory Bulb. Here, the dimensionality of the EEG was found to be low and was more oscillatory in nature.

The term "low dimension" states that these EEG series are pure brain waves, which are not corrupted by artifacts (ocular, muscular, cardiac, etc.) or by external noise (instrument noise, etc.) Changes made by Freeman correlates to cutting off the influence from Prepyriform Cortex, which is present in the Cerebral Cortex of the Brain. Figure 2 gives the low-dimensional chaotic EEG with low amplitude obtained by removing the feedback connection from the OB and AON.

Figure 3 gives the high-dimensional chaotic EEG taken from the mitral cell (M1) of the OB and from the excitatory node (E1) of the AON. Here the dimensionality of the EEG was found to be high and was more random in nature. Figure 4 gives the output of PC node. The output was a high-dimensional chaotic EEG with high amplitude with more amount of randomness due to the feedback connections from the OB and AON.

High-dimensional EEG is the EEG from scalp recordings. They are termed "high dimension" as they are more erratic when compared to low dimension EEG, since they are mixed with artifacts (ocular, muscular, cardiac, etc.) and external noise. Changes made by Freeman correlates to involving the effect of the Prepyriform Cortex activity to the brain waves.

3.2 Nonlinear Analysis of Simulated Chaotic EEG

The various nonlinear methods had been used to analyze the simulated EEG patterns of low and high dimension. The results are given in Table 1. The fractal dimension of a normal EEG lies between the values 1 and 2. Among the available fractal dimensions methods HFD gives the accurate results. It quantifies the complexity and the transient nature of an EEG signal. These results in Table 1 give the complexity of EEG signal under the low and high dimensions. More transient detections are found under high dimension EEG signal resulting in high fractal dimension.

The Lyapunov exponent method determines the dynamics of a system. The negative value of Lyapunov exponent for low-dimensional EEG indicates that the system is stable and has a limit cycle attractor. The trajectories of the system tend to a common fixed point. The Lyapunov exponent is positive for high-dimensional EEG showing that the system has a chaotic attractor. The Lyapunov exponent also gives the information of the long-term behavior of the EEG that the high dimension EEG is hardly predictable as compared with the low-dimensional EEG.

The Detrended fluctuation analysis is the method that is used to estimate the long-term temporal correlation of a dynamic system. It calculates the mean Hurst exponent. The Hurst exponent obtained for the simulation of low- and high-dimensional EEG is found to be less than 0.5. This indicated the negative correlation of the simulated EEG time series. This resembled that there have been a switching between high and low values and also stronger is the tendency for the

Table 1 Nonlinear analysis of the simulated EEG

Methods	Low dimension EEG	High dimension EEG
Higuchi fractal dimension	1.0597	1.7044
Lyapunov exponent	−2.8985	0.5307
Detrended fluctuation analysis	0.0385	0.1193

time series to revert to its long-term mean value. The system is anti persistent in nature. The DFA is a high-dimensional EEG indicating the more chaotic or non-linear nature of human brain under active state.

The above nonlinear analysis helped to analyze the chaotic EEG being simulated in the olfactory system using the Freeman K-set model. These methods also paved a better way to analyze the low-dimensional and the high-dimensional EEG.

4 Conclusion and Future Scope

The Freeman K-set model based on the olfactory system is used to simulate low-dimensional and high-dimensional EEG activity patterns. This study has enabled us to understand how brain processes the information and decodes it. The nonlinear analysis was done on the simulated EEG to validate and thereby to understand the complexity of brain functioning from one mental state to another. The simulation and validation using nonlinear techniques can help in the study of disorders such as Alzheimer's, epilepsy, sleep apnea and various mental states of brain.

References

1. Eledath, D., Ramachandran, S., Pradhan, N., Asundi, S.: Power spectral scaling and wavelet entropy as measures in understanding neural complexity. Annu. IEEE India Conf. (INDICON) pp. 1–6 (2015)
2. Skarda, C.A., Freeman, W.J.: How brains make chaos in order to make sense of the world. Behav. Brain Sci. **10**, 161–195 (1987)
3. Freeman, W.J.: Simulation of chaotic EEG patterns with a dynamic model of the olfactory system. Biol. Cybern. **56**, 139–150 (1987)
4. Klonowski, W.: Chaotic dynamics applied to signal complexity in phase space and in time domain. In: Chaos, solitons and fractals, vol. 14, pp. 1379–1387 (2002)
5. Bermudez, R.G., Pedro, J., Laencina, G.: Analysis of EEG signals using nonlinear dynamics and chaos: a review. In: Applied mathematics & information sciences, vol. 9, pp. 2309–2321 (2015)
6. Poza, J., Gómez, C., Bachiller, A., Hornero, R.: Spectral and Non-Linear analyses of spontaneous magneto-encephalographic activity in alzheimer's disease. J. Healthc. Eng. **3**, 299–322 (2012)
7. Kunhimangalam, R., Joseph, P.K., Sujith, O.K.: Nonlinear analysis of EEG signals: surrogate data analysis. IRBM **29**(4), 239–244 (2008)
8. Li, Z., Hopefield, J.J.: Modeling the olfactory bulb and its neural oscillatory processings. Biol. Cybern. **6**, 379–392 (1989)
9. Kozma, R., Freeman, W.J.: Encoding and recall of noisy data as chaotic spatio-temporal memory patterns in the style of the brains. Int. Joint Conf. Neural Networks **5**, 33–38 (2000)
10. Chang, H.J., Freeman, W.J.: Parameter optimization in models of the olfactory neural system. Neural Networks **9**, 1–14 (1996)
11. Eisenberg, J., Freeman, W.J., Burke, B.: Hardware architecture of a neural network model simulating pattern recognition by the olfactory bulb. Neural Networks **2**, 315–325 (1989)

12. Denis, R., Piazenti, M., Rosa, J.L.: A simulator for freeman K-sets in JAVA. Int. Joint Conf. Neural Networks (IJCNN), 1–8 (2015)
13. Ilin, R., Kozma, R.: Stability conditions of the full KII model of excitatory and inhibitory neural populations. Int. Joint Conf. Neural Networks, pp. 3162–3167 (2005)
14. Obayashi, M., Koga, S., Feng, L., Kuremoto, T., Kobayashi, K.: Handwriting character classification using Freeman's olfactory KIII model. Artif. Life Robot. **17**(2), 227–232 (2012)
15. Zhang, J., Lou, Z., Li, G., Freeman, W.J.: Application of a novel neural network to face recognition based on DWT. Int. Conf. Biomed. Robot. Biomechatronics, pp. 1042–1046 (2006)
16. Yang, X., Fu, J., Lou, Z., Wang, L., Li, G., Freeman, W.J.: Tea classification based on artificial olfaction using bionic olfactory neural network. Advances in Neural Networks— ISNN 2006. Lect. Notes Comput. Sci. **3972**, 343–348 (2006)
17. Obayashi, M., Sud, R., Kuremoto, T., Mabu, S.: A class identification method using Freeman's olfactory KIII model. J. Image Graph. **4**, 130–135 (2016)
18. Yao, Y., Freeman, W.J.: Model of biological pattern recognition with spatially chaotic dynamics. Neural Networks **3**, 153–170 (1990)
19. Bermudez, R.G., Pedro, J., Laencina, G.: Analysis of EEG Signals using nonlinear dynamics and chaos: a review. In: Applied mathematics & information sciences, vol. 9, pp. 2309–2321 (2015)
20. Esteller, R., Vachtsevanos, G., Echauz, J., Litt, B.: A comparison of waveform fractal dimension algorithms. In: IEEE transactions on circuits and systems. fundamental theory and applications, vol. 48, pp. 177–183 (2001)
21. Pradhan, N., Sadasivan, P.K.: The nature of dominant Lyapunov exponent and attractor dimension curves of EEG in sleep. Comput. Biol. Med. **26**(5), 419–428 (1996)
22. Wolf, A., Swift, J.B., Swinney, H.L., Vastano, J.A.: Determining Lyapunov exponents from a time series. Physica D **16**, 285–317 (1985)
23. Das, A., Das, P., Roy, A.B.: Applicability of LyapunovExponent in EEG data analysis. Complex. Int. **9**, 1–8 (2002)
24. Marton, L.F., Brassai, S.T., Bako, L., Losonczi, L.: Detrended fluctuation analysis of EEG signals. In: 7th international conference interdisciplinary in engineering (INTER-ENG2013), vol. 12, pp. 125–132 (2014)
25. Bachmann, M., Suhhova, A., Lass, J., Aadamsoo, K., Vohma, U., Hinrikus, H.: Detrended fluctuation analysis of EEG in depression. In: XIII mediterranean conference on medical and biological engineering and computing, vol. 41, pp. 694–697 (2013)
26. Zorick, T., Mandelkern, M.A.: Multifractal detrended fluctuation analysis of human EEG: preliminary investigation and comparison with the wavelet transform modulus maxima technique. PLoS ONE **8**, 1–7 (2013)

Novel Approach to Segment the Pectoral Muscle in the Mammograms

Vaishali Shinde and B. Thirumala Rao

Abstract The X-ray technique is widely used to detect the breast cancer. The X-ray image contains the breast part along with the pectoral muscles. The pectoral muscles are similar to breast tissue in terms of texture and appearance but it is not a part of breast tissue. Hence pectoral muscles removal is an essential task for breast tumor detection. In the first phase of the proposed approach, the three existing pectoral muscles segmentation methods, region growing, thresholding, and k-mean clustering has been implemented. In a later phase, machine learning-based approach to segment out the pectoral muscle has been implemented. The proposed system provides the promising results on the MIAS database.

Keywords K-means clustering · Machine learning · Pectoral muscle removal
Region growing · Thresholding

1 Introduction

Breast cancer is one of the leading causes of the death among the females worldwide. As per the survey carried out by the Breastcancer.org [1], 12% females worldwide suffer from the breast cancer in her lifetime. 28% cancer among all females in the USA suffered from the breast cancer. That means one-eighth of the females living in the USA may develop breast cancer. The detection and prevention of the breast cancer are difficult, but it helps to decrease the death rate by detecting breast cancer at an earlier stage. The mammogram screening technique is helpful to detect cancer at an earlier stage. This may help to save the life of the patient. The radiologists examine the mammogram images for abnormalities as the tumor is malignant or benign. Sometimes manual identification may result in false positive.

V. Shinde (✉) · B. Thirumala Rao
KL University, Vijayawada, Andhra Pradesh, India
e-mail: svaishu11@gmail.com

B. Thirumala Rao
e-mail: thirumail@yahoo.com

© Springer Nature Singapore Pte Ltd. 2019
P. K. Mallick et al. (eds.), *Cognitive Informatics and Soft Computing*,
Advances in Intelligent Systems and Computing 768,
https://doi.org/10.1007/978-981-13-0617-4_22

Fig. 1 Anatomy of the breast

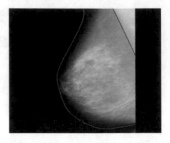

To improve the accuracy of the system, there is a necessity of automatic abnormality detection algorithm. To detect the breast tumor from the mammogram, some preprocessing techniques have to be applied to the image.

The mammogram images contain artifacts such as pectoral muscles and labels. The pectoral muscles removal is a crucial task because the texture and appearance of the pectoral muscle are nearly similar to the breast cancer tissue. The location of the pectoral muscles is found on the top-right or top-left quarter part of the mammogram image which is shown in Fig. 1.

2　Literature Survey

Sreedevi et al. [2] proposed the approach for pectoral muscles. In this approach, discrete cosine transforms (DCT) based Non-Local Mean (NLM) filter are used for noise removal. This method is most effective for removal of the Gaussian and impulse noise. Thresholding-based approach is used to segment the pectoral muscles. This system achieves the overall accuracy of 90.06% MIAS database [3].

Lakshmanan et al. [4] proposed an algorithm for pectoral muscles boundary detection. In this approach, the image is first decomposed by the Laplacian pyramids then the edges were detected using canny edge detection finally the boundaries of pectoral muscle is extracted using orientation and eccentricity of the elliptical bounding box. It achieves a false positive rate of 0.28% and false negative rate of 3.67%.

Vikhe et al. [5] proposed intensity based pectoral muscles boundary detection. The pectoral region of the mammogram images is enhanced with the novel enhancement filter. This technique achieves the acceptance rate of 96.56%.

Shrivastava et al. [6] present the accurate pectoral muscle segmentation method based on the dispersed region growing and sliding window technique. This approach uses Dispersed Region Growing Algorithm for detection of cancerous tissue. This system achieves the acceptance rate of 91.3%.

K-means clustering segmentation algorithm is used by Alam et al. [7] for removal of pectoral muscles from the digital mammogram images. The images are enhanced by the morphological operations like erosion and dilation. The image is finally segmented using K-means clustering and region growing algorithm.

In [8], global thresholding method is used to segment out the mammogram images. In global thresholding method, the histogram is used for image segmentation. Deciding the threshold value by considering the histogram, the image can be segmented into two regions.

3 Existing Pectoral Muscles Removal Algorithms

There are different approaches are attempt by the researcher for the pectoral muscle segmentation. Some of the existing methods are explained below.

3.1 Region Growing

Region growing is the segmentation method which works in the region-based properties. It can be used as a pixel-based segmentation by selecting arbitrary seed point. Based on the intensity of the seed point, it grows if the nearest pixels if the intensity of the pixel is nearer to it [9]. If the area of the image is greater the time required to segment out the image is more. Detailed steps for the region growing algorithm is as explained below.

1. Select the arbitrary seed point from the pectoral region.
2. Nearest intensity pixel of the seed point is added to the region.
3. Repeat step 2 for each of the newly added pixels.
4. When the growth of the one region stops then chooses another seed point which does not belong to the same segmented part.
5. The process is continued until all pixels in the region does not segmented.

3.2 Thresholding

Thresholding is one of the segmentation method based on the grayscale value. There are two different methods of thresholding, i.e., local and global thresholding. In this approach, Otsu global thresholding method is used for calculation of threshold value.

3.3 K-Means Clustering

The K-means clustering algorithms are used to segment out the pectoral muscles from the mammogram images. The value of K defines the image is divided into K

numbers of the region. In the proposed approach, the digital mammogram image is dividing into three clusters. The region of the first cluster contains pectoral tissue, second contains the breast lobules and third contains fatty tissue and ligaments [7]. The algorithm for k-means clustering algorithm is as below:

1. Calculate the distribution of the intensities in an image
2. Calculate the centroid of each k-random intensities
3. Repeat step 2 until cluster label finish

Calculate the distance of the intensity of each point from the centroid intensity and make a cluster of nearest intensity pixel.

$$C^i = \arg_j, \min \left\| x^{(i)} - \mu_j \right\|^2 \tag{1}$$

Recomputed the centroid for newly created clusters

$$\mu_i = \frac{\sum_{i=1}^{k} \left\{ C(j)^j \right\} x^{(i)}}{\sum_{i=1}^{k} \left\{ C(i) = j \right\}}, \tag{2}$$

where, K is a number of the cluster formed, i is the iteration performed over the all intensities and j iterate over the all centroid intensities μ_i.

4 Proposed System

The proposed block diagram is as shown in Fig. 2.

In the proposed system, three existing algorithms (region growing, thresholding, and k-means clustering) are implemented. By analyzing the segmented results of the existing three algorithm (region growing, thresholding, and k-means), it is observed that region growing algorithms give good acceptance rate but it fails for some of the images of MIAS database. But the other two algorithms give better results for the non-accepted image of region growing algorithms. Hence to select the proper segmentation algorithm, the machine learning-based algorithm is developed. The detailed steps of the proposed algorithm are shown in Fig. 2.

The process flow of proposed pectoral muscle removal algorithm is divided into training and testing part. Each step of the algorithm is explained in Fig. 2.

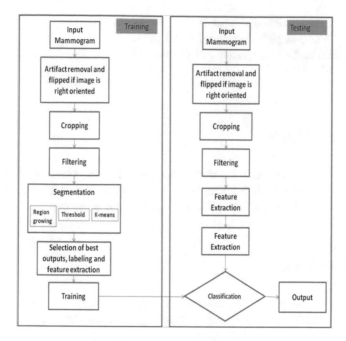

Fig. 2 Process flow of proposed automatic pectoral muscle removal algorithm

4.1   Input Images from MIAS Database

The database is created by an organization of UK research groups [10]. It consists of 320 mammogram images of right and left breast. Out of 320, 208 are normal, 63 images are benign and 51 images are malignant.

4.2   Artifacts and Label Removal

The input mammogram images contain labels, high and low-intensity artifacts. To remove the artifacts from the mammogram image the detailed algorithm is shown below:

   Algorithm: label and artifact removal from digital mammogram images
   Input: digital mammogram image

1. Apply median filter
2. Apply Otsu's global Thresholding Method and binarize to get multiple blobs
3. Select the largest blob and reject another

4. Apply morphological operation to get proper object
5. Select the largest blob area and discard the other
6. Labels and artifacts are successfully removed

   End

4.3   Image Flipping

The images of the mammogram are captured from both the breast hence images are left and right oriented both. For simplification of the process, the images need to be oriented in one side. For this the flipping operation is necessary.

4.4   Segmentation

In the proposed approach three segmentation techniques are used, i.e., region growing, thresholding, and k-mean clustering. The detailed explanation of the segmentation algorithm is given in Sect. 3.

4.5   Feature Extraction

The image can be described in the form of features. The texture of the pectoral muscles and the breast region is different hence it can be well defined by the texture features. In this approach, 14 features are extracted using Gray Level Co-occurrence Matrix [11] (Table 1).

4.6   Classification

Support vector machine is a supervised machine learning algorithm [12]. 250 images from the MIAS database are selected for training and 56 images are selected for testing. In the labeling process, the outputs of the three segmentation algorithm are analyzed. The label is assigned according to the best output provided by the respective segmentation method. SVM classifier classifies the image using linear function as

$$f(x) = W^T X + b, \tag{3}$$

Table 1 GLCM features and formulas

S. no.	Features	Formulas
1	Contrast	$\text{Contrast} = \sum_{i,j} \lvert i - j \rvert^2 p(i,j)$
2	Homogeneity	$\text{Homogeneity} = \sum_{i,j} \frac{1}{1+(i-j)^2} p(i,j)$
3	Correlation	$\text{Correlation} = \sum_{i,j} \frac{(i,j)p(i,j)-\mu_x\mu_y}{\sigma_x\sigma_y}$
4	Sum of average	$\text{Sum of Average} = \sum_{i=0}^{2(N-1)} i * p_{x+y}(i)$
5	Sum of variance	$\text{SOV} = \sum_{i,j}(1-\mu)^2 P(i,j)$
6	Standard deviation	$\text{SD} = \sqrt{\sum_{i,j=0}^{N-1} P(i,j)(i-\mu_i)^2}$
7	Autocorrelation	$\text{Autocorr} = \sum_{i,j} p(i,j)/log(p(i,j))$
8	Dissimilarity	$\text{Dissimilarity} = \sum_{i,j} \lvert i-j \rvert p(i,j)$
9	Energy	$\text{Energy} = \sum_{i,j} p(i,j)^2$
10	Entropy	$\text{Entropy} = -\sum_{i,j} p(i,j)\log(p(i,j))$
11	Difference of variance	$\text{Diff var} = \left(\sqrt{\frac{\sum \lvert p(i,j)-p(i,j)^2 \rvert}{n}}\right)^2$
12	Difference of entropy	$\text{Diff Entropy} = \sum_{i,j} \lvert i-j \rvert^2 p(i,j)$
13	INV	$\text{INV} = -\sum_{i,j} p(i,j)/(1+\lvert (p(i,j)) \rvert)$
14	INN	$\text{INN} = \sum_{i,j} p(i,j)/\left(1+\frac{\lvert p(i,j) \rvert}{i,j}\right)$

where,

X is the training samples
W is the weight assigned
b is bias or offset.

The linear SVM classifier is worthwhile for the nonlinear classifier to map the input pattern into higher dimensional feature space. The data which can be linearly separable can be examined using hyperplane and the data which is linearly non-separable those data are examined methodically with kernel function like higher order polynomial. SVM classification algorithm is based on different kernel methods, i.e., Radial basic function (RBF), linear and quadratic kernel function. The RBF kernel is applied on two samples x and x′, which indicate as feature vectors in some input space and it can be defined as,

$$K(\mathrm{x}, \mathrm{x}') = \exp\left(\frac{||\mathrm{x} - \mathrm{x}'||^2}{2\sigma^2}\right) \tag{4}$$

The value of K varies according to distances. The training data features with its label are provided to the SVM classifier to create a training model and the testing features are a test of the trained model.

5 Result

The results of the proposed system are presented in a Qualitative and quantitative way.

5.1 Qualitative Analysis

The output of the segmentation results for each segmentation algorithms is shown in Fig. 3.

From the above qualitative analysis, it is observed that for the Mdb180 image the region growing and k-means clustering algorithm shows good segmentation results but thresholding algorithm fails. Likewise, for the Mdb208 image, region growing algorithm fails to correctly segment the pectoral muscle while other two shows good results. Hence to select the proper segmentation algorithm, machine learning algorithm based on the GLCM texture feature is introduced in the proposed algorithm.

5.2 Quantitative Analysis

The performance of the proposed system is evaluated on the basis of acceptance parameter. The acceptance accuracy of each segmentation algorithm and the proposed algorithm is tabulated in Table 2.

The graphical analyses of the different methods for segmentation of the pectoral muscles are as shown in Fig. 4.

From the graphical analysis, it is observed that the proposed algorithm is superior to other algorithms.

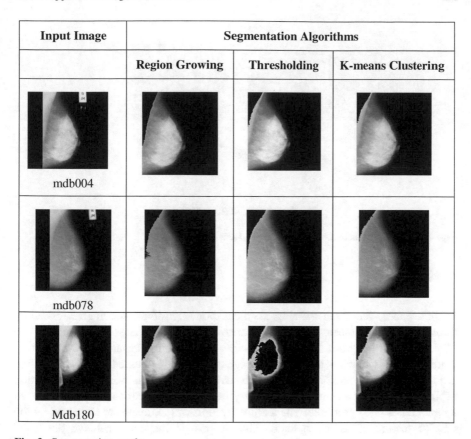

Fig. 3 Segmentation results

Table 2 Quantitative analysis

S. no.	Algorithms	Total no. of images	Acceptance	Accuracy (%)
1	Region growing	318	272	85.53
2	Thresholding	318	231	72.64
3	K-means clustering	318	212	66.67
4	SWA and DRGA [6]	322	294	91.30
5	Intensity based [5]	320	279	87.19

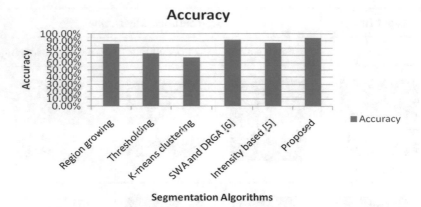

Fig. 4 Comparative analysis of performance of the segmentation of pectoral muscles

6 Conclusion

In this paper, the three existing segmentation algorithm viz. region growing, thresholding and k-means clustering algorithm used to segment the pectoral muscle of digital mammogram images. From the segmentation results of the existing algorithms, it is concluded that some images properly segmented by one algorithm, did not segment by another algorithm. To overcome this disadvantage the selection of an algorithm for an image is necessary hence in this approach, the SVM based approach has been proposed. The texture features of the image are extracted by the GLCM feature extraction method. The proposed algorithms achieve the good acceptance accuracy of 93.71% on the mini-MIAS database.

References

1. http://www.breastcancer.org/
2. Sreedevi, S., Sherly, Elizabeth: A novel approach for removal of pectoral muscles in digital mammogram. Int. Conf. Inf. Commun. Technol. (ICICT), Procedia Comput. Sci. **46**, 1724–1731 (2015)
3. Sucklin, J.: The mammographic image analysis society digital mammogram database exerpta medica. Int. Congr. Ser. **1069**, 375–378 (1994)
4. Lakshmanan, R., Thomas, S.T.P.V., Jacob, S.M., Pratab, T.: Pectoral muscle boundary detection-A preprocessing method for early breast cancer detection. In: World automation congress (WAC), Waikoloa, HI, pp. 258–263 (2014)
5. Vikhe, P.S., Thool, V.R.: Intensity-based automatic boundary identification of pectoral muscle in mammograms. In: The 7th international conference on communication, computing and virtualization, pp. 262–269 (2016)

6. Shrivastava, A., Chaudhary, A., Kulshreshtha, D., Prakash Singh, V., Srivastava, R.: Automated digital mammogram segmentation using dispersed region growing and sliding window algorithm. In: 2nd international conference on image, vision and computing (ICIVC), Chengdu, pp. 366–370 (2017)
7. Alam, N., Islam, M.J.: Pectoral muscle elimination on mammogram using K-means clustering approach. Int. J. Comput. Vision Sig. Process. 4(1), 11–21 (2014)
8. Brzakovic, D., Luo, X., Brzakovic, P.: An approach to automated detection of tumors in mammograms. IEEE Trans. Med. Imag. 9(3), 233–241 (1990)
9. Kamdi, S., Krishna, R.K.: Image segmentation and region growing algorithm. Int. J. Comput. Technol. Electron. Eng. (IJCTEE) 2(1), 103–107 (2012)
10. http://www.mammoimage.org/databases/
11. Albregtsen, F.: Statistical texture measures computed from gray-level co-occurrence matrices. In: Image Processing Laboratory Department of Informatics University of Oslo November 5, (2008)
12. Vapnik, V.: The nature of statistical learning theory. Springer, N.Y. (1995). ISBN 0-387-94559-8

Performance Analysis of TCP Variants Using Routing Protocols of MANET in Grid Topology

Namita Das, Sukant Kishoro Bisoy and Sanjukta Tanty

Abstract TCP is a transport layer protocol used for reliable transmission of data packets from one end to other. One of the important functionality of TCP is to control congestion in the network. Congestion can be controlled through the window based mechanism of TCP. TCP comes with many variants Newreno, Vegas and FullTCP (two-way TCP). In this, TCP variants such as Newreno and Vegas including FullTCP protocol is analyzed using ad hoc on demand distance vector (AODV), dynamic source routing (DSR) and destination sequenced distance vector (DSDV) routing protocols using grid topology. The simulation result using NS2 shows that performance of FullTCP protocol is better than Newreno and Vegas irrespective routing protocol used.

Keywords AODV · DSDV · DSR · Newreno · Vegas · FullTCP

1 Introduction

Mobile ad hoc network (MANET) comprised of random network by consisting of mobile nodes which communicates with each other over wireless path. Ad hoc network is more appropriate for a scenario where we do not have existing networking infrastructure. In this network, every device acts as end nodes as well as a router. In order to route the packets from one end to other many routing protocols

N. Das (✉)
Department of Computer Science and Engineering, Temple City Institute
of Technology and Engineering, Bhubaneswar, Odisha, India
e-mail: nmt.das@gmail.com

S. K. Bisoy
Department of Computer Science and Engineering, C.V. Raman College
of Engineering, Bhubaneswar 752054, Odisha, India

S. Tanty
Department of Computer Science and Engineering, DAMITS, Rourkela,
Odisha, India

© Springer Nature Singapore Pte Ltd. 2019
P. K. Mallick et al. (eds.), *Cognitive Informatics and Soft Computing*,
Advances in Intelligent Systems and Computing 768,
https://doi.org/10.1007/978-981-13-0617-4_23

available in MANET such as AODV [1], DSR [2] and DSDV [3] which are standardized by IETF MANET working groups [4]. AODV and DSR are two reactive routing protocols. Destination sequenced distance vector (DSDV) [3] is proactive routing protocol. Data can be transferred reliably using TCP [5] protocol. It can work well when the network scenario is combination of both wired and wireless. Many variants of TCP protocols are Tahoe, Reno, Newreno, Sack, Vegas. These protocols were proposed to rectify the congestion control mechanism. In last two decades several works has done on both TCP and its variants using routing protocols of MANET. Authors in [6] analyzed the performance of routing protocol using TCP as transport layer protocol.

In this work, we analyze the performance of TCP variants with FullTCP in terms of packet delivery ratio, normalized routing load, packet loss to explore suitable TCP variants for the ad hoc network environments.

The remaining portion of the paper is organized as follows. In Sect. 2 the work related our paper is presented. Section 3 explains TCP variants. Section 4 gives description of simulation set up used for the work and Sect. 5 gives performance analysis. At last the conclusion of the work is given in Sect. 6.

2 Existing Work

As the delayed acknowledgement technique has significant impact on the performance of TCP, the author in [7] analyzed the impact of delayed acknowledgement strategy on TCP variant protocols. The result indicates that Vegas protocol achieves better performance than Newreno. The Intra-protocol fairness between Vegas and Reno is analyzed in [8] using some existing routing protocols. Then, impact of delayed acknowledgement technique on fairness property of TCP variants is analyzed. The result demonstrates that fairness is better using destination sequenced distance vector (DSDV) protocols. The observation indicates that use of delayed acknowledgement technique enhances the fairness. In [9], the author analyzed the interaction between routing protocols of MANET and TCP variants and results suggested that optimized link state routing (OLSR) achieve better performance than destination sequence distance vector (DSDV) and ad hoc on demand distance vector (AODV).

3 TCP Variants

One of the main functions of TCP protocol is to deliver the data reliably over network layer. Therefore, many research work is carried out on TCP and its variants [10, 11]. In [12], the author analyzed the TCP variants for a network scenario where wired and wireless network is combined using active queue management techniques.

3.1 Newreno [13]

The Newreno has been developed with minor change in previous version Reno. In this fast recovery algorithm is modified and improved to recover multiple loss of packets in a single window and to minimize transmission timeout events [14]. It is more appropriate in the MANET environment where packet loss is huge.

3.2 Vegas

Vegas [15] adopt different mechanism in controlling the congestion. It detects congestion before any loss occurs based on the measured RTT values of the each data packets sent. It depends greatly on accuracy of estimating the available bandwidth. Based on the measured value the window size is updated to control the flow rate of source.

3.3 FullTCP

FullTCP is two-way TCP agent and currently it is only implemented with Reno congestion control. It adopts differs mechanism than others in the following ways:

- connections may be established and town down (SYN/FIN packets are exchanged).
- The data transfer is two-way (bidirectional).
- Bytes is used in sequence numbers (rather packets).

4 Simulation Set up

We study the performance of Newreno, Vegas with FullTCP using some routing protocols namely AODV, DSR, and DSDV creating a grid topology. We have considered different network sizes and evaluated the performance using NS2 network simulator [16].

Simulation tool and Parameter

In this, the performance Newreno, Vegas and FullTCP(Two-way TCP) is analyzed using routing protocols DSR, DSDV, AODV. We arranged all the nodes in a grid topology. The snapshot of the topology for 25 nodes is shown in Fig. 1. Table 1 shows the values of the parameters used in this simulation.

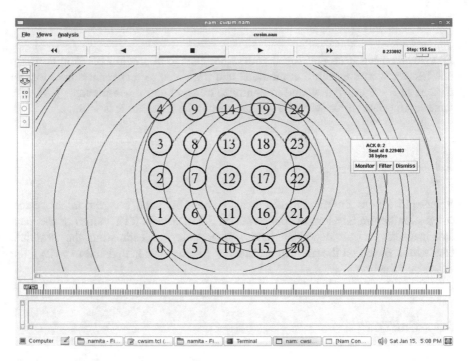

Fig. 1 Snapshot of the grid topology with 25 nodes

Table 1 Parameter values

Parameter	Value
Channel type	Wireless channel
MAC	802.11
Routing protocol	AODV, DSDV, DSR
TCP variants	Newreno, Vegas and FullTCP
Number of nodes	4, 9, 16, 25
Size of interface queue	50
Packet type and size	FTP and 512 bytes
Topography	1000 m × 1000 m
Time of simulation	20 s

5   Performance Analysis

The performance of the protocol is analyzed using packet delivery ratio (PDR), normalized routing load (NRL) and packet loss. The performance of the protocol is shown in Table 2. As the table value depicts the PDR of FullTCP is higher than Newreno and Vegas irrespective of the routing protocols are used. However, it is better at DSDV protocols.

Table 2 Packet delivery ratio (PDR) of NewReno, Vegas and FullTCP

Grid topology	Newreno			Vegas			FullTCP		
PDR (%)	AODV	DSDV	DSR	AODV	DSDV	DSR	AODV	DSDV	DSR
4 node	98.3	98.3	98.3	99.7	99.7	99.7	100	100	100
9 node	92	96.6	67.9	98.2	98.9	92.9	97.2	100	100
16 node	87.7	95	48.8	95.3	96.8	52.7	95.3	100	98.4
25 node	85.3	49.2	38.5	92.3	93.8	31.5	99	100	99

Table 3 Normalized routing load (NRL) of NewReno, Vegas and FullTCP

Grid topology	Newreno			Vegas			FullTCP		
NRL	AODV	DSDV	DSR	AODV	DSDV	DSR	AODV	DSDV	DSR
4 node	0.0032	0.0072	0.0030	0.0046	0.0073	0.0024	2	2.125	0.5
9 node	0.183	0.019	0.629	0.172	0.016	0.2	0.742	2.611	0.722
16 node	0.773	0.036	11.363	0.758	0.037	10.68	12.78	2.656	0.777
25 node	2.278	0.060	13.954	1.756	0.058	10.5	18.21	2.52	0.828

Then normalizing routing load (NRL) of these protocols are measured in percentage (%) and shown in Table 3. Normalizing routing load is defined as the ratio of total number of routing packets transmitted divided by total number of data packet received. As the table value suggest, the NRL of Vegas of is lower than others. It is achieved using DSDV routing protocols.

Then packet loss of each TCP variants is measured for different routing protocols mentioned earlier and shown in Figs. 2, 3 and 4. The packet loss for AODV routing protocol is shown in Fig. 2. It is observed that the packet loss of FullTCP protocol is quite lower than others. As compared to Vegas and FullTCP, Newreno losses more packets using AODV protocol. Moreover, the packet loss of each is increasing with the number of nodes.

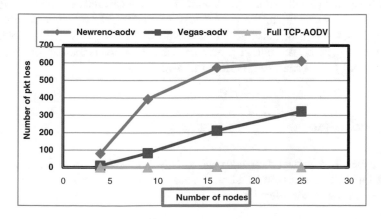

Fig. 2 Packet loss for AODV

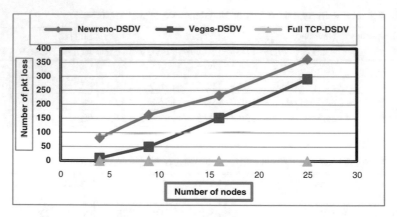

Fig. 3 Packet loss for DSDV

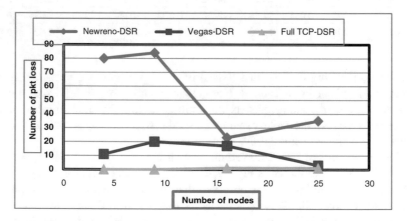

Fig. 4 Packet loss for DSR

Similarly, the packet loss of DSDV and DSR are measured and shown in Figs. 3 and 4. As shown in figure the packet loss of FullTCP lower than others. From Figs. 2, 3 and 4 it is demonstrated that the packet loss of FullTCP is lower than other irrespective of the routing protocols used. Moreover, the packet loss of all protocols is lower while using DSR routing protocols.

6 Conclusions

In this paper, TCP variant protocols such as Newreno and Vegas with two-way TCP named FullTCP is analyzed in wireless network creating a grid topology. For the analysis, we have considered three routing protocols such as AODV, DSR and DSDV.

The results using NS2 network simulator shows that FullTCP performs better than Newreno and Vegas with respect to packet delivery ratio and packet loss irrespective of routing protocol used. Moreover, the performance is better while using DSDV protocols.

References

1. Perkins, C.E., Belding-Royer, E., Das, S.R..: Ad-hoc on-demand distance vector (AODV) routing, IETF RFC 3561 (2003)
2. David, B., Johnson David, A., Maltz Yih-Chun, Hu.: The Dynamic Source Routing for Mobile Ad Hoc Networks. [online] Available: http://www.ietf.org/internet
3. Parkins, C.E., Bhagwat, P.: Highly dynamic destination sequence distance vector routing (DSDV) for mobile computers. In: Proc of ACM SIGCOMM'94, London, UK (1994)
4. Internet engineering task force.: Manet working group charter, http://www.ietf.org/html. charters/manet-charter.html
5. Postel, J.: Transmission control protocol. RFC 793 (1980)
6. Papanastasiou, S., Ould-Khaoua, M.: Exploring the performance of TCP Vegas in Mobile Ad hoc Networks. Int. J. Commun. Syst. 17(2), 163–177 (2004)
7. Bisoy, S.K., Das, A., Pattnaik, P. K., Panda, M.R.: The impact of delayed ACK on TCP variants protocols in wireless network. In: proceedings of IEEE international conference on high performance computing and application (ICHPCA), pp. 1–6, 22–24 December (2014)
8. Bisoy, S.K., Pattnaik, P.K.: Throughput of a network shared by TCP reno and TCP vegas in static multi-hop wireless network. In: proceedings of Springer ICCIDM, vol. 1, pp. 471–481, December (2015)
9. Bisoy, S.K., Pattnaik, P.K.: Interaction between internet based TCP variants and routing protocols in MANET. In: international conference on frontiers of intelligent computing: theory and applications (FICTA), pp. 423–433 (2013)
10. Jacobson, V.: Congestion avoidance and control. Comput. Commun. Rev. 18(4), 314–329 (1988)
11. Jacobson, V.: Modified TCP Congestion Avoidance Algorithm. Technical report (1990)
12. Hoe, J.: Start-up dynamics of TCP's congestion control and avoidance scheme, Master's thesis, MIT (1995)
13. Bisoy, S.K., Pati, B., Panigrahi C.R., Pattnaik, P.K.: Analysis of TCP variant protocol using active queue management techniques in Wired-Cum-Wireless networks. In: Proceeding of springer ICCIDM 2017, pp. 439–448 (2017)
14. Floyd, S., Henderson, T., Gurtov, A.: The new reno modification to TCP's fast recovery algorithm. RFC 3782 (2004)
15. Brakmo, L., O'Malley, S., Peterson, L.: TCP vegas: new techniques for congestion detection and avoidance. In: Proceeding of ACM SIGCOMM, pp. 24–35, New York, USA (1994)
16. Information Sciences Institute, The Network Simulator NS-2, http://www.isi.edu/nanam/ns/

An Ensemble Feature Selection Framework of Sonar Targets Using Symmetrical Uncertainty and Multi-Layer Perceptron (SU-MLP)

Sai Prasad Potharaju, M. Sreedevi and Shanmuk Srinivas Amiripalli

Abstract In Data Mining applications, feature selection methods has become popular area of investigation. It has become priority strategy to get significant knowledge about any chosen domain. In literature many ways are existed for feature selection (Filter, Wrapper, Embedded). With the availability of large characters of data, the feature selection has become necessary in the process of Data Mining. In this article, we presented and examined a new feature selection technique on the basis of Symmetrical Uncertainty (SU) and Multi-Layer Perceptron (MLP), then we estimated the performance of rule-based (Jrip), tree-based (J48), lazy (KNN) algorithm with ensembling techniques such as boosting, bagging on SONAR (sound navigation and ranging) data. In our proposed framework, we divided the most prominent features derived by SU into finite number of clusters. Each cluster formed by our technique has unique features. As an initial step, MLP is applied on all clusters and decided the foremost cluster as per classifier accuracy. The finest cluster of features are examined with Jrip, J48, KNN with ensembling approaches and compared with existing feature selection techniques Gain Ratio Attribute (GR), Information Gain(IG), Chi Squared Feature selection (CHI). Cluster of features originated by the proposed framework has recorded improvement with the most of the classifiers than traditional methods.

Keywords Bagging · Boosting · Multi-layer perceptron · Symmetrical uncertainty · SMOTE

S. P. Potharaju (✉) · M. Sreedevi · S. S. Amiripalli
Department of Computer Science and Engineering, KL University, Guntur 522502, Andhra Pradesh, India
e-mail: psaiprasadcse@gmail.com

M. Sreedevi
e-mail: msreedevi_27@kluniversity.in

S. S. Amiripalli
e-mail: shanmuk39@gmail.com

© Springer Nature Singapore Pte Ltd. 2019
P. K. Mallick et al. (eds.), *Cognitive Informatics and Soft Computing*,
Advances in Intelligent Systems and Computing 768,
https://doi.org/10.1007/978-981-13-0617-4_24

1 Introduction

Since few years, all business organizations, educational institutes, manufacturing agencies are getting the assistance of data mining. The generation of data sets has been rapidly increasing by the various sections of organizations. In such cases, data mining stages like preprocessing, feature selection, classification plays vital role. Data Mining involved in several disciplines to get the unexplored intelligent patterns from native dataset. Recently, with the promotions in technology in several disciplines such as health care, marketing, e-commerce, bio-medicine, finance, manufacturing and many, size of the data increasing in horizontally (records), and vertically (attributes).

The extraction of desired knowledge from the gigantic amount of data is an arduous job [1]. One of the important subject for data analysis is availability of more number of features associated with any real data set, i.e., high dimensionality. This directs to a diverse disadvantages such as performance degradation, high processing time, consumption of more amount of memory during the training phase of classifier. There are two important methods available in literature. First approach is transforming the actual feature space to a new lower dimensional space. Second approach is nominating a strong subset of features from original features, so that irrelevant and redundant features can be removed, this is called feature selection.

In occurrence of high dimensionality problems, feature selection has proved as an effective technique which boosts up the accuracy and also procure the fast predictors. But, selection of most suitable method for a given task is critical job for any given domain. In general, three categories of feature selection approaches are existed. Those are: Filter, Wrapper, and Embedded [2]. In filter method, ranking technique is used as criterion function to select an appropriate number of features. An effective ranking criterion is used to select the features and a threshold is used for discarding the features which has the less threshold value. Ranking to a variable can be given using several techniques, which includes Chi square, Symmetric Uncertainty(SU) and Information Gain. Wrapper mode uses the predictive models such as Genetic Algorithms, PSO, Naive Bayes to acquire the performance by the target algorithms [3]. Embedded mode amalgamates filter and wrapper methods.

In this current study, selected features with the proposed technique are investigated with rule. Tree, lazy based algorithms with an ensembling techniques. Ensembling techniques can increase the classification accuracy by taking the advantages or strengths of chosen classifier, or it may reduce the classification performance if weak classifier is chosen. Following are commonly used ensembling techniques which are employed for the current study.

Bagging: It stands for Bootstrap Aggregation. It is the mechanism for diminishing the prediction variance. In order to diminish the variance prediction, multiple sample data sets will be produced from original data set using random amalgamations with repetitions for training.

Boosting: It is a double-step mechanism; first step uses subsets of the original data to generate multiple averagely performing models and then accelerates (boosts) their performance by joining them together using a particular cost function. It will be repeated until the satisfied model obtained.

Our aim for the current study is to present a novel framework for deriving the best subset of features of SONAR data set. Sonar system is used to test the underwater waves which are coming from different sides. Those waves are detected by a sensor and classifies the target of side as either Rock or Mine. For this study, we mainly considered SMOTE, which is an over sampling method to address the imbalanced issue, SU for deriving the top features, MLP to measure the best subset.

2 Literature Review

Feature selection methods are assembled into supervised, unsupervised and semi-supervised according to the data set is labeled or not. In the procedure of feature selection, an important operation is, on what basis an individual feature is clearly differentiated. For the differentiating the features, there have been various methods introduced, in which CHI, SU, IG are popular [4].

In micro array gene expression data, mutual information gain is used for feature selection. Many researchers employed diverse logic in order to select the best features, e.g., Sequential random K-Nearest neighbor is considered for selection of best features from high dimensional data [5]. FAST method is introduced for feature selection of high characters data set, in which minimum spanning tree clustering method is employed for feature selection [6]. Feature unionization technique is proposed for dimension reduction. In unionization technique, instead of reducing the features, multiple features are combined for better result. SU is used for selection of best features of data set [7].

Feature selection is difficult job for analyzing high-dimensional data sets which is generating from the various application domains such as image processing, text processing, bio-informatics, etc. Many researches have drawn strong and weak points of various feature selection methods, but the best methods depends on the problem it could be applied. Depending on the problem at hand, various techniques are continuously coming up with different strategies. Examples of such strategies includes, ensembling the output of various feature selectors [8], merging feature selection with feature extraction [9]. Most of the available research focus on specific application area, such as genomic analysis, image classification, text classification, software defect prediction.

There have been number of studies conducted on a synthetic data to calculate the accuracy of selection methods. The evaluation of selection methods conducted under various circumstances, e.g., class imbalance, noise consideration, redundancy of features. In case of imbalance class label, SMOTE is applied to balance the class labels. Performance of classification algorithm is increased by applying SMOTE on Kidney disease data sets [10]. Ensembling approaches also applied to improve the prediction of kidney disease [11]. For the present study also, we employed SMOTE to address the class imbalance problem. A rotation forest ensemble decision tree algorithm is applied and it is wrapped with best search technique for feature selection [12].

Data integration is familiar to hike the accuracy of sonar signal classification. For classification of underwater mines, Gaussian mixture model (GMM) is introduced to enhance the classification outcome [13]. For the underwater passive sonar signals classification, four categories of neural network classifiers have been applied. Those classifiers are Probabilistic Neural Network (PNN), Adaptive Kernel Classifier (AKC), Learning Vector Quantization (LVQ), and Multilayer Perceptron (MLP) [14].

3   Proposed Methodology

Proposed framework is on the basis of two constituent elements: 1. Symmetric Uncertainty (SU) 2. Multi-Layer Perceptron (MLP).

SU
Symmetrical Uncertainty is a statistical estimation, it draws the SU score of a feature then assigns the rank (position) to every feature. SU score with maximum feature has top position and less score has least position. For feature selection, top positioned features can be chosen as best features depending on the need, and kind of the problem on which it is being applied. It can be defined as below.

$$SU = 2 * IG/(H(F_1) + H(F_2)),$$

where IG is Information Gain; $H(F_1)$ is Entropy of F_1 ...;$H(F_2)$ is Entropy of F_2.

SU score will be in the boundary of [0, 1]. SU score 1 indicates one feature can predict entirely others, 0 shows two features are uncorrelated. For introduced framework, feature which has SU score 0 is ignored as it cannot guide the learning model.

MLP
MLP is a most popular classifier used in various applications which is based on neurons and perceptrons. As our framework divides the initial feature space into set of clusters, to decide the strong cluster, as an initial step MLP is applied on each cluster.

Based on these two elements, the proposed method is elucidated below.

Algorithm
Input: D, C, TF, ListTF
D: Balanced data set, C:# clusters to be formed, TF: Total # features (SU score > 0)
ListTF : List of features whose SU score > 0
Output: F = {$f_1, f_2, \ldots f_n$} (Reduced Feature set)

1. Test the class imbalance of the original data set, if found imbalanced, then employ the SMOTE and find the balanced data set (D).
2. Apply SU on D and find out TF, arrange all features in ListTF in decreasing order of its position (Position 1 has high priority).
3. Form the F, such that.

    a. Order the first/next "C" features from ListTF in left to right direction, such that first feature would be pushed in first cluster, second feature would be in second cluster and so on. Get next 'C' feature from ListTF for next iteration.
    b. Order the next "C" number of features from ListTF in right to left direction, such that first feature would be pushed in last cluster, second feature would be in second last cluster and so on. Get next "C" number of feature from ListTF for next iteration.

4. Repeat step 3 (a) and 3 (b) until all features are organized.
5. Merge all vertically first-order features into first cluster (F_1), second-order features into second cluster (F_2), and so on till last cluster (F_n).
6. If all clusters are arranged with equal number of features, then stop. Otherwise, delete the last feature from the cluster which has an extra feature. (As it cannot influence the learning model).
7. Apply MLP on each cluster to know the strong cluster.
8. Derive the top "N" features by the traditional methods.(N is equal to the # features formed by proposed method).
9. Apply classifiers and measure the strength of each subset and compare the results.

Example

Assume T = 12 (total # features); TF = 10 (# features whose SU score > 0); C = 3 (# clusters); ListTF = {a1, a2, a3, a4, a5, a6, a7, a8, a9, a10}

As per the algorithm proposed, features in each cluster will be positioned as per Table 1.

Table 1 Cluster of features

First order (C1)	Second order (C2)	Third order (C3)	Direction
a1	a2	a3	Left to right
a6	a5	a4	Right to left
a7	a8	a9	Left to right
		a10	Right to left

Cluster 1 a1, a6, a7; *Cluster 2* a2, a5, a8; *Cluster 3* a3, a4, a9
Note a10 will be ignored as per step 6

4   Experiment

Implementation

The proposed framework is applied on SONAR data set which is gathered from UCI machine learning repository. Initial data set has 60 features, 2 classes [Rock-R and Mine (metal cylinder)], 208 instances. Rock has 97 instances and Mine has 111 instances. For better accuracy, these data set need to be balanced. So, SMOTE is applied on the data set and balanced it. As, SMOTE is based on KNN, for balancing the data set $K = 5$ is considered. As a result of this, 218 instances are formed out of which Rock has 107 and Mine has 111 instances. After this preprocessing, SU is applied on balanced data set and recorded the total # features (TF) whose SU score > 0, then defined the # clusters (C) to be formed. Table 2 gives the rank of each feature derived by various traditional methods (Note: Top 12 features details are given).

If there is a requirement to select "N" best features, generally any of the filter method can be applied, and top 'N' features can be considered for classification. Depending on "N", features will be varied.

After deriving feature rank, the proposed framework is employed to form the cluster of features. Then, MLP is applied to know the strong cluster. For this implementation, we formed 2, 3, 4 clusters. Cluster of features formed by proposed method, and strong cluster as a result of MLP is given in Table 3. From the Table 2, it is clear that, there are 25 features whose SU score is >0.

Top "N" features from the traditional filter methods are extracted (Refer Table 4) and analyzed with ensembling approaches bagging and boosting. Jrip, J48, KNN classifiers are considered for ensembling. Those captured results are given in Table 5. This experiment is carried out with WEKA machine learning tool with default settings.

Table 2 SU score of each feature, rank derived by SU and other traditional methods (IG, CHI, GR)

Rank	SU score	S	I	C	G
1	0.2242	11	11	11	11
2	0.2007	12	12	12	12
3	0.1636	9	9	9	58
4	0.1518	10	10	10	44
5	0.137	13	13	13	9
6	0.1167	45	48	48	54
7	0.1153	48	49	49	45
8	0.1116	44	45	52	13
9	0.1109	49	52	51	10
10	0.1006	54	51	47	2
11	0.0983	47	47	21	28
12	0.0973	28	21	4	48

S Feature ID derived by SU, *I* Feature ID derived by IG, *C* Feature ID derived by CHI, *G* Feature ID derived by GR

Table 3 Cluster of features formed by proposed method and strong cluster

#C	Cid	#N	Features in it	Strong cluster (accuracy)
2	C21	12	11, 10, 13, 44, 49, 28, 52, 5, 21, 46, 58, 43	C21 (82.56)
	C22		12, 9, 45, 48, 54, 47, 51, 4, 36, 2, 20, 8	
3	C31	8	11, 45, 48, 28, 52, 36, 2, 43	C31 (81.65)
	C32		12, 13, 44, 47, 51, 21, 46, 8	
	C33		9, 10, 49, 54, 4, 5, 58, 20	
4	C41	6	11, 44, 49, 5, 21, 43	C41 (74.77)
	C42		12, 48, 54, 4, 36, 8	
	C43		9, 45, 47, 51, 2, 20,	
	C44		10, 13, 28, 52, 46, 58	

#C Number of Clusters, *#Cid* Cluster ID, *#N* Number of Features in each cluster

Table 4 Top "N" features derived by traditional methods (refer Table 2)

#C	Method	N	Top features in It
2	IG	12	11, 12, 9, 10, 13, 48, 49, 45, 52, 51, 47, 21
	CHI		11,12, 9, 10, 13, 48, 49, 52, 51, 47, 21, 4
	GR		11,12, 58, 44, 9, 54, 45, 13, 10, 2, 28, 48
3	IG	8	11,12, 9,10,13, 48, 49, 45
	CHI		11, 12, 9, 10, 13, 48, 49, 52
	GR		11, 12, 58, 44, 9, 54, 45, 13
4	IG	6	11, 12, 9, 10, 13, 48
	CHI		11, 12, 9, 10, 13, 48
	GR		11, 12, 58, 44, 9, 54

N Number of feature

Table 5 Classification performance of two clusters

CID	Bagging			Boosting		
	Jrip	J48	KNN	Jrip	J48	KNN
C21[a]	77.06	77.06	82.56	72.93	76.60	82.56
IG	78.44	77.52	79.81	78.89	75.68	77.52
CHI	74.77	76.60	81.19	77.98	76.60	80.27
GR	74.77	76.60	81.19	77.98	76.60	80.27

[a]Performance of the top cluster formed by proposed method

5 Results and Discussion

In this section, performance of cluster of features formed by proposed method and top "N" features derived by traditional method with ensembling approach is presented (for # clusters are 2, 3, 4 refer Tables 5, 6 and 7 respectively).

Table 6 Classification performance of three clusters

CID	Bagging			Boosting		
	Jrip	J48	KNN	Jrip	J48	KNN
C31[a]	77.98	79.35	80.37	81.65	76.60	79.81
IG	73.39	71.55	78.44	70.18	72.01	75.22
CHI	72.01	72.01	73.39	70.64	74.93	72.47
GR	76.60	74.77	78.89	71.1	73.85	79.35

[a]Performance of the top cluster formed by proposed method

Table 7 Classification performance of four clusters

CID	Bagging			Boosting		
	Jrip	J48	KNN	Jrip	J48	KNN
C41[a]	75.68	74.31	72.01	73.39	75.68	72.01
IG	72.01	72.93	73.39	71.1	69.72	71.55
CHI	72.01	72.93	73.39	71.1	69.72	71.55
GR	72.93	72.01	76.6	70.64	68.34	76.14

[a]Performance of the top cluster formed by proposed method

C21 cluster of feature formed by proposed method has performed better than traditional CHI and GR methods when Jrip, J48 is applied with Bagging. It has recorded better accuracy than all traditional methods when KNN is applied with bagging. J48 and KNN recorded boosted accuracy than all traditional methods with Boosting. Graphical representation of classification performance is given in Fig. 1.

According to Table 6, C31 cluster, displayed high accuracy than all existing methods with all classifiers when applied with bagging and boosting. Graphical representation of classification performance is given in Fig. 2.

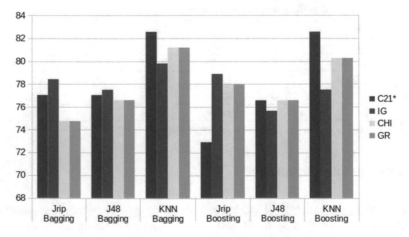

Fig. 1 Performance with two clusters

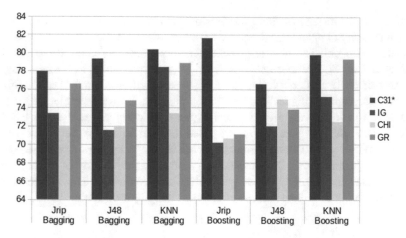

Fig. 2 Performance with three clusters

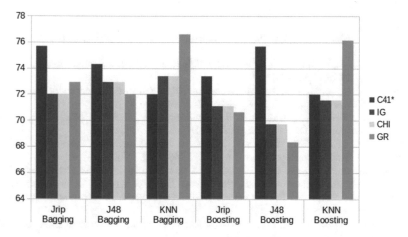

Fig. 3 Performance with four cluster

C41 cluster outperforms than all existed methods with Jrip, and J48 by bagging and boosting (refer Table 7) Graphical representation of classification performance is given in Fig. 3.

6 Conclusion

In this article, a novel cluster of feature selection framework based on Symmetrical Uncertainty (SU), and Multilayer Perceptron(MLP) was proposed. SU is used to derive the position of each feature. The new approach could generate finite clusters, in which each cluster has finite number of features without duplication. MLP is applied on each cluster to find out the strong cluster. All the cluster of features is evaluated with existing feature selection methods such as Gain Ratio Attribute Evaluator, Chi Square Feature selection, Information Gain. For evaluating the accuracy of each cluster rule based (Jrip), tree based (J48) and lazy (KNN) are applied using ensembling techniques such as Bagging, Boosting. Sonar data sets are considered to evaluate the proposed technique. After complete analysis, it has been noticed that, clusters formed by proposed method is competing with regular methods. With this framework, maximum 50, 33, 25% features can be selected if number of clusters are 2, 3, 4 respectively. However, this study is carried out on SONAR data sets. But, this can be generalized to all data sets. This method can be tested with Support vector machine to test the strength of each cluster, which is our future study.

References

1. Kumar, V., Minz, S.: Feature selection: a literature review. Smart Comput. Rev. **4**(3), 211–229 (2014)
2. Saeys, Y., Inza, I., Larrañaga, P.: A review of feature selection techniques in bioinformatics. Bioinform. **23**(19), 2507–2517 (2007)
3. Akay, M.F.: Support vector machines combined with feature selection for breast cancer diagnosis. Expert Syst. Appl. **36**(2), 3240–3247 (2009)
4. Hariri, S., Yousif, M., Qu, G.: A New Dependency and Correlation Analysis for Features. IEEE Trans. Knowl. Data Eng. **17**, 1199–1207 (2005)
5. Park, C.H., Kim, S.B.: Sequential random k-nearest neighbor feature selection for high-dimensional data. Expert Syst. Appl. **42**, 2336–2342 (2015)
6. Song, Qinbao, Ni, Jingjie, Wang, Guangtao: A fast clustering-based feature subset selection algorithm for high-dimensional data. IEEE Trans. Knowl. Data Eng. **25**, 1–14 (2013)
7. Jalilvand, A., Salim, N.: Feature unionization: a novel approach for dimension reduction. Appl. Soft Comput. **52**, 1253–1261 (2017)
8. Latkowski, T., Osowski, S.: Data mining for feature selection in gene expression autism data. Expert Syst. Appl. **42**, 864–872 (2015)
9. Bharti, K.K., Singh, P.K.: Hybrid dimension reduction by integrating feature selection with feature extraction method for text clustering. Expert Syst. Appl. **42**(6), 3105–3114 (2015)
10. Prasad Potharaju, S., Sreedevi, M.: An improved prediction of kidney disease using SMOTE'. Indian J. Sci. Technol. **9**. https://doi.org/10.17485/ijst/2016/v9i31/95634 (2016)
11. Potharaju, S.P., Sreedevi, M.: Ensembled rule based classification algorithms for predicting imbalanced kidney disease data. J. Eng. Sci. Technol. Rev. **9**, 201–207 (2016)
12. Ozcift, A., Gulten, A.: A robust multi-class feature selection strategy based on rotation forest ensemble algorithm for diagnosis of Erythemato-squamous diseases. J. Med. Syst. **36**, 941–949 (2012)
13. Kotaria, V., Changa, K.C.: Fusion and gaussian mixture based classifiers for SONAR data. In: SPIE defense, security, and sensing, pp. 80500U–80500U. International Society for Optics and Photonics (2011)
14. Chen, C.H., Lee, J.D., Lin, M.C.: Classification of underwater signals using neural networks. Tamkang J. Sci. Eng. **3**(1), 31–48 (2000)

Autonomous Path Guiding Robot for Visually Impaired People

Rajesh Kannan Megalingam, Souraj Vishnu, Vishnu Sasikumar
and Sajikumar Sreekumar

Abstract This paper introduces an intelligent and efficient path guidance robot to assist the visually impaired people in their movements. This is a novel device for the replacement of strenuous guide dogs. The robot has the capability to move along multiple paths and then remember as well as retrace all of them, thus making it a perfect substitute for a guide dog which is often a luxury for the ninety percent of blind people living in low income settings. The robot is capable to guide the user to travel places which cannot be traced using GPS. Since most of the navigation systems that are developed so far for the blind people employ a complex conjunction of positioning systems, video cameras, location mapping and image processing algorithms, we have designed an affordable low-cost prototype navigation system which serves the purpose of guiding the visually impaired people both in indoor as well as outdoor environment.

Keywords Blind guiding · Obstacle avoiding · Navigation · Retrace

1 Introduction

Data from World Health Organization surveys shows that there are around 285 million people who are visually challenged worldwide, of whom 39 million are blind and 246 million have other kinds of visual challenges and these numbers will

R. K. Megalingam (✉) · S. Vishnu · V. Sasikumar · S. Sreekumar
Department of Electronics and Communication Engineering,
Amrita University, Amritapuri, India
e-mail: rajeshkannan@ieee.org

S. Vishnu
e-mail: sourajvishnup@gmail.com

V. Sasikumar
e-mail: vishnusasikumarpp@gmail.com

S. Sreekumar
e-mail: sksreekumar94@gmail.com

© Springer Nature Singapore Pte Ltd. 2019
P. K. Mallick et al. (eds.), *Cognitive Informatics and Soft Computing*,
Advances in Intelligent Systems and Computing 768,
https://doi.org/10.1007/978-981-13-0617-4_25

rise to 75 million by the year 2020. It is not unknown to us the efforts and struggles put forth by a physically disabled person in their day-to-day life. The situation is even more complicated when it comes to a visually challenged person. It is needless to discuss the troubles they face when it comes to self-independent mobility, even if it is a small area like their backyard, neighbourhood, etc. They either depend on a cane or a guide dog.

Guide dogs need care throughout the day which means you must fit your life around their schedule as much as they do around yours. Guide dog training is a strenuous and time-consuming task. Moreover, Guide dogs only have a working life of around 8 years and then you have to start all over again with a new dog. Hence guide dogs are not a suitable choice to the 90% visually challenged people living in low income settings. A white cane though affordable does not provide adequate safety to the user. Other types of assistive devices like ETA's (Electronic Travel Aids) along with various other devices are limited in functionality to act as mere obstacle sensing and warning devices. The aim of this research paper is to implement a low-cost autonomous path guiding robot which is capable of assisting the blind and visually challenged people without the assistance of a sighted person and test the efficiency of the proposed prototype system. The paper is organized as follows: Sect. 1.2 describes the current research going on in this field. The system architecture and design approach are explained in Sect. 1.3. This is followed by the working of the robot in Sect. 1.4. Section 1.5 describes experiments, results, and discussions. Conclusion, future work and acknowledgement are presented in subsequent sections.

2 Related Work

A lot of ideas have been proposed in the field of robotics regarding travel aids for the blind people. Some of these are mentioned below. One such project is smart vision [1] which uses a camera fixed at the chest height, a portable computer and one earphone which helps the user to detect the path borders and obstacles along the path. In the intelligent guide stick [2] project, intelligent guide sticks which consist of displacement sensors were used. In the project i-path [3], a path guiding robot was developed. The robot is able to follow any particular path on the floor and is also smart enough to sense the motion of the user. In the work done on obstacle avoidance robot [4] at Amrita University, a three-wheeled self-navigating and obstacle avoiding robot was developed. The system further incorporates computer visioning using an on board computer and web camera. The project titled "High Precision Blind Navigation System Based on Haptic and Spatial Cognition" [5] used a system which combines both haptic and blind spatial perception. It is the ultra-wideband wireless positioning technology which gives them the precise positioning. The project "A Blind Guidance System Based on GCADSF Image Enhancement and Music Display" [6] address the problem of massive data, very low mapping efficiency and sound coding in two distinct way, one by introducing a system based on GCADSF image enhancement and music display. The project "Electronic Escort for the Visually Challenged" [7] uses a smart phone and a wearable RF module. The research work [8] introduces a blind cane which incorporates bluetooth module, haptic module, etc.

3 System Architecture

This prototype has been developed with emphasis of integrating advanced technologies which are economically affordable so that their benefits help to aid the visually challenged community. As shown in Fig. 1, the design of the system mainly consists of twelve parts. We used ATmega1280 Microcontroller. Matrix keypad is used as an input device through which different mode of operation is selected. A 4 × 4 matrix keypad has been used which provides the option to use 16 different keys. Joystick is used to navigate the robot. Joystick converts the X and Y axes into voltages. These values are then stored in an SD card. Arduino only has a flash memory of 256 KB which is inadequate for the robot to store data generated by the joystick. SD card helps in mass storage and data logging. The driving unit of the robot consists of a motor driver, two 12 V 45 rpm motors and four wheels. Motor driver also acts as the voltage regulator converting the 12 V received from the external rechargeable lead acid battery into a 5 V output which is in turn fed to the Arduino board. This navigational system uses sensors like ultrasonic sensors and infrared sensors. Ultrasonic sensor help in accurate measurement of distance within a range of 2 cm–3 m. Ultrasonic sensor is mounted on a servo motor in order to provide a 180° scan of the surroundings to detect the presence of obstacles. IR sensors mounted at the bottom of the robot body detects the presence of edges

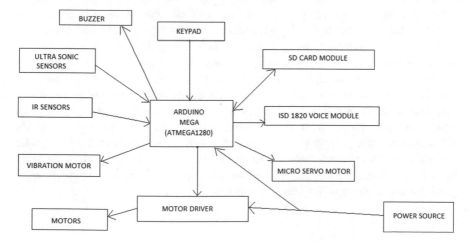

Fig. 1 System architecture

and pits below the horizontal level the robot. The audio module used is ISD 1820. Recordings are stored into on chip non-volatile memory, providing zero-power message storage. Figure 6 contains the design of the robot.

4 Working

The robot works in three basic modes. User can choose the modes with keypad which is mounted on the control panel. In the first mode, the robot can be driven through any path using the joystick attached on the control pad by a trainer. In this mode, the microcontroller processes the data from the joystick and directs the movement of the robot. Simultaneously the values are passed into an SD card where it is stored in an array format. The robot stops saving the path once the push button on the joystick is pressed. The second mode primarily focuses on the retrace of the path by retrieving the coordinates stored in the secure digital card. The values are taken one at a time from the array and passed to the microcontroller where it is processed to determine the direction of travel of the robot. Ultrasonic sensors acts as obstacle detectors while infrared sensors acts as depth and edge detectors. Ultrasonic sensors are located at a distance of 2–40 cm from the ground, each of these ultrasonic sensors can detect obstacles within a range of 20–100 cm respectively. One of the ultrasonic sensors takes a 180° sweep of the immediate surroundings.

Another useful function has been incorporated to the algorithm, which provides the user the freedom to stop in the middle of the process of path retracing and to come back to the starting point. By pressing a specific key on the keypad the user can instruct the system to go back to the starting point. As soon as the instruction reaches the system the further retrieval of values from the array ceases and then from that point in the array retrieval of values start to occur in the reverse direction till the beginning of the array. Since this retrieval of values happen in the reverse direction, the algorithm is developed in such a manner that wherever the instruction to turn left is encountered, the system is made to turn right and vice versa. As the last value stored in the SD card has been retraced and corresponding actions are executed the robots stops, by then it would have directed the user to the exact location. The robot also provides the user with an option for free navigation which can be accessed by pressing suitable key in the keypad. As mentioned in the other modes, movement of the robot is controlled by the joystick. As the robot moves the ultrasonic and edge sensors give simultaneous feedbacks to the robot. This enables the robot to avoid obstacles and edges and thus enhances the freedom of the user. Thus in this case the robot acts as an obstacle avoiding robot. The software algorithm for the working of the robot is given as follows.

Fig. 2 System processing in
learning mode

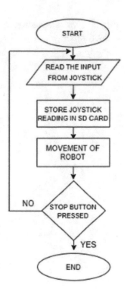

1. Start.
2. Prompt user to select the mode through keypad.
3. Obtain mode "n".
4. If n = 1, execute "learning mode".
5. If n = 2, execute "retrace mode".
6. If n = 3, execute "free mode".
7. Quit.

The robot works in three modes namely learning mode, retrace mode, and free moving mode. The working of the system when it is in either of the above three mentioned modes, i.e., learning, retrace or free mode is represented in Figs. 2, 3 and 4.

The analog joystick is made of two potentiometers, one each for the x-axis and y-axis. The input values from the joystick can vary from 0 to 1023 on both the axis. The center (x:512, y:512) is the rest position. If the joystick is moved along the x-axis from the two extreme points on the left and right, values ranging from 0 to 1023 are obtained, similar thing happens with the y-axis. Thus, the position of the joystick can be obtained from the (x, y) values. The joystick consists of a push button on the top which acts as a switch with an input either 1 or 0. The diagrammatic representation of the above description is given in Fig. 5.

5   Experiments, Results, and Discussions

The main function of the robot is to guide a visually impaired person through indoor as well as outdoor environment. The robot was tested in different scenario. The first environment was on a tar road. Tar road has rough and uneven surface.

Fig. 3 System processing in
retrace mode

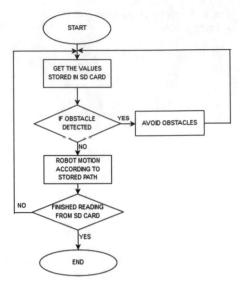

Fig. 4 System processing in
free mode

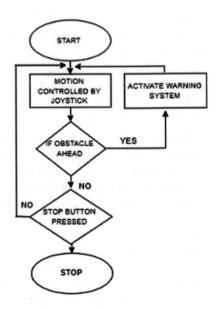

The wheels of the robot should have high amount of torque in order to overcome
the friction caused by the tar road. A user who was given a training course on the
usage of the robot was instructed to walk towards a wall from a random distance.
The user was instructed to inform as soon as he feels a vibration in his hands and
then continue walking forward till he hears a warning sound. The threshold value
that is given for the vibration of handle of the robot was the maximum range of the
ultrasonic sensor, while twice of this value was given to generate a sound in order to

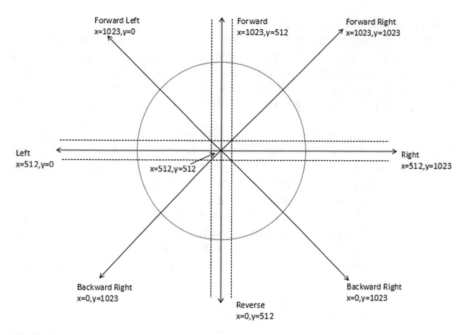

Fig. 5 Joystick position coordinates

Fig. 6 Robot design

indicate the risk of obstacles in the path of the robot. The user in this phase was able to infer the distance of obstacle from the vibratory and sound information. The user was instructed to go towards a pit which had a depth of 10 m. The robot was successful in detection of the pit. The robot stopped moving forward as soon as it detected the presence of the pit. At the end of this experiment, the user was asked to turn the robot left and right for purpose of detecting the presence of obstacles while turning.

For the second environment, the robot was tried on indoor environments like tiled floors and cement floors in the laboratory. Successful results were obtained in this experiment. In order to test the efficiency of the retracing mechanism, the robot was made to travel through a path ABCDEFG. A to B was a straight path of 50 m, after point B the robot was turned through an angle of 30° and moved 25 m to reach point C, path C to D was a semicircular path of circumference 35 m. From D the robot is moved through a straight path to reach E. At E robot is turned through an angle of 90° and moved 25 m to reach point F. From F the robot takes a left turn and travels a distance of 50 m to reach the destination G. Now the robot was switched to retrace mode and placed at point A. As expected the robot successfully moved along the path and reaches the point E. In order to test the efficiency of the sensors, an obstacle was placed along the path BC and a pit was dug along the path CD, the robot was able to detect the obstacle from a distance of 80 cm from it. Similar results were obtained when the robot encountered the pit on the path. The entire path was 60 cm broad.

To evaluate the response time of the system, two different paths were chosen. Path 1 was a 10 m long straight line path with a width of 50 cm, while path 2 was a straight line having a length of 15 m and width of 50 cm. The time taken by the robot to travel along the paths was recorded for the learning mode as well as the retracing mode. The observed values have been tabulated and shown in Tables 1 and 2 respectively. The mean response time for the learning mode in path 1 = 22.1919 s and in path 2 = 32.7878 s. The mean response time for the retrace mode in path 1 = 29.5458 s and in path 2 = 44.3501 s. Difference in response time of the two modes in path 1 = 7.3539 s and path 2 = 11.5623 s.

The deviation of the robot from a mean path was observed in the retrace mode for a number of trials and the variation in values have been plotted in the graph (Fig. 7) based on observed values given in Table 3. The analysis of the graph

Table 1 Calculation of response time in learning mode	S. No	Time (s)	
		Distance = 10 m	Distance = 15 m
	1	21.22178	31.8326
	2	19.70598	29.559
	3	20.4639	30.6958
	4	24.2527	36.379
	5	25.31525	35.47287

Table 2 Calculation of response time in retrace mode

S. No	Time (s)	
	Distance = 10 m	Distance = 15 m
1	21.22178	31.8326
2	19.70598	29.559
3	20.4639	30.6958
4	24.2527	36.379
5	25.31525	35.47287

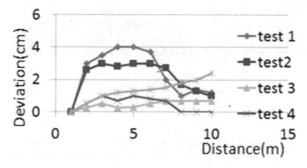

Fig. 7 Plot of distance versus deviation

Table 3 Calculation of deviation

Distance from origin in m	Deviation in cm from mean path				
	Test 1	Test 2	Test 3	Test 4	Test5
1	0	0	0	0	0
2	3	2.6	0.25	0.5	0.5
3	3.5	3	0.5	1	1
4	4	2.8	0.25	0.7	1.2
5	4	3	0.3	1	1.3
6	3.7	3	0.5	0.82	1.4
7	2	2.7	0.7	0.68	1.5
8	1	1.7	0.7	0	1.8
9	1.4	1.3	0.7	0	2
10	1.2	1	0.7	0	2.4
Mean deviation	2.38	2.11	0.46	0.47	1.31

shows the maximum deviation to be 4 cm from the mean value, the minimum deviation was observed to be 0 cm from the mean path and the maximum average deviation from the mean path was observed as 2.38 cm and the minimum average deviation was observed to be 0.46 cm from the mean path.

6 Conclusion and Future Work

Throughout our experiments and research we have tried to see how a simple technique of making a robot remember and retrace a path can be effectively applied in the field of humanitarian robotics to design a compact and user-friendly robot to assist the movements of a visually challenged person. With the use of SD card module, the memory of the system has been enhanced, so as to provide the provision to save multiple paths to the system. The safety of the user has been reinforced by the proper use of multiple sensors and simultaneous warning systems, which help the user to avoid any obstacles or similar threats along the path.

Acknowledgements We would like to thank all-powerful god for aiding us in the successful completion of our project. We recognize the help delivered to us by Amrita Vishwa Vidyapeetham and the Humanitarian Technology Lab for aiding us throughout the course of this project.

References

1. José, J., Farrajota, M., Rodrigues, J.M.F., du Buf, J.M.H.: The smart vision local navigation aid for blind and visually impaired persons. Int. J. Digital Content Technol Appl. **5**(5) 362–375 (2011)
2. Kang, S.J., Kim, Y.H., Moon, I.H.: Development of an intelligent guide-stick for the blind. In: Proceedings of the IEEE 2001 International Conference on Robotics 8, Automation, Seoul, 21–26 May 2001
3. Toha, S.F., Yusof, H.M., Razali, M.F., Halim, A.H.A.: Intelligent path guidance robot for blind person assistance. In: Informatics, Electronics and Vision (ICIEV), Fukuoka, 15–18 July 2015
4. Balasubramanian, K., Arunkumar, R., Jayachandran, J., Jayapal, V., Chundatt, B.A., Freeman, J.D.: Object recognition and obstacle avoidance robot. In: Control and Decision Conference (CCDC), 17–19 June 2009
5. Ma, J., Zheng, J.: High precision blind navigation system based on haptic and spatial cognition. In: 2nd International Conference on Image, Vision and Computing (ICIVC), Chengdu, China, 2–4 July 2017
6. Yanan, T., Hongwei, L., Jinghong, L.: A blind guidance system based on GCADSF image enhancement and music display. In: Control and Decision Conference (CCDC), Chongqing, China, 28–30 May 2017
7. Rao, S.N., Aswathy, V.R.: An electronic for the visually challenged. In: 2016 IEEE International conference on Computational Intelligence and computing research (ICCIC), 15–17 Dec 2016
8. Megalingam, R.K., Nambissan, A., Thambi, A.: Sound and touch based smart cane: better walking experience for visually challenged. In: IEEE Canada International Humanitarian Technology Conference, Montreal, Canada, 1–4 June 2014

Analysis of Diabetes for Indian Ladies Using Deep Neural Network

Saumendra Kumar Mohapatra and Mihir Narayan Mohanty

Abstract Diabetes is a common disease in recent years. Almost 50–60% of human being in the society suffer from diabetes respective of new born baby to elderly people including male and female. Due to the food habits and stressful life the disease occurs as per the suggestion of the physicians. Also it is found that for most of the cases the disease is hereditary. In this work authors have considered the Indian ladies for this disease analysis. Proper diagnosis requires accurate measurement that helps the ladies to give birth the healthy children. The data has been collected from PIMA Indian Diabetes database and are analyzed. As the database has vast the concept is based on Data Mining techniques and Big data analysis. Deep Neural Network (DNN) is used to analyze the data and disease. The result for different data set is shown in the result section. The measurement is done based on Root mean square error (RMSE).

Keywords KDD · Data mining · Healthcare data mining · Deep neural network Root mean square error

1 Introduction

Due to the improvement of computer science and data base technology, the amount of data in the data bases increases at an exponential speed. It is important to extract useful information from this large amount of data. The information extraction or knowledge extraction from the data base is known as KDD (Knowledge discovery in the data base). The KDD has different process to extract knowledge from the huge amount of data base. Data mining is a process of knowledge discovery where

S. K. Mohapatra · M. N. Mohanty (✉)
Department of Electronics and Communication Engineering,
Siksha 'O' Anusandhan University, Bhubaneswar, Odisha, India
e-mail: mihir.n.mohanty@gmail.com

S. K. Mohapatra
e-mail: saumendra.mohapatra.27@gmail.com

© Springer Nature Singapore Pte Ltd. 2019
P. K. Mallick et al. (eds.), *Cognitive Informatics and Soft Computing*,
Advances in Intelligent Systems and Computing 768,
https://doi.org/10.1007/978-981-13-0617-4_26

potentially useful and previously unknown information is extracted from large volume of data base by using different algorithms. The traditional data base technology is unable to face the problems of "excess data" and "information explosion" [1]. Nowadays different techniques are used to solve data mining task. Neural Network is one of the widely used technique which works on the concept of the Biological Neural Network. It consists of an interconnected group of Neurons and process an information that enters into it. Neural Networks have been employed in an extensive kind of supervised and unsupervised learning applications [2–5].

In the recent years most of the researchers are using a new advanced technique of neural network called DNN which can be applied in data mining, image processing, speech enhancement, pattern recognition, biomedical informatics and many more. Deep learning network is a kind of artificial neural network system with various data representation layers that learn representation by expanding the level of deliberation from one layer to another layer. The neurons of the DNN utilize all the similar activation function like unipolar sigmoid function. The performance results of the proposed method vary according to the different activation function [6].

The human brain has a massive amount of various natural neurons that all have distinctive functionalities. Thus, all organic neurons do not have the same mathematical function because they are not completely actually and organically equal. To impersonate the organic neural networks in human brain, more composite DNN with different neurons are built. That makes use of different mathematical activation functions as a substitute of having a conventional DNN. These Networks relatively with conventional ANN, additionally called shallow neural network (SNN) [7]. We can apply deep learning to lean from adequately large amount of unsupervised data. The DNN algorithms are the design of continuous layers in which each layer applies a nonlinear change on its input and gives a depiction in its output. The advantage of this algorithm is that the weights and bias are globally optimized between the neurons and the activation function [8–10].

It is more important to extract useful information from the medical databases which stores health care information like patient's record, disease details, etc. Different tools can be implemented in the medical data bases for diagnosis and prognosis of diseases [11]. Diabetes disease is a steady area of vitality to social protection bunch. In United states around 8.3% of the aggregate masses have diabetes impact and around 79 million people have been resolved to have pre-diabetes. Different techniques can be applied on the diabetes data set for the better prediction of this disease from very earlier stage. Different tools like Artificial neural network (ANN), WEKA, RST, etc., have been used for the accuracy level prediction of many diseases like cancer, heart, diabetes, etc. Different prediction level has been generated from different tools [12, 13].

The rest of the paper is organized as follows. In Sect. 2 the proposed methodology is presented. The experiment result discussion is presented on Sect. 3 the summarization and conclusion part is described on Sect. 4.

2 Methodology

Deep learning is a structure of the primary credit assignment problem which means modifying the weights that influence the neural network demonstrate the needed performance. A key property of these models is that they can separate complex factual conditions from data [8]. Both supervised and unsupervised learning are integrated together in the proposed data mining model based on deep learning. Here the input data means training data for network that have an explicit label. The learning model is adjusted until the prediction result reaches an expected accuracy. The architecture for deep learning is presented in Fig. 1.

In our purposed model we are taking the training and testing data set from the original data set which applied in MATLAB 2015 for the calculation of performance result using DNN. Figure 2 describes our proposed architecture of data mining model based on deep learning for diabetes data set.

3 Result Discussion

Experiments were led utilizing Pima Indians female diabetes data set [14]. Table 1 describes the dataset of the diabetes disease. An aggregate number of 6912 data were chosen for training and staying 768 data were utilized for testing the different DNNs.

The training data set was the collection of the all the information of diabetes diseases. It includes no. of times pregnant, Plasma glucose concentration (a 2 h in an oral glucose tolerance test), Diastolic blood pressure, Triceps skin fold thickness, serum insulin, Body mass index, Diabetes pedigree function, Age and a class variable (contain 0 for tested negative and 1 for tested positive). Figures 3, 4, 5, 6, 7, 8, 9 and 10 presents the data samples for the deep neural network.

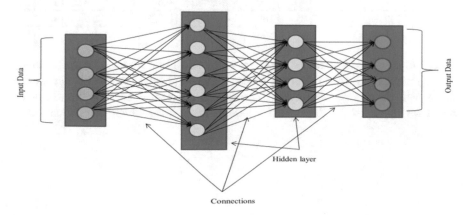

Fig. 1 Architecture of deep neural network

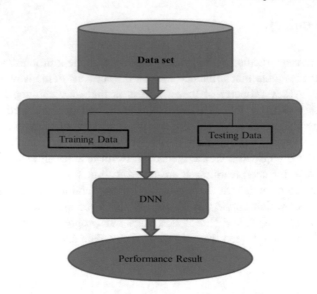

Fig. 2 Architecture of data mining model based on deep learning for diabetes data set

Table 1 Training and testing dataset for the experiment

Training data								Testing data
Pregnant	Plasma	Blood pressure	Skin	Insulin	Body mass index	Diabetes pedigree function	Age	Class
1	85	66	29	0	26.6000	0.3510	31	0
1	89	66	23	94	28.1000	0.1670	21	0
5	116	74	0	0	25.6000	0.2010	30	0
6	148	72	35	0	33.6000	0.6270	50	1
5	183	64	0	0	23.3000	0.6720	32	1
0	137	40	35	168	43.1000	0.2880	33	1

The topological structure of the DNN was critical to estimation of the accuracy of the training and testing data set. In our experiment total number of four hidden layers with four nodes are taken. To measure the performance of the estimated result we employed the Root mean square error (RMSE) which was defined as:

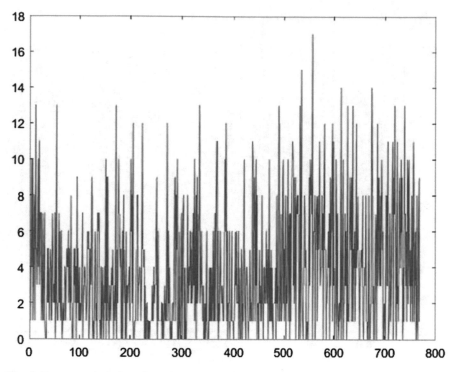

Fig. 3 Representation of number of times pregnant attribute of diabetes data set

$$\text{RMSE} = \sqrt{\frac{\sum_{i=1}^{n} (v_i - \hat{v}_i)^2}{n}},$$

where v_i is defined as estimated output and $\hat{v}_i$ is the corrected output of the ith input data for n number of data. The RMSE interprets the standard deviation which is the difference between estimated values and observed values. In our experiment we observed the root mean square value for the diabetes data set is 0.4762.

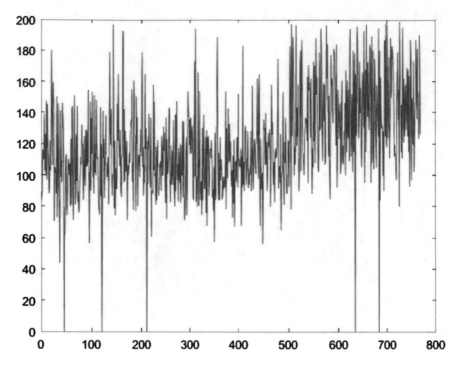

Fig. 4 Representation of plasma glucose concentration for diabetes data set

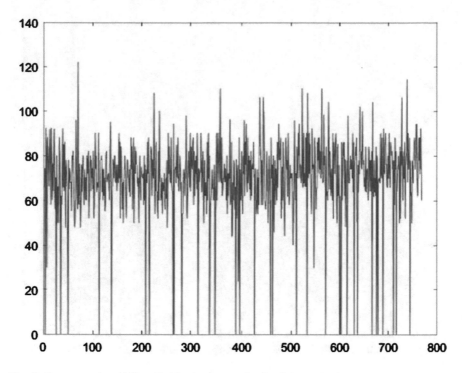

Fig. 5 Representation of diastolic blood pressure for the diabetes data set

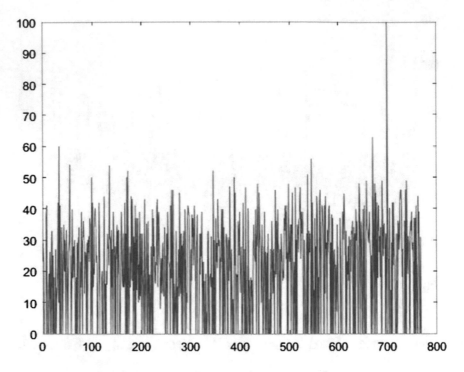

Fig. 6 Representation of triceps skin fold thickness for the diabetes data set

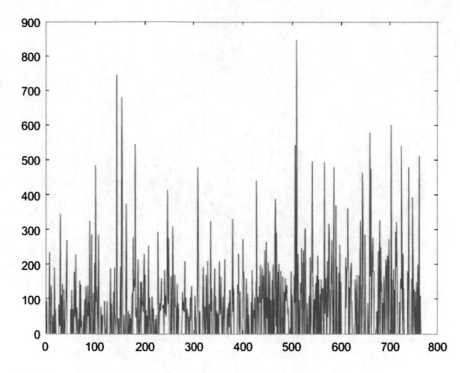

Fig. 7 Representation of 2-h serum insulin for the diabetes data set

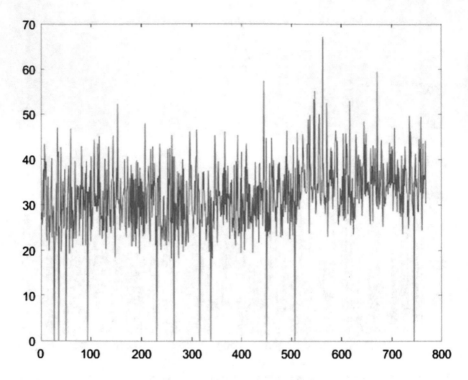

Fig. 8 Representation of body mass index for the diabetes data set

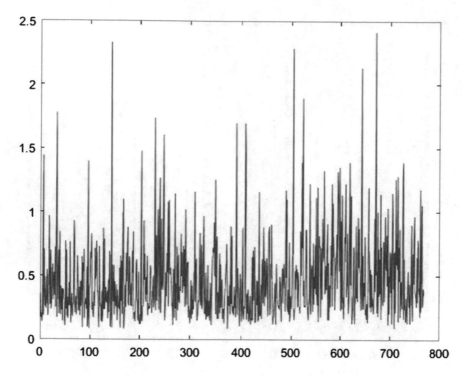

Fig. 9 Representation of diabetes pedigree function for the diabetes data set

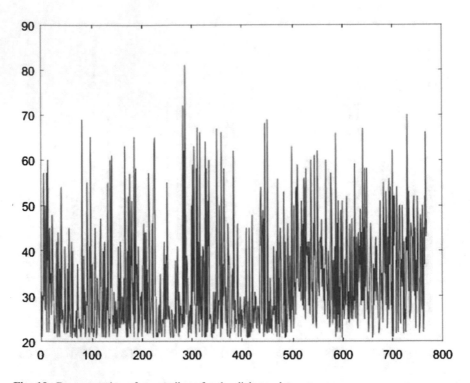

Fig. 10 Representation of age attribute for the diabetes data set

4 Conclusion

Deep learning has demonstrated to be an efficient tool for data analysis in case of Data mining and big data. Deep learners were mainly successful in trouble that in the past proved to be very complicated in Artificial Intelligence research work. In this paper DNN based data mining method has been performed for women suffering from diabetes. The Pima Indian Diabetes data set is introduced for training the deep neural network. Using 768 samples from the data base DNN is performing its result by using different learning algorithm. In future other models may be used for even accurate result and also optimization techniques may be applied for comparison.

References

1. Ma, Y., Tan, Y., Zhang, C., Mao, Y.: A data mining model of knowledge discovery based on the deep learning. In: IEEE Conference on Industrial Electronics and Applications (ICIEA), pp. 1212–1216 (2015)
2. Craven, M.W., Shavlik, J.W.: Using neural networks for data mining. Future Gener. Comput. Syst. **13**(2-3), 211–229 (1997)

3. Radhimeenakshi, S.: Classification and prediction of heart disease risk using data mining techniques of support vector machine and artificial neural network. In: IEEE International Conference on Computing for Sustainable Global Development, pp. 3107–3111 (2016)
4. Lu, H., Setiono, R., Liu, H.: Effective data mining using neural networks. IEEE Trans. Knowl. Data Eng. **8**(6), 957–961 (1996)
5. Girija, D.K., Shashidhara, M.S., Giri, M.: Data mining approach for prediction of fibroid disease using neural networks. In: IEEE International Conference on Emerging Trends in Communication, Control, Signal Processing and Computing Applications, pp. 1–5 (2013)
6. Zhang, L.M.: Genetic deep neural networks using different activation functions for financial data mining. In: IEEE International Conference on Big Data, pp. 2849–2851 (2015)
7. Setlak, G., Bodyanskiy, Y., Vynokurova, O., Pliss, I.: Deep evolving GMDH-SVM-neural network and its learning for Data Mining tasks. In: IEEE Conference on Computer Science and Information Systems, pp. 141–145 (2016)
8. Wlodarczak, P., Soar, J., Ally, M.: Multimedia data mining using deep learning. In: IEEE International Conference on Digital Information Processing and Communications, pp. 190–196 (2015)
9. Najafabadi, M.M., Villanustre, F., Khoshgoftaar, T.M., Seliya, N., Wald, R., Muharemagic, E.: Deep learning applications and challenges in big data analytics. J. Big Data **2**(1), 1 (2015)
10. He, M., He, D.: Deep learning based approach for bearing fault diagnosis. IEEE Trans. Ind. Appl. **53**(3), 3057–3065 (2017)
11. Pandey, S.C.: Data mining techniques for medical data: a review. In: IEEE International Conference on Signal Processing, Communication, Power and Embedded System, pp. 972–982 (2016)
12. Saxena, K., Sharma, R.: Diabetes mellitus prediction system evaluation using c4. 5 rules and partial tree. In: IEEE International Conference on Reliability, Infocom Technologies and Optimization, pp. 1–6 (2015)
13. Sharma, R., Singh, S.N., Khatri, S.: Medical data mining using different classification and clustering techniques: a critical survey. In: IEEE International Conference on Computational Intelligence and Communication Technology (CICT), pp. 687–691 (2016)
14. https://archive.ics.uci.edu/ml/datasets/pima+indians+diabetes

Deep Neural Network Based Speech Enhancement

Rashmirekha Ram and Mihir Narayan Mohanty

Abstract Enhancement of the speech signal is an essential task in the adverse environment. Several algorithms have been designed from several years to improve the quality. Mostly Neural Network and its variants are utilized for classification purpose. This paper exhibits the speech enhancement method based on the Deep Neural Network (DNN) to improve the quality and to increase the Signal-to-Noise Ratio of the speech signal. Different hidden layers are set to test the results. The audio features are extracted by using the short time Fourier transforms. The use of audio features improves the speech enhancement performance of DNN. Segmental Signal-to-Noise Ratio (SegSNR) and Perceptual Evaluation of Speech Quality (PESQ) are measured to test the results.

Keywords Deep neural network · Adaptive linear neuron · Perceptual evaluation of speech quality · Segmental signal-to-noise ratio · Neural network · Speech enhancement

1 Introduction

During the last decades, speech enhancement attracts the research attention due to its practical challenges in real time applications: mobile telephony, speech recognition, hearing aids design, etc. The main objective of speech enhancement method is to increase the voice quality of the deteriorated speech signal without distorting the signal. Many algorithms have been designed for this purpose. However, the performance is not always satisfactory in real-world environment [1, 2].

R. Ram · M. N. Mohanty (✉)
Department of Electronics and Communication Engineering,
Siksha 'O' Anusandhan University, Bhubaneswar, Odisha, India
e-mail: mihir.n.mohanty@gmail.com

R. Ram
e-mail: ram.rashmirekha14@gmail.com

© Springer Nature Singapore Pte Ltd. 2019
P. K. Mallick et al. (eds.), *Cognitive Informatics and Soft Computing*,
Advances in Intelligent Systems and Computing 768,
https://doi.org/10.1007/978-981-13-0617-4_27

Boll proposed the spectral subtraction method for decreasing the impacts of acoustical added noise in speech signal. But the occurrence of musical noise and the use of noisy phase degrade the quality of the synthesized speech signal [3]. Lots of modifications have been made to achieve the desired accuracy. Nonlinear spectral subtraction algorithm, Multiband spectral subtraction algorithm, Minimum mean square spectral subtraction, extended spectral subtraction algorithms are some of them. Various adaptive algorithms, subspace algorithms, statistical model-based algorithms are also proposed for speech enhancement [4]. A two-step noise reduction (TSNR) method and harmonic regeneration noise reduction methods are proposed by Siddala Vihari et al. These methods provide better results compared to spectral subtraction, Wiener filtering, and Decision directed approach [5]. Least mean squares algorithm (LMS), Recursive least squares (RLS) algorithm, State-space RLS (SSRLS) algorithm are well suited in nonstationary environment. Better SNR and MSE are achieved in SSRLS method [6].

Recently, machine learning approaches have attracted more attention in the field of speech enhancement. Adaptive linear neuron (ADALINE) network is applied as an adaptive filter for the purpose of speech enhancement. The weights and bias of the network are adapted by LMS algorithm [7]. The feed forward network and Wavelet are applied to ADALINE for enhancing the speech signal. The power spectral density proves the ADALINE filtered signal provides better enhanced signal [8]. The fractional coefficients of Discrete Cosine transform are used in ADALINE for speech enhancement [9].

In noisy environment, understanding of speech is difficult for cochlear implant (CI) users. To improve the intelligibility of the speech, neural network based speech enhancement is proposed in the current year. The decomposed time-frequency bins extract the auditory features and fed to the neural network to acquire the high SNRs of information. This estimation attenuates the noise and preserves the speech signals [10]. An overview of the neural network is presented in [11]. Different applications are mentioned in this field. The artificial neural network and its variations are also mentioned.

A regression approach based Deep Neural Network (DNN) is proposed by Yong Xu et al. A mapping function is calculated between the noisy speech signal and the clean speech signal of DNN. The comparison is made in between the proposed algorithm and the minimum mean square error (MMSE) noise reduction algorithm. Both the objective and subjective tests prove better result in DNN-based speech enhancement method [12]. The linear predictive (LP) parameter estimation is measured between speech and noise signals based on DNN. Multiple layers of DNN is designed using LP coefficients. An improved LMS adaptive filter (ILMSAF) combines with the DNN for speech enhancement. The adaptive filter coefficients are estimated through Deep belief Network (DBN) and afterward the enhanced speech signal is obtained by ILMSAF [13, 14]. The design of hearing aids instruments are also possible by using the DNN. Different feature sets are considered to feed to the Deep Neural Network [15].

The paper organization is as follows: Sect. 1 provides the Introduction of the work. Section 2 presents DNN-based speech enhancement method. Section 3 discusses the results of different speech signals. Finally Sect. 4 concludes the work.

2 Speech Enhancement Using Deep Neural Network

The block diagram of DNN for speech enhancement method is shown in Fig. 1. This method is separated into two phases: training phase and testing phase. A group of noisy and clean speech samples are developed in the training stage. All the signals are carried to the frequency domain by employing Short Time Fourier Transform (STFT). The logarithmic magnitude spectra of the noisy signals are considered. These features are fed to the DNN to generate the log-power spectra features of the enhanced signal. The noisy phase is computed from the deteriorated speech signals but not considered for training. To learn the deep neural network of noisy log spectra, the multiple restricted Boltzmann machines (RBMs) are arranged [16]. The input and the output features are obtained from the noisy-clean log spectra.

Steps in DNN for Speech enhancement:

- Speech signals are processed into a sequence of frames.
- The frame length is 512 samples with an overlap of 50%.
- The sampling frequency of the speech signal is 8 kHz.
- STFT is employed to calculate the FFT of each and every overlapped-windowed frame.
- Subsequently 164 dimensions of log-power spectrum features train the DNN.
- To test the DNN, clean signals and some randomly chosen noisy signals are considered.
- Inverse FFT and Overlap add methods are applied for signal reconstruction.

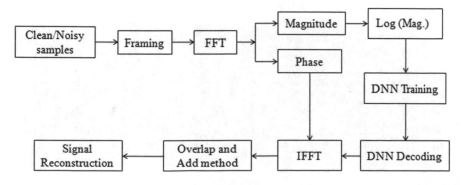

Fig. 1 Block diagram of DNN-based speech enhancement method

3   Results and Discussion

In this section, four types of noise signals, i.e., AWGN, Babble, Train and Restaurant are considered for speech enhancement using DNN. These noise signals are synthesized the noisy speech training samples. All 1024 utterances are taken from the Loizou databases and the four noise levels of SNR 0, 5, 10, and 15 dB are added to build the training set. One sample of clean data and all noisy training data sets are applied to train the DNN model. Another 250 arbitrarily selected utterances from the database are taken to build the test data set for each combinations of noisy types and noise levels. Two different sorts of noise signals, Street and Car are considered to evaluate the mismatch condition.

Figure 2 shows the clean signal and Fig. 3 shows the Noisy signal (affected by AWGN) spectrograms. The spectrogram of the enhanced signal presents in Fig. 4.

Two objective speech quality measures, i.e., SegSNR and PESQ are calculated to estimate the quality of enhanced speech signal. Also, the listening tests are performed for comparison. In this work, only some of the results are shown due to limited space.

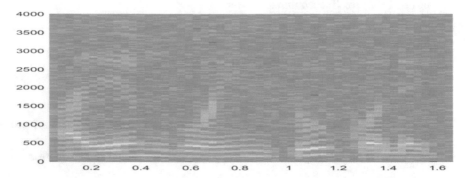

Fig. 2 Spectrogram of the clean signal

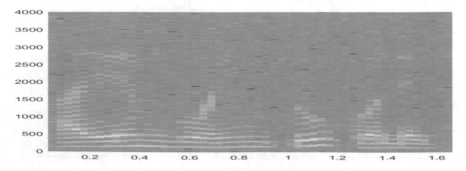

Fig. 3 Spectrogram of the noisy signal (affected by AWGN)

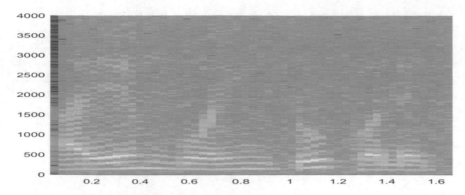

Fig. 4 Spectrogram of the enhanced signal after DNN

For each layer of RBM, the number of epoch is set as 20. For pre-training the learning rate parameter is 0.005 whereas for fine tuning the learning rate is 0.2. 80 epochs are considered for the mini batch size 100. All the features of DNN model are normalized to zero mean and unit variance.

Figure 5 represents the SegSNR results of four various noise types across different SNRs of training data set on the test set. 1024 hidden units and 15 frames are chosen for DNN. For 0 and 5 dB the Babble noise provides better SegSNR whereas for 10 and 15 dB the Restaurant noise results better.

Different hidden layers (DNN_L) are considered for test the results. Where L is the hidden unit and $L = 1, 2$, and 3 are used. Table 1 shows the PESQ scores of different noise levels and hidden layers. A maximum value of PESQ is 3.65 obtained for DNN_3 at SNR 10 dB. The DNN-based speech enhancement method enhances the speech signals of improved quality.

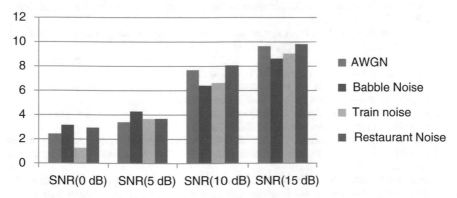

Fig. 5 SegSNR results of different training data on the test data (different SNRs of four noise types)

Table 1 PESQ scores on the test set of different SNRs

	SNR (0 dB)	SNR (5 dB)	SNR (10 dB)	SNR (15 dB)
Noisy	1.68	1.93	2.42	2.51
DNN_1	1.52	2.54	2.78	2.26
DNN_2	2.76	2.79	3.41	3.35
DNN_3	2.02	3.23	3.65	3.47

DNN_* represents the hidden layer number

4 Conclusion

DNN-based speech enhancement method is proposed in this paper. The prominent training data set is difficult to handle the complex configuration of DNN. By using preprocessing of the speech signals and the acoustic context of the information, the enhanced speech signal is less distorted. The objective measures show the highest PESQ score 3.65 in SNR 10 dB. The score is improved when the hidden layers are increased. All the experiments are performed by considering the NOIZEOUS database in MALAB environment. Results demonstrate that the DNN-based speech enhancement method outperforms better in both objective and subjective measures. In future, the real-time speech data sets are considered to perform noise adaptation for DNN-based speech enhancement method.

References

1. Loizou, P.: Speech Enhancement: Theory and Practice. CRC Press, Boca Raton (2007)
2. Haykin, S.: Adaptive Filter Theory, 3rd edn. Prentice Hall, Upper Saddle River (1996)
3. Boll, S.F.: Suppression of acoustic noise in speech using spectral subtraction. IEEE Trans. ASSP **27**, 113–120 (1979)
4. Chaudhari, A., Dhonde, S.B.: A review on speech enhancement techniques. In: International Conference on Pervasive Computing (ICPC) (2015)
5. Vihari, S., Murthy, A.S., Soni, P., Naik, D.C.: Comparison of speech enhancement algorithms. Procedia Comput. Sci. **89**, 666–676 (2016)
6. Ram, R., Mohanty, M.N.: Performance analysis of adaptive algorithms for speech enhancement applications. Indian J. Sci. Technol. **9**(44), 6 (2016)
7. Fah, L.B., Hussain, A., Samad, S.A.: Speech enhancement by noise cancellation using neural network. In: IEEE Conference on TENCON (2000)
8. Daqrouq, K., Abu-Isbeih, I.N., Alfauori, M.: Speech signal enhancement using neural network and wavelet transform. In: International Multi-Conference on Systems, Signals and Devices (2009)
9. Ram, R., Mohanty, M.N.: Fractional DCT ADALINE method for speech enhancement. In: International Conference on Machine Learning and Computational Intelligence (2017) (Communicated)
10. Goehring, T., Bolner, F., Monaghan, J.J.M., Dijk, B., Zarowski, A., Bleeck, S.: Speech enhancement based on neural networks improves speech intelligibility in noise for cochlear implant users. Hear. Res. **344**, 183–194 (2017)

11. Prieto, A., Prieto, B., Ortigosa, E.M., Ros, E., Pelayo, F., Ortega, J., Rojas, I.: Neural networks: an overview of early research, current frameworks and new challenges. Neurocomputing **214**, 242–268 (2016)
12. Xu, Y., Du, J., Dai, L., Lee, C.: A regression approach to speech enhancement based on deep neural networks. IEEE/ACM Trans. Audio, Speech, Lang. Process. **23**(1), 7–19 (2015)
13. Li, Y., Kang, S.: Deep neural network based linear predictive parameter estimations for speech enhancement. IET Signal Process. **11**(4), 469–476 (2017)
14. Li, R., Liu, Y., Shi, Y., Dong, L., Cui, W.: ILMSAF based speech enhancement with DNN and noise classification. Speech Commun. **85**, 53–70 (2016)
15. Goehring, T., Yang, X., Monaghan, J.J.M., Bleeck, S.: Speech enhancement for hearing-impaired listeners using deep neural networks with auditory-model based features. In: European Signal Processing Conference (2016)
16. Hou, J.C., Wang, S.S., Lai, Y.H., Lin, J.C., Tsao, Y., Chang, H.W., Wang, H.M.: Audio-visual speech enhancement using deep neural networks. In: IEEE Signal and Information Processing Association Annual Summit and Conference (APSIPA), pp. 1–6 (2016)

Wind Power Forecasting Using Support Vector Machine Model in RStudio

Archana Pawar, V. S. Jape and Seema Mathew

Abstract Wind energy has gained a lot of importance in few decades, as it is the cleanest form of renewable energy and available at free of cost. Wind power generation is increasing rapidly due to concerns about global warming and financial incentives from government. With the increased percentage of wind power in power grid, wind power forecasting has become essential to system operator for electric power scheduling and power reserve allocation. Wind power producers can get benefit of wind power prediction while bidding in the electricity market. This paper presents the wind power forecasting model using Support Vector Machine technique. In this model, wind speed, wind direction, air temperature, air pressure and air density are taken as input parameters to build accurate and efficient model. New statistical computing software called RStudio is used to develop prediction model. The model is trained and tested by using data of Dutch Hill Wind Farm available on website of National Renewable Energy Laboratory (NREL). According to results, the model performs significantly better than linear SVM and other regression models of SVM. The model is used to implement one hour ahead wind power forecasting and the results are validated with dataset.

Keywords Support vector machine (SVM) · Wind power forecasting (WPF)
RStudio · Numerical weather prediction (NWP) · Radial basis function (RBF)
Epsilon regression SVM (ε-SVM) · Nu regression SVM (υ-SVM)

A. Pawar (✉) · V. S. Jape · S. Mathew
PES's Modern College of Engineering, Pune, India
e-mail: archanadpawar@gmail.com

V. S. Jape
e-mail: jape_swati@yahoo.co.in

S. Mathew
e-mail: seema_mathew47@rediffmail.com

© Springer Nature Singapore Pte Ltd. 2019   289
P. K. Mallick et al. (eds.), *Cognitive Informatics and Soft Computing*,
Advances in Intelligent Systems and Computing 768,
https://doi.org/10.1007/978-981-13-0617-4_28

1 Introduction

Non-conventional energy sources are inexhaustible, available at free of cost and do not contribute to global warming. Considering the advantages of non-conventional or renewable energy sources, many countries have started utilizing them for power generation. Wind is one of the most important renewable energy sources and is being utilized for power generation. The intermittent nature of wind has created challenges in wind power penetration into power grid. Due to uncertainty of wind power, power system operator has to tackle problems like economic load scheduling, reserve allocation, etc. So, wind power prediction has become very essential for power system operators also for wind power producers to get incentives from government for producing it as well as to get benefit by bidding in electricity market.

There are various techniques available and research is going on to build the best and accurate model for wind power Forecasting. Many statistical research models for wind power prediction are available in the literature. Bhaskar and Singh [1] proposed adaptive wavelet neural network (AWNN) model to forecast wind speed by using wavelet decomposition of wind series which is further used as input to feed forward neural network (FFNN) to forecast wind power. Sideratos and Hatziargyriou [2] proposed an advanced statistical method for WPF, in which artificial intelligence and fuzzy logic technique is used. This model uses NWP along with historical wind data and can give 1–48 h ahead prediction. Zeng and Qiao [3] developed a short tern WPF model by using wavelet support vector machine, in which wind speed is predicted using SVM with wavelet kernel and then wind power is found from wind profile for predicted wind speed. This model performs well for short term forecasting but it failed to capture the trend of wind variation. Recent research has shown that, SVM-based model outperform the ANN based model in WPF, as it takes less computational time than ANN-based model [4].

The paper proposes a WPF model which uses Support Vector Machine technique. The model considered uses the different parameters like speed and directions of wind and characteristics of air like temperature, pressure and density as an input, as these parameters do have impact on wind power generation. SVM with the combinations of various kernel functions and types of regression is implemented in statistical analysis software called RStudio. It is an integrated development environment for R programming language. SVM with radial kernel function and nu regression performs well compare to other combinations. The model is validated with data of Dutch Hill Wind Farm available on website of National Renewable Energy Laboratory (NREL) [5]. The data is available after every 5 min interval.

2 Factors Affecting Wind Power

Wind power generation by wind turbine can be calculated by Eq. (1),

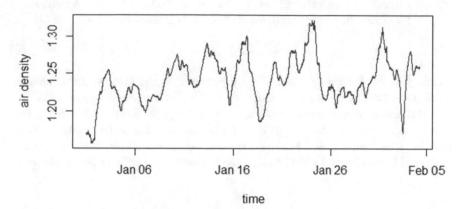

Fig. 1 Variation of air density of Dutch Hill Wind Farm available online on website of National Renewable Energy Laboratory

$$\text{Power} = kC_p \frac{1}{2} \rho A V^3 \tag{1}$$

where, P is power output (kW), C_p is maximum power coefficient, ρ is air density (lb/ft^3), A is rotor swept area (ft^2), V is Wind speed (mph) and k is 0.000133 a constant.

In the wind power calculations the air density is usually taken as constant in time. But the density of moist air is not constant and is a function of air temperature, air pressure and relative humidity. Figure 1 shows the air density of Dutch Hill Wind Farm varies with respect to time. By taking variations of air density into account the potential RMS error decreases by up to 16% and the accuracy of WPF is increased [6]. So, it is very essential to consider air characteristics like density, pressure and temperature along with wind speed and wind direction while forecasting wind power generation. In the proposed WPF model, all these parameters are taken as input.

3 Support Vector Machine

Support vector machine is supervised learning technique of machine learning. SVM finds an optimal hyperplane to separate data by using selected subset of data. Data points in this subset of data are called as support vectors. The hyperplane which has maximum margin is the optimal hyperplane. It is obtained by solving the constrained optimization problem using optimization technique. If the data is not linearly separable, it is mapped into a high dimensional space by a nonlinear mapping. After mapping data, SVM performs a linear regression in this high dimensional space.

The SVM for wind power forecasting can be expressed by Eq. (2):

$$\hat{y}_t = \omega^T \varnothing(x_t) + b \tag{2}$$

where, $\hat{y}_t$ is the forecasted value of wind power, x_t is the input vector containing input parameters and b is a bias term. The term ω^T is the weight vector and Φ is a kernel function, which maps input vector x_t to a higher dimensional vector $\Phi(x_t)$. Then linear regression is used to train SVM and to forecast wind power. Three types of SVM: linear, epsilon regression and nu regression are implemented and compared here. Linear SVM is epsilon regression with the linear kernel function.

3.1 Epsilon Support Vector Machine (ε-SVM)

In the ε-SVM, selection of ε is very crucial, as lesser value of it increases number of support vectors increasing computational cost and large value of it leads to lower number of SVs but increase in permissible error. The SVM parameters in (2) can be obtained by solving the following problem (3).

$$\min \quad \frac{1}{2}\omega^T\omega + \gamma \sum_{i=1}^{l}(\zeta_i + \zeta_i^*)$$

$$\text{Subject to } (\omega^T\varnothing(x_i) + b) - y_i \le \varepsilon + \zeta_i^* \tag{3}$$

$$y_i - (\omega^T\varnothing(x_i) + b) \le \varepsilon + \zeta_i$$

$$\zeta_i, \zeta_i^* \ge 0, \ i = 1, \ldots, l$$

where, ζ_i and ζ_i^* are the nonnegative slack variables. γ is a regularization parameter. $\Phi(x_i)$ is the RBF kernel function.

The ε-SVM of (2) can be given by Eq. (4):

$$\hat{y}(x_t) = \sum_{i=1}^{l}(\alpha_i - \alpha_i^*)K(x_t, x_i) + b \tag{4}$$

where, α_i and α_i^* are nonnegative Lagrange multipliers. The samples associated with non-zero coefficients $(\alpha_i - \alpha_i^*)$ have approximation errors equal to or larger than ε and those samples are known as the support vectors (SVs).

3.2 Nu Support Vector Machine (υ-SVM)

As it is difficult to select an appropriate ε, a new parameter υ is introduced which controls the number of SVs as well as training errors. υ has a range of [0, 1] and the resulting model is known as υ-SVM. The constrained optimization problem is given by Eq. (5).

$$\min \quad \frac{1}{2}\omega^T\omega + \gamma\left(\upsilon\varepsilon + \frac{1}{l}\sum_{i=1}^{l}(\zeta_i + \zeta_i^*)\right)$$

$$\text{Subject to} \quad (\omega^T + \varnothing(x_i) + b) - y_i \le \varepsilon + \zeta_i \quad (5)$$

$$y_i - (\omega^T\varnothing(x_i) + b) \le \varepsilon + \zeta_i^*$$

$$\zeta_i, \zeta_i^* \ge 0, \quad i = 1, \ldots, l, \quad \varepsilon \ge 0$$

Then υ-SVM of (2) can be given by Eq. (6) as follows,

$$\hat{y}(x_t) = \sum_{i=1}^{l}(\alpha_i^* - \alpha_i)K(x_i, x_t) + b \quad (6)$$

4 Proposed Model

Figure 2 shows the block diagram of proposed model. In this model input parameter values are normalized first. The model is developed in RStudio Software using R programming language. The software specially used for statistical analysis. In SVM, function data is scaled internally to zero mean and unit variance. Feature representation of input data is done by kernel function and further SVM prediction model is developed by using υ regression SVM method (given by Eq. 6). De-normalization of target value is also done internally and predicted value is generated.

The model is built using historical dataset of one month (January 1, 2011–January 31, 2011) of Dutch Hill Wind Farm available on website of National Renewable Energy Laboratory (NREL). Data samples are available at the sampling rate of one sample per 5 min. 2/3rd of data is utilized to train the model and one-third of data is utilized to test model.

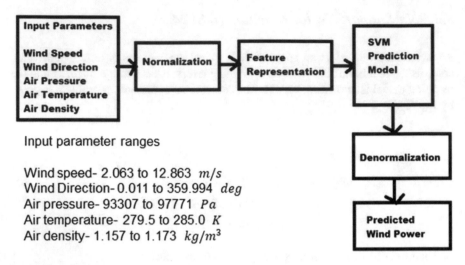

Fig. 2 Block diagram of proposed model

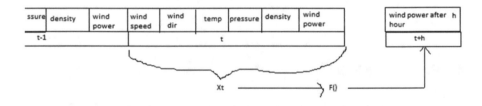

Fixed step prediction scheme

Fig. 3 Prediction scheme of h hour ahead wind power forecasting

4.1 Prediction Scheme

Figure 3 shows the prediction scheme of 1 h ahead WPF model considered here. In this scheme, current values of input parameters are taken as input and the value of wind power after an hour is forecasted.

5 RStudio

The three models are implemented in RStudio. It is freely available and open source integrated development software, where statistical techniques can be implemented for various applications [7]. Due to its open source nature, people working in statistics, keep on adding their work to it in the form of packages. These packages can be installed and used while doing any project in it. After installing a package one need to include library of that package so that functions available in that package can be used in program. RStudio is enriched with such a various packages. The packages installed here for implementing considered model are given below,

1. e1071
   This package offers an interface to the software called Libsvm which is used for classification, regression, and distribution estimation by sSVM.
2. Rminer
   This package contains functions which are useful to calculate errors of classification and regression models.
3. MLmetrics
   It is the collection of various metrics which evaluate loss, score and performance of classification and regression models.
4. Forecast
   It has functions to calculate accuracy, autocorrelation, and correlations. Autocorrelation is useful in analyzing data.

6 Simulation Results

Model is trained with the various lengths of training data like 15, 30 and 60 days. Model with 30 days data as a training length is suited best for time and complexity. Time taken by model to get trained is around 2 min. The parameters of linear SVM, ε-SVM and υ-SVM model are given in Table 1.

Figure 4 gives comparison of the actual and the forecasted wind power for the linear SVM, ε-SVM, and υ-SVM model. The predicted values of linear model have much deviation from the actual observations. The predicted value follows closely the actual wind power in ε-SVM model. In υ-SVM model, the predicted value follows almost the actual observation of wind power. It is clear that, υ-SVM model is the best forecasting model among the three models.

The residuals from a fitted model are the differences between the observed and the forecasted value of response variable. If the errors are independent, prediction

Table 1 Parameters of linear SVM, ε-SVM and υ-SVM model

SVM parameters	Linear SVM	ε-SVM	υ-SVM
Type	Eps-regression	Eps-regression	Nu regression
Kernel	Linear	Radial	Radial
cost	4	4	4
Gamma (γ)	0.5	0.5	0.5
Epsilon (ε)/Nu (υ)	$\varepsilon = 0.1$	$\varepsilon = 0.1$	$\upsilon = 0.5$
No. of support vectors	3782	153	3510

True and predicted wind power (all models)

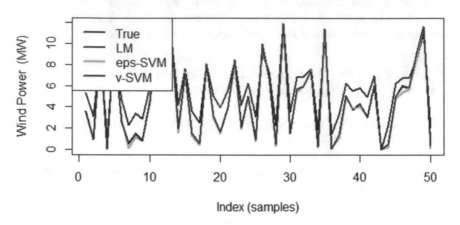

Index (samples)

Fig. 4 Comparison of performance of linear, ε-SVM and υ-SVM model

model fits perfectly. That means, residual (error) should have less or no correlation with its lagged value. Figure 5 shows the autocorrelation of residuals of linear, ε-SVM and υ-SVM model respectively. The errors (residuals) of linear SVM and ε-SVM models are correlated with their lagged values of residuals. Residuals of υ-SVM model are less correlated. So, υ-SVM model fits well compared to other two models.

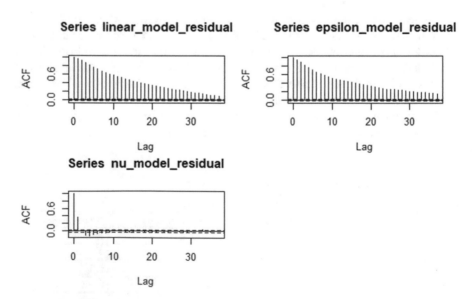

Fig. 5 Correlation of residuals of linear, ε-SVM and υ-SVM model

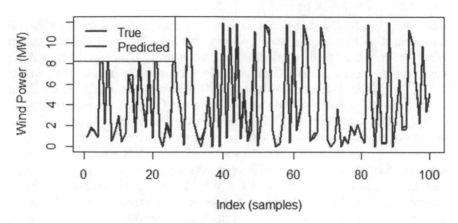

Fig. 6 Performance of 1 h ahead prediction by v-SVM model

Table 2 Performance measures of prediction models and 1 h ahead forecasting (v-SVM) model

Prediction model	Mean absolute error (MAE)	Mean absolute percentage error (MAPE) (%)	Root mean squared error (RMSE)
ε-SVM	0.1921661	1.912728	0.2258683
v-SVM	0.01402083	0.1266071	0.02194454
Linear model	0.5792569	11.44996	0.7263405
1 h ahead v-SVM	0.27645	0.4861	0.5879

Figure 6 gives the comparison of actual and predicted wind power by 1 h ahead wind power forecasting scheme using v-SVM model. The performance measures of it are given in Table 2.

Table 2 gives the comparison of three predictive models. Nu regression SVM (v-SVM) model has very less mean absolute error and mean absolute percentage error compare to other two models i.e. 0.014 and 0.127% respectively. Linear SVM model has highest values of MAE and MAPE compare to other two models. So, v-SVM Model out performs and gives the accurate forecasting of wind power generation. So, v-SVM model is used further to implement 1 h ahead forecasting scheme which has acceptable errors.

7 Conclusion

This paper has presented a SVM model, in which v-regression algorithm is implemented to forecast the wind power generation. The model considered is trained using wind speed, wind direction, air density, air temperature and air pressure, unlike other models which use only wind speed to predict wind power, Due to this, accuracy of model is increased. Three prediction models using linear SVM, epsilon regression (ε-SVM) and nu regression (v-SVM) are implemented. In case of epsilon regression model, one has to compromise either in computational cost or permissible error. But, in case of nu regression model, v (nu parameter) controls margin errors and number of support vectors. v-SVM gives us control over number of support vectors which will control the computational cost. The model is trained and tested using data of Dutch Hill Wind Farm which is available online on website of National Renewable Energy Laboratory. The model is validated with the data and results show that v-SVM Model performs better than other two models. It has 0.01402083 mean absolute error (MAE) and 0.1266071% mean absolute percentage error (MAPE). The model is implemented in statistical analysis software 'RStudio' which is free and open source integrated development environment (IDE) for R programming language. Further, the model is used to implement 1 h ahead wind power forecasting and the results are validated with the dataset. The future scope of this project is to develop a web based application for wind power forecasting using the proposed model in RStudio.

References

1. Bhaskar, K., Singh, S.N.: AWNN-assisted wind power forecasting using feed-forward neural network. IEEE Trans. Sustain. Energy 3(2) (2012)
2. Sideratos, G., Hatziargyriou, N.D.: An advanced statistical method for wind power forecasting. IEEE Trans. Power Syst. 22(1) (2007)
3. Zeng, J., Qiao, W.: Short-term wind power prediction using a wavelet support vector machine. IEEE Trans. Sustain. Energy 3(2) (2012)
4. Sreelakshmi, K., Ramakanth Kumar, P.: Performance evaluation of short term wind speed prediction techniques. IJCSNS Int. J. Comput. Sci. Netw. Secur. 8(8) (2008)
5. Draxl, C., Hodge, B.M., Clifton, A., McCaa, J.: The Wind Integration National Dataset (WIND) toolkit. Appl. Energy 151, 355366 (2015)
6. Farkas, Z.: Considering Air Density in Wind Power Production. Eötvös University, Hungary (2011)
7. R Core Team: R: A Language and Environment for Statistical Computing. R Foundation for Statistical Computing, Vienna, Austria. (2017). URL https://www.R-project.org/

A Survey: Classification of Big Data

Dinesh Kumar and Mihir Narayan Mohanty

Abstract In the current decades large data sets are mostly available from the source, extraction and analysis of data is an interesting and challenging task. Big Data relate to expansive bulk size, developing datasets that are intricate and have numerous self-ruling spring. Prior advances were not ready to deal with capacity and handling of enormous dataset in this manner Big Data idea appears. This is a monotonous employment for clients to distinguish precise data from enormous unstructured data. Along these lines, there ought to be some system which characterize unstructured data into sorted out shape which causes client to effectively get to required data. Arrangement systems over big value-based database give expected dataset to the clients from huge datasets further straightforward way. There are two primary arrangement procedures, administered and unsupervised. In this paper we concentrated on to investigation of various administered characterization methods. Encourage this paper demonstrates use of every system and their points of interest and confinements.

Keywords Big data · Classification · Structured · Unstructured

1 Introduction

Big Data consists of large Data sets that cannot be managed efficiently by the common data base management systems. These datasets range from terabytes to Exabyte. Mobile phones, credit cards, Radio Frequency Identification (RFID)

D. Kumar
Department of CSE, Sri Sai College of Engineering and Technology,
Badhani, Pathankot, India
e-mail: dineshgarag82@gmail.com

M. N. Mohanty (✉)
Department of Electronics and Communication Engineering,
Siksha 'O' Anusandhan University, Bhubaneswar, India
e-mail: mihir.n.mohanty@gmail.com

© Springer Nature Singapore Pte Ltd. 2019
P. K. Mallick et al. (eds.), *Cognitive Informatics and Soft Computing*,
Advances in Intelligent Systems and Computing 768,
https://doi.org/10.1007/978-981-13-0617-4_29

Fig. 1 Feature of big data

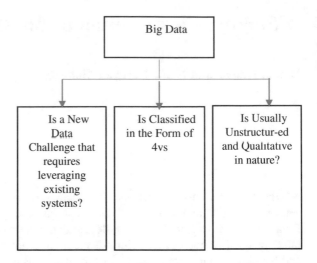

Fig. 1 Feature of big data

device, and social networking platforms create huge amounts of data that may reside unutilized at unknown servers for many years. Similarly, healthcare industries have generated and kept in storage vast amount of data which are very useful for diagnosis and disease prediction [1–4].

Advancement of Internet benefits, lists and questioned substance are developing very quickly. This put a demand on internet surfer organizations to tackle the problems associated with handling considerable huge information. Google has developed programming models such as the GFS [3] and the MapReduce to accommodate the issues and difficulties associated to handling of information, administration, examination concerned to Internet scale. Likewise, substance produced by clients, sensors, and different omnipresent information sources additionally fueled the staggering information streams, and needs a basic change on the registering design and extensive scale information preparing instrument.

This is truly an information age where data is being generated at an alarming rate. This large number of data is often termed as Big Data. The term BigData is now widely used, Particularly in the IT industry, where it has generated various job opportunity (Fig. 1).

2 Types of Big Data

At the Davos Forum in Switzerland, a report on "Big data, Big Impact" was introduced in early 2012. There, it was declared that huge information has turned into another sort of financial resources, much the same as gold or cash. Gartner, a worldwide famous research organization, issued the much talked Hype Cycles between 2012 and 2013, that characterized enormous information registering, social examination, and put away information investigation into 48 rising advancements

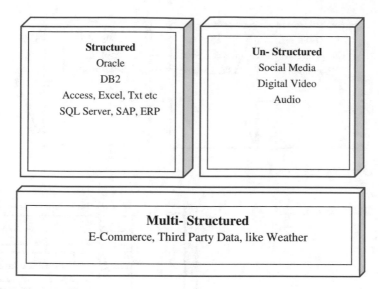

Fig. 2 Big data difference

that merit most attention. Hadoop is generally utilized as a part of huge information applications related to business such as spam sifting, arrange looking, clickstream investigation, and, social proposal. Also, significant scholastic research is currently in light of Hadoop.

There are three Types of Big data

(a) **Social Data**: Social Data refers to the information collected from various social networking sites and online portals like Facebook, LinkedIn, and Twitter etc.
(b) **Machine Data**: Machine Data refers to the Information generated from RFID chips, Bar code Scanners, and sensors. For example RFID Chips, GPS results.
(c) **Transactional Data**: Transactional Data Refers to the information generated from online shopping sites, retailers and B2B Transactions like ebay, flipkart, amazon etc. (Fig. 2).

Semi-Structure Data + Structured Data + Unstructured Data = Big Data

In the real-world scenario the unstructured data is larger than Semi-structured data and structured data. Structured data may be considered as the data that has repeating patterns. On the other hand, the unstructured data is a set of data that might or might not have any logical or repeating patterns. And semi-Structured data, also known as having a scheme less structure, in other words, data is stored inconsistently in rows and columns of a database (Figs. 3 and 4).

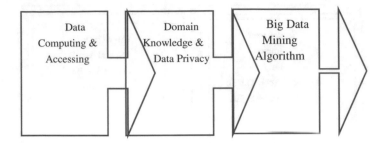

Fig. 3 Tier of big data

Fig. 4 Classification of big
data

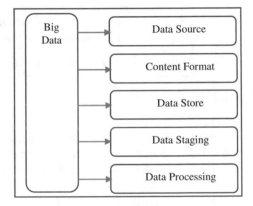

3 Classification of Big Data

Classification has been a data mining method which characterizes the unstructured
data into the organized class and gatherings. It can serve the clients for learning
revelation and, future arrangement. It gives a knowledge basic leadership [5–9].
There are basically two stages in Classification such as learning process stage
wherein a gigantic preparing informational index are provided and investigation
happens then standards and examples are made. At that point the execution of
second stage begins that is assessment or trial of informational collections and
chronicles the exactness of an arrangement designs. This area quickly portrays the
managed characterization strategies, for example, Decision Tree and Support
Vector Machine. There ought to be some system which arranges unstructured
information into sorted out shape which encourages client to effectively get to
required information. Characterization systems over huge value-based database
give expected information to the clients from substantial datasets more straightforward way. K closest neighbors (kNN) is a proficient lethargic learning calculation and has effectively been created in genuine applications. It is extremely
normal to scale the past kNN technique to the vast scale datasets [10–14].

There are two types of classification:

(a) Human-Driven
(b) Machine-Driven

Human-Driven: This data is the record of human encounters, beforehand recorded in books and gems, and later in photos, sound and video. Human-sourced data is presently completely digitized and put away wherever from PCs to interpersonal organizations. Information are inexactly organized and frequently ungoverned.

Machine-Driven: got from the wonderful development in the quantity of sensors and machines used to gauge and record the occasions and circumstances in the physical world. The yield of these sensors is machine-produced information, and from straightforward sensor records to complex PC logs, it is all around organized. As sensors multiply and information volumes develop, it is turning into an undeniably vital segment of the data put away and handled by numerous organizations. Its very much organized nature is reasonable for PC handling, yet its size and speed is past conventional methodologies.

There is a contrast between Data put away on customary Data Bases verses Big Data, and it depends of its inclination, we can describe five kinds of sources

1. Sensors/meters and activity records using electronic contraptions: This kind of information is conveyed steadily. In this case, both the number as well as the periodicity of recognitions impression varies every so often so that relies on a lap of time or occasion of certain event or activity (per case passing of an auto by the camera vision edge). In other case, it relies on the manual control (Strictly speaking, it can be comparable to the occasion of an event). Such source generally depends on the farthest point of the corresponding sensor mostly to take exact estimations.

2. Social collaboration: It is the information made as a result of human associations using certain framework, such as the Internet. One of the broadly perceived information is due to casual associations of human being. Such information proposes subjective having the quantitative edges often involve some form of energy need to be computed. Such quantitative perspectives are less requesting to gage than subjective edges, beginning ones recommends, checking the number of recognitions accumulated either by land or transient traits. The idea corresponding to the second ones generally depends upon the precision of the counts associated with expel the significance of the substance which are consistently found as unstructured substance written in trademark vernacular, instances of examination that are created utilizing this information are feeling examination, incline focuses examination, etc.

3. Business trades: Data conveyed in view of business function may be recorded either in sorted out or unstructured databases. Exactly while recording on composed information, the most broadly perceived issue to separate information to get authentic markers happens to be the huge volume of information. Its periodicity of age in light of the fact that these information are periodically made at a fast pace and an expansive number of records are conveyed in a minute.

This is true even when enormous associations such as the supermarket chains are involved in recording arrangements. Such information is not by and large made in positions which may particularly secured in social databases. An electronic receipt is one such instance of this example of source. It is essentially a structure. However if it is required to put the information available in a social database, we need to apply a system that suits the information on different tables so as to institutionalize the information similarly with the social database theory. It isn't in plain substance and may be a PDF, a photograph, Excel record, etc. One issue that need to be considered here is the technique that needs time and is effectively expressed. Since the information potentially is being conveyed too snappy, so we would require unmistakable procedures to effectively use such information, setting it up without making available on a social database, at the same time discarding a couple of observations or criteria and parallel dealing with such system. The nature of such information conveyed with respect to the business trades is immovably associated with the capacity to get operator recognitions and similarly process them.

4. Electronic Files: The files are unstructured reports, either statically or dynamically made secured or conveyed as electronic records, such as the Internet pages, accounts, sounds, PDF archives, etc. They have substance of phenomenal interest yet are independent in which various procedures may be used such as the substance mining, plan affirmation, and so forth. Nature of the estimations often relies mostly on the capacity to evacuate and viably decipher all the delegate information from those reports.

5. Broadcastings: Mainly suggested video and sound made on steady, getting true information from the substance of such electronic information is too much complex now a days and derives huge computational as well as exchanges control. Once handled such issues of evolving over "cutting edge basic" substance to "modernized information" substance, we will have practically identical burdens like the ones we encounter on social joint efforts.

Choice Tree (DT) Decision Tree demonstrates is a progressive recursive detachment of the info space into class locales. It has choice hubs and clears out. Innocent Bayes classifier remains a straightforward probabilistic classifier based on Bayes' hypothesis with gullible freedom suppositions [5]. KNN calculation is the least difficult calculation of characterization calculation and straightforward. The closest neighbor (NN) run recognizes the classification of obscure information point on the premise of its closest neighbor whose class is as of now known [6]. However the arrangement correctnesses that the SVM methods give are fantastic. Along these lines embracing SVM methods is very favored. Presently the test is to discover the answer for enhance the SVM strategy [13, 14].

4 Conclusion

Expository models ought to likewise be operationally productive. This alludes to the endeavors expected to gather the information, preprocess it, assess the model, and sustain its yields to the business application (e.g., crusade administration, capital estimation). Particularly in a real-time web based scoring condition (e.g., misrepresentation discovery) this might be a pivotal trademark. Information are key elements for any diagnostic exercise. Henceforth, it is imperative to completely consider and list all information sources that are of potential enthusiasm before beginning the investigation. The govern here is the more information, the better. Be that as it may, genuine information can be messy as a result of irregularities, deficiency, duplication, and combining issues. All through the systematic displaying steps, different information separating components will be connected to tidy up and decrease the information to a reasonable and pertinent size. It is absolutely critical that each datum preprocessing step is deliberately supported, completed, approved, and recorded before continuing with advance examination. In this paper, we talked about the distinctive managed characterization procedures on Big Data, its highlights, sorts and order.

References

1. Wu, X., Zhu, X., Wu, G., Ding, W.: Data mining with big data. IEEE Trans. Knowl. Data Eng. **26**(1), 1041–4347 (2014)
2. Wang, D., Liu, X., Wang M.: A DT-SVM strategy for stock futures prediction with big data. In: IEEE International Conference on Computational Science and Engineering. 978-0-76955096, pp. 1/13 (2013)
3. Ghemawat, S., Gobioff, H., Leung, S.-T. The Google file system. In: ACM SIGOPS Operating Systems Review, ACM, vol. 37, pp 29–43 (2003)
4. Dean, J., Ghemawat, S.: Mapreduce: simplified data processing on large clusters. Commun. ACM **51**(1), 107–113 (2008)
5. Kesavaraj, G., Sukumaran, S.: A study on classification techniques in data mining. In: IEEE ICCCNT, pp. 4–6 (2013)
6. Suthaharan, S.: Big Data Classification: Problems and Challenges in Network Intrusion Prediction with Machine Learning. University of North Carolina at Greensboro, USA (2012)
7. Kotsiantis, S.B.: Supervised Machine Learning: A Review of Classification Techniques. vol. 31, pp. 249–268, Informatica, USA (2007)
8. Yu, H., Yang, J., Han, J.: Classifying large data sets using SVMs with hierarchical clusters. In: SIGKDD '03 Washington, DC, 1581137370/03/0008 (2003)
9. Piao, Y., Park, H.W., Jin, C.H., Ryu, K.H.: Ensemble Method for Classification of High-Dimensional Data. IEEE. 978-1-4799-3919-0/14 (2014)
10. Yenkar, V., Bartere, M.: Review on data mining with big data. Int. J. Comput. Sci. Mob. Comput. **3**(4), 97–102 (2014)
11. Mohammed, G.H., Zamil, A.L.: The application of semantic-based classification on big data. In: IEEE International Conference on Information and Communication Systems (ICICS). 978-1-4799-3023 4/14 (2014)

12. Dai, W., Ji, W.: A map reduce implementation of C4. 5 decision tree algorithm. Int. J. Database Theor. Appl. SERSC **7**(1), 49–60 (2014)
13. Mohanty, M.N., Kumar, A., Routray, A., Kabisatpathy, P.: Evolutionary algorithm based optimization for PQ disturbances classification using SVM. Int. J. Control Autom. Syst. **8**(6), 1306–1312 (2010)
14. Jeyakumar, V., Li, G., Suthaharan, S.: Support vector machine classifiers with uncertain knowledge sets via robust convex optimization. Optim. J. Math. Program. Oper. Res. 1–18 (2012). https://doi.org/10.1080/02331934.2012.703667

Review Paper: Licence Plate and Car Model Recognition

R. Akshayan, S. L. Vishnu Prashad, S. Soundarya, P. Malarvezhi
and R. Dayana

Abstract The main intention of this study paper is to explore and analyze the numerous methods that are used for license plate extraction and identification. Analysis is done by detailed study of the prevailing methodologies and their drawbacks were duly notified. Various concepts such as OCR reading, bounding method and other computational techniques were thoroughly studied and are recorded below. Thus, with the knowledge gained from this review, a pristine and a noble method were developed for the same by utilizing the concept of API. This not only involves extraction of characters from the license plate but also aids in the identification of the vehicle model, as a whole, with the help of image processing. This method is not merely effective for detection but also holds good for certain amount of automation with the aid of the upheld database. These techniques of detection backboned by databases can prove to be very effective in a wide range of use-cases in the current world. Such identification as well as compilation of useful data can be implemented in applications such as automatic toll system, fee collection in parking areas, traffic disciplinary maintenance, vehicle monitoring inside a premise and much more.

Keywords License plate · Vehicle identification · Automatic toll system
Automated parking fee collection · API

R. Akshayan · S. L. Vishnu Prashad · S. Soundarya · P. Malarvezhi (✉)
R. Dayana
SRM Institute of Science and Technology, Kattankulathur, Chennai, India
e-mail: malar83@gmail.com

R. Akshayan
e-mail: akshayanaki@gmail.com

S. L. Vishnu Prashad
e-mail: vishnuprashad3@gmail.com

S. Soundarya
e-mail: sssoundaryas@gmail.com

R. Dayana
e-mail: dayana.r@ktr.srmuniv.ac.in

© Springer Nature Singapore Pte Ltd. 2019   307
P. K. Mallick et al. (eds.), *Cognitive Informatics and Soft Computing*,
Advances in Intelligent Systems and Computing 768,
https://doi.org/10.1007/978-981-13-0617-4_30

1 Introduction

License plate extraction and model identification [1–4] of a vehicle has immense importance in various issues. Implementation of such system is mainly done in automatic toll collection [1], parking fee collection, etc. The earliest version of automobile identification was done with microwave and infrared system [1]. Over the course of time, this process was further extended and using a CCD camera, image segmentation of license plate was processed [2]. With advent of capturing technology, a method called Optical character recognition was devised to extract plate information under various circumstances [5]. This includes places where an image of a vehicle in traffic has to be processed or in cases, where a vehicle has to be identified within a mixture of other objects. Texture analysis is applied along with statistical methods [6] to provide information that is obtained from traffic camera images. Usage of Tesseract engine and neural networks [7] has given more efficient character segmentation and identification. Based on prior knowledge on the images, an automated detection which detects only the edges of license plate was performed [8]. This process helps to identify merely the license plate with the help of above-mentioned process. Reduction of noise can be avoided by the process of converting a colored image to a gray scale image. To allow recognition of each character, segmentation can be done by yet another method [9]. In the same method, an image was converted to YDbDr format to specifically detect characters [10]. The automated fee calculation was thus done using these recognition techniques, along with certain prescribed rules that was controlled via a program [11].

1.1 Various Techniques Involving Extraction of Characters from License Plate

Extraction of license plate characters has been extensively researched for various purposes. Some of the effective theories have been put forth in this section.

1.1.1 Optical Character Recognition [OCR]

As put forth in papers mentioned [7, 8, 11], the OCR technology is used for segmentation of the characters that are available in the license plate. The process involves capturing and extraction of the image pertaining only to the license plate, as shown in Fig. 1. The image so obtained is segmented by character, which is then recognized individually, using the OCR process.

The process of such image capturing technology is different with those mentioned in [7]. Here, all the operations are programmed in a mobile device. This mobile device, which either runs on android or iOS, is used to capture the image with its camera. Then on, with the help of OpenCV libraries and image processing

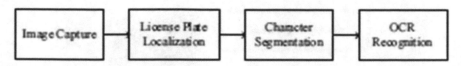

Fig. 1 System using OCR recognition

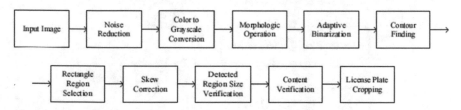

Fig. 2 License plate localization

techniques, further process in the system can be performed. In paper [8], for better recognition purpose, the process of machine learning is utilized. By using supervised machine learning, various types of plates are studied whose data are stored in XML. This process could facilitate the system from differentiating the license plate from other objects. Nevertheless, the main disadvantage of this method lies in its usage of XML. Since, they are large data storage units, they take up high disk space, which in turn overburden the system. Moreover, analyzing paper [11], this indicates a similar process except for the use of gray scaling, given by the Eq. (1) as,

$$Y = 0.299R + 0.587G + 0.114B \tag{1}$$

Whereas in case of [8], binary form of image is obtained, as shown in Fig. 2. This results in possibility of error due to improper detection of details.

1.1.2 Using Bounding Method

Analyzing this method [5], implies that, even though OCR is applied for preliminary process, the recognition block, which identifies the edge of the license plate, varies. The image obtained in this case is extracted from the overall vehicular image. Thus, instead of processing whole image as in [9], only the license plate is isolated for further processing. This method is illustrated in Fig. 3. As a result, the computational error can be minimized, which can be seen from the Table 1. Despite these advantages, this method cannot be applied for complex image detection due to the presence of multiple edges. Thus, additional complexity is introduced due to this process.

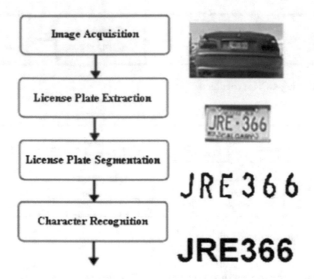

Fig. 3 Using bounding method

Table 1 Results of the system [9]

Units	Number of accuracy	Percentage of accuracy (%)
Extraction	42/45	93.33
Segmentation	39/45	86.67
Recognition	42/45	93.33

1.1.3　Extraction Process for Characters

The system suggested in paper [2] provides the process of masking the unwanted components obtained from the image through camera.

This system, as shown in Fig. 4, consists of a simple unit wherein the arrival of the car is identified with the help of IR sensor, upon which the gate would be closed. Thus, the image of the plate can be captured and processed. The quality of the image is maintained by fixing a threshold value to eliminate noise components. The threshold value may vary with different parameter such as,

W_A: width-threshold; H_A: height-threshold; A_A: area-threshold$_{max}$

Modified Adaptive thresholding is utilized for segmentation process. This method proves its effectiveness in terms of lesser processing time and memory. However, it produces lesser computational power when compared to OCR.

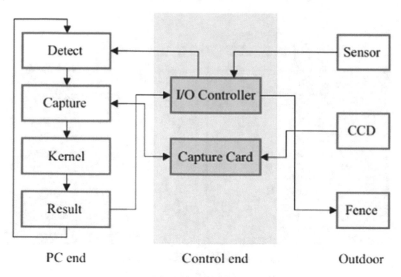

Fig. 4 Process of extraction using sensor

1.1.4 Computation for Multiple License Plates

All the above suggested techniques are effective and adaptable for a clear image of single license plate. But in a real-time scenario, a traffic camera can obtain images with certain resolution, only for a certain number of images. In such cases, the process, become tedious and this may raise the complexity of the processes, thereby making the system prone to error. This paper [10] provides a novel method for extraction of characters from numerous license plates at a single time. From the Fig. 5, it is shown that the process involves converting a RGB image to YDbDr format. This system was designed specifically for China where the number plates are of either blue or yellow (for heavy vehicles). Thus, with such high precision, it can be implemented in various important places. Nevertheless, the major short-coming in this method is the requirement of high quality image thereby increasing the cost of the system. Likewise, in yet another paper [6], similarly, the image from the traffic cam is identified by probabilistic feature detection. But the main draw-back in this method is to achieve the correct probability and to overcome the increased complexity when dealing with multiple plates.

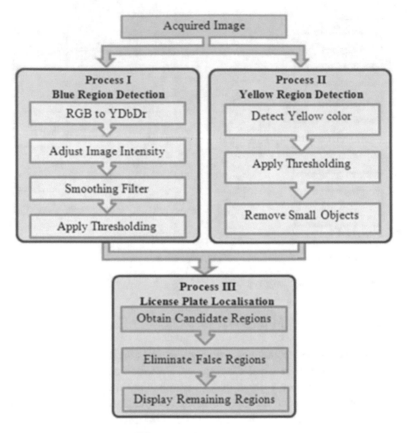

Fig. 5 Detection of multi-colored license plate

2 Suggested Method for License Plate Character Extraction

A suggested method, as shown in Fig. 6, involves integration of API with a motion detector sensor. With advancement of technology in recent years, Google engine have let out various improvised version of resources for innumerable purposes. Utilizing these resources has made computational analysis with stored database much simpler. With the aid of motion sensor, the arrival of vehicle can be detected and with optimum resolution, an image is taken from the front and rear end of the car. These images are further processed with the help of Google API, which in turn provides the required result with more accuracy. This is done by integrating the camera with Raspberry pi, which processes and then sends this image to Google-Vision-API. By suitable programming, the text can be retrieved from the

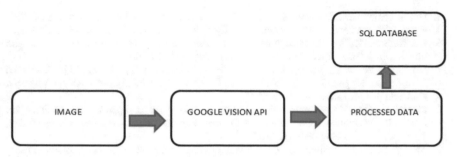

Fig. 6 Proposed method of detection and extraction

image upon which the characters are extracted. Using a SQL database, the data is stored. When required, the data can be extracted, deleted or inserted as and when required. Unlike XML, the data is not redundantly large and thus computation space is further conserved.

3 Summary

Thus, various methods were analyzed pertaining to license plate detection and recognition. Each method was studied thoroughly and were estimated and compared on their performance basis. By considering all the shortcomings in the subsisting systems, a novel method of image recognition was put forth for license plate detection along with car model recognition. The proposed method was tested in terms of performance, operating speed, cost and its effectiveness to be implemented in real-life situations. Under all these criteria, the proposed method of detection involving API provides precise output with a sturdy database to support in various applications. With the help of faster processing system by raspberry pi, computational error can also be minimized. Thus, by means of simpler programming structure, it can be made adaptable to various use-cases.

References

1. Foote, R.S.: Automatic vehicle identification: tests and applications in the late 1970's. IEEE Trans. Veh. Technol. **29**(2), 226–229 (1980)
2. Wu, B.-F., Lin, S.-P., Chiu, C.-C.: Extracting characters from real vehicle license plates out-of-doors. IET Comput. Vis. **1**(1), 2–10 (2007)
3. Liu, G., Ma, Z., Du, Z., Wen, C.: The calculation method of road travel time based on license plate recognition technology. In: Proc. Adv. Inform. Tech. Educ. Commun Comput. Inform. Sci., vol. 201, pp. 385–389 (2011)
4. Kranthi, S., Pranathi, K., Srisaila, A.: Automatic number plate recognition. Int. J. Adv. Tech. **2**(3), 408–422 (2011)

5. Du, S., Ibrahim, M., Shehata, M., Badwy, W.: Automatic License Plate Recognition (ALPR): a state-of-the-art review. IEEE Trans. Circ. Syst. Video Technol. **23**(2), 311–325 (2013)
6. He, H., Shao, Z., Tan, J.: Recognition of car makes and models from a single traffic-camera image. IEEE Trans. Intell. Transp. Syst. **16**(6), 3182–3192 (2015)
7. Do, H.N., Vo, M.T., Vuong, B.Q., Pham, H.T., Nguyen, A.H., Luong, H.Q.: Automatic license plate recognition using mobile device. In: 2016 International Conference on Advanced Technologies for Communications (ATC)
8. Miyata, S., Oka, K.: Automated license plate detection using a support vector machine. In: 2016 14th International Conference on Control, Automation, Robotics & Vision Phuket, Thailand, 13–15th Nov 2016 (ICARCV 2016)
9. Mahesh Babu, K., Raghunadh, M.V.: Vehicle number plate detection and recognition using bounding box method. In: 2016 International Conference on Advanced Communication Control and Computing Technologies (ICACCCT)
10. Asif, M.R., Chun, Q., Hussain, S., Fareed, M.S.: Multiple licence plate detection for Chinese vehicles in dense traffic scenarios. IET Intell. Transp. Syst. **10**(8), 535–544 (2016)
11. Yimyam, W., Ketcham, M.: The automated parking fee calculation using license plate recognition system. In: International Conference on Digital Arts, Media and Technology (ICDAMT) (2017)

Adoption of Big Data Streaming Techniques for Simultaneous Localization and Mapping (SLAM) in IoT-Aided Robotics Devices

Nyasha Fadzai Thusabantu and G. Vadivu

Abstract The evolution of low-powered devices through the Internet of Things (IoT) has enabled the technology community to come up with solutions to problems faced by pervasive networks. IoT enables communication between devices, i.e., "machine to machine" communication. With this regard, the evolution of IoT-aided Robots was birthed. Robotic devices constantly need to communicate and share their location and the surrounding environment, a concept known as Simultaneous Localization and Mapping (SLAM). Normally this data is shared through traditional techniques, but with the exploding data universe, there is need to come up with an alternative, fast, and efficient methods for management and transfer of data. This research proposes the adoption of big data streaming techniques to manage data transfer and communication during SLAM. Ultimately, big data streaming techniques will be used in critical applications where the analytic process has to happen in real time and decisions need to be made within a short time.

Keywords SLAM · Big data · Streaming · IoT · Robotics · Localization

1 Introduction

IoT has evolved from smart objects. These include micro-services, micro-electromechanical systems (MEMS) and the convergence of wireless technologies. An object is regarded as "smart" when it can exchange data with other devices on the internet [1]. Any sensor can be an object. Consider an example where the temperature and pressure of an environment need to be monitored, the object should send the monitored data by connecting to the Internet [1].

N. F. Thusabantu (✉)
Big Data Analytics, SRM University, Chennai 603203, India
e-mail: nyasha.fadzai@gmail.com

G. Vadivu
Department of Information Technology, SRM University, Chennai 603203, India
e-mail: vadivukar@ktr.srmuniv.ac.in

© Springer Nature Singapore Pte Ltd. 2019
P. K. Mallick et al. (eds.), *Cognitive Informatics and Soft Computing*,
Advances in Intelligent Systems and Computing 768,
https://doi.org/10.1007/978-981-13-0617-4_31

In this case, we are dealing with robotic objects. In robotics, the mapping or generation and familiarization of environment are handled by a concept called Simultaneous localization and mapping (SLAM). SLAM system major trait is that it should operate in real time not offline map-generation which means all processes need to complete before the next camera frame is available [2, 3]. However, the issue is not about simply algorithms faster, because as it explores the area SLAM system is continuously updating and expanding its map, therefore, storage becomes a major problem [4, 5]. The alternative would be to introduce strategies to limit the map points which can potentially be matched. This, however, introduces further problems, the major one being restricting the ability of the system to identify and recognize places it has maneuvered before [6]. This makes the tracking more prone to many failures. In such scenarios, this is then where the adoption of big data streaming techniques come into play. These challenges can be overcome by utilizing cache memory window in big data streaming technologies such as Kafka, Flume, Storm, Spark, etc.

Problem statement: SLAM system operates in real time making it very crucial to process frames before the next one arrives, but the challenge is any IoT robotic devices should be optimized to use low power and this compromises on process speed and storage space when the data being received becomes large.

This research explores Kafka architecture as a means to stream the data from the IoT-aided robotic devices to their environment during the process of SLAM.

2   Literature Review

Massive amounts of data comes from IoT "things" that send data in real time. These things can be devices or sensors. There are three main steps followed in processing large volumes of data from the sensor at scale: ingestion, storage, and analytics [7, 8]. The biggest concern is how these huge amounts of IoT data can be analyzed and stored in a timely manner, which does not result in losing any information [9].

Basically, at each layer, there is a need for a broker that maintains a balance between the data produced and the data that is consumed. For instance, if millions of devices are sending data at regular intervals to an IoT platform, [9] it needs to buffer the data before it can process it. Consider there are multiple consumers of this device data. The question is how do we connect the data to the specific right channel for the specific right kind of analysis?

There are protocols used in IoT for each purpose [10, 11]. These can range from basic infrastructure protocols to data protocols. Some protocols use a many-to-many paradigm and the broker decouples the publisher to the subscriber and acts as a message router.

The best suited for this is an IoT protocol that is called Message Queuing Telemetry Transport (MQTT). It uses publish-subscribe messaging system thereby ensuring quality of service (QoS), secure communication, and persistence [12].

Fig. 1 Kafka environment

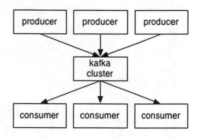

Currently, SLAM faces major challenges when the data being sent is greater than the speed at which it is being dispensed. Kafka is more focused on transmitting data to completion in the shortest possible time rather than the reliability of that data [13] (Fig. 1).

3 Design and Methodology

The ideal solution would be adopting big data streaming techniques that do not result in any data losses while receiving live streams of data from the devices. Each frame/stream has to be processed at the fastest possible time and persistently stored for a reasonable amount of time until it is proven by any statistical model that the likelihood or probability of that data being invoked in the shortest possible time is minimal to zero. This data can be analyzed and transformed into memory then it can be transmitted to disk for storage. Also, cloud-based parallel implementation for results that need to stay longer in memory [11, 12] (Fig. 2).

Simulation of SLAM adopted from python based FastSLAM application while the data to be simulated in an IoT environment is obtained from data generators. This process is continuously fed into the MQTT-Kafka Bridge while monitoring various parameter using DataDog dashboard (Fig. 3).

Real-time analysis obtained by creating an IoT environment using Raspberry Pi 3 Model B and testing model on various sensors.

4 Experimental Results

The hypothesis is that big data streaming technologies by nature improve the time complexity and speed in processing of large and real-time data.

Equation 1 Hypothesis of result

$$\lim_{n \to \infty} \left(1 + \frac{1}{n}\right)^n I * ST = \uparrow S,T \tag{1}$$

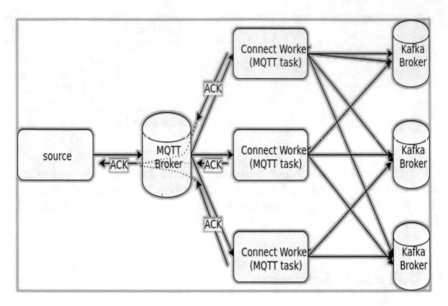

Fig. 2 MQTT-Kafka bridge

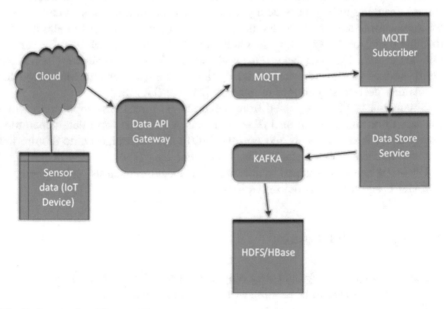

Fig. 3 Proposed architecture diagram

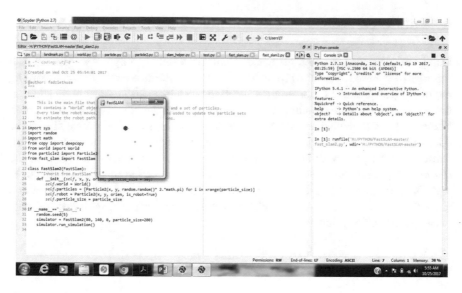

Fig. 4 FastSLAM simulation

I   Input (unstructured data)
ST   Streaming technique
S   Speed
T   Time.

The above screenshots were taken from the continuous monitoring of the MQTT-Kafka bridge during SLAM. Image processing technique used was canny edge detection from OpenCV library (Fig. 4).

The Kafka results generated and compared:

617,425 records/sec—8623 record faster
78.12 MB/sec—8 MB/sec faster.

5 Conclusion

Data pushed from the sensors is usually unstructured therefore this requires it to be continuously analyzed, transformed and moved to memory before it can be stored on disk. Ultimately, adoption of big data streaming techniques will increase the processing speed and improves efficiency in critical applications where the analytic process has to happen in real time and decisions need to be made within a short time.

6 Future Directions

Using any streaming technologies or mechanisms would be of great advantage as it improves on speed. We are proposing the use of Kafka among other streaming methodologies because of its ability to keep the data in a persistent state. In addition to that, IoT works with certain protocols depending on the layer of implementation. Since Kafka uses publish-subscribe model it is only logical to adopt MQTT as the data protocol because of its operation that is modeled around the same publish subscribe principle. With this in mind, it is of paramount importance that in the future a deeper research be done with respect to authentication of the messages from a Kafka broker. Bridging MQTT and Kafka opens a door of vulnerabilities and since this research is directed towards critical data applications security of that data becomes the major priority.

References

1. Ray, P.: A survey on internet of things architectures. EAI Endorsed Trans. Internet Things **2**(5), 151714 (2016)
2. Chatterjee, A., Rakshit, A., Singh, N.N.: Simultaneous Localization and Mapping (SLAM) in mobile robots. In: Vision Based Autonomous Robot Navigation. Studies in Computational Intelligence, vol. 455. Springer, Berlin (2013)
3. Yuan, X., Zhao, C.: 3D-SLAM based on point-plane matching. ROBOT **33**(2), 215–221 (2011)
4. Begum, M., Mann, G., Gosine, R.: An evolutionary algorithm for simultaneous localization and mapping (SLAM) of mobile robots. Adv. Robot. **21**(9), 1031–1050 (2007)
5. Williams, H., Browne, W., Carnegie, D.: Learned action SLAM: sharing SLAM through learned path planning information between heterogeneous robotic platforms. Appl. Soft Comput. **50**, 313–326 (2017)
6. Dissanayake, M., Newman, P., Clark, S., Durrant-Whyte, H.F.: A solution to the simultaneous localization and map building (SLAM) problem. IEEE Trans. Robot. Autom. **17**(3), 229–241 (2001)
7. Google Cloud Platform: Architecture: Real-Time Stream Processing for IoT| Architectures | Google Cloud Platform. [online] Available at: https://cloud.google.com/solutions/architecture/real-time-stream-processing-iot (2017). Accessed 30 July 2017
8. Hortonworks: Real World Examples: Real-time Data From Internet of Things. [online] Available at: https://hortonworks.com/blog/real-world-examples-real-time-data-internet-things/ (2017). Accessed 03 Aug 2017
9. Kamburugamuve, S., Christianse, L., Fox, G.: A framework for real time processing of sensor data in the cloud. J. Sens. **2015**, 1–11 (2015)
10. Postscapes.com. IoT Standards & Protocols Guide | 2017 Comparisons on Network, Wireless Communications, Security, Industrial. [online] Available at: https://www.postscapes.com/internet-of-things-protocols/ (2017). Accessed 10 Sep 2017
11. Survey Paper for IOT: Capacity planning and cloud technologies. Int. J. Sci. Res. (IJSR) **5**(4), 1812–1815 (2016)
12. Kim, S., Kim, D., Oh, H., Jeon, H., Park, H.: The data collection solution based on MQTT for stable IoT platforms. J. Korea Inst. Inf. Commun. Eng. **20**(4), 728–738 (2016)
13. Bendre, M., Thool, V.: Analytics, challenges and applications in big data environment: a survey. J. Manage. Analytics **3**(3), 206–239 (2016)

A Proposed Framework for Evaluating the Performance of Government Initiatives Through Sentiment Analysis

P. S. Dandannavar, S. R. Mangalwede and S. B. Deshpande

Abstract The main goal of any Government is to secure the basic rights of its citizens, promoting the welfare (in general) and economic growth while maintaining domestic tranquilly and achieving sustainable development. Government and its agencies introduce several initiatives for the welfare of its citizens and to improve the quality of public services. The performances of all such initiatives need to be evaluated with the involvement of the citizens at large, to draw insights into the public acceptance of such initiatives. These insights gained can be used to restructure/transform the government initiatives to make them more successful. Routinely citizens use social networks to post their opinions, views, and comments. Analyzing social content media is thus very important for Governments to take decisions. Sentiment Analysis, as a tool can be used to analyze the citizens feedback expressed on all such social media.

Keywords Sentiment analysis · Lexicon based methods · Machine learning
Twitter · Government initiative

1 Introduction

Sentiment Analysis (SA) deals with extraction of sentiment (+ve/−ve/neutral) from a piece of text (document/sentence/phrase) towards an entity (person/idea/topic/sub-topic). Though the field of SA emerged way back in the 1990s, it gained major

P. S. Dandannavar (✉) · S. R. Mangalwede · S. B. Deshpande
Department of CSE, KLS Gogte Institute of Technology,
Belagavi, Karnataka, India
e-mail: padmad@git.edu

S. R. Mangalwede
e-mail: mangalwede@git.edu

S. B. Deshpande
e-mail: sbdeshpande@git.edu

© Springer Nature Singapore Pte Ltd. 2019   321
P. K. Mallick et al. (eds.), *Cognitive Informatics and Soft Computing*,
Advances in Intelligent Systems and Computing 768,
https://doi.org/10.1007/978-981-13-0617-4_32

attention from 2004 onwards. The importance and value of social media monitoring, and SA as a means to achieve it, was noticed by researchers and analysts around 2007 [1]. SA has been a hot topic of research for quite some time. Researchers have used SA for analyzing product reviews, movie reviews, poll predictions, i.e., for commercial purposes. Individuals seek the "online reputation" of a product before making a product purchase, rather than gather feedback from friends/relatives. SA helps organizations to analyze their customers. Though SA has been used by organizations for a decade now, its value for Government/ Government agencies was not very apparent until 2010, when Gartner suggested the Open Government Maturity Model and at the fourth level of maturity, SA was proposed as a means to achieve collaboration for Governments [1].

Growing importance of SA coincides with the growth of social media (reviews, blogs, micro blogs, online discussion forums, twitter, Face book etc.) [2]. People express their opinions on these social platforms. The opinions are not only about products/services/movies but also about various topics and issues from the social domain [3]. Citizens around the world are also using on line platforms for expressing their opinions on various government missions/policies/initiatives. These opinions reflect the sentiment of the user expressing it. Government agencies can use SA as a Social Media Analytics tool to analyze such citizen sentiment and get feedback of their newly launched missions and policies [4]. Social media can provide deep insights into what citizens want [5]. Analyzing social content is thus very important for Government and their agencies. This can help in transformation or restructuring of the Government efforts to make the Government Initiatives more successful and sustainable [4]. **In short, Governments can make use of SA to measure public service satisfaction**.

The organization of the remainder of the paper is as follows: Sect. 2 identifies the benefits of measuring public service satisfaction. Section 3 highlights the importance of Twitter as a possible and important source of data for analysis. Section 4 presents the summary of related work carried out in the area of analyzing the government initiatives. Section 5 presents an overview of the proposed framework.

2 Benefits of Measuring Public Service Satisfaction

Twitter SA has been used in several areas—product reviews, movie reviews, poll predictions, etc. However, not much attention has been given to using citizen SA for smart city governance. With the exponential growth of social media in the last few years, Governments have realized that SA can be a great tool to understand their citizens better. Larger populations of citizens are living in cities than in rural areas and cities are getting smarter than ever before. The common challenge that Governments across the world are facing, is to ensure that cities are governed and managed in an efficient way. It is thus inevitable for public agencies and Governments to move closer to their citizens to better understand their sentiment.

Ignoring citizen sentiment can impact Governments [5]. For smart city governance and monitoring, efforts are being made by Governments across the world to move more closer to their citizens. Twitter SA is opening new opportunities to help Governments move more closer to their citizens.

Several initiatives have been introduced by Governments and its agencies aimed at improving the quality of services. Societal involvement in evaluation of government initiatives is necessary. Evaluation of such initiatives, using SA, makes it possible to identify public opinion—whether +ve/−ve/neutral. The results of such evaluation can help in restructuring the government efforts to make the initiatives more successful. If all such restructuring efforts are linked to the performance evaluation by the society at large, the transformation in public services quality will be successful and sustainable [5]. The opinions of people, about different government schemes, can be analyzed using SA. The results of such analysis can be used to predict whether a scheme has been successful at people's level or not.

Social SA (SSA) can serve as a useful tool and be used by Government and its agencies to regularly keep a tab on the pulse of its citizens, which can pave way for better governance [5]. SSA helps in identifying—the most talked about programs—whether good/bad; citizens feelings about such programs and policies; the most positively talked about aspects of the programs, etc. Government agencies can then use answers obtained for such questions to finetune their policies to address specific citizens concerns.

3 Twitter—Source of Data for Analysis

Twitter is a micro-blogging platform for users to voice their opinion publicly about government issues, current affairs, policies, product items, movies, sports and so on. The large volume of data that is available at one place can be mined and useful results produced. Twitter data can also be analyzed to show people's opinion on just about anything—a product, a service, a company, a movie, a particular topic etc. The results of such analysis can help the public opinion reach the concerned in a better and organized manner.

Another aspect that works in favor of Twitter data is that—twitter data (tweets) is publicly and freely available. Tweets can be collected using twitter API which is relatively more simple compared to scraping blogs from the web.

4 Previous Related Work

The work of Prerna Mishra, Ranjana Rajnish, and Pankaj Kumar [3] focused on analyzing the performance of the "Digital India" campaign. 500 related tweets were collected and analyzed using a sentence—level, dictionary-based approach to classify the sentiments as +ve, −ve or neutral. The results of analysis were in favor of the initiative—50% positive, 20% negative and 30% neutral.

Endah Susilawati's [6] chose "Electrical Services in Indonesia" as a specific case study to measure public service's satisfaction using SA. A crawling program was built which was used to collect the tweets used in the work from a collection of Indonesian tweets. Naive Baye's algorithm was used to classify the tweets. Analysis of the results so obtained revealed that the electrical service in Indonesia performed poorly in the citizen's perspective. This was evident in the number of tweets classified: +ve (12), −ve (162) and neutral (31).

The authors [4] in their work attempted to analyze the impact of different government programs viz., Swachh Bharat, Digital India, Beti Bachao Beti Padao. They used the official government site (www.my.gov) to extract the data needed for analysis. The authors developed a system that not only determined the response of individuals but also was capable of predicting whether the scheme will be successful or not. The work on SA was based on two important parts viz., (1) Data Extraction and preprocessing and (2) Classification. The performance of classification algorithms was evaluated based on their precision and recall.

Sahil Raj, Tanveer Kajla [7] focused on Swachh Bharat Abhiyan—a cleanliness campaign launched by the Government of India. The authors: (a) analyzed the pattern of the tweets to identify the citizen's perception regarding this initiative and (b) determined the popularity of the mission region wise by extracting the geolocations of the tweets. This information can be used to identify and apply strategies to make the campaign gain popularity in the lesser known regions of India. The work concluded that the campaign is a success among the people of India.

In the opinion of Ravi and Sandipan [5], Government agencies can benefit by monitoring and analyzing the citizen sentiment from social media. Such kind of analysis can also help Government's to take decisions. The authors proposed a model for the same and performed SA using an IBM system—CCI (Cognos Consumer Insight). A social benefits organization in the US was chosen to carry out the work. The scope of the study included not only understanding how the various programs of the organizations were perceived by citizens, but also identifying the root causes leading to the perception. This key information was then used prepare an actionable roadmap to reduce negative perceptions.

The research gap identified in the work [1] was the lack of twitter SA based on location. Work in this direction led to a novel approach of filtering tweets based on location. This, the author felt, is useful in determining what kind of sentiment is expressed by each part of the country towards a particular initiative. The author chose to mine twitter data related to a very recent government initiative—"demonetization", with an aim of concluding whether this kind of region wise classification can yield more specific results.

Mengdi Li et al.'s work [8] on twitter SA resulted in a sentiment model that can be of help for Government agencies—to monitor the moods of its citizens and also track the fluctuations in citizens moods, using mapping techniques. A sentiment classifier was built using the multinomial Naive Baye's classifier, which employed a variety of features including "emojis". 7.5 million tweets generated in New York

City were retrieved, using Twitter's streaming API, for the purpose of study. A framework was proposed by the authors to map citizen sentiment from Twitter. A specific micro-blogging feature, emoji, was considered for feature extraction to build the sentiment classification system. The results of experiments showed that the sentiment model built was effective.

Payal Yadav et al. [9] work focused on SA based on text and emoticons and more on showing the importance of emoticons in SA. The authors discuss about the importance of emoticons in SA and how the SA results are affected by their presence. Visual cues (like smiling/crying/laughing) help in identifying the sentiment expressed in a day-to-day face-to-face communication. People use emoticons as an alternative to visual cues in text communication. The polarity of a statement can change if emoticons are used. Focusing on the relationship that exists between text and emoticons helps in classifying the polarity of text in an exact manner.

In the work [10] twitter API and the open-source R tool were used to collect, pre-process and analyze social data using SA. The aim was to determine citizen's opinion about the scheme's/initiatives announced by the central government. The authors presented a case study of eight Indian Government initiatives to emphasize the importance of analyzing online opinions generated by users. The Naive Baye's machine learning algorithm was applied to the datasets collected and sentence level SA was performed using PMI, which was used to calculate the semantic orientation of phrases and words. Results were depicted in terms of +ve/−ve and neutral tweets for eight different government schemes. The results indicated how the polarities vary with different schemes.

Monika Sharma's [11] work on SA considered the government initiative announced on November 8, 2016 i.e., demonetization, for analysis. The aim was to find out the opinions of citizens on this initiative through tweets collected from twitter. Using the AFINN directory, for each such word its rating (from +5 to −5) was extracted. Using the rating of each word, the average rating of the tweet was calculated as either +ve/−ve.

Author [12] opines that Government's can use big data to promote the public good, just like businesses use big data to pursue profits. Governments can use big data to help them overcome national challenges and serve their citizens. The author in his work: (a) compared the two sectors (business & government) with respect to decision-making processes, goals etc (b) examined several current applications in technologically advanced countries and (c) examine big data applications and initiatives in the business sectors that can be implemented by governments.

5 Overview of the Proposed Framework

After the relevant tweets have been extracted (using hash tags to streamline and increase the effectiveness of the search), these tweets must be brought into proper format before performing SA. The various steps to be followed—from collection of tweets to their classification—are shown in the proposed framework shown as Fig. 1.

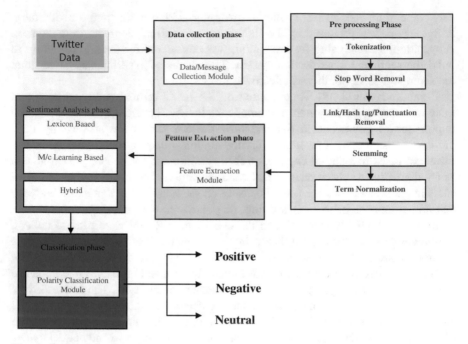

Fig. 1 Proposed framework

5.1 Micro-Blog

There are various sources of data for performing SA. Twitter makes available tremendous amount of data (tweets) as a public timeline for easy gathering. Hence twitter will act as the source of input to the system.

5.2 Data Collection Module

Data collection is an important part of SA. This module performs data fetching operation. Twitter allows researchers to collect tweets by using a twitter API. To collect tweets one must create a developer account in twitter to obtain twitter credentials (i.e., API key, API secret key, Access Token and Access Token Secret). These keys and tokens are then used to obtain twitter authorization to extract tweets. Data can be retrieved as XML/JSON format.

5.3 Data Preprocessing Module

Data that is collected is raw in nature—often containing lot of errors and incon-sistencies. Before such data is used it should undergo cleaning operations. Preprocessing aims to process and present the tweets in an organized format and increase the machine understanding of the text, as most tweets are in unstructured text form [12].

Preprocessing includes

- Subject Capitalization—for the easiness of feature selection and in order to remove the difference between "Text" and "text", all the words are converted either into lower or upper case.
- Tokenization—a tweet is segmented into tokens using word delimiters, such as space and punctuation marks. All such tokenized words form a bag of words.
- Stop-word removal—words that carry that no meaning and do not help in determining the sentiment of text such as "a", "an", "the", "has", "have", etc., are removed from the bag of words.
- Link and hash tag removal—external and address links do not help in detecting polarity and features and hence must be removed. Hash tag's are also removed, as the hash tagged item ma not refer to the subject [12].
- Punctuation removal—punctuation marks also carry no meaning for the analysis and hence they can be removed from the input text.
- Stemming-some words have similar roots but differ only in affixes, i.e., words are constant at the front and vary at the end. For eq., connection, connective, connected, connecting have the same root—connect. Taking off the variable part is called stemming and the residual part is called the stem.
- Term normalization—involves changing the form of verb/adjective back to base form. For eq., the verb "talking" is replaced with "talk". Abbreviation's are replaced with its meaning.
- Slang handling—involves normalizing unstructured tweets by removing the repeated characters. For eq., the word "gooooooood" is reduced to "good".

5.4 Feature Extraction/Selection Module

After preprocessing, relevant feature extraction for classification and feature extraction/selection (FS) is the difficult task in SA. The main task of FS is to decrease the dimensionality of the feature space and thus the computational task. FS techniques reduce the feature vector length by mapping the high-dimensional data on lower dimensional space. FS techniques can help improve the classification accuracy and also provide better insight into important class features, resulting in a better understanding of sentiments.

Some of the features include

(a) Term presence and frequency: It usually consists of n-grams of words and their frequency counts.
(b) Parts of speech tagging: words in the text are tagged with their respective parts of speech in order to extract adjectives nouns verbs which add meaning to the sentiment.
(c) Opinion words and phrases: words or phrases that indicate opinion of the text
(d) Negation: presence of words like "not", "nor", "neither" may reverse the sentiment of whole sentence. E.g.: "not good"
(e) Twitter specific features: Presence of emoticons in tweets, positive or negative hash tags are all twitter specific features which add meaning to the sentiment.

5.5 Sentiment Analysis Module

The task of SA can be performed using either the Lexicon/Lexical based methods, Machine Learning based methods or hybrid methods.

5.5.1 Lexicon Based Methods

The vocabulary-based approach utilizes supposition lexicon with sentiment words and match them with the information for deciding extremity. There are three systems to develop a vocabulary: manual development, corpus-based strategies and word reference based techniques. The manual development is a troublesome and tedious assignment. Corpus-based strategies can deliver feeling words with moderately high exactness. In lexicon-based methods, the thought is to first gather a little arrangement of conclusion words physically with known information, and after that to develop this set via hunting in the WordNet word reference down their equivalent words and antonyms.

5.5.2 Machine Learning Based Methods

Machine learning (ML), is a branch of artificial intelligence.

ML for SA needs two data sets viz., a training and a test set. A supervised learning classifier uses the training set to learn and train itself with respect to the differentiating attributes of text, and the performance of the classifier is tested using test dataset. SA using ML starts with collection of dataset containing labeled tweets. This dataset should be first preprocessed using various Natural Language processing (NLP) techniques. Then features that are relevant for SA need to be extracted and finally the classifier is trained and tested on unseen data. Several machine learning algorithms like Maximum Entropy (ME), Naive Baye's (NB) and Support Vector Machines (SVM) are usually used for classification of text (tweets).

5.5.3 Hybrid Method

Finally in the hybrid approach, the blend of both the ML and the dictionary based methodologies can possibly enhance the opinion characterization execution.

5.6 Polarity Classification Module

This step includes analyzing preprocessed tweets and assigning polarity (i.e., a number ranging from 1 to −1) to each word present in the tweet and thus the polarity or sentiment of entire tweet is determined. Once the polarities are added it becomes easy to understand sentiment or opinion of users. Depending on the polarity the tweets are classified into positive or negative or unbiased.

- Positive Tweets—are the tweets which demonstrate a decent or positive reaction towards something. For instance tweets such as: "It was an inspiring movie!!!" or "Best movie ever".
- Negative Tweets—can be classified as the tweets which show a negative response or oppose towards something.
  For example tweets such as—"Waste of time" or "Worst movie ever".
- Neutral tweets—can be classified as the tweets which neither show a support or appreciate anything nor oppose or depreciate it. It also includes tweets which are facts or theories. For example a tweet such as—"Earth is round".

6 Conclusions

By analyzing citizen's opinions, using SA, of its various schemes/initiatives, the Government and/or its agencies stand to benefit in the following few ways:

(1) Summarization of opinions—so that Governments can have participation of citizens by knowing their opinion, in formulation and implementation of various policies [3].
(2) Governments can gain insights into citizen's opinions about their public initiatives/services [5].
(3) Government can keep tab on the pulse of its citizens which will pave way for better governance.
(4) Agencies can obtain deep insights into how citizens feel. The insights so gained can be used for smart city monitoring and governance.

(5) Proves helpful for evaluation of Government performance from citizen's perspective instead of undertaking exhaustive people's survey, which is both expensive and time consuming.

(6) Using citizen's opinion expressed on social media, the Government can find out the popular schemes and feedback of citizens.

References

1. Gupta, F., Singal, S.: Sentiment analysis of the demonetization of the economy 2016, India region wise. In: IEEE 7th International Conference on Cloud Computing, Data Science and Engineering—Confluence 2017 (978-1-5090-3519-9/17 © 2017 IEEE)
2. Kim, G.-H., Trimi, S., Chung, J.-H.: Big-data applications in the government sector. Contributed Article Commun. ACM **57**(3) (2014)
3. Mishra, P., Rajnish, R., Kumar, P.: Sentiment analysis of twitter data: case study on digital india. In: IEEE 2016 International Conference on Information Technology (InCITe)—The Next Generation IT Summit (978-1-5090-2612-8/16 ©2016 IEEE)
4. Mahajan, A., Divyavir, R., Kumar, N., Gade, C., Deshpande, L.A.: Analysing the impact of government programmes. Int. J. Innovative Res. Comput. Commun. Eng. **4**(3) (2016, March)
5. Arunachalam, R., Sarkar, S.: The new eye of government: citizen sentiment analysis in social media. In: IJCNLP 2013 Workshop on Natural Language Processing for Social Media (SocialNLP), pp. 23–28, Nagoya, Japan, 14 Oct 2013
6. Susilawati, E.: Public services satisfaction based on sentiment analysis. case study: electrical services in Indonesia. In: IEEE 2016 International Conference on Information Technology Systems and Innovation (ICITSI) Bandung—Bali, 24–27 Oct 2016 (978-1-5090-2449-0/16 ©2016 IEEE)
7. Raj, S., Kajla, T.: Sentiment analysis of Swachh Bharat Abhiyan. Int. J. Bus. Analytics Intell. **3**(1) (2015) (ISSN: 2321-1857)
8. Li, M., Ch'ng, E., Chong, A., See, S.: The new eye of smart city: novel citizen sentiment analysis in twitter. In: International Conference on Audio, Language and Image Processing 2016 (ICALIP 2016) (978-1-5090-0654-0/16 ©2016 IEEE)
9. Yadav, P., Pandya, D.: SentiReview: sentiment analysis based on text and emoticon. In: IEEE International Conference on Innovative Mechanisms for Industry Applications (ICIMIA 2017) (978-1-5090-5960-7/17 ©2017 IEEE)
10. Naiknaware, B.R., Kawathekar, S., Deshmukh, S.N.: Sentiment analysis of indian government schemes using twitter datasets. In: International Conference on Recent Advances in Computer Science, Engineering and Technology, IOSR Journal of Computer Engineering (IOSR-JCE) (e-ISSN: 2278-0661, p-ISSN: 2278-8727), pp. 70–78
11. Sharma, M.: Twitter sentiment analysis on demonetization an initiative government of India. Int. J. Recent Trends Eng. Res. (IJRTER) **03**(04) (2017) (ISSN: 2455-1457)
12. Chong, W.Y, Selvaretnam, B., Soon, L.-K.: Natural language processing for sentiment analysis. In: IEEE 4th International Conference on Artificial Intelligence with Applications in Engineering and Technology (978-1-4799-7910-3/14 © 2014 IEEE)

The Extent Analysis Based Fuzzy AHP Approach for Relay Selection in WBAN

Subarnaduti Paul, Arpita Chakraborty and Jyoti Sekhar Banerjee

Abstract A revolutionary technology in the field of healthcare monitoring system to manage illness for maintaining wellness by concentrating on prevention and early detection of disease are popularly known as Wireless Body Area Networks (WBAN) which is highly localized wireless networks along with different sensors placed in the human body or surface mounted on the particular places of the body. Though WBAN is a specially designed sensor network to implement ubiquitous and affordable health care autonomously, anytime and anywhere, it faces numerous challenges like frequent link loss due to postural body movement, size, and complexity of the sensors, channel condition, and power consumption. This letter provides FAHP using the Extent Analysis scheme for relay node selection that prioritizes the vagueness of the decision-makers during the relay node selection procedure.

Keywords Relay selection · Wireless body area network · Two-hop extended star topology · Fuzzy AHP · Triangular fuzzy number

1 Introduction

Constant monitoring of a patient by concentrating on prevention and early detection of disease is the foremost objective of WBAN [5] which has excelled due to increasing exploration of wireless communication and semiconductor technologies concurrently. To achieve ubiquitous and affordable healthcare monitoring system

S. Paul · A. Chakraborty · J. S. Banerjee (✉)
Department of ECE, Bengal Institute of Technology, Kolkata 700150, India
e-mail: tojyoti2001@yahoo.co.in

S. Paul
e-mail: paulsubarna98@gmail.com

A. Chakraborty
e-mail: chakraborty_arpita2006@yahoo.com

© Springer Nature Singapore Pte Ltd. 2019
P. K. Mallick et al. (eds.), *Cognitive Informatics and Soft Computing*,
Advances in Intelligent Systems and Computing 768,
https://doi.org/10.1007/978-981-13-0617-4_33

autonomously, anytime and anywhere, WBAN has been designed specially and has been standardized by IEEE in 2007. This network contains mainly sensor nodes and a Hub or coordinator. Sensor node provides body related information. However achieving reliable communication and consumption of less battery power are the major two constraints of WBAN. As the well-being of human lives is the supreme criteria of medical services, hence reliable communication is the most demanding factor of WBAN. WBAN environment is directly function of the network conditions as postural body movements, i.e., movement of the sensor node produces link loss between the coordinator node and the sensor node which in turn create network instability and network partitioning.

The power of the device enables mobility to the sensor nodes and thus consumption of less battery power is very important issue in WBAN. As WBAN comprises of wearable devices or sensors, it appears to be quite strenuous to recharge or replace the devices customarily and hence demands less power consumption evidently. Therefore to meet our objectives as well as consume less device power, these sensor devices need to be organized in such a way that it exhausts less power. The solution to this problem is average transmission power consumption is to be minimized by setting low duty cycle i.e., sampling rate and is specified by IEEE 802.15.6 in technical requirement document (TRD) publication.

Another component which is responsible for calculating the communication power is the distance between the sensor nodes; greater is the distance more will be the communication power which is commonly even higher than the processing power. Here relay nodes are the best possible accessible option to reduce power consumption rate as well as few other severe drawbacks faced in WBAN.

Presently, plenty of research is aimed at how to choose the desired relay node in cognitive radios [1, 2], cooperative communications and sensor networks [5]. In this letter, authors deal with IEEE 802.15.6 (WBAN) where presence of multiple relays is accepted.

The notable findings of this correspondence are discussed below. First, for WBAN, authors have employed the extent analysis based Fuzzy AHP approach for relay selection which supports vagueness of the decision makers during the relay node selection procedure. Second, the proposed scheme can intensively meet both of the requirements of WBAN, i.e., enhanced reliability of the communication systems and optimum consumption of power.

The paper has been organized as follows. The system models along with problem formulation are discussed in Sect. 2. Section 3 provides the Relay Node Selection parameters. Section 4 provides the Extent Analysis based Fuzzy AHP approach for Relay Selection procedure. Section 5 emphasizes on the results and discussions. Finally the conclusion is described in Sect. 6.

2 System Model

The IEEE 802.15.6 illustrates the extension of two-hop type star network to provide an advanced communication connection using another helper or cooperating node, which is generally called as Relay Node or Carrier Node. In Fig. 1a, Node 2 and Node 1 act as a Relay node, which connect the Nodes (4 and 5) and Node 6 respectively to the Coordinator. When any node finds its connection with the hub is week then the communicating node or the relay node begins the procedure of two-hop communication. During the first hop, the node transmits its own frames to the carrier node and in next hop the relay node relays the same frames to the coordinator. There are two basic approaches for relay finding and choice method in the star network; one is Coordinator centric and the other is Relay node centric approach.

Coordinator-Centric Approach: In case of a weak communication connection linking the sender and the coordinator, the sender initiates a searching process for a suitable relay node to establish a connection with hub via relay. The procedure is executed by eavesdrop Management and ACK messages exchanging in between the carrier and the hub, which specifies that the connection between the sender and the hub is possible to establish via relay.

Relay Node Centric Approach: Unlike the previous approach, this approach attempts to cover both the connected or disconnected nodes to establish a connection with the suitable relay node for the purpose of communication with the coordinator. This is executed by sending a Broadcast Message generating from a relay node and targeted to the connected or disconnected nodes.

In both of these approaches, the carrier node which responds first with an Acknowledgement message gets selected. But in this paper, we deal with such type of wireless body area networks which supports the best carrier choice out of multiple carriers.

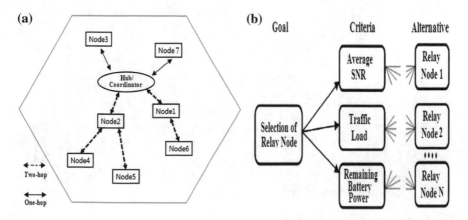

Fig. 1 **a** IEEE 802.15.6 two-hop star topology for WBAN. **b** FAHP hierarchy for advanced relay choice method

Figure 2 demonstrates FAHP hierarchy for proposed carrier node choice method. Initially when a sender is unable to connect to the coordinator due to loss of its direct link, the process to discover the candidate relay nodes is commenced by exchanging few management messages. Next, from the pool of available relay nodes the sender node determines the eligible one by using extent analysis based FAHP. Ultimately, the redemption of the direct connection between the sender node and the hub is performed to save the unnecessary wastage of energy and transmission latency

When an ACK frames or a BEACON is not received by the Sender node from the coordinator, it is assumed that the direct link is lost somehow and *RelayNodeDiscovery* message is transmitted to adjacent carrier nodes and starts Discovery Timer. On receiving the *RelayNodeDiscovery* message, each available relay node transmits an information frame (*RelayInfo*) containing average Signal-to-Noise Ratio, relay node incurred traffic load, and remaining battery power information. Here the FAHP-based carrier choice procedure has been adopted to find out the best carrier node out of the carrier pool to establish the link with the coordinator via the carrier or relay node. Due to random change of network condition, the direct link can be recovered through a holistic approach to save consumption of redundant energy and transmission latency. Hence, the Direct Link Recovery algorithm is considered.

Figure 2 shows the direct link recovery process which can either be coordinator-based- or sender node-based initiative.

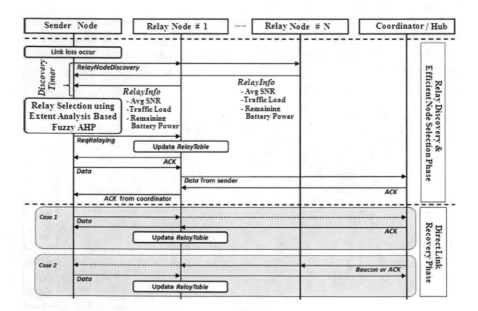

Fig. 2 Proposed relay selection scheme for the IEEE 802.15.6

3 Relay Node Selection Parameters

3.1 Average Signal-to-Noise Ratio (Avg SNR)

Average SNR for ith node (Avg SNR_i), denotes SNRs of the sender node and the hub respective to a candidate carrier or relay node i, which is formulated by

$$\text{Avg SNR}_i = \frac{\text{SNR}_{i \to \text{node}} + \text{SNR}_{i \to \text{cord}}}{\text{SNR}_{i \to \text{node}} \times \text{SNR}_{i \to \text{cord}}},$$

where $\text{SNR}_{i \to \text{node}}$ is the Signal-to-Noise Ratio from the ith candidate carrier node to the sender and $\text{SNR}_{i \to \text{cord}}$ is the Signal-to-Noise Ratio from the hub to the ith candidate carrier node.

3.2 Relay Node Incurred Traffic Load

The entire traffic load incurred on the ith relay candidate $\left(\text{TL}_i^{\text{total}}\right)$ is defined by

$$\text{TL}_i^{\text{total}} = \frac{1}{\text{TL}_i + \sum \text{TL}_{\text{Relaying Node}}},$$

where TL_i refers to the ith relay candidate's own traffic and to indicate the traffic of the different nodes communicating with the hub through ith candidate relay, $\text{TL}_{\text{Relaying Node}}$ is considered.

3.3 Remaining Battery Power

The remaining Battery Power of ith node $\left(E_i^{\text{ratio}}\right)$ is described in below-mentioned manner,

$$E_i^{\text{ratio}} = \frac{E_i^{\text{Rest}}}{E_i^{\text{Init}}},$$

where E_i^{Rest} and E_i^{Init} denote the remaining and initial power of ith node.

4 Extent Analysis Based FAHP Approach for Relay Selection

FAHP is a fuzzy extension of the traditional AHP process which finds its application in the various multi-criteria decision-making process. Unlike AHP [2, 7, 9] Fuzzy AHP takes into account the vagueness of the decision makers with the help of the fuzzy logic associated with it [1, 8]. It involves linguistic variables in association with the triangular fuzzy numbers (TFN) to evaluate the square matrix of both criteria and alternatives. Although a number of propositions have been made by different researchers regarding the implementation of Fuzzy AHP process and its triangular numbers, in this letter authors have considered Extent Analysis procedure, which is proposed by Chang [3] to evaluate the relative significance weights of both the alternatives and criteria by introducing S_i i.e. synthetic extent value of the evaluation matrix.

Step 1: Criteria and alternatives are compared and assigned a value using the Linguistic terms (see Table 1). For every TFN, its membership function is defined within the interval [0, 1] which can be stated as:

$$\mu_M(x) = \begin{cases} \frac{x}{m-l} - \frac{l}{m-l}, & l \leq x \leq m, \\ \frac{u}{u-m} - \frac{x}{u-m}, & m \leq x \leq u, \\ 0, & \text{otherwise,} \end{cases} \tag{1}$$

where $l \leq m \leq u$, u and l denote upper and lower value respectively and m is the modal value.

Step 2: The primary objective of FAHP is to find out the relative significance of each pair of factors by using TFN. The comparison matrix $A = \left(a_{ij}\right)_{n \times n}$ of n criteria or alternatives using TFNs is represented as follows:

Table 1 Triangular fuzzy numbers and linguistic representation

Saaty scale	Definition	Triangular fuzzy scale	Triangular fuzzy reciprocal scale
1	Equal significance	(1, 1, 1)	(1, 1, 1)
3	Moderate significance	(2, 3, 4)	(1/4, 1/3, 1/2)
5	Essential or strong significance	(4., 5, 6)	(1/6, 1/5, 1/4)
7	Very strong significance	(6, 7, 8)	(1/8, 1/7, 1/6)
9	Extreme significance	(9, 9, 9)	(1/9, 1/9, 1/9)
2	In between values of the two adjacent judgments	(1, 2, 3)	(1/3, 1/2, 1)
4		(3, 4, 5)	(1/5, 1/4, 1/3)
6		(5, 6, 7)	(1/7, 1/6, 1/5)
8		(7, 8, 9)	(1/9, 1/8, 1/7)

$$A = \begin{bmatrix} 1 & \tilde{a}_{12} & \cdots & \tilde{a}_{1n} \\ \tilde{a}_{21} & 1 & \cdots & \tilde{a}_{2n} \\ \cdots & \cdots & \cdots & \cdots \\ \tilde{a}_{n1} & \tilde{a}_{n2} & \cdots & 1 \end{bmatrix} \tag{2}$$

Step 3: Suppose $x = \{x_1, x_2, \ldots x_n\}$ and $g = \{g_1, g_2, \ldots g_n\}$ are assumed as an object set and goal set, correspondingly. As per extent analysis method, every object is judged and executes extent analysis (EA) to obtain every goal g_i. Hence m numbers of values are produced for every object with following notations:

$$M_{g_i}^1, M_{g_i}^2, \ldots, M_{g_i}^m \quad \text{where} \quad i = 1, 2, \ldots n,$$

The synthetic extent value for the ith object is described as

$$S_i = \sum_{j=1}^m M_{g_i}^j \otimes \left[\sum_{i=1}^n \sum_{j=1}^m M_{g_i}^j \right]^{-1} \quad \text{where } j = 1, 2, \ldots m \tag{3}$$

To calculate $M_{g_i}^j$, we need to execute the fuzzy addition process of the m numbers of EA values for a specific matrix so that,

$$\sum_{j=1}^m M_{g_i}^j = \left[\sum_{j=1}^m l_j, \sum_{j=1}^m m_j, \sum_{j=1}^m u_j \right] \tag{4}$$

$$\text{and } \sum_{i=1}^n \sum_{j=1}^m M_{g_i}^j = \left[\sum_{i=1}^n l_i, \sum_{i=1}^n m_i, \sum_{i=1}^n u_i \right] \tag{5}$$

The inverse of the Eq. 5 is defined as

$$\left[\sum_{i=1}^n \sum_{j=1}^m M_{g_i}^j \right]^{-1} = \left(\frac{1}{\sum_{i=1}^n u_i}, \frac{1}{\sum_{i=1}^n m_i}, \frac{1}{\sum_{i=1}^n l_i} \right) \tag{6}$$

Step 4: The Degree of possibility of $M_1 = (l_1, m_1, u_1) \geq M_2 = (l_2, m_2, u_2)$ is stated as

$$V(M_1 \geq M_2) = \sup_{x \geq y} [\min(\mu_{M_1}(x), \mu_{M_2}(y))] \tag{7}$$

A pair (x, y) which satisfies both the conditions $x \geq y$ and $\mu_{M_1}(x) = \mu_{M_2}(y) = 1$, results $V(M_1 \geq M_2) = 1$. As M_1, M_2 are convex TFNs, it is presented as $V(M_1 \geq M_2) = 1$ provided $m_1 \geq m_2$,

Fig. 3 Intersection point D

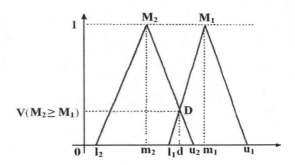

$$V(M_2 \geq M_1) = \text{hgt}(M_1 \cap M_2) = \mu_{M_1}(d), \tag{8}$$

Here D is considered as the highest intersection point between μ_{M_2} and μ_{M_1} (see Fig. 3). $V(M_1 \geq M_2)$ and $V(M_2 \geq M_1)$ are equally required for further computation.

Step 5: Degree of possibility for $M_i (i = 1, 2, \ldots, k)$ is stated in following manner where a TFN M to be greater than k convex fuzzy numbers

$$\begin{aligned} V(M &\geq M_1, M_2, \ldots, M_K) \\ &= V[(M \geq M_1) \text{ and } (M \geq M_2) \text{ and } \ldots \text{ and } (M \geq M_K)] \\ &= \min \quad V(M \geq M_i), \quad i = 1, 2, \ldots k. \end{aligned} \tag{9}$$

Suppose that

$$d'(A_i) = \min \ V(S_i \geq S_k) \quad \text{where } k = 1, 2, \ldots n, \& \quad k \neq i \tag{10}$$

Weight vector W' is represented in following manner;

$$W' = (d'(A_1), d'(A_2), \ldots, d'(A_n))^{\mathrm{T}} \tag{11}$$

Finally, normalized weight vectors are obtained using normalization process

$$W = (d(A_1), d(A_2), \ldots, d(A_n))^{\mathrm{T}} \tag{12}$$

Here W can be treated as a non-fuzzy number

5 Results and Discussion

Authors describe the relay selection FAHP process in wireless body area networks WBAN using the following example as (Table 2);

Where the three decision factors are expressed by A, B, and C. For this paper, authors consider avg Signal-to-Noise Ratio (A) as the least important decision

Table 2 Comparison matrix for criteria

Criteria	A	B	C
A	(1, 1, 1)	(1/3, 1/2, 1)	(1/5, 1/4, 1/3)
B	(1, 2, 3)	(1, 1, 1)	(1/3, 1/2, 1)
C	(3, 4, 5)	(1, 2, 3)	(1, 1, 1)

factor, Relay load incurred TL (B) as the second and remaining battery power (C) as the most significant decision factor. The service provider analyzed the situation and prioritized the criteria in the ratio 1:2:4. Hence, the pairwise comparison of three possible pairs are expressed as C:A = 4:1, A:B = 1:2 and B:C = 1:2. The normalized relative weights are calculated as $W = \{0.00, 0.33, 0.67\}$ with the help of Eq. 12, which differs from the true weight vector $W = \{0.143, 0.286, 0.571\}$ (see Fig. 4). It is obvious that the weights obtained from extent analysis are different from the true weights [10].

The weights of the relay nodes with respect to each criterion must be calculated after calculating the average weights of the criteria. Thus, the relay node with the highest weight (Relay 2 possess the highest value) is considered as the selected relay (see Fig. 5a, b). In this letter, the results are also verified through an online

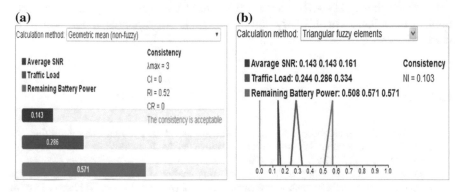

Fig. 4 **a** normalized weights of criteria using non-fuzzy method **b** triangular fuzzy elements

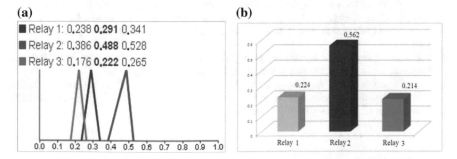

Fig. 5 **a** ranking of the relay nodes using Geometric mean method **b** extent analysis method

technical computing software [4] described by Ramík [6] which has considered geometric mean method (see Fig. 4a). We have simulated a scenario where three criteria, i.e., average SNR, traffic load and remaining battery power along with three relaying nodes have been considered. Different methods of FAHP may lead different decisions but in our problem both the methods (Extent analysis and geometric mean) converge to a same solution or decision.

6 Conclusion

WBAN has drawn much attention and became the most promising technology at medical service facilities due to its varied range of utility and important role to improve the human well-being. But researchers find different practical difficulties like frequent link loss, network instability, higher power consumption, and unreliable communication, while implementing WBAN in major application areas. Keeping the above criterion in mind, the proposed Extent Analysis based Fuzzy AHP Relay Node Selection scheme can successfully cope up with the problem of vagueness among the decision makers and the other above-discussed problems. Additionally, Extent Analysis based Fuzzy AHP algorithm performs better than other Fuzzy AHP schemes with respect to time complexity and hence implicit exploitation of this very algorithm in this correspondence turn out to be more effective.

References

1. Banerjee, J.S., Chakraborty, A., Chattopadhyay, A.: Fuzzy based relay selection for secondary transmission in cooperative cognitive radio networks. In: Proceedings of OPTRONIX, pp. 279–287, Springer (2017)
2. Banerjee, J.S., et. al.: Relay node selection using analytical hierarchy process (AHP) for secondary transmission in multi-user cooperative cognitive radio systems. In: Proceedings of ETAEERE, pp. 745–754, Springer (2018)
3. Chang, D.: Applications of the extent analysis method on fuzzy AHP. Eur. J. Oper. Res. **95**(3), 649–655 (1996)
4. Holecek, P., Talašová, J.: A free software tool implementing the fuzzy AHP method
5. Paul, S., Chakraborty, A., Banerjee, J.S.: A fuzzy AHP-Based relay node selection protocol for wireless body area networks (WBAN). In: IEEE Proceedings of OPTRONIX 2017 (Press), (2017)
6. Ramík, J., Korviny, P.: Inconsistency of pair-wise comparison matrix with fuzzy elements based on geometric mean. Fuzzy Sets Syst. **161**(11), 1604–1613 (2010)
7. Saha, O., Chakraborty, A., Banerjee, J.S.: A decision framework of IT-based stream selection using analytical hierarchy process (AHP) for admission in technical institutions. In: IEEE Proceedings of OPTRONIX 2017 (Press) (2017)

8. Saha, O., Chakraborty, A., Banerjee, J.S.: A fuzzy AHP approach to IT-based stream selection for admission in technical institutions in India. In: Proceedings of IEMIS (Accepted), AISC-Springer (2018)
9. Saaty, T.L.: How to make a decision: the analytic hierarchy process. Eur. J. Oper. Res. **48**(1), 9–26 (1990)
10. Wang, Y.M., Luo, Y., Hua, Z.: On the extent analysis method for fuzzy AHP and its applications. Eur. J. Oper. Res. **186**(2), 735–747 (2008)

Fingerprint Classification by Filter Bank Approach Using Evolutionary ANN

Annapurna Mishra and Satchidananda Dehuri

Abstract This paper presents an evolutionary artificial neural network for classification of fingerprints in the area of biometric recognition. An efficient way for feature extraction from the fingerprints using a Gabor filter bank has been studied very rigorously and extracted potentially useful features. Here five classes of fingerprints have been taken into consideration. We have conducted experimental study to prove the effectiveness of the method on NIST-9 database. It is evident from the results that the method is effective in classifying the fingerprints with a varying degree of accuracy vis-à-vis to the different parameters setting.

Keywords Classification · Gabor filter · EANN · Feature extraction
Evolutionary

1 Introduction

Classification of fingerprint refers to assigning a class label to an unclassified fingerprint. This is perhaps the most important stage in AFIS system because it introduces mechanism for indexing and facilitates matching over a large database. For a known class of fingerprint, the comparison is to be done with a stored database class, which reduces the time complexity of the identification process significantly [1]. Various approaches of fingerprint classification are available in the literature including the model based technique proposed by Edward Henry (1900)

A. Mishra (✉)
Department of Electronics and Communication Engineering,
Silicon Institute of Technology, Silicon Hills, Patia,
Bhubaneswar 751024, Odisha, India
e-mail: annapurnamishra12@gmail.com

S. Dehuri
Department of Information and Communication Technology,
Fakir Mohan University, Vyasa Vihar, Balasore 756019, Odisha, India
e-mail: satchi.lapa@gmail.com

© Springer Nature Singapore Pte Ltd. 2019
P. K. Mallick et al. (eds.), *Cognitive Informatics and Soft Computing*,
Advances in Intelligent Systems and Computing 768,
https://doi.org/10.1007/978-981-13-0617-4_34

—a very popular classifier. It uses the number and locations of singular points (cores and deltas). A set of rules are designed by researchers for detection of singular points and classifies them in five classes [2, 3] namely left loop, right loop, whorl, arch, and tented arch. Here classification is done using an evolutionary ANN. This neural network uses genetic algorithm, which is an evolutionary computing technique for its weight update. This technique is less sensitive towards noise since it does not depend on the detection of singular points. So this is a more robust and accurate technique for fingerprint classification.

As shown in Fig. 1 structure based method categorizes images using approximated orientation field and partitions the field into homogeneous portions by relational graphs [3]. Maio and Maltoni [4] have adopted this method of image splitting.

Some other approaches used are frequency-based and hybrid approach. In frequency-based approach the frequency spectrum of fingerprints are used to classify the images [4]. Hybrid approach is an approach which combines two or more ways to classify fingerprints [5, 6]. These methods show some indication but not been applied on large fingerprint databases [4–6].

The proposed algorithm is based on the multichannel representation of fingerprints using a Gabor filter bank tuned in different orientation and a multi-stage evolving artificial neural network (EANN) classifier (Fig 2).

(a)   (b)

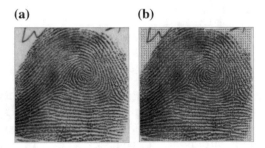

Fig. 1 Gray scale image and orientation field (**a**) and (**b**)

(a)   (b)   (c)   (d)   (e)

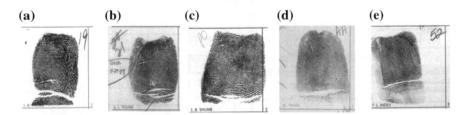

Fig. 2 Five fingerprint classes from NIST-9 database (**a**) right loop (**b**) left loop (**c**) whorl (**d**) arch (**e**) tented arch

The steps involved in the adopted classification algorithm [3, 4] are:

- Fix a reference point in the fingerprint image and spatial tessellation is performed around the region taking the reference point as center.
- Segment the tessellated image into a set of small informative images which contains certain ridge structure; and then calculate the standard deviation in each sector to generate the feature vector.

2 Steps of Feature Extraction

Global ridge and furrow structure contains information about the class of a fingerprint [7]. Capturing these global features will lead to an effective classification [4, 5].

The steps for feature extraction algorithm are as follows:

- Separate the region of interest from the rest of the image by detecting a valid reference point and defining a spatial tessellation around the reference point that is core point.
- Decompose the region of interest into a set of sub-images, which preserves global ridges and furrows structures.
- Calculate the average absolute deviation of intensity values of pixels in each sector from its mean intensity value to form the feature vector.

2.1 Feature Extraction

Let $C_{i\theta}(x, y)$ is the component sub image with respect to the direction θ for sector S_i. For $i = 0, 1, 2, \ldots, 36$ and $\theta \in [0°, 45°, 90°, 135°]$, a feature is the standard deviation $F_{i\theta}$ is given as

$$F_{i\theta} = \sqrt{\sum_{k_i} (C_{i\theta}(x, y) - M_{i\theta})^2},$$

(1)

where k_i is the number of pixels in S_i and $M_{i\theta}$ is the mean pixel intensity in $C_{i\theta}(x, y)$. So we have a ($36 \times 4 = 144$) dimensional feature vector.

3   Classification

Fingerprint classification is a difficult task for the presence of small interclass variations and large intra-class variations within the fingerprint classes, which is to be considered [8, 9]. EANN is an artificial neural network based on evolutionary learning. The idea behind this is the fact that the success of an individual not only depends on his personal knowledge or skills which he has acquired through learning (concept of neural network) but it also depends on his genetic factors (concept of genetic algorithm) [2, 10, 11]. Genetic algorithm (GA) is a function optimization algorithm [12]. It is a search algorithm based on the mechanics of natural selection and natural genetics. It simulates the process of evolution based on the survival of fittest concept (Darwinian theory). It searches for global optimum (maximum or minimum) of a particular function very effectively. A function having local and global minima is shown in Fig. 3. Point 1 and 3 represent the local minimum points of the function in region 1 and 3, whereas 2 is the global minima. Other traditional searching algorithms get struck in the local optimum points but GA can search for the global optimum. The process of optimization using GA is, a population of $2n$ to $4n$ trial solutions are used, where n is the number of variables in the function that we want to optimize [13, 14]. Each solution is done usually by a string of binary or real coded variables, corresponding to chromosomes in genetics. Here we have implemented real coded genetic algorithm (RCGA) and combined genetic algorithm with neural network. So our neural network learns through evolution. Figure 4 shows the flowchart of network training using GA.

The objective function for our optimization problem is the mean square error (MSE) and the parameters which are to be optimized are network weights and biases. The first stage of the classifier is trained to classify fingerprints into loops and non-loops. According to the classification result of first stage classifier one of the two classifiers from the second stage is selected which are trained with two classes of fingerprint each. Further a classifier in third stage classifies arch from tented arch. The structure of our classifier is shown in Fig. 5.

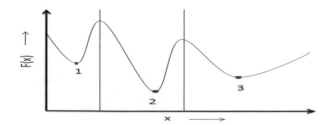

Fig. 3 A function with several local minima and a global minima

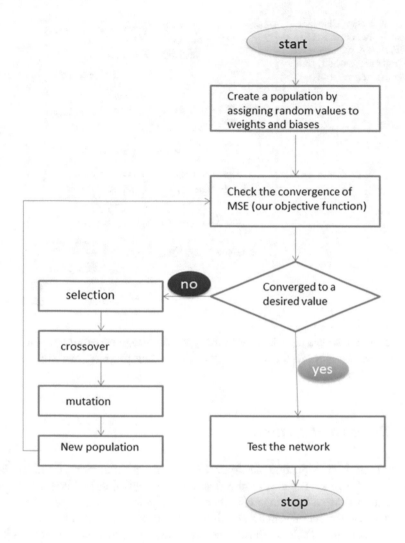

Fig. 4 Flowchart of neural network training using GA

4 Experimental Work

4.1 Dataset and Environment

Our observations of fingerprint classification algorithm is based on NIST-9 database for the five class classification problem [12, 15]. NIST-9 database consists of 5400 fingerprints from 2700 fingers with their natural frequency of occurrence where loops and whorls are majority. Each fingerprint is labeled into one of the five classes (R, L, W, A, and T). However, here 1648 images are used, out of which

Fig. 5 Two class Classifier
to classify five classes (LL,
RL, W, A, TA represents left
loop, right loop, whorl, arch
and tented arch respectively)

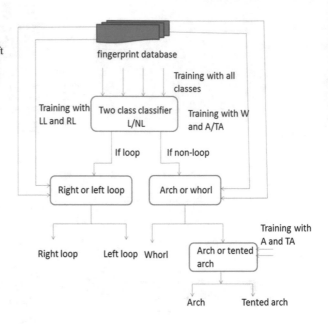

80% i.e., 1318 images are taken for training the first stage classifier and 800 images
to train each classifier in the second stage. For testing purpose, rest 20%, i.e., 330
images are used.

4.2 Results and Analysis

The four class classification accuracy by altering the number of hidden layer
neurons is shown in Table 1. Figure 6 shows the confusion matrix for four class
classification using a two-stage classifier. Classification accuracy also changes by
tuning the parameters of genetic algorithm like crossover, mutation, and recombination percentage. Table 2 shows the variation of classification accuracy by
changing the above parameters. Classification accuracy is reported to be increasing
by incorporating rejection option in the feature extraction stage. If some of the
images for which a valid tessellation cannot be defined are rejected. In this case
classification accuracy will increase as shown in Table 3.

Table 1 Classification accuracy obtained by changing the number of hidden layer neuron

S. No.	Number of hidden layer neuron	Classification accuracy (%)
1	16	78.56
2	18	81.84
3	20	85.12
4	22	83.23

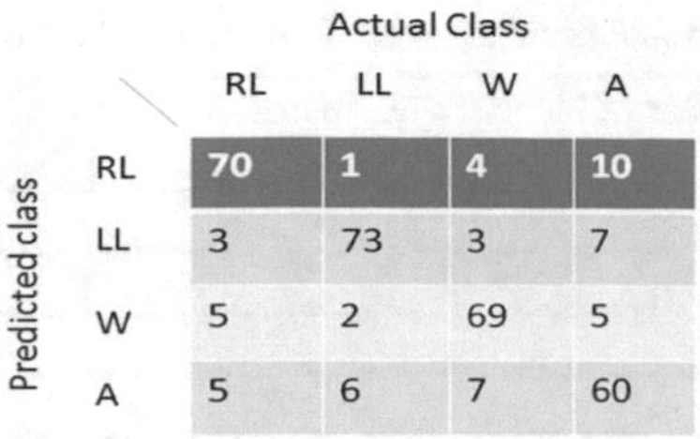

Actual Class

		RL	LL	W	A
Predicted class	RL	70	1	4	10
	LL	3	73	3	7
	W	5	2	69	5
	A	5	6	7	60

Fig. 6 Confusion matrix of four class classification

Table 2 Classification accuracy obtained by changing parameters of genetic training process

S. No.	Crossover (%)	Mutation (%)	Recombination (%)	Classification accuracy (%)
1	40	30	30	74.25
2	50	30	20	80
3	50	20	30	83.23
4	60	20	20	80.45

Table 3 Classification accuracy with % of rejected fingerprints

S. No.	Percentage of rejected fingerprint	Classification accuracy (%)
1	10	85.54
2	20	88.32
3	30	90.12

From Table 1 it can be seen that 20 hidden neuron produces the best result. A few number of hidden neuron leads to under-fitting in which the neural network misclassifies. Excess number of hidden neuron leads to over-fitting in which the network remembers things, it fails to generalize. From Table 2 it is clear that excess mutation leads to loss of genetic properties so a mutation percent of 20 yields better result. Table 3 shows increase in classification accuracy with increase in rejection rate. It is maximum when around 30% of fingerprints are rejected due to invalid tessellation. A further increase in rejection rate decreases the training database of the classifier which leads to misclassification due to improper training of the network. For five class classification, we use a third stage of EANN classifier which is trained with arch and tented arch to separate these two classes. Five classes classification accuracy by altering the number of hidden neurons is shown in Table 4. Table 5 shows classification accuracies obtained by tuning the genetic parameters and Table 6 shows classification accuracies obtain by varying the rejection rate. Also incorporating rejection option in the classifier training stage improves the performance of the classifier.

Table 4 Five class classification accuracy with different numbers of hidden layer neuron

S. No.	Number of hidden layer neuron	Classification accuracy (%)
1	16	70.12
2	18	76.50
3	20	81.45
4	22	75.40

Table 5 Five class classification accuracy obtained by tuning the genetic parameters

S. No.	Crossover (%)	Mutation (%)	Recombination (%)	Accuracy (%)
1	40	30	30	73.12
2	50	30	20	74.28
3	50	20	30	81.45
4	60	20	20	76.15

Table 6 Five class classification accuracy obtained by incorporating different rejection rates

S. No.	Rejection (%)	Classification accuracy (%)
1	0	81.45
2	10	83.14
3	20	85.34
4	30	86.78

Table 4 is obtained by keeping crossover, mutation, and recombination percentages to 50, 20, and 30, respectively and with 0% rejection. Table 5 is obtained by keeping number of hidden layer neuron to 20 and with 0% rejection. For Table 6, the crossover, mutation and recombination percentages to 50, 20, and 30 are fixed respectively and rejection percent to 0%.

5 Conclusion

The classification of fingerprints into five broad classes using a multichannel filter based feature extraction algorithm and EANN, gives 82% accuracy with no rejection. Our classifier needs less training samples than general feed-forward, back-propagation neural network. It takes about 6 seconds to classify one fingerprint to its respective class. However our algorithm suffers from inaccurate detection of core point. A more enhanced performance can be easily promised by improving the accuracy of core point location. Since image decomposition (convolution with Gabor filter) takes almost 90% of the total computational time, special purpose hardware for convolution can significantly decrease the overall time for classification.

This work has scope of improvement in various directions, which includes developing more efficient and robust classifiers by harnessing the best attributes of particle swarm optimization (PSO), Ant colony optimization (ACO), and support vector machine (SVM) for fingerprint recognition. Our future research bag also includes principal component analysis (PCA) for dimensionality reduction without losing significant information. Our algorithm is sensitive towards the location of core point, so, a more robust core point estimation algorithm will improve the performance of the classifier. This proposed algorithm doesn't handle rotation; hence attempt can be made in this direction by approximating a set of reference of the fingerprint.

Acknowledgements The first author would like to thank the technical support of Department of Information and Communication Technology, Fakir Mohan University, Vyasa Vihar, Balasore.

References

1. Wayman, J.L., Jain, A.K., Maio, D., Maltoni, D.: An introduction to biometric authentication systems. In: Biometric Systems (2005)
2. Jain, A.K., Prabhakar, S., Hong, L., Pankanti, S.: Filterbank based fingerprint matching. IEEE Trans. Image Process. **9**, 846–859 (2000)
3. Jain, A.K., Prabhakar, S., Hong, L.: A multichannel approach to fingerprint classification. IEEE Trans. Pattern Anal. Mach. Intell. **21**(4), 348–359 (1999)
4. Cappeli, R., Maio, D., Maltoni, D., Wayman, J.L., Jain, A.K.: Performance evaluation of fingerprint verification systems. IEEE Trans. Pattern Anal. Mach. Intell. **28**, 3–18 (2006)
5. Cappeli, R., Maio, D., Lumini, A., Maltoni, D.: Fingerprint image reconstruction from standard templates. In: IEEE Transactions on Pattern Analysis and Machine Intelligence, vol. 29 (2007)
6. Maltoni, D., Maio, D., Jain, A.K., Prabhakar, S.: Handbook of fingerprint recognition. Springer Science and Business Media (2009)
7. Jain, A.K., Prabhakar, S., Hong, L., Pankanti, S.: Filterbank based fingerprint matching. IEEE Trans. Image Process. **9**(5), 846–859 (2000)
8. Cappeli, R., Lumini, A., Maio, D., Maltoni, D.: Fingerprint classification by directional image partitioning. IEEE Trans. Pattern Anal. Mach. Intell. **21**, 402–421 (1999)
9. Jain, A.K., Ross, A., Prabhakar, S.: An introduction to biometric recognition. IEEE Trans. Circuits Syst. Video Technol. **14**, 4–20 (2004)
10. Bhanu, B., Lin, Y.: Genetic algorithm based feature selection for target detection in SAR images. Image Vis. Comput. **21**, 591–608 (2003)
11. Vieira, H., Borges, D.L.: Fingerprint classification with neural networks. In: Brazilian Symposium on Neural Networks, pp. 66–72 (1997)
12. Watson, C.I., Wilson, C.L.: NIST special database 4, fingerprint database. Nat'l Inst. Stand. Technol. (1992)
13. Wilson, C.L., Candela, G.T., Watson, C.I.: Neural network fingerprint classification. J. Artif. Neural Networks **1**(2), 203–228 (1993)
14. Candela, G.T., Grother, P.J., Watson, C.I., Wilkinson, R.A., Wilson, C.L.: PCASYS—a pattern-level classification automation system for fingerprints. Technical report NISTIR 5647 (1995)
15. Jain, A.K., Hong, L., Pankanti, S., Bolle, R.: An identity authentication system using fingerprints. Proc. IEEE **85**(9), 1365–1388 (1997)

A Sequence-Based Cellular Manufacturing System Design Using Genetic Algorithm

C. R. Shiyas, B. Radhika and G. R. Vineetha

Abstract This paper is presented with an algorithm for manufacturing cell system design and part family identification. The model is suitable for establishing a good division of machine cells and part families considering operation sequence data. The aim of this model is the maximization of group technology efficiency value which is mostly used for measuring the worth of cellular configurations when route matrix data is considered in design. Allocating machines to different machine cells is carried out using a randomized procedure based on genetic algorithm. Five situations based on four problems were subjected to comparison based on Group Technology Efficiency (GTE) with two other methods from the literature and it is observed that the new algorithm is either outperforming the other methods or giving the best results obtained from them.

Keywords Group technology efficiency · Cellular manufacturing
Genetic algorithm · Sequence data

1 Introduction

The cellular manufacturing system (CMS) is a type of plant layout based on group technology (GT) philosophy and has developed in the last two decades as an effective manufacturing layout model under medium variety and volume production conditions. A CMS tries to achieve the efficiency of a flow line production system, at the same time retain the flexibility of a functional layout. The objective in a CMS

C. R. Shiyas (✉) · B. Radhika · G. R. Vineetha
Cochin University College of Engineering Kuttanadu, Pulincunnoo,
Alappuzha, Kerala, India
e-mail: crshiyas@rediffmail.com

B. Radhika
e-mail: bradhika79@gmail.com

G. R. Vineetha
e-mail: chachu.gr@gmail.com

© Springer Nature Singapore Pte Ltd. 2019   353
P. K. Mallick et al. (eds.), *Cognitive Informatics and Soft Computing*,
Advances in Intelligent Systems and Computing 768,
https://doi.org/10.1007/978-981-13-0617-4_35

design is the satisfactory partition of the manufacturing facility into cells in which similar components for manufacturing and its associated machines are arranged together for processing. Such a division of manufacturing system into cells often leads to reduced production time, reduced inventory levels, easy tooling, less setup time, timely delivery, less rework and waste materials, improved managerial control, and better quality [1].

In most of the methods available for cell formation, it is seen that the use of part-machine-incidence matrix (PMIM) is the type of data set used. In this category the methods include procedures based on array management, hierarchical grouping, non-hierarchical grouping, mathematical modeling approaches, graph theory based approaches, heuristic approaches, etc. Heuristic approaches generally rely on procedures based on Genetic algorithm [2], Tabu search [3], Simulated annealing [4] etc. They are also widely used for developing solution method for mathematical models in CMS design.

In the literature only very few models are seen considering the route matrix which contains operation sequence data of parts visiting the machines. Using sequence data in design will directly affect the number of intercell movement in the system and intercell movement is one of the parameter considered in many models which has to be minimized. The intercell movement may be more if we consider a route matrix compared to a PMIM. Hence considering sequence data will lead to more sensible design of CMS. CASE algorithm developed by Nair and Narendran [5] is one of the early models in CMS design that considers sequence of operations. They introduced a similarity coefficient that uses the operation characteristics of the PMIM and a non-hierarchical grouping algorithm has been proposed. Park and Suresh [6] proposed an new method to compare the performance of a fuzzy adaptive resonance theory (ART) neural network, which is a recently developed neural network method, with the performance of a conventional hierarchical grouping technique by considering operation sequence data in design. Won and Lie [7] in their approach for minimizing intercell moves used two types of input data matrices in which the operation sequences of parts and volume of production of parts are included. Shiyas and Pillai [8] integrated intracell layout also into the cell design problem. Integer programming models are also very common in cell design approaches [9].

There are many performance measures for checking the block diagonal form obtained as a result of different methods of design which include grouping efficiency and grouping efficacy [10]. But, all these measures suitable only if the input is a part-machine incident matrix. This cannot be used for checking the efficiency of the model where operation sequence data is considered in design. Hence, Group Technology Efficiency (GTE) proposed by [11] can be taken as a measure of performance considering sequence of parts. The proposed model tries to achieve good machine cells by maximizing Group Technology Efficiency.

2 The Proposed Model and Algorithm

Group Technology Efficiency (GTE) proposed by Harhalakis et al. [11] is a measure of performance of CMS considering sequence of parts. It is given by

$$\text{GTE} = \frac{(I_{\max} - I_{\text{need}})}{I_{\max}},$$

where $I_{\max}$—Maximum intercell movement possible in the system and I_{need}—Intercell movement needed by the system.

In the present model our objective is to maximize the Group Technology Efficiency (GTE) which is a measure of the quality of cells obtained. The different steps of the algorithm are given below. The genetic algorithm details are provided in the next section.

Step 1   Input the route matrix to the model.

Step 2   Enter the values of algorithm parameters.

Step 3   Initial population which represent the manufacturing cells are generated randomly.

Step 4   Assigning of parts to the above machine cells based the customized law.

Step 5   Compute the Group Technology Efficiency value for all population which will give the fitness function of the algorithm.

Step 6   Form subsequent generations through doing operations such as reproduction, cross over operation, mutation operation, replacement strategy and do part allocation in the new generation.

Step 7   Assess the algorithm and is stopped when the termination criteria is reached and choose the design with maximum Group Technology Efficiency value else return to 6.

The algorithm is coded in MATLAB.

3 Solution Methodology

The solution procedure for the model is based on genetic algorithm (GA) which will arrive at a manufacturing cell division by maximizing GTE. Hence the above algorithm uses GA as its core procedure for the formation of manufacturing cells. A heuristic is applied for part assignment for part family formation. Hence the algorithm developed uses GA and part assignment heuristic for cell formation, and part family identification.

Genetic algorithm is a randomized search process where the survival of the fittest principle is used. GA starts with a combination of solution called the initial population and each entity in it is termed as a chromosome, which represents a solution for the given input fulfilling all constraints. The size of population generally depends on the length of the chromosome. During the run of the new chromosomes are obtained through consecutive iterations which are called generations. Chromosomes are randomly selected and are subjected to many genetic operators which include reproduction operation, crossover operation, mutation, etc., during successive generation. The presence or absence of a chromosome, in the following generation, is based on its fitness value which is calculated based on the objective function value.

Chromosome representation: For any GA based procedure, the first and most important stage is the chromosome coding. Each chromosome has been made up of a sequence of genes. In the present model, each gene indicates a manufacturing machine cell (number) and the place of each gene in the chromosome indicates the type of machine. The number of types of machine will be the length (Total genes) of all chromosomes.

For example, a chromosome "312342" indicates a four cell configuration as follows:

Cell No 1   machine 2.
Cell No 2   machine 3 and 6.
Cell No 3   machines 1, 4.
Cell No 4   machine 5.

Population Initialization: The initialization is done randomly. The initial solution is the old population when the first generation of the genetic algorithm is formed. Generally, the size of each population is determined with respect to the length of the chromosome. A general rule applied is; the size of the population is made equal to that of the chromosome length. Here population size is set as 2.5 times the chromosome length [1].

Evaluation Function: Each solution in the population pool has to be assessed for its objective function value to find the fitness value. In this model, the objective function is the maximization of GTE.

Genetic algorithm operators: Three basic genetic operators are used here, i.e., reproduction operator, crossover operator, and mutation. The function of reproduction is the selection of parents for the coming generation. In reproduction, strings are chosen probabilistically based on its fitness using stochastic-sampling without replacement method [1]. The crossover operation carried out with a crossover probability and makes offspring by the exchange of information enclosed in the parents. Here two-point crossover is used and the points of cross over are selected randomly. Mutation is performed for avoiding a chromosome being unchanged any time. Mutation is desirable because sometimes they may lose some potentially valuable genetic material through other genetic operators. The mutation done based on a small probability.

Replacement Strategy: Poor off-springs based on fitness value are replaced in the new generation through this strategy. The off-springs are evaluated for the fitness value. The aim of this process is to have improved generations than the previous generation. This is done by admitting off-springs in the new generation only to members with fitness better than that of the current population. This is carried out by the comparison of the fitness values of the old and new generations. The fittest will become the population for the next generation.

Terminating the Genetic Algorithm: The genetic algorithm iterates, and as the process proceeds, the subsequent generations include chromosomes with higher fitness function values. Termination criterion is used to stop the iteration. Two termination criteria are used here namely of population convergence, and the maximum number of generations.

The GA parameters used are given below

Crossover Probability is taken as 0.8.
Mutation Probability is taken as 0.1.
Number of generations for stopping as150.
Maximum number of generations as 2000.

4 Illustrative Example

The algorithm is coded in MATLAB. In this study, a model based on genetic algorithm is suggested for cell design model considering production sequence of parts. Then the results are compared to some of the previously available methods which consider sequence data. GTE is taken as a measure for comparing the quality of solutions obtained. The results obtained from the new method is compared to the results given using CASE algorithm [5], ART1 approach [12] as given in Table 1.

The comparison is done by selecting configurations with same number of cells from other methods and new method. This is required because the GTE will be high for less number of cells and its maximum value is obtained when a single cell is formed. The number of problems selected for comparison is four and total five situations based on number of cells are analyzed. The problem size ranges from 7 machines, 7 parts to 25 machines, 40 parts. Table 1 show the comparative results based on GTE, exceptional elements and number of intercell moves. Out of the 5 situations tested the proposed method outperformed other two methods in three situations based on GTE and obtained equal GTE for the remaining cases. This indicates the superiority of the new method over the other two in forming

Table 1 Comparison table

Problem size	No. of cells	CASE			ART1			Proposed method		
		EE	IM	GTE	EE	IM	GTE	EE	IM	GTE
7 × 7	2	2	4	69.25	2	4	69.25	1	2	84.62
20 × 8	3	10	17	58.54	0	17	58.54	10	17	58.54
20 × 20	4	NA	NA	NA	2	15	74.58	14	16	77.78
20 × 20	5	15	19	67.8	6	19	69.49	20	19	73.61
40 × 25	5	7	NA	NA	6	22	72.04	26	22	72.04

Table 2 Input for problem 1

Machine	Part						
	Part 1	Part 2	Part 3	Part 4	Part 5	Part 6	Part 7
Machine 1	1	1					
Machine 2	3	2					
Machine 3	2		1	1		2	
Machine 4			2	2			3
Machine 5					1	4	1
Machine 6					2	3	2
Machine 7					3	1	4

Table 3 Output

		Part						
		P1	P2	P3	P4	P5	P6	P7
Machine	M1	1	1					
	M2	3	2					
	M3	2		1	1		2	
	M4			2	2			3
	M7					1	4	1
	M5					2	3	2
	M6					3	1	4

manufacturing cells considering sequence. Table 2 shows the input and Table 3 shows the result obtained by the new method for the first problem situation in Table 1 where an improved solution is obtained. Hence overall the new method is better that the other two based on the problems analyzed.

5 Conclusion

In the present model, an algorithm has been developed to design cellular manufacturing systems considering sequence of operation of parts. For solving the proposed model, genetic algorithm based procedure is used. The objective of the model is the maximization of Group Technology Efficiency. The model efficiency is verified with example problems of various sizes taken from the literature. The configurations obtained are compared with results given by other two methods considering sequence data. Five situations based on four problems were subjected to comparison based on group technology efficiency and it is observed that the new algorithm is either outperforming the other methods or giving the best results of the other two methods. The model may be further extended by incorporating parameters like production volume, processing time, alternate routes, dynamic demand, etc., for better adherence to real situations

References

1. Wemmerlo, V.U., Johnson, D.J.: Cellular manufacturing at 46 user plants: implementation experiences and performance improvements. Int. J. Prod. Res. 1(35), 29–49 (1997)
2. Pillai, V.M., Subbarao, K.A.: Robust cellular manufacturing system design for dynamic part population using a genetic algorithm. Int. J. Prod. Res. 46(1), 5191–5210 (2008)
3. Adenso-Diaz, B., Lozano, S.: A model for the design of dedicated manufacturing cells. Int. J. Prod. Res. 46, 301–319 (2008)
4. Chen, C.L., Cotruvo, N.A., Baek, W.: A simulated annealing solution to the cell formation problem. Int. J. Prod. Res. 33, 2601–2614 (1995)
5. Nair, J.G., Narendran, T.T.: CASE: A clustering algorithm for cell formationwith sequence data. Int. J. Prod. Res. 36, 157–179 (1998)
6. Park, S., Suresh, N.C.: Performance of Fuzzy ART neural network and hierarchical clustering for part machine grouping based on operation sequences. Int. J. Prod. Res. vv. 41(14), 3185–3216 (2003)
7. Won, Y., Lee, K.C.: Group technology cell formation considering operation sequences and production volumes. Int. J. Prod. Res. 39, 2755–2768 (2001)
8. Shiyas, C.R., Madhusudanan, Pillai V.: An algorithm for intra-cell machine sequence identification for manufacturing cells. Int. J. Prod. Res. 5, 2427–2433 (2014)
9. Alijuneidi, T., Bulgak, A.: A: designing a cellular manufacturing system featuring remanufacturing, recycling, and disposal options: a mathematical modeling approach. CIRP J. Manufact. Sci. Technol. 19, 25–35 (2017)
10. Kumar, C.S., Chandrasekharan, M.P.: Grouping efficacy: a quantitative criterion for goodness of block diagonal forms of binary matrices in group technology. Int. J. Prod. Res. 28, 233–243 (1990)
11. Harhalakis, G., Nagi, R., Proth, J.M.: An efficient heuristic in manufacturingcell formation for group technology applications. Int. J. Prod. Res. 28, 185–198 (1990)
12. SudhakaraPandian, R., Mahapatra, S.S.: Manufacturing cell formation with production data using neural networks. Comput. Ind. Eng. 56, 1340–1347 (2009)

Multimodal Biometrics Authentication Using Multiple Matching Algorithm

Govindharaju Karthi and M. Ezhilarasan

Abstract Biometric recognition system is the popular technique used for authentication applications. Multimodal biometrics are more likely used for biometric recognition system since it has more advantages. This paper proposes a new multimodal biometric system which combines two feature extraction algorithms in fingerprint recognition system to ensure the optimal security. A fingerprint image is applied to two different algorithms and the matching process was carried out. The algorithms are distance method and template-based method. In the distance method, the center point and the ridge points of the fingerprint image were captured, each ridge point was connected with the center point of the fingerprint and the distance was calculated. In template-based matching method, the set of ridges were extracted and the template was generated based on ridges and minutiae set and compared with the template from the database. Finally the results are combined at the decision level fusion; the user is authenticated and the proposed algorithm focuses on the accuracy, universality, and ease of use.

Keywords Distance method · Templates · Authentication · Multimodal biometrics · Fingerprint recognition

1 Introduction

Biometrics is the individual's behavioral or physiological pattern which is used for individual's identification or verification. The biometrics could be broadly classified based on their pattern as physiological or behavioral biometrics. In the digital

G. Karthi (✉)
Department of Computer Science and Engineering, Pondicherry Engineering College,
Puducherry, India
e-mail: karthi.govindharaju@gmail.com

M. Ezhilarasan
Department of Information Technology, Pondicherry Engineering College, Puducherry, India
e-mail: mrezhil@pec.edu

© Springer Nature Singapore Pte Ltd. 2019
P. K. Mallick et al. (eds.), *Cognitive Informatics and Soft Computing*,
Advances in Intelligent Systems and Computing 768,
https://doi.org/10.1007/978-981-13-0617-4_36

environment, the biometrics technologies could be used for authentication, confidentiality, and various security applications. In the case of behavioral biometrics, the traits are gait keystroke dynamics, signature and so on. The physical traits are face, voice pattern, DNA, fingerprint, and palm. The biometric recognition process gets feature vector from the biometric traits which is used to identify or recognize the individual. The biometric recognition process start with the feature extraction process, matching and it is stored as template in the database for the purpose of matching

Fingerprint [1] is the most user-friendly biometric trait widely used in authentication, access control, and in forensics. Fingerprint made up of edges and furrows which is called as ridges and valleys respectively. These ridges and valleys patterns form the features of fingerprint. These features are of two major categories global and local. The global feature includes ridge orientation, singular points, number of ridges, etc. The features such as minutiae points and ridge ends belong to local features.

The performance metrics used for the biometric recognition are as follows:

(1) FAR: The overall percentage of the biometric system which falsely validates the invalid user.
(2) FRR: The average percentage of the biometric system which falsely rejects the genuine user.

In the proposed system, a new multimodal biometrics model is used. In general, the multimodal biometric system depends on various scenarios. Based on this the following are some of the various scenarios of multimodal biometric recognition system [2].

Multiple sensors: single biometric trait is recorded by multiple sensors. So we get different set of biometric features for different sensors.

Multiple classifiers: In this scenario, multiple classifiers are used for the feature set extracted from the single biometric trait.

Multiple instances: The biometric trait is attained for multiple times by the same sensor.

Multiple traits: Here more than one biometric trait of an individual is used for recognition. For example, any two combinations of the biometric trait iris, fingerprint, face, palm, hand geometry, etc., is used.

In the proposed system, the second scenario which uses multiple matching algorithms is used to produce better performance of the biometric recognition system.

The organizations of the papers is as follows, the first section gives the introduction of the work, the section two covers the related works done and the section three discuss about the proposed algorithms needed for feature extraction and matching. The section four deals with the experimental results. The section five concludes the work.

2 Related Works

In this section related works of multimodal biometrics is discussed. Afsar et al. [3] designed an Automatic Fingerprint Identification System (AFIS) technique. A new thinning algorithm gives an overall better performance of the fingerprint system. Ng et al. [4] proposes new Adjacent Orientation Vector (AOV) features for fingerprint matching. This feature gives better matching performance to overcome the distortion.

Qi et al. [5] the feature extraction is based on global orientation features and the matching performed by Euclidean distances. Thai et al. [6] gives a standardized fingerprint model which uses a transformation after extraction process in order to reduce the effect of noise. Jain et al. [7] gives a latent fingerprint identification algorithm for solving large nonlinear distortion images. An enhanced matching algorithm [8] is proposed, which combines two different algorithms to improve the local correlation score. The local enhancement algorithm and the gradient decent algorithm are used to improve the matching performance.

A two-step process of fingerprint extraction is proposed [9]. In the first step, a Fast Fourier transform is applied to the fingerprint image for enhancement and in the second step, the image is applied with thinning algorithm. Finally, the features are extracted and the similarity index is calculated. In [10], a newly designed multimodal biometric system is proposed, which used three different traits and they uses different features extraction algorithm. Finally all the features are combines at the feature level to produce fused features as template.

In [11], a hybrid fingerprint matching algorithm is proposed which uses a local minutiae matching algorithm and a consolidated final stage multiple minutiae matching algorithms.

3 Proposed System

The proposed work is elaborated in the diagram given in Fig. 1. The proposed system starts with samples of the fingerprint images taken from the same finger using the same sensor. The given fingerprint image is performed with fingerprint enhancement techniques namely thinning, orientation estimation for extraction of level one features. In this approach, both the global as well as the minutiae features are extracted. Core point detection and the minutiae features are extracted and they are applied with different matching algorithms. Template matching method is applied for both the features (local and global) whereas; the distance method uses only the minutiae features. These features are stored in the database as templates.

With this the registration or enrollment process is fulfilled in the fingerprint recognition process. For the verification process, the same process is done and finally matching module is done. The output of the biometric recognition verification process is the matching score of the real template with the existing template

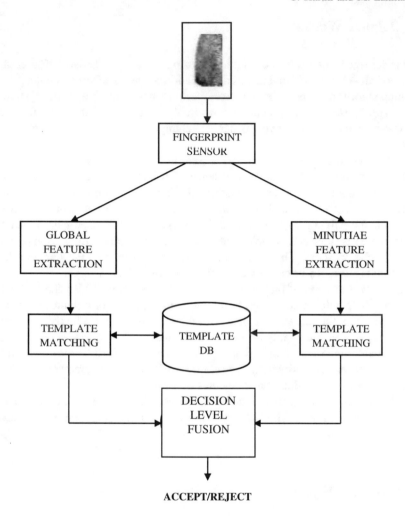

Fig. 1 Fingerprint recognition using different matching algorithm

in the database. Since we have employed two different features algorithms the matching had to be done in an adaptive manner.

The performance measure is calculated with two parameters; FAR and FFR for biometrics system. The obtained features are forced to generate the matching templates, i.e., matching template of the biometric data of the user obtained during the registration phase. The matching scores of the template from the data base with the scores found from the finger print image. In decision level fusion, the user was identified as Imposter or Genuine based on the output of the two matching modules. The decision level fusion uses weighted sum rule for taking final decision.

3.1 Finger Print Recognition Using Template Matching

The template matching algorithm compares the level one features and the core point of the fingerprint. The level one features are of the stored template is compared verses the given template for matching. The core point is detected for the aligning purpose of inconsistent pairs of fingerprint. In Fig. 2, an inconsistent fingerprint pairs is given.

The matching process uses both the global and local features for the identification process. The level one features mainly used for the classification process and the level 2 minutiae points are used for identification.

3.2 Fingerprint Recognition Using Distance Method

The distance method can be done with the help of the Principle Component analysis. In this method, a core point is calculated as in the case of template matching algorithm and from the core point the neighboring minutiae points distance is calculated using the Eq. 1.

The distance between the core and each real ridge was calculated by the following Eq. 1.

$$D_i = \sqrt{(C - x_i)^2 + (C - y_i)^2},\qquad (1)$$

where, end was the real ends identified from the binary fingerprint image; D_i is distance between the core point and the minutiae points in the fingerprint image. The X and Y are the minutiae points of the fingerprint image respectively. Figure 3 shows the distance relationship between the ridge points.

Fig. 2 Inconsistent fingerprint pairs

Fig. 3 Core and ridge points connected FP image

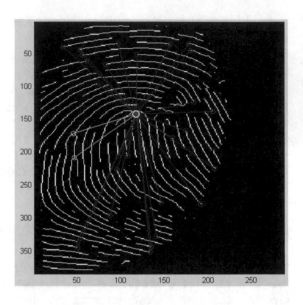

4 Results and Discussion

The experiment environment is core i5 with CPU speed of 4.16 GHz. A total of 1200 matches is performed to calculate the performance of the recognition system. The fingerprint image is taken from FVC 2004 database. The FVC 2004 database consists of four data sets named as DB1, DB2, DB3, and DB4. Each dataset has it won characteristics. FVC2004 for each dataset has 800 fingerprints of 100 fingers and total of eight impressions for each finger as shown in Fig. 4. The experiment is performed by taking two random fingerprints from each data set for a matching process. From each fingerprint image, the feature extraction process is applied and both the template and distance matching algorithm is applied and finally the matching level fusion process is applied. From the experiment a better result is obtained compared with the existing algorithms. The performance is calculated

The False Acceptance Rate (FAR) was identified by Eq. 2.

$$FAR = \frac{\sum_{n=1}^{N} FAR(n)}{N} \tag{2}$$

And the Genuine Acceptance rate (GAR) was identified by Eq. 3,

$$GAR = \frac{\sum_{n=1}^{N} GAR(n)}{N} \tag{3}$$

The FAR and GAR for the fingerprint is (3.57–7.14)% and (95–100)% respectively.

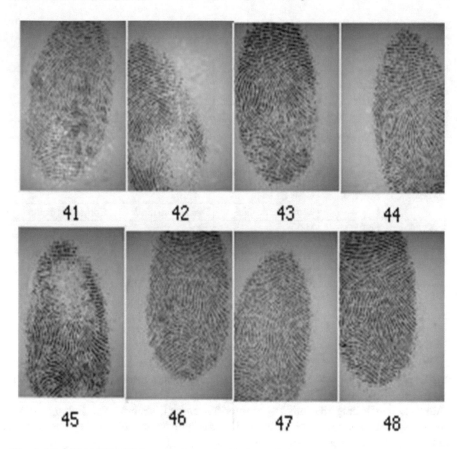

Fig. 4 DB4 FVC 2004 FP images of user 4 with 8 samples

Tables 1 and 2 gives the comparison of results with the existing and the proposed work. In the existing system any one of the matching method is considered and the performance results where given. As far as the proposed system is concerned, both the matching techniques have applied and the following result has been obtained. The sample 1 gives the average performance of the existing and proposed system, the first set of samples is taken from the DB1 FVC 2004. The proposed system provides 1.25% better results than the existing algorithm. For the second set of sample is taken from DB2 dataset of FVC 2004 is taken and the average performance shows that 1.48% better performance and for the third set of sample the fingerprint images is taken from the dataset DB3 of FVC 2004. The results obtained gives a superior results for the proposed system by 1.30% and finally the DB4 dataset is used for last set of samples and the output gives a good results an increase of 2.01% compared to existing system.

Table 1 FAR results of fingerprint of FVC 2004 dataset

Database	Dataset	FAR (%)		
		Template method	Distance method	Combined method
FVC 2004	DB1	2.14	1.97	1.71
	DB2	2.12	1.68	1.43
	DB3	1.37	1.11	1.05
	DB4	1.57	1.05	0.78

Table 2 GAR obtained from the fingerprint FVC 2004 dataset

Database	Dataset	GAR (%)		
		Template method	Distance method	Combined method
FVC 2004	DB1	100	100	100
	DB2	98.99	99.54	99.71
	DB3	97.98	98.64	99
	DB4	96	99.14	99.75

5 Conclusion

In this paper, two different matching algorithms is combined in order to provide an improved security in critical systems. This paper makes use of two classifiers that provide multimodal biometrics for the purpose of providing secured authentication. Thus it is concluding that this paper provides enough security for authentication. Finally the results are combined at the decision level fusion; the user was authenticated and focused on the accuracy, universality, and ease of use. This model provides the advantages of using the multimodal biometrics. From the performance, it is clear that the proposed system gives better results when compared with the existing system. For all sample set taken it is observed that the FAR and GAR of the proposed system is better than the existing system,.

References

1. Jain, A.K., Pankanti, S.: Fingerprint Classification and Matching, pp. 57–62. IBM T. J. Watson Research Center (1979)
2. Rahal, S.M., Aboalsamah, H.A., Muteb, K.N.: Multimodal Biometric Authentication System —MBAS, pp. 1026–1030. Information and Communication Technologies (2016)
3. Afsar, F.A., Arif, M., Hussain, M.: Fingerprint Identification and Verification System using Minutiae Matching. Conference on Emerging Technologies (2004)
4. Ng, G.S., Tong, X., Tang, X., Shi, D.: Adjacent orientation vector based fingerprint minutiae matching system. In: Proceedings of the 17th International Conference on Pattern Recognition, vol. 1, pp. 528–531 (2004)

5. Qi, J., Wang, Y.: A robust fingerprint matching method. In: Pattern Recognition, vol. 38, pp. 1665–1671 (2005)
6. Thai, L.H.: Fingerprint recognition using standardized fingerprint model. (IJCSI) Int. J. Comp. Sci. Issues, **7** (2010)
7. Jain, A.K., Feng, J.: Latent fingerprint matching. IEEE Trans. Pattern Anal. Mach. Intell. **33**, 88–100 (2011)
8. Li, J., Tulyakov, S., Govindaraju, V.: Improved local correlation method for fingerprint matching. In: International Symposium on Computing and Networking, pp. 560–562 (2014)
9. Patel, H., Asrodia, P.: Fingerprint matching using two methods. Int. J. Eng. Res. Appl. **2**, 857–860 (2012)
10. Priya, B.L., Rani, M.P.: Authentication of identical twins using tri modal matching. In: Computing and Communication Technologies (CCT), pp. 30–33 (2017)
11. Tran, M.H., Duong, T.N., Nguyen, D.M., Dang, Q.H.: A local feature vector for an adaptive hybrid fingerprint matcher. In: International Conference on Information and Communications (ICIC), pp. 249–253 (2017)

Kinship Verification from Facial Images Using Feature Descriptors

Aarti Goyal and T. Meenpal

Abstract Kinship Verification via facial images is an emerging research topic in the field of biometrics, pattern recognition, and computer vision. It is motivated by the findings that individuals with some genetic relations have certain similarities in their facial appearances. These similarities in the facial appearance is a result of inherited facial features from one generation to next generation, especially from parents to children. The researchers use these inherited facial features to perform Kinship Verification and validate their results. Most of the existing methods in Kinship Verification are based on metric learning which aims to improve the verification rate ignoring the effect of salient facial features. This paper aims to learn the effect of extracted facial features between pair of facial images to perform Kinship Verification. This paper proposes different feature descriptors to describe salient facial features and support vector machine classifier to learn these extracted facial features. To validate the accuracy of the proposed methods, experiments are performed on KinFaceW-I dataset. The results obtained outperformed previous results and would encourage researchers to focus on this emerging topic.

Keywords Kinship verification · Genetic similarity · Feature extraction
Feature classification · Support vector machine (SVM)

1 Introduction

Kinship is a broad term with distinct meanings based on the kind of associated framework. In anthropology, kinship is defined as the network of relationships between people in the society [1–3]. Whereas in biology, kinship is defined as the

A. Goyal (✉) · T. Meenpal
Department of Electronics and Telecommunication,
NIT Raipur, Raipur, India
e-mail: aarti29wasnik@gmail.com

T. Meenpal
e-mail: tmeenpal.etc@nitrr.ac.in

© Springer Nature Singapore Pte Ltd. 2019
P. K. Mallick et al. (eds.), *Cognitive Informatics and Soft Computing*,
Advances in Intelligent Systems and Computing 768,
https://doi.org/10.1007/978-981-13-0617-4_37

coefficient of relationship or degree of genetic relatedness between individuals of the same species.

Recent psychological studies have come across the fact that genetic relatedness is much higher for biologically related individuals compared to non-biologically related individuals. Humans with some common genes have similarities regarding their appearance, gestures, behavior, voice and many more [4, 5]. Among all these, human facial appearance is the most important factor for measuring genetic similarities. Motivated by these studies, many computer vision scientists and researchers have studied the problem of finding genetic similarities between individuals via facial image analysis [6].

The individuals with any kind of genetic similarities are termed as kins and their relationship is termed as kinship relationship. The process of finding this relationship via correlated visual resemblance is called Kinship Verification. Kinship Verification is the task of finding whether any kinship relationship exists or does not exist between a given pair of facial images.

Kinship Verification via facial images has immense applications in solving issues around criminal sectors like human trafficking, child smuggling, adoptions and finding missing child [7, 8]. It may also assist in developing albumin software which would distinguish families from impostors. This would also be helpful in collecting soft information to improve the accuracy of different face recognition algorithms. Kinship Verification may also be used as an additive assistance to complex and costly DNA verification process [9].

Kinship Verification can be used to find similarities between some possible candidates in DNA verifications and then DNA testing can be used to get the final verdict. Figure 1 shows some basic visual attributes which are usually used for genetic similarity measures between pair of facial images.

This paper mainly focuses on determining whether there exists any kinship relation in a given pair of facial images. Specifically, four kinds of kinship relations are examined in his paper. These kinship relations are father–son (F–S) relation, mother–son (M–S) relation, father–daughter (F–D) relation and mother–daughter (M–D) relation. To validate the accuracy of the proposed method, experiments are performed on KinFaceW-I dataset. KinFaceW-I dataset has a wide range of all possible four kinds of relations.

Fig. 1 Basic visual features to determine similarity between pair of facial images in Kinship verification

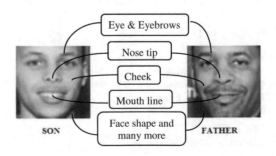

2 Related Work

The first attempt in Kinship Verification by Fang et al. [10] employs distinct local facial features such as histogram of gradient, skin color to recover cues from facial images. DAISY descriptor [11] proposes to match distinct facial features with Spatial Gaussian Kernels. Prototype-based discriminative feature learning (PDFL) [12] proposes to learn discriminative middle features to characterize similarities between facial images. Later some research papers have been published in Kinship Verification based on either feature-based method or learning-based method. Some of these feature based methods are self-similarity based representation (SSR) [11], Gabor gradient orientation pyramid (GGOP) [13] and semantic related attributes [14]. While some of these learning-based methods are transfer subspace learning [15, 16], metric learning such as neighborhood repulsed metric learning (NRML) [7] and discriminative multi-metric learning (DMML) method [6]. Most of these existing metric learning methods learn conventional Euclidian or Mahalanobis distance for Kinship Verification.

3 Kinship Datasets

Kinship dataset is a collection of facial images describing different kind of kinship relations. The main kinship relations are father–son (F–S) relation, mother–son (M–S) relation, father–daughter (F–D) relation, mother–daughter (M–D) relation, brother–brother (B–B) relation, sister–sister (S–S) relation, brother–sister (B–S) relation, father–mother–son (FM–S) relation, and father–mother–daughter (FM–D) relation. At present, there are six kinship datasets available describing these kinship relations. A complete study of all available datasets is given in this section.

The first kinship dataset is Cornell KinFace dataset [10] that contains RGB facial images collected from internet. This dataset includes 150 image pairs of parents-child but due to some security reasons, seven parent–child pairs have been removed from the dataset. UB KinFace dataset [8, 15, 16] is the second publicly available dataset that contains three types of images. First is the image of the children, second is the image of the parents when they are young and third is the image of parents when there are old. Images of the old parents in the dataset act as intermediate data between children and young children. Total 600 facial images present in the dataset corresponds to 200 images of children, 200 images of young parents and 200 of old parents.

Kinship Face in the Wild dataset commonly known as KinFaceW [7] is divided into two different datasets namely, KinFaceW-I dataset and KinFaceW-II dataset. Each of these two datasets have all the possible four kinds of kinship relations, namely father–son (F–S) relation, mother–son (M–S) relation, father–daughter (F–D) relation, and mother–daughter (M–D) relation. The difference between these two datasets is that, KinFaceW-I dataset contains facial image pairs from different photos or sources under uncontrolled conditions while KinFaceW-II contains facial image pairs cut from the same photo resulting in biased Kinship Verification accuracy. This is the reason why KinFaceW-I dataset is used in this paper to validate the accuracy of the proposed methods.

Another dataset is Family 101 dataset [17] which is the first largest kinship dataset. Family 101 dataset contains 206 core families with 607 individuals. The total number of images in this dataset is approximately 14,816. Another dataset is UvA NEMO Smile dataset [9, 18] which is the only kinship database with face videos and still images.

The last kinship dataset is Tri-subject Kinship Face Dataset commonly known as TSKinFace dataset [19]. This dataset contains two different kinds of kinship relations namely, parent–daughter (FM–D) relation and parent–son (FM–S) relation. A comparative evaluation of all available kinship datasets is given in Table 1.

Table 1 Summary of available kinship datasets

Data set		Resolution	Types of relationships	Total kinship groups	Total image	Uncontrolled condition
Cornell KinFace [10]		100 × 100	Parent–child	143	287	Yes
UB KinFace [8, 15, 16]		89 × 96	Child–young parent	200	600	Yes
			Child–old parent	200		
KinFace W	KinFaceW-I [7] KinFaceW-II [7]	64 × 64	F–S	156	1066	Yes
			F–D	134		
			M–D	127		
			M–S	116		
		64 × 64	F–S	250	2000	Yes
			F–D	250		
			M–D	250		
			M–S	250		
Family 101 [17]		120 × 150	Family	206	14,816	Yes
UvA-NEMO Smile [9, 18]		1920 × 1080	F–S, M–S, F–D, M–D, S–S, B–B, B–S	1240	400	No
TSKinFace [19]		64 × 64	FM–D	502	787	Yes
			FM–S	513		

4 Methodology

This paper proposes an effective method to determine kinship relations between given pair of facial images using feature descriptors and learning SVM classifier. The feature descriptors are used to extract the salient facial features. These extracted facial features are then concatenated to create a high-dimensional feature vector. Support vector Machine (SVM) learns these high-dimensional feature vectors to classify facial images based on feature similarities.

In this paper, KinFaceW-I dataset is used to validate the Kinship Verification accuracy. Figure 2 gives a sample of some positive kinship pairs of KinFaceW-I Dataset. Positive kinship pair corresponds to real or own parent–child pair. While negative kinship pair corresponds to pair of one's parent with another's child.

4.1 Feature Extraction

Preprocessing is performed for each facial image in the KinFaceW-I subset for unbiased comparison between proposed methods. The various steps performed in preprocessing of each facial image are, (i) rotating each image to vertically align, (ii) scaling each image to align the eye positions, (iii) cropping facial region to remove unwanted background if needed, (iv) resize image into 64 × 64 pixel and convert RGB to gray scale image. The processed image is then used for feature extraction.

In the proposed method, the high-dimensional feature vectors techniques, Histogram of Gradient and Local Binary Patterns are used to describe facial appearance.

Histogram of Gradient (HOG): In HOG feature extraction method, each image of size 64 × 64 is divided into 8 × 8 overlapping blocks. This generates 196 blocks of size 8 × 8. Each 8 × 8 block is then represented by 9-dimensional uniform pattern histogram. These extracted features are then concatenated in a single block to form 1764-dimensional feature vector for final facial appearance. Figure 3 shows a facial image and its corresponding HOG features.

Fig. 2 Sample of some positive kinship pairs of KinFaceW-I dataset

INPUT IMAGE

HOG FEATURES

Fig. 3 Input image and corresponding HOG features. For 64 × 64 image, the HOG feature dimension is 1764

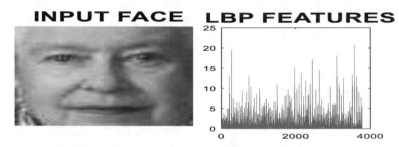

Fig. 4 Input Image and corresponding LBP features. For 64 × 64 image, the LBP feature dimension is 3776

Local Binary Patterns (LBP): In LBP feature extraction method, each image of size 64 × 64 is divided into 8 × 8 non-overlapping blocks. This generates 64 blocks of size 8 × 8. Each block is represented by a 59-dimensional uniform pattern LBP feature. These extracted LBP features are then concatenated in a single block to a 3776-dimensional feature vector. Figure 4 shows a facial image and its corresponding LBP features histogram.

4.2 Kinship Classification

After feature extraction process is performed and a high-dimensional feature vector is obtained to represent each image, the next stage is classification. The feature vectors of each of the two feature extraction methods (HOG and LBP) are used for supervised classification using support vector machine. SVM classifier is trained for the label-specific features and obtain the corresponding attributes vectors for result comparison.

5 Experimental Results

The proposed method is evaluated on KinFaceW-I dataset for all the four possible relations. The output of feature extraction method yields feature vectors of length 1764 and 3776 for HOG and LBP descriptors respectively. These feature vectors then train SVM classifier. The resultant is HOG descriptor trained SVM classifier and LBP descriptor trained SVM classifier. The result of HOG-SVM method for some samples of positive and negative pairs of KinFaceW-I dataset is demonstrated in Fig. 5. While the result of LBP-SVM method is demonstrated in Fig. 6. The verification rate (%) of the proposed methods on different subsets of KinFaceW-I dataset is tabulated in Table 2.

HOG-SVM

Correct match   False Match

Fig. 5 Sample images of results of HOG-SVM method. The first image in each pair is the image of son from (FS KinfaceW-I) and the correct matched image is the respective father of query image

LBP-SVM

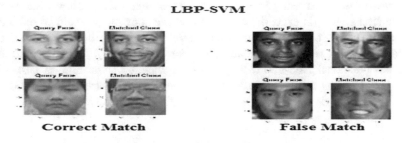

Correct Match   False Match

Fig. 6 Sample images of results of LBP-SVM method

Table 2 Verification rate (%) of the proposed methods on different subsets of KinFaceW-I dataset

Approach		F–S	F–D	M–D	M–S	Mean
Feature	Classifier					
HOG	SVM	73.3	75.6	75.6	68.9	73.35
LBP	SVM	75.6	77.8	75.6	73.3	**75.57**

The mean accuracy of LBP-SVM method is 2.22% higher than HOG-SVM method is mentioned in bold

Table 3 Comparison list of verification accuracy (%) of KinFaceW-I datasets with other state-of-art methods (All the results are compared for LBP feature extraction)

Method accuracy (%)	Datasets				
	F–S	F–D	M–D	M–S	Mean
Lu et al. [7]	72.5	66.5	66.2	72	69.9
Yan et al. [6]	74.5	69.5	69.5	75.5	72.25
Yan et al. [12]	73.5	67.5	66.1	73.1	70.1
Liu et al. [21]	73.39	71.70	71.14	77.57	73.45
Wang et al. [22]	71.8	62.7	66.4	66.6	66.9
Zhou et al. [23]	81.7	71.1	69.6	74.3	74.1
Proposed approach	75.6	77.8	75.6	73.3	75.57

The experiment results show that LBP-SVM method has performed better than HOG-SVM method for all the subsets of the dataset. The mean accuracy of LBP-SVM method is 2.22% higher than HOG-SVM method. The experiment results of LBP-SVM method are compared with different state-of-the-art methods to validate the performance accuracy of the proposed method. Table 3 shows that LBP-SVM method has achieved better results than all the state-of-art methods for each subset of KinFaceW-I dataset.

6 Computational Time

The proposed methods are implemented in Intel Core i5 processor of 2.71-GHz CPU and 8 GB RAM using MATLAB 2015 version. Table 4 shows the average computational time (in seconds) for proposed methods on one fold of subsets of KinFaceW-I dataset. The computational time for testing in both HOG and LBP methods are approximately same and less. For training phase, LBP method takes more computational time due to larger size of feature vector. But as LBP gives much better results compared to HOG and the training phase is performed only once in the methods and can be done offline, 30 s of time can be accepted.

Table 4 The average computational time (in seconds) for proposed methods on one fold of subsets of KinFaceW-I dataset

Method	Training (s)	Testing (s)
Pendar [20]	353.81	0.1
HOG-SVM	66.447	4.186
LBP-SVM	**91.27**	4.314

The average computational time for training of LBP-SVM method is mentioned in bold

7 Discussion

In this section, some potential applications of Kinship Verification are presented. Kinship Verification helps to determine kinship relations which would assist in solving issues like human trafficking, child smuggling, child adoptions and finding missing children. When Kinship Verification is performed with high accuracy, it could be used for labeling of facial images to identify the family tree.

Kinship Verification would be an additive assistance to complex and costly DNA verification process. Currently, DNA testing is a dominant approach to verify kinship relation between individuals but is costlier and require tedious legal approval. Hence if Kinship Verification accuracy is improved, it can be used to find similarities between some possible candidates in DNA verifications and then DNA testing can be used to get the final verdict.

8 Conclusion

This paper has discussed various aspects of Kinship Verification. A complete structure of all available kinship datasets has been presented. The proposed methods have been evaluated for all the four subsets of KinFaceW-I dataset. Experimental results showed that LBP-SVM method with mean accuracy 75.57% outperformed HOG-SVM based method with mean accuracy 73.35%. LBP-SVM method has achieved better results than different state-of-art methods for each subset of KinFaceW-I dataset. This demonstrates the validity and efficiency of the proposed method. The experiment results would inspire researchers to focus on Kinship Verification so that it may be helpful in various discussed applications. In future works, the proposed method would be employed for other kinship datasets to validate the verification accuracy.

Acknowledgements The research work is supported by Science and Engineering Research Board (SERB), Department of Science and Technology (DST), Government of India for the research grant. The sanctioned project title is "Design and development of an Automatic Kinship Verification system for Indian faces with possible integration of AADHAR Database." with reference no. ECR/2016/001659.

References

1. Maloney, L.T., Dal Martello, M.F.: Kin recognition and the perceived facial similarity of children. J. Vis. **6**(10), 4 (2006)
2. Dal Martello, M.F., Maloney, L.T.: Where are kin recognition signals in the human face? J. Vis. **6**(12), 2 (2006)
3. Dal Martello, M.F., Maloney, L.T.: Lateralization of kin recognition signals in the human face. J. Vis. **10**(8), 9 (2010)

4. Kaminski, G., Dridi, S., Graff, C., Gentaz, E.: Human ability to detect kinship in strangers' faces: effects of the degree of relatedness. Proc. R. Soc. Lond. B **276**(1670), 3193–3200 (2009)
5. Oda, R., Matsumoto-Oda, A., Kurashima, O.: Effects of belief in genetic relatedness on resemblance judgments by japanese raters. Evol. Hum. Behav. **26**(5), 441–450 (2005)
6. Yan, H., Lu, J., Deng, W., Zhou, X.: Discriminative multimetric learning for kinship verification. IEEE Trans. Inf. Forensics Secur. **9**(7), 1169–1178 (2014)
7. Lu, J., Zhou, X., Tan, Y.P., Shang, Y., Zhou, J.: Neighborhood repulsed metric learning for kinship verification. IEEE Trans. Pattern Anal. Mach. Intell. **36**(2), 331–345 (2014)
8. Xia, S., Shao, M., Luo, J., Fu, Y.: Understanding kin relationships in a photo. IEEE Trans. Multimedia **14**(4), 1046–1056 (2012)
9. Dibeklio̅ glu, H., Salah, A.A., Gevers, T.: Are you really smiling at me? Spontaneous versus posed enjoyment smiles. In: European Conference on Computer Vision, pp. 525–538. Springer (2012)
10. Fang, R., Tang, K.D., Snavely, N., Chen, T.: Towards computational models of kinship verification. In: 17th IEEE International Conference on Image Processing (ICIP), pp. 1577–1580. IEEE (2010)
11. Guo, G., Wang, X.: Kinship measurement on salient facial features. IEEE Trans. Instrum. Meas. **61**(8), 2322–2325 (2012)
12. Yan, H., Lu, J., Zhou, X.: Prototype-based discriminative feature learning for kinship verification. IEEE Trans. Cybern. **45**(11), 2535–2545 (2015)
13. Zhou, X., Lu, J., Hu, J., Shang, Y.: Gabor-based gradient orientation pyramid for kinship verification under uncontrolled environments. In: Proceedings of the 20th ACM International Conference on Multimedia, pp. 725–728. ACM (2012)
14. Xia, S., Shao, M., Fu, Y.: Toward kinship verification using visual attributes. In: 21st International Conference on Pattern Recognition (ICPR), pp. 549–552. IEEE (2012)
15. Shao, M., Xia, S., Fu, Y.: Genealogical face recognition based on ub kinface database. In: IEEE Computer Society Conference on Computer Vision and Pattern Recognition Workshops (CVPRW), pp. 60–65. IEEE (2011)
16. Xia, S., Shao, M., Fu, Y.: Kinship verification through transfer learning. In: IJCAI, pp. 2539–2544 (2011)
17. Fang, R., Gallagher, A.C., Chen, T., Loui, A.: Kinship classification by modeling facial feature heredity. In: 20th IEEE International Conference on Image Processing (ICIP), pp. 2983–2987. IEEE (2013)
18. Dibeklioglu, H., Ali Salah, A., Gevers, T.: Like father, like son: facial expression dynamics for kinship verification. In: Proceedings of the IEEE International Conference on Computer Vision, pp. 1497–1504 (2013)
19. Qin, X., Tan, X., Chen, S.: Tri-subject kinship verification: understanding the core of a family. IEEE Trans. Multimedia **17**(10), 1855–1867 (2015)
20. Alirezazadeh, P., Fathi, A., Abdali-Mohammadi, F.: A genetic algorithm-based feature selection for kinship verification. IEEE Signal Process. Lett. **22**(12), 2459–2463 (2015)
21. Liu, Q., Puthenputhussery, A., Liu, C.: Inheritable fisher vector feature for kinship verification. In: IEEE 7th International Conference on Biometrics Theory, Applications and Systems (BTAS), pp. 1–6. IEEE (2015)
22. Wang, M., Li, Z., Shu, X., Tang, J., et al.: Deep kinship verification. In: IEEE 17th International Workshop on Multimedia Signal Processing (MMSP), pp. 1–6. IEEE (2015)
23. Zhou, X., Shang, Y., Yan, H., Guo, G.: Ensemble similarity learning for kinship verification from facial images in the wild. Inf. Fusion **32**, 40–48 (2016)

A Feature Averaging Method for Kinship Verification

Niharika Yadav, Aarti Goyal and T. Meenpal

Abstract Over a past few years, kinship verification using facial images has been gaining significant attention by different researchers. Kinship verification is motivated by the human inheritance and psychological findings that a child resembles his/her parent more than any other person in terms of facial appearance. In this paper, a new Feature Averaging method is proposed for kinship verification. This method creates an average vector that maximizes the variance between kinship pairs and minimizes the variance between non-kinship pairs. Multiple features are extracted using different feature descriptors for each facial image to create a high-dimensional feature vector. Principal Component Analysis (PCA) is used for dimension reduction of the high-dimensional feature vector by selecting apposite features which give useful information. The resultant PCA reduced feature vector characterizes each facial image with salient features. Experimental results are presented for different kinship datasets and the obtained verification accuracy demonstrates the performance of the proposed method.

Keywords Kinship verification · Genetic similarity · Feature averaging
Feature descriptor

1 Introduction

Human inherit traits from their ancestors due to genetic relatedness. Like other traits, visual resemblance is due to gene heredity which passes from parents to children [1, 2]. Many psychological studies aim to understand how humans visually

N. Yadav (✉) · A. Goyal · T. Meenpal (✉)
Electronics and Telecommunication, NIT Raipur, Raipur, India
e-mail: niharika9714@gmail.com

T. Meenpal
e-mail: tmeenpal.etc@nitrr.ac.in

A. Goyal
e-mail: aarti29wasnik@gmail.com

© Springer Nature Singapore Pte Ltd. 2019
P. K. Mallick et al. (eds.), *Cognitive Informatics and Soft Computing*,
Advances in Intelligent Systems and Computing 768,
https://doi.org/10.1007/978-981-13-0617-4_38

perceive and identify relations based on appearance, gestures, behavior and voice [3, 4]. These studies show that amongst all these, facial appearance substantiate genetic similarity between parents and children [5, 6].

Human inheritance and many psychological findings have motivated, the idea of kinship which has now become an active area of research [2, 3, 6, 7]. Kinship is a broad term with distinct meanings based on the kind of associated framework. In anthropology, kinship is defined as the network of relationships between people in the society [8] whereas in biology, kinship is defined as the coefficient of relationship or degree of genetic relatedness between individuals of the same species [9]. The individuals with any kind of genetic similarities are termed as kins and their relationship is termed as kinship relationship. The process of finding this relationship via correlated visual resemblance is called kinship verification. In computer vision, kinship verification is the task of training a machine to evaluate if any kinship relation exists between a given pair of facial image based on feature extraction and classification.

Kinship verification plays a vital role in the field of social network analysis. It may be helpful in solving issues like human trafficking, child smuggling, child adoptions, surveillance, criminal investigations, and finding missing children. Kinship verification would be an additive assistance to complex and costly DNA verification process. Currently, DNA testing is a dominant approach to verify kinship relation between individuals but is costlier and requires tedious legal approval. Hence if kinship verification accuracy is improved, the overall cost is significantly reduced.

In this paper, a new Feature Averaging method is proposed which creates an average vector that maximizes the variance between kinship pairs and minimizes that between non-kinship pairs. First, multiple features are extracted using different feature descriptors for each facial image in the kinship dataset. A high-dimensional feature vector is obtained that contains some discriminative relevant features while other features are redundant. Principal Component Analysis (PCA) is hence used for reduction of the high-dimensional feature vector by selecting apposite features.

The proposed Feature Averaging method then uses this reduced feature vector to create an average vector by normalization. The average vector then minimizes the distance between kinship pairs and maximizes the distance between non-kinship pairs. The proposed method is evaluated on different kinship datasets, KinFaceW-I, KinFaceW-II, and Cornell KinFace to validate the performance accuracy.

2   Related Work

This section describes the related work and available datasets of kinship verification.

2.1 Kinship Verification

The pioneer attempt of kinship verification is originated from Bressnan et al. [10], in which the authors evaluated phenotype matching. These traits and facial landmarks are highly correlated in a kinship pair. Despite, no efforts have been made by researchers to develop algorithm for kinship verification until Fang et al. [11] came up with the first Kinship database named Cornell KinFace database in 2010.

This paper categorizes existing methods of kinship verification into three different classes: (i) Feature based, (ii) Metric Learning based (iii) Other methods.

Feature-Based methods: Feature-based kinship verification methods focus on characterizing extracted local features rather than holistic features. Fang et al. [11] evaluated a set of 22 low-level image features and show that computer vision techniques can be effectively employed for kinship verification. Guo et al. [12] proposed DAISY descriptor to extract and describe salient facial features and developed a dynamic scheme to stochastically combine familial traits. Zhou et al. [13] proposed spatial pyramid learning-based feature descriptor (SPLE) and Liu et al. in 2016 [14] presented an innovative inheritable Fisher vector feature method (IFVF) for kinship verification.

Learning-Based methods: Learning-based kinship verification methods aim to reduce the large divergence of distribution between parent and child. Xia et al. [15] proposed an extended transfer subspace learning method (TSL) that aims to reduce the large similarity gap between child and old parent. Lu et al. [16] proposed neighborhood repulsed metric learning method (NRML) for kinship verification via facial images. They further proposed a multiview NRML method (MNRML) to improve the usage of multiple feature descriptors. Other metric learning methods are discriminative multimetric learning method (DMML) [17], neighborhood repulsed correlation metric learning method (NRCML) [18] and Ensemble similarity learning method (ESL) [19].

Other methods: Jhang et al. [20] proposed convolution neural network (CNN) architecture composed by two convolution max pooling layers for kinship verification. Wang et al. [21] presented a novel deep Kinship Verification model (DKV) for Kinship Verification. In this model, the authors integrated the excellent deep learning architecture into metric learning aiming to improve the performance. Zhou et al. [22] proposed a new relative symmetric bilinear model (RSBM) and Xu et al. [23] proposed a novel Structured Sparse Similarity Learning method (S3L).

2.2 Kinship Datasets

There are six available databases for kinship verification. Cornell KinFace, KinfaceW-I/II, UB KinFace, TSKinFace, Family 101 and UvA-NEMO Smile. A complete description of these kinship databases is given in Table 1. Out of all these kinship datasets, this paper evaluates the performance of KinFaceW-I,

Table 1 Summary of available Kinship datasets

Data set		Kinship relation	Total image pairs
Cornell KinFace [11]		Parent–child	143
UB KinFace [15]		Child–young parent	200
		Child–old parent	200
KinFaceW [16]	KinFaceW-I	F–S	156
		F–D	134
		M–D	127
		M–S	116
	KinFaceW-II	F–S	250
		F–D	250
		M–D	250
		M–S	250
Family 101 [24]		Family	206
UvA-NEMO Smile [25]		F–S, F–D, M–D, M–S, S–S, B–B, B–S	1240
TSKinFace [22]		FM–D	502
		FM–S	513

KinFaceW-II, and Cornell KinFace datasets. These datasets consist of four fundamental relationships existing in the society, namely, Father and his Son (F–S), Father and his Daughter (F–D), Mother and her Son (M–S) and Mother and her daughter (M–D).

3 The Proposed Feature Averaging Method

Let P be positive sample subset and N be negative sample subset. Positive samples are image pairs with kinship relations (example, fs_001_1 and fs_001_2 corresponds to positive sample of father–son subset in KinFaceW-I dataset) and negative samples are image pairs with non-kinship relations (fs_001_1 is randomly combined with any of the son's image except fs_001_2).

For a kinship dataset with n pairs of kinship relation, P and N subsets are represented as

$$\begin{cases} P = \{(x_i, y_i) | i = 1, 2, \ldots, n\} \\ N = \{(x_i, y_j) | j = 1, 2, \ldots, n\} \end{cases}, \tag{1}$$

where, x_i and y_i belongs to m-dimensional feature vector of ith parent and child representing kinship pair. While, x_i and y_j belongs to m-dimensional feature vector of ith parent and jth child representing non-kinship pair.

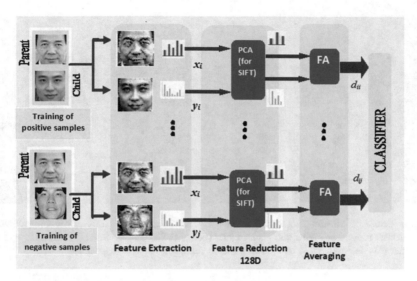

Fig. 1 Flowchart of the proposed method for training stage

In the proposed method, an average vector is obtained by normalizing m-dimensional feature vector. The resultant is then used to minimize the distance between x_i and y_i kinship pairs and maximizing that between x_i and y_j non-kinship pairs by the following equation,

$$\text{Minimize } d_{ii} = \|A_i \text{ and } B_i \|$$
$$\text{Maximize } d_{ii} = \|A_i \text{ and } B_j \|, \tag{2}$$

where, $A_i, B_i, \ldots, B_j$ corresponds to averaged values of x_i, y_i and, y_j feature vectors respectively.

A flowchart of the proposed method for training stage is given in Fig. 1. Sample of kinship pairs for KinFaceW-I, KinFaceW-II and Cornell KinFace is shown in Fig. 2.

4 Experiment Evaluation

4.1 Datasets Evaluated

The proposed method is evaluated on KinFaceW-I, KinFaceW-II, and Cornell KinFace to validate the performance accuracy. A brief description of these datasets is given as

KinFaceW-I: KinFaceW-I dataset has 533 pairs of RGB facial images. Out of these, 156 pairs belong to father–son relation, 134 pairs belong to father–daughter relation, 116 pairs belong to mother–son relation and 127 pairs belong to mother–

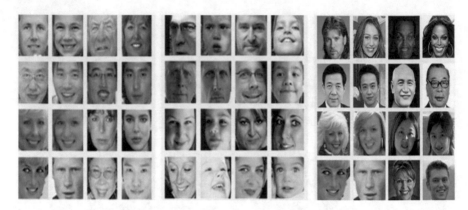

Fig. 2 Sample of kinship pairs for **a** KinFaceW-I, **b** KinFaceW-II, **c** Cornell KinFace. First row in each dataset represents Father and his Daughter (F–D), Father and his Son (F–S), Mother and her daughter (M–D) and Mother and her Son (M–S) relation row-wise respectively

daughter relations. All the images of KinFaceW-I dataset are already aligned and cropped to 64 × 64 pixels in coherence to eye centroid, hence only RGB to Gray conversion is needed while preprocessing.

KinFaceW-II: KinFaceW-II dataset is larger than KinFaceW-I dataset with 1000 RGB images pairs. This dataset is equally divided into 250 image pairs for father–son, father–daughter, mother–son relation and mother–daughter relations. All the images of KinFaceW-II dataset are also aligned and cropped to 64 × 64 pixels, requiring only RGB to gray conversion for preprocessing.

Cornell KinFace: Cornell KinFace is a very small dataset with 143 RGB image pairs of resolution 100 × 100 pixels. This dataset has 64 pairs for father–son, 33 pairs for father–daughter,18 pairs for mother–son and 28 pairs for mother–daughter relations.

Cornell KinFace dataset is the first dataset introduced for kinship verification with a very small number of kinship pairs, for example, mother–son relation has only 18 pairs. Therefore, most of the previous papers on kinship verification has no evaluated this dataset. But, in this paper, we have evaluated the performance accuracy of Cornell KinFace dataset to present the accuracy of the first introduced kinship dataset. Hence an effort is done to validate the performance accuracy of proposed method for this dataset.

4.2 Experimental Setup

Preprocessing: In this paper, a small amount of preprocessing is performed on facial images of kinship dataset. Each facial image is converted from RGB to grayscale and histogram equalization is performed to mitigate the illumination factor. For KinFaceW-I, KinFaceW-II, and Cornell KinFace, each facial image is cropped to 64×64 to remove any available non-facial region and left with the required facial region for kinship verification.

Feature Descriptors: After preprocessing, salient features are extracted from the image using feature descriptors. It is important that the features must be invariant to scale transformations and be more dominant traits on display. Hence, SIFT [26] and SURF [27] feature descriptors are used in this paper to extract salient features from facial images. The basic steps involved in these two feature descriptors are given as

SIFT(Scale Invariant Feature Transform)

1. Create a scale space of key points for each facial image $I(x, y)$ using gaussian kernel with different values of in each octave.

$$L(x, y, \sigma) = G(x, y, \sigma) * I(x, y), \qquad (3)$$

where, G is gaussian function and * defines convolution.

2. For SIFT, key points are detected using Difference of Gaussian (DOG) given by equation

$$D(x, y, \sigma) = (G(x, y, k\sigma) - G(x, y, \sigma)) * I(x, y) \qquad (4)$$

3. Local maxima and minima are extracted for $D(x, y, \sigma)$ and corresponding key points are detected.
4. For each key point, local gradient in 3×3 neighborhood centered at the key point is evaluated based on image gradient magnitude and orientation.
5. Feature vector of each key point is generated with 4×4 array location grid and eight orientations.

SURF (Speed-Up Robust Features)

All the basic steps involved in SURF descriptor is similar to SIFT. The only difference is that, SURF calculates the determinant of Hessian matrix of each pixel point to detect the local extreme point.

4.3   Feature Extraction

We have conducted feature extraction using SIFT and SURF feature descriptors. We have performed feature reduction using PCA on SIFT feature vectors by deducting redundant features. While, no PCA reduction is needed for SURF descriptor and this is the first time SURF descriptor is used in kinship verification.

SIFT: In the experiment, 64×64 facial image is divided into non-overlapping cell of size 16×16 with grid space of 4 pixels. A 4×4 gradient window with eight orientations give a feature vector of length $128D$ for each SIFT descriptor. Hence, multiple SIFT descriptors each of length $128D$ is obtained in which some descriptors represent discriminative relevant features while other represent redundant features. PCA is used as dimensional reduction technique to remove redundant features and maintaining stable features which had high contrast and eliminating those with low contrast and poor localization. In our work, PCA is used to eliminate redundant features resulting in 128×6 SIFT descriptors. Finally, each facial image is represented by a $768D$ feature vector.

SURF: In this paper, SURF extract key points using Hessian matrix. A 64×64 facial image is divided into non-overlapping cell of size 16×16 with grid space of 4 pixels. A 4×4 gradient window give a vector of length $64D$. Hence, each feature descriptor is of length $64'D$. The major advantage of SURF descriptor is that, it obtain less and more relevant features with higher accuracy. Hence, the size of overall feature vector is less and PCA is not needed for dimension reduction.

4.4 Performance Evaluation

To evaluate the performance of the proposed method, the fivefold cross validation is used on kinship datasets. In fivefold cross validation, we divide each subset of kinship datasets equally so that each fold has nearly equal numbers of positive with kinship relations and negative without kinship relations samples of the facial images. Four folds positive samples combined with four folds negative samples act as the training set for the classifier. The remaining folds of positive and negative samples act as the testing set. Table 2 shows the range of different folds of kinship datasets used in the experiment.

KNN classifier with Euclidean distance is used with neighbors selected between 3 and 6, keeping 1 as the standard value.

5 Experimental Results

The performance accuracy of proposed method for different datasets are given as

Results for KinFaceW-I: Table 3 shows the result for KinFaceW-I dataset for SIFT and SURF descriptors. The performances of all tested local features for four

Table 2 Range of different folds of the Kinship datasets

Dataset		Folds				
		1	2	3	4	5
KinFaceW-I	F–S	1–31	32–63	64–95	96–127	128–156
	F–M	1–27	28–55	56–83	84–107	108–134
	M–S	1–23	24–47	48–71	72–95	96–116
	M–D	1–25	26–50	51–75	76–100	101–127
KinFaceW-II (All datasets)		1–50	51–100	101–150	151–200	201–250
Cornell KinFace	F–S	1–13	14–26	27–39	40–53	54–64
	F–M	1–7	7–14	15–21	22–28	29–33
	M–S	1–4	5–8	9–12	13–15	16–18
	M–D	1–6	7–12	13–18	19–24	25–28

Table 3 Verification rate (%) of the SIFT and SURF descriptors on different subsets of KinFaceW-I dataset

Method	Feature representation	F–S	F–D	M–S	M–D	Mean
Feature averaging	SIFT	57.8	53	54.16	50	53.74
	SURF	57.7	51.9	54.3	52	**53.97**

The bold value shows that the corresponding feature descriptor gives the best result (mean verification rate) for the dataset

Table 4 Verification rate (%) of the SIFT and SURF descriptors on different subsets of KinFaceW-II dataset

Method	Feature representation	F–S	F–D	M–S	M–D	Mean
Feature averaging	SIFT	53	48	56	54	52.75
	SURF	62	54	51	55	**55.5**

The bold value shows that the corresponding feature descriptor gives the best result (mean verification rate) for the dataset

Table 5 Verification rate (%) of the SIFT and SURF descriptors on different subsets of Cornell KinFace dataset

Method	Feature representation	F–S	F–D	M–S	M–D	Mean
Feature averaging	SIFT	43.6	45.4	48	45.4	43.3
	SURF	57.7	64.3	50	50	**55.5**

The bold value shows that the corresponding feature descriptor gives the best result (mean verification rate) for the dataset

subsets are boosted by feature averaging method. The mean accuracy of SURF outstands SIFT descriptor by 0.23%

Results for KinFaceW-II: Table 4 shows the result for KinFaceW-II dataset for SIFT and SURF descriptors. The performances of all tested local features for four subsets are boosted by feature averaging method. The mean accuracy of SURF outstands SIFT descriptor by 2.75%

Results for Cornell KinFace: Table 5 shows the result for Cornell KinFace dataset for SIFT and SURF descriptors. The performances of all tested local features for four subsets are boosted by feature averaging method. The mean accuracy of SURF outstands SIFT descriptor by 12.2%.

6 Conclusion

The aforementioned research concludes the performance of a new Feature Averaging method for kinship verification. The proposed method creates an average vector that maximizes the variance between kinship pairs and minimizes that between non-kinship pairs. The obtained experimental results demonstrated that averaging method has achieved the competitive performance on KinfaceW-I, KinFaceW-II and Cornell KinFace datasets.

For multiple feature extraction, we have used SIFT of feature dimension 768D and SURF of feature dimension 128D. The obtained results showed that SURF descriptor has performed better than SIFT descriptor for the proposed method even with a very small-sized feature vector. In future works, we will employ more efficient feature descriptors along with Feature Averaging method for kinship verification and would try to obtain better verification accuracy.

Acknowledgements This research work is supported for the research grant by Science and Engineering Research Board (SERB), Department of Science and Technology, Government of India. The sanctioned project title is "Design and development of an Automatic Kinship Verification system for Indian faces with possible integration of AADHAR Database." with reference no. ECR/2016/001659.

References

1. Dal Martello, M.F., Maloney, L.T.: Where are kin recognition signals in the human face? J. Vis. **6**(12), 2 (2006)
2. DeBruine, L.M., Smith, F.G., Jones, B.C., Roberts, S.C., Petrie, M., Spector, T.D.: Kin recognition signals in adult faces. Vis. Res. **49**(1), 38–43 (2009)
3. Dal Martello, M.F., Maloney, L.T.: Lateralization of kin recognition signals in the human face. J. Vis. **10**(8), 9 (2010)
4. Wu, H., Yang, S., Sun, S., Liu, C., Luo, Y.J.: The male advantage in child facial resemblance detection: behavioral and ERP evidence. Soc. Neurosci. **8**(6), 555–567 (2013)
5. Alvergne, A., Oda, R., Faurie, C., Matsumoto-Oda, A., Durand, V., Raymond, M.: Cross-cultural perceptions of facial resemblance between kin. J. Vis. **9**(6), 23 (2009)
6. Kaminski, G., Dridi, S., Graff, C., Gentaz, E.: Human ability to detect kinship in strangers' faces: effects of the degree of relatedness. Proc. R. Soc. Lond. B **276**(1670), 3193–3200 (2009)
7. Kaminski, G., Ravary, F., Graff, C., Gentaz, E.: Firstborns disadvantage in kinship detection. Psychol. Sci. **21**(12), 1746–1750 (2010)
8. International Encyclopedia of the Social Sciences (2008) Kinship
9. International Encyclopedia of the Social Sciences (1968) Kinship
10. Bressan, P., Dal Martello, M.: Talis pater, talis filius: perceived resemblance and the belief in genetic relatedness. Psychol. Sci. **13**(3), 213–218 (2002)
11. Fang, R., Tang, K.D., Snavely, N., Chen, T.: Towards computational models of kinship verification. In: 17th IEEE International Conference on Image Processing (ICIP), pp. 1577–1580. IEEE (2010)
12. Guo, G., Wang, X.: Kinship measurement on salient facial features. IEEE Trans. Instrum. Measur. **61**(8), 2322–2325 (2012)
13. Zhou, X., Hu, J., Lu, J., Shang, Y., Guan, Y.: Kinship verification from facial images under uncontrolled conditions. In: Proceedings of the 19th ACM International Conference on Multimedia, pp. 953–956. ACM (2011)
14. Liu, Q., Puthenputhussery, A., Liu, C.: Inheritable fisher vector feature for kinship verification. In: IEEE7th International Conference on Biometrics Theory, Applications and Systems (BTAS), pp. 1–6. IEEE (2015)

15. Xia, S., Shao, M., Luo, J., Fu, Y.: Understanding kin relationships in a photo. IEEE Trans. Multimedia **14**(4), 1046–1056 (2012)
16. Lu, J., Zhou, X., Tan, Y.P., Shang, Y., Zhou, J.: Neighborhood repulsed met-ric learning for kinship verification. IEEE Trans. Pattern Anal. Mach. Intell. **36**(2), 331–345 (2014)
17. Yan, H., Lu, J., Deng, W., Zhou, X.: Discriminative multimetric learning for kinship verification. IEEE Trans. Inf. Forensics Secur. **9**(7), 1169–1178 (2014)
18. Yan, H., Zhou, X., Ge, Y.: Neighborhood repulsed correlation metric learning for kinship verification. In: Visual Communications and Image Processing (VCIP), pp. 1–4. IEEE (2015)
19. Zhou, X., Shang, Y., Yan, H., Guo, G.: Ensemble similarity learning for kinship verification from facial images in the wild. Inf. Fusion **32**, 40–48 (2016)
20. Zhang12, K., Huang, Y., Song, C., Wu, H., Wang, L., Intelligence, S.M.: Kinship verification with deep convolutional neural networks (2015)
21. Wang, M., Li, Z., Shu, X., Tang, J., et al.: Deep kinship verification. In: 17th International Workshop on Multimedia Signal Processing (MMSP), pp. 1–6. IEEE (2015)
22. Qin, X., Tan, X., Chen, S.: Tri-subject kinship verification: understanding the core of a family. IEEE Trans. Multimedia **17**(10), 1855–1867 (2015)
23. Xu, M., Shang, Y.: Kinship measurement on face images by structured similarity fusion. IEEE Access **4**, 10280–10287 (2016)
24. Fang, R., Gallagher, A.C., Chen, T., Loui, A.: Kinship classification by modeling facial feature heredity. In: 20th IEEE International Conference on Image Processing (ICIP), pp. 2983–2987. IEEE (2013)
25. Dibeklioglu, H., Ali Salah, A., Gevers, T.: Like father, like son: Facial expression dynamics for kinship verification. In: Proceedings of the IEEE International Conference on Computer Vision, pp. 1497–1504 (2013)
26. Lowe, D.G.: Distinctive image features from scale-invariant keypoints. Int. J. Comput. Vis. **60**(2), 91–110 (2004)
27. Bay, H., Tuytelaars, T., Van Gool, L.: Surf: Speeded up robust features. In: Computer vision—ECCV, pp. 404–417 (2006)

An Artificial Intelligence Technique for Low Order Harmonic Elimination in Multilevel Inverters

Ritika Yadav and Birinderjit Singh

Abstract This paper presents elimination of lower order harmonics in multilevel electrical converters with ascend DC sources utilizing distinctive computations. To wish out the chosen lower prepare harmonics from the yield voltage or current wave of a multilevel electrical converter, acceptable amendment angles were calculated and thus the facility change amendment devices were switched as required be. The amendment angles were brindled from the non-coordinate conditions got from the Fourier course of action of the yield voltage of the converter. A genetic rule was sent to lightweight nonlinear conditions. With the projected approach, the specified amendment angles square measure with capably handled utilizing genetic to want out the lower organize harmonics from the convertor voltage waveform for varied modulation indexes. Different strategies likewise gave great answer for these conditions. Moreover many derivatives of different algorithms have also been used. Different algorithms have its own pros and cons. A few calculations take much computational time yet give better outcome and the other way around.

Keywords Multilevel inverter · Harmonic elimination · Proposed rules

1 Introduction

Electric power quality is a term which has discovered extending thought in charge or power designing in the present years. Notwithstanding the way that this subject has constantly been imperative to control engineers, it has expected impressive enthusiasm for the 1990's. Electric power quality means various things for other people's individuals [1]. To most electric power plans, the term implies a particular satisfactorily high audit of electric organization yet past that there is no far reaching

R. Yadav (✉) · B. Singh
Electrical Engineering, Chandigarh University, Chandigarh, India
e-mail: ritika18sep1993@gmail.com

B. Singh
e-mail: birinderjit@msn.com

© Springer Nature Singapore Pte Ltd. 2019
P. K. Mallick et al. (eds.), *Cognitive Informatics and Soft Computing*,
Advances in Intelligent Systems and Computing 768,
https://doi.org/10.1007/978-981-13-0617-4_39

assention. The live of vitality quality relies upon the needs of the equipment that is being given. What's great power quality for an electrical motor may not be good enough for a personal computer. Typically the administration quality implies keeping up a bended influx of transport voltages at evaluated voltage and recurrence [2]. The waveform of electrical power at age make is totally bended and free from any damage. A large number of the power transformation and utilization hardware are additionally intended to work under unadulterated sinusoidal voltage waveforms. In any case, there are numerous gadgets that contort the waveform. These bends may engender wherever all through the electrical framework. As of late, there has been associate degree dilated utilization of non-straight masses that has caused associate degree dilated a part of non-sinusoidal streams and voltages in electrical network [3] Plan of vitality quality zones may be made by the wellspring of the issue, for instance, converters, attractive circuit non linearity, circular segment heater or by the wave state of the flag, for example, sounds, glimmer or by the recurrence range (radio recurrence obstruction) [4]. The wave wonders connected with management quality could be delineated into synchronous and non synchronous wonders. Synchronous wonders guide those in synchronizing with AC wave at management repeat. The standard parts of electric power quality may be requested as:-

1. Fundamental concepts
2. Sources
3. Instrumentation
4. Modeling
5. Analysis
6. Effects

Figure 1 demonstrates a portion of the average voltage aggravations [5].

2 Literature Review

Harmonic decrease is a standout amongst the most difficult issues identified with multilevel inverters. As per various methodologies on finish of harmonics of a fell construction supply electrical converter. Here the fundamental idea is to dispose of

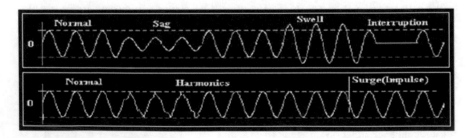

Fig. 1 Average voltage representation [5]

particular harmonic with a legitimate decision of switch angles. During this paper genetic algorithm rule improvement procedure is use to accomplish applicable switch angle for decrease the all harmonic and finish of a particular lower compose harmonics with real portions at the pined for regards. This paper displays an outline of different topologies, administration strategies and tweak courses used by these electrical converters. Regenerative and impelled topologies are additionally examined. Table 1 demonstrates that related work with low request harmonics diminished in construction electrical converter. This paper contains five areas. Segment II exhibits a theoretical diagram of set up and case setup. One amount of a multilevel inverter or construction electrical converter is talked concerning Sect. 3.

Table 1 Related work for harmonic elimination in multilevel inverter

S/N	Authors	Proposed approaches
1	Habelter et al. (1999)	Cascade inverter is a characteristic fit for expansive car every single electric drive since it utilizes a few levels of DC voltage sources [4]
2	Tolbert et al. (2000)	Multilevel inverter structures have been produced to beat deficiencies in strong state switching devices evaluations so they can be connected to high-voltage electrical frameworks [6]
3	Yen Shin Lai et al. (2002)	Comparison with the specific harmonic end PWM strategy, for a similar number of killed low request harmonics, the exhibited method gives the benefits of both lower THD and less exchanging [7]
4	Chiaason et al. (2003)	It is indicated however the exchanging end during a construction electrical converter may be accomplished a needed key voltage and not produces explicit higher request harmonics [8]
5	Keith Jeremy et al. (2004)	A brought together approach is introduced to comprehend the harmonic disposal conditions for the greater part of the different exchanging plans [9]
6	Hew Wooi Ping et al. (2010)	It proposes associate optimum pulse width modulation technique that uses answer up to region criteria and harmonic infusion [10]
7	Tarafdar et al. (2010)	It can be connected to take care of the issue in a more straightforward way, notwithstanding when the quantity of exchanging edges is expanded and the assurance of these edges utilizing the resultant hypothesis approach isn't conceivable [11]
8	Damoun et al. (2011)	Compared with existing techniques, the planned strategy doesn't embrace complicated condition gatherings and is considerably easier to be used on account of expansive variety of exchanging edges, or numerous exchanging edges per voltage level in construction inverters [12]
9	Saha et al. (2013)	The planned strategy is re-enacted on a seventeen levels and fifteen levels electrical converter and ideal exchanging edges square measure resolved to lose low harmonics [13]

SHE-PWM is mentioned in Sect. 4. Concerning projected calculation is analyzed in Sect. 5. Concerning projected approach is introduced in Sect. 6. In conclusion, discourses and finishing up comments are introduced in Sect. 7.

3 Cascaded Multilevel Inverter

The idea of electrical converters has been presented since 1975. The term multilevel began with the three-level gadget [14]. Multilevel or structure inverters are actuation certain the profound world and additionally business within the current decade for high-voltage and medium-voltage vitality management. It combines a sought after voltage from a many levels of dc voltages as inputs. By taking sufficient number of dc sources, a nearly sinusoidal voltage waveform can be synthesized [14]. A multilevel converter not only achieves high power ratings, but also enables the use of renewable energy sources [15]. Structure inverters have changed into a productive and right down to earth announces expanding the facility and diminishing the hints of AC waveforms. Stood out from the standard two-level voltage supply electrical converter, the stepwise yield voltage are those the genuine favorable position of multilevel inverter.

This advantage results in higher power quality, better electromagnetic compatibility, lower switching losses, higher voltage capability, and needlessness of a transformer at distribution voltage level, thereby reducing the costs. Plentiful multilevel converter topologies have been proposed during the last two decades. Multilevel inverters are generally divided into three configurations: diode-clamped, flying capacitor, and cascaded H-bridge multilevel inverters [12]. Among these electrical converter topologies, electrical converter accomplishes the upper yield voltage and power levels and furthermore the higher enduring quality in view of its measured structure. Electrical inverters depend on an arrangement association of a few single stage full scaffold inverters. This structure is supplied for achieving medium yield voltage levels utilizing simply normal low voltage segments [12]. A structure electrical converter, an influence electronic convenience, is equipped for giving wanted substituting voltage level at the yield utilizing various lower level DC voltages as data. Electrical converter appeared in Fig. 2 has been decided for the execution of firefly control by virtue of its deliberate quality, effortlessness of administration, require less assortment of components, general less weight and cost once appeared differently in relation to exchange styles of electrical converter [11]. To allow a curved yield, a couple of H-bridge structure converter is associated in arrangement. Every cell contains one H-expansion and yield voltage made by electrical converter is absolutely the entire of the respectable variety of voltages created by each cell i.e. on the off likelihood that there are k cells in electrical

Fig. 2 Cascaded multilevel
inverter [12]

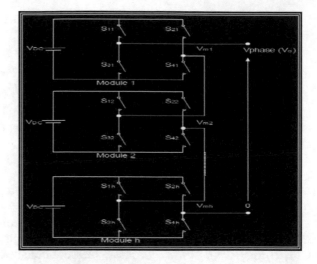

converter then the amounts of yield voltage levels would be 2k + 1 [9]. 11-level electrical converter is decided for execution of firefly run the show.

4 Selective Harmonic Elimination

In case of SHE, selected lower order harmonics are eliminated while remaining harmonic components are reduced to minimize THD [16]. In this paper bring down request sounds i.e. i.e. fifth, seventh, eleventh, thirteenth are wiped out. The articulation wants key voltage V_1 in below specify equation. Additionally, the connection between the essential and also the greatest approachable voltages is given by modulation index (m_{11}) is portrayed on the grounds that the extent of the focal yield voltage V_1 to the foremost extreme potential major voltage V_{1max} [17]. The greatest essential voltage is noninheritable once when the entire switch angles are zero i.e.

$$V_{1max} = \frac{4V_{dc}}{\pi} \tag{1}$$

In this manner the articulation for m_{11}

$$m_{11} = \frac{\pi V_1}{4V_{dc}} \quad (0 \leq m_{11} \leq 1) \tag{2}$$

Numerical harmonic eliminated issue can be defined as

$$
\begin{aligned}
&\cos(\theta_1) + \cos(\theta_2) + \cdots + \cos(\theta_m) = m_{11}M \\
&\cos(5\theta_1) + \cos(5\theta_2) + \cdots + \cos(5\theta_m) = 0 \\
&\cos(7\theta_1) + \cos(7\theta_2) + \cdots + \cos(7\theta_m) = 0 \\
&\cos(11\theta_1) + \cos(11\theta_2) + \cdots + \cos(11\theta_m) = 0 \\
&\cos(13\theta_1) + \cos(13\theta_2) + \cdots + \cos(13\theta_m) = 0 \\
&\cdots \\
&\cdots \\
&\cos(n\theta_1) + \cos(n\theta_2) + \cdots + \cos(n\theta_m) = 0
\end{aligned}
\tag{3}
$$

The condition (3) will be an arrangement of supernatural condition, known as SHE condition. From the condition obscure exchanging point or switching angle $\theta_1, \theta_2, \theta_3, \theta_4, \theta_5, \theta_6, \theta_7, \ldots, \theta_m$ square measure designed with the assistance of the given estimation of m_{11} (from zero to one) for trigger semiconductor switches [18].

5 Proposed Approach

As per proposed approach right off the bat to examine different course multilevel inverters. At that point ascertain the estimation of harmonics at various levels. After that actualize the genetic calculation on an eleventh level and ideal exchanging edges are resolved to kill low request harmonics and to limit THD.

6 Simulated Results

As indicated by proposed approach actualize the genetic calculation on an eleventh level and ideal switching angle are resolved to wipe out low request harmonics and to limit THD in MATLAB. The outcomes for ideal switching angles $\theta_1, \theta_2, \theta_3, \theta_4$ and θ_5 are planned in Fig. 3 demonstrates the best switch angle once this strategy is connected to an 11-level electrical converter to limit the harmonic distortion.

For a three-arrange load, the line-to-line voltage harmonic distortion is imperative, since the triplen harmonics, appear within the stage voltage, square measure exhausted from the line voltage.

For the switch angles nonheritable by genetic calculation the stage voltage and line-voltage harmonic mutilation is enrolled and arranged in Fig. 4 separately. Table 1 demonstrates the ideal exchanging point and aggregate harmonic contortion for three stage eleventh level inverter (Table 2).

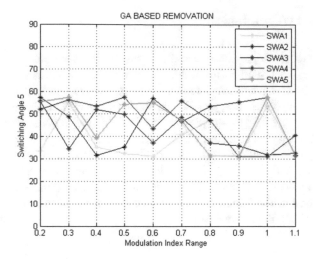

Fig. 3 Switching angle w.r.t modulation index

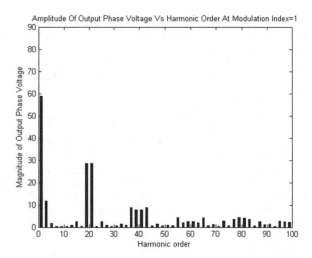

Fig. 4 THD

Table 2 Optimum switching angle and THD

Optimum switch angle and harmonic elimination	Previous rule	Proposed rule
θ_1	6.39	5.01
θ_2	18.9	16.1
θ_3	2.68	27.90
θ_4	44.98	42.18
θ_5	62.08	59.01
THD	6.83	6.05

7 Summary/Conclusion

As per past methodologies harmonic diminishment is a standout amongst the most difficult issues identified with multilevel inverters. By philosophies numerous heuristic methods have been actualized to get the arrangement of supernatural conditions. Along these lines, genetic calculation is reproduced on an eleventh level and ideal switching angles are resolved to take out low request harmonic and to limit Total Harmonic Distortion (THD). Low request harmonics in the H-Bridge multilevel inverter or construction electrical converter are limited utilizing particular harmonics end through genetic calculation. As per proposed calculation an effective and advanced framework is built up for least harmonics mutilation contrasted with past methodologies. Promote new system can be used for various levels in same H-Bridge multilevel inverter.

References

1. Nawaz, F., Yaqoob, M., Ming, Z., Ali, M.T.: Low order harmonics minimization in multilevel inverters using firefly algorithm. In: Power and Energy Engineering Conference (APPEEC), 2013 IEEE PES Asia-Pacific, pp. 1–6. IEEE (2013)
2. Nisha, G.K., Ushakumari, S., Lakaparampil, Z.V.: Harmonic elimination of space vector modulated three phase inverter. In: Proceedings of the International Multi conference of Engineer and Computer Scientists, vol. 2 (2012)
3. Manjrekar, M.D., Steimer, P.K., Lipo, T.A.: Hybrid multilevel power conversion system: a competitive solution for high-power applications. IEEE Trans. Ind. Appl. 36(3), 834–841 (2000)
4. Tolbert, L.M., Peng, F.Z., Habetler, T.G.: Multilevel converters for large electric drives. IEEE Trans. Ind. Appl. 35(1), 36–44 (1999)
5. Kumbha, V., Sumathi, N.: Power quality improvement of distribution lines using DSTATCOM under various loading conditions. Published in International Journal of Modern Engineering Research (IJMER), vol. 2, issue 5, pp. 3451–3457, Sep.–Oct. 2012
6. Tolbert, L.M., Peng, F.Z.: Multilevel converters as a utility interface for renewable energy systems. In: Power Engineering Society Summer Meeting, vol. 2. IEEE (2000)
7. Shyu, F.-S., Lai, Y.-S.: Virtual stage pulse-width modulation technique for multilevel inverter/converter. IEEE Trans. Power Electron. 17(3), 332–341 (2002)
8. Chiasson, J., et al.: Eliminating harmonics in a multilevel converter using resultant theory. In: 2003 IEEE 33rd Annual Conference on Power Electronics Specialists PESC '02, vol. 2. IEEE (2003)
9. McKenzie, K.J.: Eliminating harmonics in a cascaded H-bridges multilevel inverter using resultant theory, symmetric polynomials, and power sums. Dissertation, University of Tennessee, Knoxville (2004)
10. Ahmed, R.A., Mekhilef, S., Ping, H.W.: New multilevel inverter topology with reduced number of switches. In: Proceedings of the 14th International Middle East Power Systems Conference (MEPCON'10), Cairo University, Egypt, pp. 19–21 (2010)
11. Taghizadeh, H., Hagh, M.T.: Harmonic elimination of cascade multilevel inverters with nonequal DC sources using particle swarm optimization. IEEE Trans. Ind. Electron. 57(11), 3678–3684 (2010)

12. Ahmadi, D., et al.: A universal selective harmonic elimination method for high-power inverters. IEEE Trans. Power Electron. **26**(10), 2743–2752 (2011)
13. Saha, S., et al.: Comparative analysis of selective harmonic elimination of multilevel inverter using genetic algorithm, vol 2, pp. 48–53. Published by Global Journal of Selective Harmonic Elimination of Multilevel Inverter using Genetic Algorithm (2013)
14. Akbari, M., Golkar, M.A., Tafreshi, S.M.: Firefly algorithm-based voltage and frequency control of a hybrid ac–dc microgrid. In: Proceedings of 17th Conference on Electrical Power Distribution Networks (EPDC), pp. 1–7. IEEE (2012)
15. Salehi, R., et al.: Elimination of low order harmonics in multilevel inverters using genetic algorithm. J. Power Electron. **11**(2), 132–139 (2011)
16. Peng, F.Z.: A generalized multilevel inverter topology with self voltage balancing. In: Industry Applications Conference, 2000. Conference Record of the 2000 IEEE, vol. 3, pp. 2024–2031. IEEE (2000)
17. Sirisukprasert, S.: Optimized harmonic stepped-waveform for multilevel inverter (1999)
18. Carrara, G., et al.: A new multilevel PWM method: a theoretical analysis. IEEE Trans. Power Electron. **7**(3), 497–505 (1992)

Object Classification Using SIFT Algorithm and Transformation Techniques

T. R. Vijaya Lakshmi and Ch. Venkata Krishna Reddy

Abstract Recognition of objects, as well as identification and localization of three dimensional environments is a part of computer vision. In the proposed study the objects in a war field are classified. Images extracted from the video stream are utilized to classify the objects of interest (soldier, tree and tank). Distinguishable features of the objects are extracted and these features are used to identify and classify the objects. The SIFT algorithm used to find the features from such images are processed to classify the objects such as soldier, tank, tree, etc. The key points generated using SIFT algorithm are used to build a pyramid. The features extracted from these pyramids using various transforms are further classified in this work.

Keywords Object identification · SIFT key points · Transformation techniques

1 Introduction

Research using pattern recognition techniques is very important and widely used since the last five decades. It has attracted the interest of researchers in many domains like artificial intelligence, computer engineering, nerve biology, medical image analysis, archeology, geologic reconnoitering, space navigation, armament technology and so on.

Computer vision deals with the recognition of objects, as well as identification and localization of three dimensional environments [1–3]. Biometric recognition can be based on face, finger print, iris or voice and can be combined with the automatic verification of signatures and PIN codes [4–8].

T. R. Vijaya Lakshmi (✉)
MGIT, Gandipet, India
e-mail: vijaya.chintala@gmail.com

Ch.Venkata Krishna Reddy
CBIT, Gandipet, India
e-mail: krishnareddy.chintala@gmail.com

© Springer Nature Singapore Pte Ltd. 2019
P. K. Mallick et al. (eds.), *Cognitive Informatics and Soft Computing*,
Advances in Intelligent Systems and Computing 768,
https://doi.org/10.1007/978-981-13-0617-4_40

In the proposed work, images will be extracted from the video stream (LiDAR signals) and objects of interest like soldier, tree, tank, etc. would be recognized for assisting the missile to hit the target. Assuming that the images are extracted from the video stream, large amount of images would be generated in the laboratory environment for both training and testing database sets. Distinguishable features of the objects of interest like tanks and soldiers would be extracted. These features will be used to identify and classify the objects. The instantaneous position of these objects also in the war field scene would be found out to aid the missile path. The process and time taken for identifying these 3D objects should be in a fraction of second for this particular application. Hence algorithms would be developed for the above said military application.

This work is proposed for assisting a missile to hit the target in a war field. Images will be extracted from the video stream and objects of interest (soldier, tree and tank) would be recognized. Distinguishable features of the objects are extracted and these features are used to identify and classify the objects. The instantaneous position of these objects is found to aid the missile path.

2 Motivation and Scope of the Work

With the emergence of twenty first century, there is an increased use of digital technologies focusing on automation to decrease human intervention in day to day scenarios. Image processing is gaining significance today realizing that a picture conveys a thousand words, helps in easier data compression and can be ideally implemented using digital techniques. It is playing a vital role in any autonomous system. Hence embarking object recognition methodologies facilitates the computer systems to identify the object and then take appropriate action. The applications of object recognition are endlessly diverging from self-driving cars to autonomous drones and also in security systems and medical imaging technologies.

There is scope in the field of rescue and rehabilitation of victims of natural disasters as the system is able to ever increasingly detect threats. The proposed project deals with the imaging of battlefield and recognition of objects present there and will help in guiding UAV (Unmanned Aerial Vehicles) and missiles to locate the target accurately and destroy it.

The scope of pattern recognition is immense and is being used in various fields; there are wide ranges of technologies that use this ranging from micro plate reading to robotics.

3 Literature Review

For objects of high symmetry additional processing was to be invoked since, for such objects, relational considerations invoked by the pruning module can fail to produce a unique pose transform; this additional processing was simple as it requires to select from the available labels for a small number of scene surfaces to calculate a pose transform from these labels and then verify the transform by using the labels available for the other scene surfaces.

Lam and Greenspan [9] have implemented a system for object recognition and registering repeatable interest segments from 3D surfaces.

According to Flynn and Jain [10] "Computer vision system is to identify and localize instances of predefined 3D models in images and thereby add benefit to industry and other environments".

Dumont et al. [11] have implemented a system which recognizes the video monitoring library in the camera of Air Traffic Control (ATC) System. Using this we can detect airplanes and servicing trucks on runway using cameras or thermal imagery.

4 Research Methodology

A moving aircraft captures the video of a war field scene which is shown in Fig. 1 and from that video, extraction of image will be done by controlling the frame speed. The extracted image is applied to a pattern recognition system as shown in the proposed model (refer Fig. 1), for finding the required objects of interest like tanks, soldiers. The time required for data acquisition, feature extraction and object identification should consume very less time in the order of micro or milli-seconds, apart from high recognition accuracy.

It is assumed that specific image frame can be extracted from LiDAR images by using the existing methods and controlling the frame rate of the video. After the video frame is extracted, the objects in the image are to be identified and classified based on shape, structure and if required texture. This extracted frame is an ordinary image of $M \times N$ size with M rows and N columns. These images are preprocessed by removing noise from them.

The preprocessed images extracted are further used in the feature extraction step. The Scale-Invariant Feature Transform (SIFT) features are extracted from these images to detect the object of interest. The steps involved in SIFT feature descriptor is depicted in Fig. 2.

The key points in the images are identified first using Gaussian filters. The image pyramids are built using the key points. Then the features from the pyramid images are refined to create a sub-image.

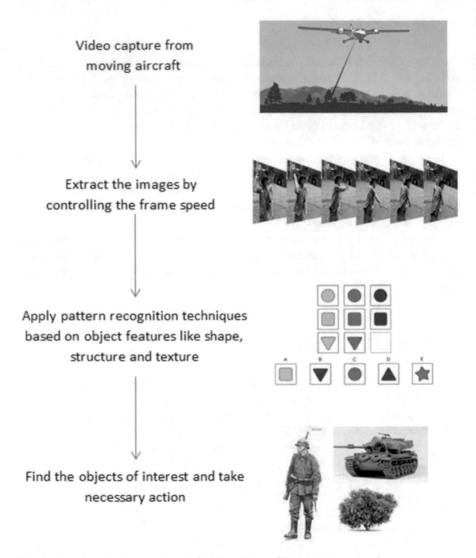

Fig. 1 Steps involved to acquire and find the objects of interest

These are further used in the classification process. In this step the points are clustered based on the distance measure. The clustered points are then used to classify the objects of interest such as tank, soldier, etc. using various transformation techniques. The results obtained with the transform techniques are reported in the next section.

Fig. 2 Steps involved in SIFT feature descriptor

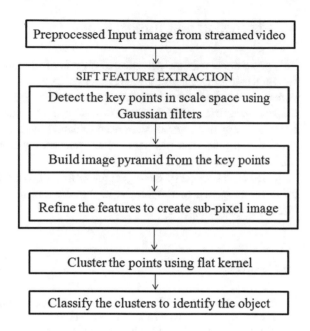

5 Experimental Results

The number of images considered in the current work for both training and testing are 10,672. They are divided into two sets. For testing 1/8 of them are used and the rest are used for training. The number of training samples used is 9338 and the number of samples tested in the current work is 1334. The classification results obtained with various transforms are tabulated in Table 1.

FFT influenced with the geometric structural properties of the patterns in the image is used to identify the patterns of interest. The classification rate obtained with FFT is 64.5%. The energy compaction property of DCT for correlated images helps better in identification of the objects. The temporal and spacial information captured by DWT identifies the objects of interest in the war field images.

As the texture of the images and the background of the images are similar few of them could not be identified correctly. Based on the shape of the object under test a few of them are correctly identified using the proposed methodology. In future using better techniques the objects of interest can be detected.

Table 1 Classification rates obtained using various transformation techniques

Transformation technique	No. of correctly recognized samples	% of classification accuracy
Fast fourier transform	861	64.5
Discrete cosine transform	947	70.9
Discrete wavelet transform	1153	86.4

6   Conclusion and Future Scope

This paper deals with the detection of objects of interest in the war field. The videos captured using missile aircraft are processed in the work to extract the images in the defense field. Performing various preprocessing operations the images are normalized to extract SIFT features from them. The shape and texture information extracted from these features are used to detect the objects of interest such as tank, soldier, trees, etc. The best recognition rate achieved with the proposed model is 86.4%.

Based on the shape of the object under test a few of them are correctly identified using the proposed methodology. In future using better techniques the objects of interest can be detected.

References

1. Stein, F., Medioni, G.: Structural indexing: efficient 3-D object recognition. IEEE Trans. Pattern Anal. Mach. Intell. **14**(2), 125–145 (1992)
2. Koley, C., Midya, B.L.: 3-D object recognition system using ultrasound. In: Proceedings of the 3rd International Conference on Intelligent Sensing and Information Processing, Bangalore, pp. 99–104 (2005)
3. Mashor, M.Y., Osman, M.K., Arshad, M.R.: 3D object recognition using 2D moments and HMLP network. In: Proceedings in International Conference on Computer Graphics, Imaging and Visualization, pp. 126–130 (2004)
4. Kim, W.Y., Kak, A.C.: 3-D object recognition using bipartite matching embedded in discrete relaxation. IEEE Trans. Pattern Anal. Mach. Intell. **13**(3), 224–251 (1991)
5. Guo, Y., Bennamoun, M., Sohel, F., Lu, M., Wan, J.: 3D object recognition in cluttered scenes with local surface features: a survey. IEEE Trans. Pattern Anal. Mach. Intell. **36**(11), 2270–2287 (2014)
6. Lowe, D.G.: Local feature view clustering for 3D object recognition. In: Proceedings of the 2001 IEEE Computer Society Conference on Computer Vision and Pattern Recognition, pp. 682–688 (2001)
7. Reddy, A.D., et al.: Quantifying soil carbon loss and uncertainty from a peatland Wild reusing multi-temporal LiDAR. Remote Sens. Environ. **170**, 306–316 (2015)
8. Kodors, S., et al.: Building recognition using LiDAR and energy minimization approach. Procedia Comput. Sci. **43**, 109–117 (2015)
9. Lam, J., Greenspan, M.: 3D object recognition by surface registration of interest segments. In: International Conference On 3D Vision, pp. 199–206 (2005)
10. Flynn, P.I., Jain, A.K.: 3-D object recognition using constrained search. IEEE Trans. Pattern Anal. Mach. Intell. **13**(10) (1991)
11. Dumont, G., Berthiaume, F., St-Laurent, L., Debaque, B., Prevost, D.: AWARE: a video monitoring library applied to the air traffic control context. In: International Conference on Advanced Video and Signal Based Surveillance, pp. 153–158 (2013)

Advancement in Transportation and Traffic Light Monitoring System

Shantam Tandon, Karthik Subramanian, Harshit Tambi
and Dhanalakshmi Samiappan

Abstract Thousands of lives are lost due of the negligence of traffic rules and overrunning of traffic lights. Modification of traffic lights is important so that no one could overrun it and a safer society is established. Instead of traditional traffic lights, an advanced traffic light system should be implemented so that a safe transportation is incorporated. Establishment of various systems is necessary for a safer road sense and good traffic knowledge. It will lead to development in the field of transportation and help enhance the quality of traffic system in India.

Keywords Traffic lights · ITS · FCD · RF-ID · Sensor · Laser alarm
Push-to-talk

1 Introduction

Traffic lights are signaling devices placed at crossroads, highways, turnings, junctions, and similar locations to control the flow of traffic in that area and manage the crowd for the smooth movement of traffic as well as people.

Traffic lights provide a correct way for the movement of traffic by displaying lights of a standard color (red, yellow, and green) following a universal color code [1].

They arc as follows:

- The green light allows traffic to move in the given direction as denoted by arrows, if there is room on the road for safe movement of traffic at the given intersection.
- The yellow light is a warning that the signal is about to change to red or green. The reactions of road users on a yellow light vary with some jurisdictions

S. Tandon (✉) · K. Subramanian · H. Tambi · D. Samiappan
Department of Electronics and Communication Engineering, SRM University,
Kattankulathur, Kancheepuram, Tamil Nadu, India
e-mail: shantamtandon96@gmail.com

© Springer Nature Singapore Pte Ltd. 2019
P. K. Mallick et al. (eds.), *Cognitive Informatics and Soft Computing*,
Advances in Intelligent Systems and Computing 768,
https://doi.org/10.1007/978-981-13-0617-4_41

Fig. 1 Conventional traffic
light system

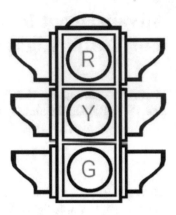

requiring drivers to stop if it is safe to do so, and others allowing drivers to go
through if safe to do so [1].

- A flashing yellow indicates that drivers may pass if no pedestrians are on the
  crossing and it is safe to do so [1].
- The red signal prohibits any traffic from moving ahead on the road as it is not
  safe and traffic on other lanes of the crossway or similar roads is in motion.

These three-basic lights are clubbed to form a system which is mounted on a
pole to control the traffic and guide it to maintain the road discipline as much as
possible.

In Fig. 1, a conventional traffic light system is should. Red color is used as a
warning signal because it can be seen from far distances due to its longer
wavelength.

2 Traffic Lights

A single traffic light pole consists of these three lights which guide the traffic. The
light pole is placed at various locations like crossways, intersections, railway tracks,
etc. according to their need. A need of the traffic pole is observed by analyzing the
location, known as aspects of the traffic light pole.

Aspect of a light pole is referred to the number of light systems that face the
driver or the number of systems that face the different roads at the intersection. The
various aspects of the light pole are

2.1 Single Aspect

A single aspect consists of a single system or a pair of systems that guides the
traffic.

2.2 Dual Aspect

A dual aspect mostly consists of two systems mounted vertically to each other. These are mainly found at railway crossings, one of which shows yellow color when no traffic is observed and the other shows red and green to maintain the passage.

2.3 Multi-aspect

A multi-aspect consists of a number of systems (three or more) that guide the traffic to stop or move in multiple directions.

Establishing the lights as per the aspects required on the road, traffic cycle signaling and all red are also important features of a traffic system.

- Traffic Cycle Signaling refers to the switching of lights from one phase to another and controlling the traffic accordingly.
- All red phase refers to the condition when all the traffic lights present on an intersection turn red, so as to clear the traffic in the middle and allow pedestrians to cross safely.

3 Present Technologies

Earlier, when the transportation system became prevalent in the urban society, traffic control officer or traffic policeman were present at crossways and intersections so as to guide the traffic and manage the jams. Slowly, the amount of traffic increased with urbanization and more and more vehicles made it a difficult task for a single officer to manage. Automation helped the transportation system to progress, which led to the formation of traffic light system, now used to control and monitor the traffic in the urban as well as rural areas.

Current technologies in the traffic system include many innovations which help in storing a disciplined traffic on the roads. These technologies include:

3.1 Lane Control

Lane Controlling includes specific traffic light boards that are installed in particular lanes of a multi-lane network to manage the traffic. It consists of multiple traffic lights that guide the driver on switching lanes accordingly.

3.2 Speed Sign Boards

Speed Signs are installed at several intervals to guide the drivers to maintain a particular speed for the smooth movement of traffic.

3.3 Programmable Visibility Signals

Initially, incandescent/halogen bulbs having low efficiency were used to signal the lights. Later on, these lights were replaced by high efficiency L.E.D lights. The problem faced by L.E.D lights were that in cold countries, snow used to cover the poles which led to road accidents. Remedy to this problem was installation of heating instrument near the L.E.D panel to melt the snow.

High visibility signals utilize light diffusion techniques to create the signal beams which pass through Fresnel lens which converge it into a beam of a certain diameter.

3.4 Camera Detection & Surveillance

High-quality cameras are installed at various junctions to monitor the moving traffic and keep a check on road accidents and other activities. Captured images are processed using software like MATLAB, to obtain the traffic density which is used for dynamic controlling of traffic light at a signal.

3.5 Promals

The Research Group in Mathematical Programming, Logistics & Simulation has designed a mechanism which uses various aspects of real time traffic to detect brakes, automatic control and driving speed and report them to every vehicle present in that area. Example, it tells the driver the distance of every vehicle near to it.

3.6 Local Time-Based Controller

A local time-based controller uses fixed time intervals or manual controlling of traffic light at a signal post. It does not require any sensors or mechanism to judge the traffic density and guide the traffic accordingly.

3.7 NEMA

National Electrical Management Association manufactured devices like TS-1 & TS-2 which are capable of controlling two phases actuated intersections and offers limited control. Traffic detection devices are installed to control the time difference at each intersection and manage the traffic.

3.8 Detection Techniques

Dynamic timing of traffic signals requires various sensors to change the time accordingly. Installation of surveillance cameras to find the vehicle density is one of the major methods. Next includes installation of inductive loops and magnetometers under the road which detect the traffic density by change in the magnetic field due to presence of magnetic substances in the vehicles.

4 Problems

Automation in the field of transportation led to a greater advancement and helped to maintain high-density traffic in extreme conditions as well. But, breaking of traffic rules is a common problem since the emergence of traffic rules. Due to breaking of rules, many problems occur which lead to chaos on roads and hamper movement with heavy loss in economy.

Some of the major problems faced nowadays due to the conventional traffic light system which are installed are:

4.1 Uncontrolled Traffic

In most of the developing countries like India, there is no limitation in purchase of vehicles and hence, a family can purchase a number of vehicles which in turn increase the density of vehicles on the roads. This increase in vehicles cause traffic jams and cause chaos on roads leading to waste of time and loss of economy. More and more vehicles increase the probability of road accidents which is a major concern in this urban transportation.

4.2 Violation of Traffic Rules

Automation in the field of transportation has led to the removal of police officer from the junctions and intersections which has provided the drivers the liberty to overrun the signal as there is no check majorly on Indian roads. This lead to road accidents and also cause loss of life and human economy.

4.3 Pedestrian Safety

A major concern in modern transportation system is the safety of people on roads. Pedestrian safety is not proper due to overrunning of signals and no check for guilty. So, people drive without any fear of fine and hence road accidents are increasing.

4.4 Encroachment

Encroachment of roads is also one of the problems which lead to traffic jams as roads are covered by shops and people which hamper traffic movement.

4.5 Emergency Vehicles and Other Aspects

In the current system, there is no provision for passage of emergency vehicles if there is a red light. This should be looked into and changed so as to give importance to the emergency systems. Proper maintenance of lanes while driving is important to maintain good traffic sense. The presence of animals on roads also hampers the movement of traffic which should be checked regularly.

5 Terminologies Used

For evolving the transportation system, an advanced traffic light pole incorporated with sensors should be installed which uses some of the pre-developed technologies. These terminologies together with the traffic pole structure form the basis of the proposal to revolutionize the transportation system. Some of the major terminologies are as follows:

5.1 Floating Car Data (FCD)

Floating Car Data (FCD) refers to the mechanism which is used to judge the traffic density in a particular area on the road. FCD uses working mobile networks of each operator in a particular area to calculate the number of people in that location and uses complex algorithms to judge the traffic density in that particular location. It is a good technique to gain the data on vehicle density so as to feed it to the system and help dynamic operation of traffic light signaling cycle [2].

5.2 Intelligent Transportation Systems (ITS)

Intelligent transportation systems (ITS) are advanced applications which use complex algorithms and artificial intelligence to maintain the decorum of the traffic on the road and control the vehicles and guide them properly.

Although ITS can be used anywhere and everywhere, it is more applicable in the field of road transport, including infrastructure, vehicle and traffic management and mobility management, as well as for interfaces with other modes of transport [3].

ITS systems also involve surveillance of the roadways, which is a priority for security. It plays a role in the mass evacuation of people in urban centers after casualties, such as natural disaster or threat [4]. Most of the infrastructure planning is done parallel to ITS technology so as to enhance homeland security and provide a better living and transporting system (Fig. 2).

Some of the basic uses of ITS include [5]

- Car Navigation System
- Traffic Signal Control System
- Container Management System
- Message Signaling
- Number Plate Recognition
- CCTV Security

Fig. 2 Flowchart for general principle used

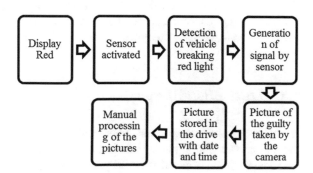

ITS also has advance applications like:

- Parking Guidance
- Weather Information
- Bridge De-Icing.

6 Proposal

In India, the situation of traffic is getting worse due to increase in the number of vehicles and degradation of roads. To improve the transportation system and bring a positive variation in this sector, several concepts should be introduced in the conventional system and some mechanisms should be established.

These technological changes in the conventional system include

6.1 Introducing RF-ID E-Cards

A Radio Frequency ID cards should be introduced for every vehicle. The card should be linked with Aadhar Card and Bank Account so that it can used to register the guilty and help speed up the process. The card should contain the details of the owner of the vehicle as well as driver details. It will also contain the vehicle details which will be a quicker detection about the details of the vehicle.

6.2 Introduction of Speed Sensors

At several intervals, speed limiting boards are displayed which define the speed limit on the road. At particular distances after these sign boards, speed sensors should be installed with cameras to detect the vehicles that are over running these limits. These sensors should also be installed at intersections and crossways to avoid accidents. The cameras can be used to capture the number plate of the vehicle and it can be identified. The newly introduced E-Cards can register the guilty easily.

6.3 Laser Light Alarming

At crossways and intersections, where traffic lights are installed to guide and control the traffic, over running of signals is a common problem. This causes accidents and also there is no check on the guilty. A laser initiated line is drawn at the signal before the zebra crossing to stop the vehicles from over running. The laser

Fig. 3 Outline of the laser alarming model

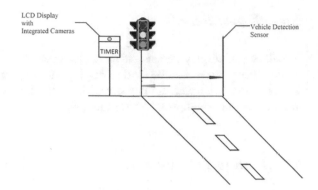

mechanism works in the similar principle as the burglary alarm; as soon as the path of laser is cut, the alarms turn on. An emitter is installed at one end of the road and is received by the receiver at the other end. When the circuit breaks, the system is activated and the guilty is captured in the camera as well as registered by the E-Card (Fig. 3).

6.4   Sound Detectors for Emergency Vehicles

Several sound sensors should be installed at various posts to judge the incoming of an emergency vehicle like an ambulance or fire truck. The data should be fed to the main system so as to control the traffic signal cycling and the path should be available so that the vehicle can pass easily without any hamper in its movement. The emergency siren generates a sound of particular frequency which will help to discriminate between emergency sound and noise.

6.5   Hologram -Guided Pedestrian Crossing

The laser light is the final stop for the vehicles on a red light. After the laser light, a curtain of red light is created by use of hologram to make a tunnel of zebra crossing for pedestrian safety. It will enable a safer crossing of roads at busy signals and crossings without subways. It will create a 3D effect which will force the incoming vehicles to stop at the signal. This will lead to less accidents at signals and pedestrian safety.

The hologram can be created by using Pepper's Ghost principle, the advantage of which is to enhance the intensity of hologram and make the image sharper and appear brighter.

6.6 Display Boards

Installation of display screens at particular intervals is necessary for guiding the traffic. These screens can be used to display shorter paths with less traffic and help the traffic to reach destination on time. They can also be used for advertisements at night so as to generate revenue for the government.

6.7 Push-to-Talk Button

A push-to-talk button should be installed near every signal so as to make the communication between the pedestrians and the officials easier. This button will directly connect the pedestrians to the substation traffic control official for any help. It can be used in case of any emergency and will be a great help for the people using the roads itself.

6.8 Check on Animals

A check on stray animals roaming on the roads is also necessary as they cause a lot of traffic jams as well as accidents. Absence of stray animals from roads will lead to safer driving and also help in cleanliness of roads.

7 Conclusion

As a whole, a traffic light pole at a single crossway should consist of a three or more light traffic system which will guide the movement of vehicle in given direction. It should have a L.E.D display screen to show the short paths and roads with less traffic to enable faster movement of traffic. It will also help for generation of revenue which can be used in public sectors or development of road infrastructure. Introduction of speed and sound sensors will also help the system to move towards advancement by decreasing the frequency of road accidents and improving the safety of pedestrians. Sound sensors will help in clearing of lanes for emergency vehicles which in turn will help resolving calamities and disasters. Introducing push-to-talk buttons will also bring the officials closer to the public and help in solving on spot problems which occur on the roads. This technique is similar to the buttons used in metro trains to communicate with driver. A hologram-initiated zebra crossing will force a stop at the moving traffic so as to ensure the safety of people walking on the roads. A laser initiated overrunning mechanism is also established to keep a check on the guilty, thus ensuring less accidents. The major

Fig. 4 Basic layout of advanced T lights monitoring system

change is the introduction of RF-ID E-Cards which will help in faster tracking process. RF-ID scanners can be installed at every crossing to check the movement of vehicles. Any vehicle guilty of any reason can be traced easily. Tolls can be cut automatically be e-banking system making it a faster process. It will also help in tracking of lost vehicles as it can be registered wherever the vehicle crosses a RF-ID sensor. Surveillance cameras will also help in this process which are pre-installed at various check posts. Dynamic controlling of traffic signal cycle is done with the data recovered from inductive loops and floating car data.

The advancement in the technology of transportation is a much-needed aspect of modern-day society. As road transport is one of the major part in transportation, road safety should be improved. Introduction of these mechanisms, for the upgradation of traffic light system will lead to faster movement of vehicles as well as safety of vehicles on the road. Pedestrian safety will also increase with check on the guilty. These differences in the modern system from the conventional system will reduce the frequency of accidents and improve the conditions of people on the roads as well as reduce the quantity of jams, hence facilitate a better, safer and faster mode of transportation (Fig. 4).

References

1. Traffic Light. Wikipedia, Wikimedia Foundation, 30 Sept 2017. https://en.m.wikipedia.org/wiki/Traffic_light
2. Floating Car Data (FCD) in the Transportation Planning—TRIMIS—European Commission. TRIMIS, 19 May 2015. https://trimis.ec.europa.eu/project/floating-car-data-fcd-transportation-planning
3. Umrao, P.K.: Intelligent Transportation System. LinkedIn SlideShare, 30 Apr 2009. www.slideshare.net/guest6d72ec/intelligent-transportation-system
4. Zhou, H.: The Internet of Things in the Cloud a Middleware Perspective. CRC Press, Boca Raton (2015)
5. Vanajakshi, L., et al.: Intelligent Transportation System (2010)

Optimized Methodology for Hassle-Free Clustering of Customer Issues in Banking

G. Naveen Sundar, D. Narmadha, S. Jebapriya and M. Malathy

Abstract The unprecedented growth of issues generated in banking sector is extremely huge. It is important to prevent customer churn by retaining existing customers and acquiring new customers so that is important for analyzing. Since data stored in the databases of banks are generally complex and are of varying dimensions such as consumer loan, debt collection, credit reporting and mortgage, the procedure for data analysis becomes very difficult. This paper presents a simplified framework for clustering the various issues by using a combination of data mining techniques. Hence in huge datasets issues from recorded by the customers are clustered using an efficient clustering algorithm. The parameters such as execution time and prediction accuracy are used to compare the results of the algorithms.

Keywords Decision tree · MapReduce

1 Introduction

Bank databases store huge volume of information such as the issues given by the customers. The maintenance of the historical records is necessary in order to provide prompt solutions to the problem. This information would help them to prevent customer churn and increase the profits of the banks. Therefore, there is a need for a

G. Naveen Sundar (✉) · D. Narmadha · S. Jebapriya
Karunya Institute of Technology and Sciences (KITS), Coimbatore, India
e-mail: naveensundar@karunya.edu

D. Narmadha
e-mail: narmadha@karunya.edu

S. Jebapriya
e-mail: jebapriya@karunya.edu

M. Malathy
VelTech High Tech (VTHT), Chennai, India
e-mail: malathianandan@gmail.com

P. K. Mallick et al. (eds.), *Cognitive Informatics and Soft Computing*,
Advances in Intelligent Systems and Computing 768,
https://doi.org/10.1007/978-981-13-0617-4_42

better way to analyze the customer issues and cluster the similar issues into groups. Moreover, banks pay more attention to marketing strategy. In any banking sector marketing strategy is deployed to provide hassle-free services to the customer. This helps the banks to retain huge success by retaining old customers and add new customers. It is extremely difficult for the banks to serve customers on timely basis. It may be efficient for the banks to optimal services if the client issues are clustered together. To overcome this challenge, this research focuses to cluster the client issues by comparing the efficiency of various clustering algorithms.

In recent times, big data mining gained considerable attention to these issues. Majority of the activities done so far was based on the profile of the customers. Any real-time applications generate volumes of data which is quite complex and cannot be handled using traditional database management systems. Hence big data analytics came to existence to manage the large amount of datasets. Data analytics is growing over the decades in leaps and bounds but big data analytics is different such that it deals with the data which is not collected is a defined form. This is a newly developing method which deals with the unstructured data and real-time data collection in large volume. The big data analytics relates on four major constituents. Volume represents the amount of data in a large dataset which is measured per day. Second, velocity that deals with the speed of data that are generated. Then, variety depends upon the data. The data can be structured or unstructured data while the big data is mostly unstructured data since it is highly scalable. Finally, veracity in dictionary is stated as truthfulness. In big data it refers to data which is not gathered in a definite form but mined into meaningful information. It deals with keeping the dirty data as clean one in the storage base. Hence it helps the processor to keep away from the dirty data.

One of the major components is Hadoop which is a key technology to handle big data, its analytics and streaming data. Distributed processing along large sets of data across a cluster is performed using Hadoop. Figure 1 shows the complete Big data analytic framework. Hadoop framework consist a lot of elements namely Hadoop Distributed File System (HDFS), MapReduce, HBase, Pig, Hive, Sqoop, Zookeeper, Avro, Cassandra, Mahout, Tez, Spark.

Categorizing of similar calls is not only done using Hadoop but Mahout also plays a major role in similarity algorithm. Various similarity algorithms are used in

Fig. 1 Big data analytic framework

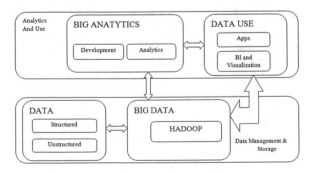

large datasets namely k-means, canopy, fuzzy based algorithms, etc. The dataset preferred is consumer complaints on banking sector, where the consumer issues are clustered efficiently using the clustering algorithms are their results are compared based on accuracy and execution time. The algorithms discussed in this research such as HFKHM (Hybrid Fuzzy K-Harmonic Means) and KM (K-Means) are compared using the parameters such as precision and time of execution of the algorithms.

2 Literature Review

Let us see the various methodology based on similarity algorithm is larger datasets.

2.1 Similarity Search in High Dimensions via Hashing

Aristides gions et al. [1], introduced Locality sensitive hashing-based algorithm. This approach provides an improvement in running time over other for searching in high-dimensional spaces based on hierarchical tree decomposition. The speed is measured by no disk blocks accessed to answer query. The disadvantage of the hashing method is that it needs additional storage overhead.

2.2 Approach to Generate New Patterns of Customer

Dr. Sankar Rajagopal (2011) introduced a methodology for grouping customers with similar issues and then to generate novel patterns for customers. This approach worked with two main stages. In the beginning stage, preprocessing is done to fill all the lost data and make it complete. In the second stage, data is clustered to find the customers with good value and low risk. This proposed approach came up with a high value of customer.

2.3 Approach to Find the Utilization of Health Care Resources

Noah lee et al. [2] introduces a methodology to mine diabetic electronic method records to explore the relationship between resources utilization of health care records and the recovery from diseases. Geometric temporal event matrix representation is used to represent the health records. It is difficult process as the data representation is incomplete.

2.4 Unsupervised Approach to Cluster Patient Records

Jacob and Ramani [3] both focused on the cluster of patient records via unsupervised clustering and compares the performance of classification algorithms on the clinical data. Hence they both modified the data mining framework into two distinct phases namely clustering and classification.

They both evaluated the performance of eight clustering algorithm and 16 classification algorithm on the dataset and compared the clustering accuracy followed by the accuracy given by the classification algorithms before and after feature selection. Hence the results of Quinlan and Random tree algorithm have less computation time compared to other six algorithms.

2.5 Approach to Determine Similarity in Medical Social Network

Klenk et al. [4] introduced an approach to determine the survival rate of the patients using medical social network. The algorithm works in terms of survival time distribution. It is not less dependent on a time to event. The advantage of this approach is to produce high performance and increase usefulness of the healthcare system.

Fahim et al. [5] gave an intuition to develop a higher version of the KM algorithm. The job of locating the set of points appears to be challenging in the traditional k-means approach. The proposed method is most suitable for large number of clusters because k-means algorithm while working on the large dataset it makes the computation process very expensive and the processing time also exceeded. Hence, for very large datasets k-means algorithm may not be quite suitable.

2.6 MapReduce-Based Approach for String Similarity in Banking

Deng et al. [6], introduced a MapReduce-based method to process ascendable string similarity joins. This approach used a Mass Join light weight algorithm which filters input data into its equivalent pairs of key values. This approach is feasible to work for large no of dissimilar pairs without an increase in the cost of the transmission. Thereby MapReduce-based approach helps to gain a greater reduction in the number of pairs of candidates produced.

2.7 Fuzzy-Based Technique

Al-Taani et al. [7] they proposed a system which composed of five stages schema. The input dataset is read and training dataset is built and a fuzzy based approach is used to find the resemblance between prepared dataset and the test data. This approach makes use of the conjunction and disjunction operators of fuzzy model to explore the similarity. Finally, membership degree of the test document is computed which takes a value between 0 and 1.

Vinicius et al. [8] proposed hierarchical-based clustering methods such as agglomerative and divisive to group the data which are initially distributed. These clusters are then iteratively clustered taking into concern of some similarity measure until all of the data points belong to a particular cluster. Even though the degree of similarity is huge during the initial processing stage but dissimilarity increases during the course of time.

Rostam Niakan Kalhori et al. [9] proposed a method which partitions the input dataset into clusters. Here separation of clusters is done by using a new measures based on credibility. The data points of a cluster are identified using possibilistic based clustering. However, this approach is susceptible to noise.

In the year 2015, Wu et al. [10] introduced an approach which is dependent on two main algorithms such as improved possibilistic C-Means Clustering and K-Harmonic means. This lead to the insight of an enhanced HFKHM algorithm. The ultimate challenge in K-harmonic means is unwanted data which is generally termed as noise. This noise factor can be greatly reduced using HFKHM. This approach shows better result in terms of accuracy when compared to other algorithm.

3 Proposed Work

Customer Relationship Management is one of the best practices followed in companies to analyze the sales, communications, and services provided to the customer. Support team of CRM plays a very vital role in proper data organization and providing prompt solutions to the customer issues. The ultimate goal of any business is to increase the number of customers and the quality of services rendered to them. Unsupervised algorithm such as clustering algorithm is used to group the issues with similar kind of approaches. Figure 2 portrays the proposed architecture for the complaints registered by the customers in banking sector. In the implementation phase, large-scale data analysis is done using Hadoop platform. The customer issues are exported in csv file format into the HDFS framework. This is led into the process of various clustering algorithms and clusters of client issues are generated.

Fig. 2 Framework for
clustering client issues

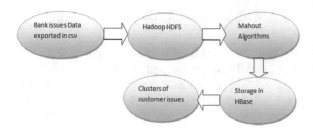

4 Results and Discussion

The performance of KM, KHM, and HFKHM is evaluated using the Consumer
complaint banking sector dataset. The performance metrics such as time and prediction accuracy are utilized to compare the performances of the algorithm. There is
a differences in the performances based on the system process. Table 1 deals with
the results of three clustering algorithm.

Figure 3 shows the pictorial representation of the comparison between the
algorithms such as k-means, K-harmonic means, and hybrid fuzzy k-harmonic
means in terms of prediction accuracy. When the number of records is 100 the
prediction accuracy is approximately 85% for k-means, 90% for KHM and 90.2%
for HFKHM. This increase in the prediction accuracy for other set of iterations
when the number of records is 500, 600, and 1000.

Figure 4 shows the pictorial representation of the difference in execution time
for various algorithms. It is clearly visible that there is an increase in execution time
for above-mentioned algorithm when compared to other algorithms, even though
there is an increase of time which is quite negligible when we compare to the
increase in the prediction accuracy.

Table 1 Prediction accuracy of algorithms

Algorithm	No of records	Clustering accuracy	Execution time (s)
KM	100	85	38
	500	84.5	389
	600	83	37
	1000	84.5	36
KHM	100	89	47
	500	89.2	48
	600	88.7	49
	1000	86.3	49
HFKHM	100	90.2	50
	500	90.3	51
	600	90.67	52
	1000	91.4	51

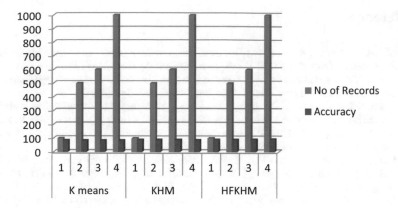

Fig. 3 No. of records versus accuracy for various algorithms

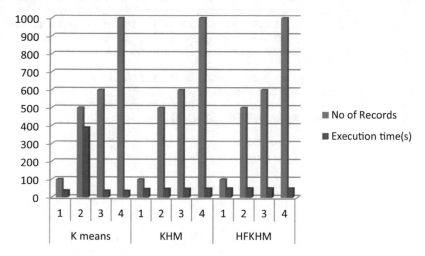

Fig. 4 No. of records versus execution time

5 Conclusion

There is an incredible increase in the data generation in many applications most specifically in banking sector. Hence, it is a need of an hour to cluster the complaints registered by the customers in the banking sector. Literature analysis was carried out to cluster the customer complaints efficiently. It is observable that HFKHM produces better result in terms of the prediction accuracy even though the execution time is huge which is negligible.

References

1. Gionis, A., Indyky, P., Motwaniz, R.: Similarity search in high dimensions via hashing. In: Proceedings of the 25th VLDB Conference, Edinburgh, Scotland (1999)
2. Jiang, J.Y., Liou, R.J., Lee, S.J.: A fuzzy self-constructing feature clustering algorithm for text classification. IEEE Trans. Knowl. Data Eng. 23(3), 335–349 (2011)
3. Jacob, S.G., Ramani, R.G.: Evolving efficient clustering and classification patterns in lymphography data through data mining techniques. Int. J. Soft Comput. (IJSC) 3(3) (2012)
4. Klenk, S., Dippon, J., Fritz, P., Heidemann, G.: Determining patient similarity in medical social networks. In: Proceedings of the MEDEX 2010 (2010)
5. Fahim, A.M., Salem, A.M., Torkey, F.A., Ramadan, M.A.: An efficient enhanced k-means clustering algorithm. J. Zhejiang Univ. Sci. (ISSN 1009–3095, ISSN 1862-1775) (2006)
6. Deng, D., Li, G., Hao, S., Wang, J., Feng, J., Li, W.S., Join, M.: A mapreduce-based method for scalable string similarity joins. In: 30th International Conference on Data Engineering (ICDE), China (2014)
7. Al-Taani, A.T., Al-Awad, N.A.K.: A comparative study of web-pages classification methods using fuzzy operators applied to arabic web-pages. World Academy of Science, Engineering and Technology. Int. J. Comput. Electr. Autom. Control Inf. Eng. 1(7) (2007)
8. Guelpeli, M.V.C., Garcia, A.C.B.: An analysis of constructed categories for textual classification using fuzzy similarity and agglomerative hierarchical methods. In: Third International IEEE Conference Signal-Image Technologies and Internet-Based System (SITIS), pp. 92–99 (2007)
9. Rostam Niakan Kalhori, M., Fazel Zarandi, M.H., Turksen, I.B.: A new credibilistic clustering algorithm. Inf. Sci. 279, 105–122 (2014)
10. Wu, X., Wu, B., Sun, J., Qiu, S., Li, X.: A hybrid fuzzy K-harmonic means clustering algorithm. Appl. Math. Model. 39, 3398–3409 (2015)

A Hybrid Optimization Algorithm Based on Ant Colony and Particle Swarm Algorithm to Address IP Traceback Problem

Amrita Saini, Challa Ramakrishna and Sachin Kumar

Abstract The Internet today is highly vulnerable to security threats. The rate of cybercrime has increased proportionately with its usage. Out of numerous possible attacks, the most precarious is Denial of Service (DoS) attack. In DoS the attacker uses the vulnerabilities of compromised hosts in a network and create an attack network called Botnet. The identity of the bots is disguised by using fake source addresses in Internet Protocol (IP) header known as address spoofing. Further, the stateless nature of IP does not allow verification of source address thus making the attack easier. The best way to handle DoS attacks is to reach the source of the attack and block it. IP traceback is a proactive and effective approach to detect the origin of the DoS attack. Once attack origin is detected attack can be blocked, routine network traffic can be restored, chances of future attacks can be prevented and most importantly the responsible attacker can be brought in front of the law. The technique of backtracking for finding an anonymous attacker on a vast network is a complex combinatorial optimization problem, which falls under NP-hard category. In this paper, we have proposed a hybrid approach by integrating Ant Colony Optimization (ACO) and Particle Swarm Optimization (PSO), to find the efficient solution of IP traceback problem. The main focus of our work is to increase the convergence rate and further reduce the computational complexity of ACO algorithm by combining the distance-based search technique used by ACO with particle velocity based search used by PSO algorithm. The performance of proposed algorithm is evaluated by simulating it on network simulator 2 and the results show that the method can successfully and efficiently detect the DoS attack path with reduced convergence time and computational complexity.

Keywords DoS · TCP/IP · ICMP · PSO · ACO

A. Saini (✉) · C. Ramakrishna
National Institute of Technical Teachers Training and Research, Chandigarh, India
e-mail: amrita.cse@nitttrchd.ac.in

S. Kumar
Snow and Avalanche Study Establishment, DRDO, Chandigarh, India

© Springer Nature Singapore Pte Ltd. 2019
P. K. Mallick et al. (eds.), *Cognitive Informatics and Soft Computing*,
Advances in Intelligent Systems and Computing 768,
https://doi.org/10.1007/978-981-13-0617-4_43

429

1   Introduction

Denial of Service (DoS) poses a severe threat to the Internet by preventing the authorized users from assessing the services and resources from the network. DoS attacks are common due to stateless nature of TCP/IP protocol suite in which packets are routed without source address verification. Thus, IP address in attack packets is generally spoofed to implement a DoS attack [1–3]. A DoS attack can reduce victim's ability to respond to the legitimate user by making victim's resources unavailable for some time or ceasing them permanently by flooding huge traffic thus depleting victim's bandwidth and processing power or both [4]. Many organizations use a combination of firewalls and intrusion detection system (IDS) to secure their network. IDS detect the malicious traffic and coordinate with firewalls to block it. Most of the work done in this area focused on pacifying the effect of the attack by sustaining it to a certain extent. Also, these techniques fail in case if falsified or spoofed IP addresses are used to launch these attacks. Such passive approaches can provide only provisional solution, which neither eradicates the root cause of the problem nor captures and produces the attacker in front of the law.

One proactive approach to deal with DoS attack is IP traceback, which is the scheme of finding the origin of the attack. It is used to detect the attack path or the path followed by attack packets. IP traceback is not a preventive measure but it begins only after the attack has actually started. Also, traceback is quite beneficial in post-attack analysis and we can block the attack origin to mitigate future possible attacks. The major challenges in IP traceback are segregating attack packets from legitimate packets and finding attackers identity with spoofed IP address packets.

The conventional IP traceback schemes require alteration of various network components. Either specific fields of the IP header of a packet are used to store router's information in the encoded form or a certain amount of the packet describing factors are stored at the routers for the purpose of IP traceback. Also, the schemes require all the routers in network to support traceback mechanism [5]. Foroushani et al. [6] described and implemented probabilistic packet marking (PPM) for performing IP traceback. The path information in the form of fingerprint is embedded in the packets randomly according to some predefined probability, which helps in detecting the origin of attack. Liu et al. [7] implemented dynamic probabilistic packet marking (DPPM) technique by using distance traveled by packets as a parameter for marking probability instead of using fixed probabilities which results in better performance. Yu et al. [8] implemented traceback by using dynamic deterministic packet marking (DDPM) where instead of marking all the nodes only the nodes involved in attack like ingress routers are marked and used to detect the attack source. Saurabh and Sairam [9] performed Internet Control Message Protocol (ICMP) based IP traceback also known as iTrace schemes. They have considered the main problem with these schemes, that is huge traffic generated by ICMP packets and they proposed a system of two bloom filters known as additive and multiplicative bloom filters, which when incorporated with reverse iTrace reduces the number of iTrace generated approximately by 100 times. Thus it prevents iTrace from becoming another DoS attack during the reflector attack. Aljifri et al. [10] proposed a novel approach using header compression for traceback. Header compression results in storing more information in the packet without affecting its size and hence proves to be an effective PPM technique for IP traceback. Fen et al. [11] proposed IP traceback by using time to live (TTL) field of IP header. The information generated by TTL is used to generate attack path thus no additional information needs to be incorporated in the IP packets.

These conventional traceback approaches need a lot of changes in infrastructure and put a lot of computational load on routers. Also, they require high memory capacity in routers and consume a lot of network bandwidth in carrying traceback information. A new way to deal with IP traceback problem is the use of meta-heuristic techniques. Meta-heuristic is a heuristic or procedure designed to generate an optimal solution to an optimization problem, especially when complete information is not available or computation capacity is limited. These techniques proves to be a better choice involving less computational effort in solving combinatorial optimization problems, from a set of large feasible solutions, as compared to simple heuristics, iterative methods or other optimization algorithms. Since IP traceback is a combinatorial optimization problem, which falls under NP-hard category, a meta-heuristic approach can provide better and fast solution as compared to conventional techniques. The goal of this work is to propose an IP traceback approach, which uses a hybrid of two meta-heuristic algorithms and existing flow information to determine the attack origin. The proposed scheme hybridizes Particle Swarm Optimization (PSO) and Ant Colony Optimization (ACO) into PSO-ACO to minimize the number of resources required and the time required to converge to the most probable attack path.

2 Related Work

Lai et al. [12] have proposed traceback approach based on ant algorithm to find out the source of DoS attack. The algorithm is inspired by food collecting behavior of ants. In an ant algorithm, all the ants (agents) cooperate with each other by exchanging necessary information in the form of stored pheromone, a chemical substance released by ants. Ants mark the path they traverse by laying pheromone in certain quantity according to the priority of the path. Rest of the ants follow the path according to a certain probability depending on the quantity of pheromone laid by ants. When multiple ants interact with each other the algorithm converges to a sub-space of many good solutions out of which the best solution can be selected as the final solution [5]. The proposed traceback approach uses flow level information to detect attack source. The approach overpowers conventional approaches by possessing two features, quick convergence and heuristics, which help in reaching the attack source quickly. The performance of approach was successfully evaluated on two simulated networks, NSFNET and DFN.

Hamzehkolaie et al. [13] have proposed a bee colony algorithm for detecting the DoS attack source. They used the quick convergence and heuristics to find the source of DoS attack. The food source exploring nature of bees forms the basis of solution to traceback problem. The traffic flow information serves as a food source for bees. The path with more traffic is more prone to fall under DoS affected path. Simulation results prove the efficiency of algorithm. The technique proves to be efficient as it does not require much storage, also it do not demand any changes in the routing infrastructure.

Wang et al. [14] proposed an improved ant colony algorithm for IP traceback problem. The algorithm focuses on improving the ability of ants to converge to attack path by implementing ant colony algorithm on smaller subgroups where each subgroup has its own pheromone update rule. Though the algorithm converges slower than conventional ant colony algorithm but it uses a local update rule along with the global update rule to update the path information of moving ants which generates a more globally optimum path.

3   Introduction to PSO and ACO

3.1   Particle Swarm Optimization

Particle Swarm Optimization is a meta-heuristic technique proposed by Kennedy and Eberhart [15] based on natural behavior of the swarm of birds. Initially, it generates a set of random solutions and with each generation, it upgrades the solution till an optimum solution is reached. Each particle in PSO has a position and a velocity and it keeps the memory of its previous best position, pbest and the best position attained by all the particles in swarm called gbest. Every move of particle is governed by the values of pbest and gbest [16]. Each particle first moves toward local optimum solution and converges with each generation. PSO algorithm optimizes the values of pbest and gbest using a fitness function. The velocity of particle is updated according to Eq. (1).

$$V_{\text{new}} = w \times V_{\text{old}} + c_1 \times r_1 \times (\text{pbest} - P_{\text{old}}) + c_2 \times r_2 \times (\text{gbest} - P_{\text{old}}) \quad (1)$$

Here P_{old} and V_{old} are the position and velocity of particle in previous iterations. V_{new} is the new value of velocity calculated for present iteration.c_1, c_2 are learning factors lying between [0,1]. r_1, r_2 are random numbers lying in range (0,1) to restrict the rise in the value of velocity to a certain permissible level and w denotes the weight or inertia factor. After updating velocity each particle updates its position according to the equation Eq. (2).

$$P_{\text{new}} = V_{\text{new}} + P_{\text{old}} \tag{2}$$

Particles updates their velocity and position till the termination conditions are met. The algorithm describes the steps involved in PSO.

- Initialize position and velocity of particles randomly;
- Repeat;
- Calculate local best and global best solution, i.e., pbest and gbest;
- Update velocity and position of particles according to Eqs. (1) and (2);
- Until termination condition is met or number of iterations are over;
- Final value of gbest gives the solution.

3.2 Ant Colony Optimization

In ant-based IP traceback [12] each ant remembers the value of fitness function, which is the best position attained by it. A distance metric storing the path lengths of all routers and a probability metric based on flow information and pheromone value of each path are maintained according to Eq. (3). Each ant makes a move depending on the values of distance and probability metrics and updates the value of pheromone deposited on the path traversed by it according to Eq. (4). The algorithm runs till all the ants converge to a single path.

$$p_{ij}(t) = \frac{[\tau_{ij}(t)]^\alpha [\eta_{ij}(t)]^\beta}{\sum_{j \in N_i} [\tau_{ij}(t)]^\alpha [\eta_{ij}(t)]^\beta} \tag{3}$$

Here $p_{ij}(t)$ is the probability of ant choosing a path from node i to node j. τ_{ij} is the pheromone concentration on the edge between nodes i and j. $\eta_{ij}(t) = \frac{1}{d_{ij}}$, where d_{ij} is the distance between node i and j. α and β are decay factors representing the fading value of pheromone with time. Pheromone value is updated according to Eq. (4).

$$\tau_{ij}(t+1) = (1 - \rho)\tau_{ij}(t) + \Delta\tau_{ij}(t) \tag{4}$$

Here ρ represents pheromone decay rate and its value lies in the interval $[0,1]$ and $\Delta\tau_{ij}(t)$ represents additional pheromone deposited over the time period t calculated as shown in Eq. (5).

$$\Delta\tau_{ij}(t) = \sum_{k=1}^{m} \tau_{ij}(k) \tag{5}$$

ACO suffers from the problem of convergence to local optimal solution. The algorithm describes the steps involved in ACO.

- Initialize position of ants and the value of pheromone trail;
- Repeat for all ants;
- Construct distance metric based on flow information and path length;
- Construct probability metric according to Eq. (3);
- Use local search to decide move of each ant on the basis of metrics;
- Update the pheromone concentration according to Eq. (4);
- Until stopping condition is met or edge router is reached.

3.3 Hybrid PSO-ACO Algorithm

The proposed hybrid algorithm integrates the distance-based search of ACO with velocity-based search of PSO to design a new meta-heuristic approach, which improves the performance of ACO algorithm in solving IP traceback problem. Both ACO and PSO suffer from the problem of convergence to local optimum due to unrealistic initialization of pheromone concentration in ACO and distance of particles in case of PSO. In ACO every move of ant is decided according to its previous best move so local update is there. In PSO-ACO the position of each ant is updated on the basis of local as well as global best value of fitness function resulting into faster convergence and a more globally optimum solution. Also PSO-ACO converges to attack path with lesser number of ants thus we can say that computational complexity of hybrid algorithm is less as compared to ACO algorithm. Equation (6) represents the local visibility of an ant k to reach its target located at (i, j).

$$\eta_{ij}^k(t) = \Omega^k / d_{is}^k(t) \tag{6}$$

The value of $\eta_{ij}^k(t)$ should be less than one and if value of $\eta_{ij}^k(t) > 1$ then it is set as $\eta_{ij}^k(t) = 1$. Ω^k is the local detection factor based on flow information and $d_{is}^k(t)$ is the distance between the ant k and the attack source. If the visibility of target is high it means the ant is near the attack source and hence the move of ant is decided by this factor. The movement is governed by updating its velocity and position. The velocity of particles in PSO-ACO algorithm is calculated according to Eq. (7) and then the position of particles is updated using Eq. (8). Here three weight factors Ψ_e, Ψ_c, Ψ_s are used in calculating the velocity of particle. Ψ_e is the explorative factor, which helps ant to gather knowledge of the environment. Ψ_c is the cognitive factor, whose value depends on the local best position attained by an ant in its journey and Ψ_s is the social factor, whose value depends on the global best position attained by all the ants in the network.

$$V_{i,d}^k(t+1) = \Psi_e \times r_e \times V_{i,d}(t) + \Psi_c \times r_c \times \left(P_c - P_{i,d}^k(t)\right) + \Psi_s \times r_s$$
$$\times \left(P_s - P_{i,d}^k(t)\right) \tag{7}$$

$$P_{i,d}^k(t+1) = P_{i,d}^k(t) + V_{i,d}^k(t+1) \tag{8}$$

Here P_c and P_s are local and global best position values and $X_{i,d}^k(t), V_{i,d}(t)$ are the position and velocity of ant k at time t on edge (i, d). r_e, r_c, r_s are random numbers whose value lie between 0 and 1. Equation (9), (10) and (11) represents the use of ACO algorithm in the optimization process. So these equations help to converge to the DoS attack path by ACO. The value of pheromone intensity of ants is updated according to Eq. (9).

$$J_{ij}^k(t+1) = \rho \times \left(J_{ij}^k(t) + T_{ij}^k\right) - (1 - \rho) \times e \times J_{ij}^k(t), \tag{9}$$

where the value of ρ lies between 0 and 1. e is the elimination parameter which represents the decay in the value of pheromone with time. The value of e is kept less than 1 so that the process of pheromone decay can be made slow as compared to pheromone deposition by the ants. The ant, which is not a source agent just propagates the pheromone it receives from source agent. Here T_{ij}^k is the value of pheromone intensity received from neighbor ants whose value is

$$T_{ij} = \begin{cases} \alpha, & \text{if source pheromone} \\ \beta, & \text{otherwise} \end{cases} \tag{10}$$

The value of α and β lies in the range $0 < \beta < \alpha < 1$. When an ant detects the path without the help of any neighboring agent, it is called a source agent. The pheromone released by source agent is called source pheromone. The target utility $\mu_{ij}^k(t)$ is calculated according to Eq. (11).

$$\mu_{ij}^k(t) = \frac{\left(k_1 \times W_{ij}^k(t) - k_2 \times J_{ij}^k(t)\right)}{R} \tag{11}$$

Here $\mu_{ij}^k(t)$ denote the target utility of ant k and $W_{ij}^k(t)$ and $J_{ij}^k(t)$ denotes the path weight and pheromone intensity respectively. The value of path weight depends on the flow information, which is an indicator of number of DoS attack packets on the path. R is the local redundancy factor, which defines the number of neighbor ants who refer to the same path. k_1 and k_2 are constants. In ACO ants follow a greedy approach and tend to choose the path with more target utility and hence leads to the trapping of ants to a local optima solution. The hybrid algorithm is effective in detecting the attack path by overcoming the problem of converging to local minima. The hybrid PSO-ACO algorithm is described below.

PSO-ACO algorithm

At t = 0, initialize N ant agents in 2D space;
Set the value of pheromone matrix as null; $\{P_{ij}\}^k = \{\}$;
For k = 1 to N
Set the value of $\psi_c = \psi_s = 0$;
Calculate $V_{i,d}^k(t+1)$ using Eq. (7);
Calculate $P_{i,d}^k(t+1)$ using Eq. (8);
End For
While (stopping condition not met or number of iterations are over)
For k = 1 to N do
Update the value of pheromone matrix $\{P_{ij}\}^k$;
For m = 1 to size $\{P_{ij}\}^k$
Find η_{ij}^k using Eq. (6);
If $\eta_{ij}^k > 1$ then set $\eta_{ij}^k = 1$;
Find J_{ij}^k using Eq. (9);
Find T_{ij} using Eq. (10);
Find $\mu_{ij}^k(t)$ using Eq. (11);
End for
           Set the value of $\psi_e \neq 0$;
           Set the value of $\psi_c \neq 0$;
           Set the value of $\psi_s \neq 0$;
Calculate $V_{i,d}^k(t+1)$ using Eq. (7);
Calculate $P_{i,d}^k(t+1)$ using Eq. (8);
End for
Set t = t + 1;

---

4 Simulation Results

Both the ACO algorithm and PSO-ACO algorithm were simulated on network simulator2 (NS2) to compare the results generated by both algorithms. N_0–N_{10} are eleven communicating nodes in a wireless scenario. Node 2 (N_2) is the victim node and node 3 (N_3) is the attacking node. We have simulated four legitimate traffic flows in the network and one DoS attack flow through the routers 3, 10, 5, 0, 4 and 2. Table 1 displays the comparison of ACO and PSO-ACO on the basis of number of ants used to detect the DoS attack path and the number of iterations each algorithm uses to reach to the final attack path and hence the attack source. The data shows that PSO-ACO converges to attack path with lesser number of ants and it finds the solution of IP traceback problem in lesser number of iterations as compared to ACO.

The results in the form of graphs are displayed in Figs. 1 and 2. Figure 1 shows that PSO-ACO takes lesser number of iterations to converge to DoS attack path as compared to ACO. Hence its convergence rate is better as compared to ACO algorithm. The path $2 \rightarrow 4 \rightarrow 0 \rightarrow 5 \rightarrow 10 \rightarrow 3$ which is the DoS attack path, is followed by maximum number of ants as shown in Fig. 2. Also the number of ants

Table 1 Performance Analysis of ACO and PSO-ACO

Algorithm used for IP traceback	Number of ants (agents) used to detect DoS attack path	Number of iterations used to perform IP traceback
Ant Colony Algorithm (ACO)	24	12
Hybrid Algorithm (PSO-ACO)	16	8

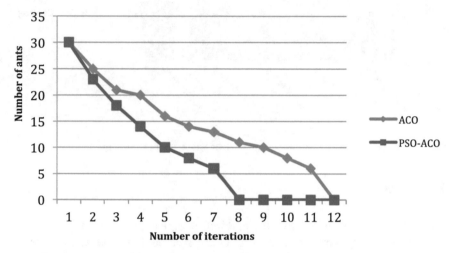

Fig. 1 Comparison of convergence rate of ACO with PSO-ACO

used to detect attack path is found to be less in case of PSO-ACO as compared to ACO as shown in Fig. 2.

The results show that though both the algorithms converges to the DoS attack path but the PSO-ACO algorithm reaches the solution faster and with the help of lesser number of ants. The computational complexity of algorithm is directly proportional to the number of ants and the convergence rate is inversly proportional to number of iterations used to detect the path. Thus, PSO-ACO algorithm has less computational complexity and better convergence rate over ACO algorithm. So, it proves to be a better approach in finding the solution to IP traceback problem.

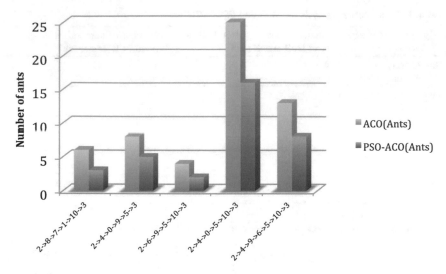

Fig. 2 DoS path detected with different number of ants by both algorithms

5 Conclusion

Both ACO and PSO suffer from the problem of convergence to local sub-optimal solution. Initial movement of ants in ACO is governed by the initial concentration of pheromone on paths and if more pheromone is there on sub-optimal paths then the algorithm gets trapped into local optima. To overcome this problem a large number of ants are needed thereby increasing the convergence time of algorithm. In PSO also initial position of particles is fixed randomly which may sometimes result in a bad sub-optimal solution. Thus PSO-ACO integrates the local update information obtained from distance metric of ACO with the global update information obtained from velocity metric of PSO (based on pbest and gbest) and selects a more reliable path. Simulation results show that the proposed algorithm outperforms ant-based traceback in terms of convergence rate and it successfully generates a more globally optimum solution. We have also observed that PSO-ACO converges to attack path in lesser number of iterations and with less number of particles or ants. From observations, we believe that the proposed hybrid approach is an effective way of finding optimal solution to traceback problem.

As a future work, methods could be devised to further improve the convergence rate of algorithm especially in case of larger networks and in more practical conditions. In large search space convergence to local optima is another major challenge, which could be looked upon as future work.

References

1. Patel, S., Jha, V.: Various anti IP spoofing techniques. J. Eng. Comput. Appl. Sci. **4**(1), 27–31 (2015)
2. Gupta, N., Dhiman, M.: A study of DDOS attacks, tools and DDOS defense mechanisms. Int. J. Eng. Res. Appl. **1**(3), 438–440 (2011)
3. Moore, D., Shannon, C., Brown, D., Voelker, G., Savage, S.: Inferring internet denial-of-service activity. ACM Trans. Comput. Syst. **42**(2), 115–139 (2006)
4. Mirkovic, J., Reiher, P.: A taxonomy of DDoS attack and DDoS defense mechanisms. ACM SIGCOMM Comput. Commun. Rev. **34**(2), 39–53 (2004)
5. Dorigo, M., Gambardella, L.M.: Ant colony system: a co-operative learning approach to traveling salesman problem. IEEE Trans. Evol. Comput. **1**(1), 53–66 (1997)
6. Aghaei-Foroushani, V., Heywood, A.: Probabilistic flow marking for IP traceback (PFM). IEEE Trans. Comput., 229–236 (2015)
7. Liu, J., Lee, Z., Chung, Y.: Dynamic Probabilistic packet marking for efficient IP traceback. Elsevier J. Comput. Netw. **51**(3), 866–882 (2007)
8. Yu, S., Zhou, W., Guo, S., Guo, M.: A feasible IP traceback framework through dynamic deterministic packet marking. IEEE Trans. Comput. **65**(5), 1418–1427 (2016)
9. Saurabh, S., Sairam, A.: ICMP based IP traceback with negligible overhead for highly distributed reflector attack using bloom filters. Elsevier J. Comput. Commun. **42**, 60–69 (2014)
10. Aljifri, H., Smets, M., Pons, A.: IP traceback using header compression. Elsevier J. Comput. Secur. **22**(2), 136–151 (2003)
11. Fen, Y., Hui, Z., Shuang, C., Xin-Chun, Y.: A lightweight IP traceback scheme depending on TTL. Elsevier J. Procedia Eng. **29**, 1932–1937 (2012)
12. Lai, G., Chen, C., Jeng, B., Chao, W.: Ant-based IP traceback. Elsevier J. Exp. Syst. Appl. **34**(4), 3071–3080 (2008)
13. Hamedi-Hamzehkolaie, M., Sanei, R., Chen, C., Tian, X., Nezhad, M.: Bee-based IP traceback. In: IEEE International Conference on Fuzzy Systems and Knowledge Discovery, pp. 968–972 (2014)
14. Wang, P., Lin, H., Wang, T.: An improved ant colony system algorithm for solving IP traceback problem. Elsevier J. Inf. Sci. **326**, 172–187 (2016)
15. Kennedy, J., Eberhart, R.: A new optimizer using particles swarm theory. In: Proceedings of Sixth International Symposium on Micro machine and Human science, IEEE Service Center, Piscataway, pp. 39–43 (1995)
16. Kuo, R., Hong, S., Huang, Y.: Integration of particle swarm optimization-based fuzzy neural network and artificial neural network for supplier selection. J. Appl. Math Model **34**, 3976–3990 (2010)

OEFC Algorithm—Sentiment Analysis on Goods and Service Tax System in India

K. Purushottama Rao, Anupriya Koneru and D. Naga Raju

Abstract Significance of the execution of paradoxical rules effects the economy of the country. Politicians need to predict the effect of any rule before it is implemented. Our rule makers introduced Goods and Service Tax recently in order to strengthen the economy of India. Nowadays, public are used to offer their opinions on Social media. There are lot of Tweets on Goods and Service Tax. To analyze the opinion of public we proposed an algorithm called Opinion Extraction using Favorites Count. This algorithm is applied on the twitter tweets to extract the opinion of public on Goods and Service Tax. The performance of this algorithm is compared with general sentiment analysis method.

Keywords Goods and service tax · Tweets · Favorites count · Opinion mining
Sentiment analysis

1 Introduction

Indian economy is growing very fast. To improve the living standards of Indian citizens our government is collecting taxes. Through this revenue the economy of the country is also improved. The Indian government is collecting two kinds of taxes from the citizens. Those are Direct tax and Indirect Tax. Direct Tax would collect from an individual directly like Income tax and corporate tax. Indirect tax would collect from the citizens indirectly like sales tax which is collected when you sell a product, Service tax which is added to services provided by our country, etc. Government is also collecting some other taxes like Toll tax, Gift tax, Stamp duty,

K. Purushottama Rao (✉) · A. Koneru · D. Naga Raju
LBRCE, Mylavaram, India
e-mail: k.purushottamaa@gmail.com

A. Koneru
e-mail: smartykoneru@gmail.com

D. Naga Raju
e-mail: dnagaraj_dnr@yahoo.co.in

© Springer Nature Singapore Pte Ltd. 2019 441
P. K. Mallick et al. (eds.), *Cognitive Informatics and Soft Computing*,
Advances in Intelligent Systems and Computing 768,
https://doi.org/10.1007/978-981-13-0617-4_44

etc. In all these ways government is raising the revenue of the country which helps our next generation to walk on golden roads of India.

Indian government had announced Goods and Service tax on 01-07-2017. GST is a single tax system which replaces some of the indirect taxes collected by the state governments. GST percentages varied based on the type of the product. Many of the economist thought that GST impacts the Indian Economy positively. They appreciated the government for taking such kind of innovative decision. As general public, if we think about GST, it is differ from other taxes. On every product that reaches from manufacturer to consumer it adds many taxes which really increases the cost of the product. To control this, our Indian Government has introduced single tax system which is GST. This is collected in three different ways. Those are SGST, CGST, and IGST. GST removes cascading tax effect and has lesser compliances. It really regulates the unorganized sectors and would be a composition scheme for small businesses.

Indian Government intended to launch a monstrous campaign by sending one minister from the Union Cabinet to every state just to create awareness and also to deliver issues identified on the GST. The chosen delegates promote the advantages of GST in rural India. Publicity has been done at provincial zones, and make the normal man and business group mindful about the uses of GST. Government has conducted some introduction sections by taxation experts, for clearing questions regard to the GST.

In present-day society, the voice of the general population shapes the core of any enactment or strategy on the land. In any case, it is imperative to first acknowledge and comprehend what the public opinion truly is. There is a blended reaction over GST. Government is an open servant and does what the greater part of the public says. Many channels conducted opinion poll on GST. Numerous news articles were published by supporting GST. Social Media like Facebook, Twitter and WhatsApp is a quick and effective apparatus to share the general public opinions. So, in this paper we proposed an algorithm OEFC which works on the data collected from Twitter to extract public opinion on GST.

The rest of the paper is organized as follows. Section 3 provides the recent work on sentiment analysis. Section 4 discuss about methodology specified in the OEFC Algorithm. Section 5 deals with the experimental work of the proposed algorithm. The next section discusses the issues while implementing the algorithm along with basic assumptions which need to be considered. The last section provides conclusion and future work.

2   Related Work

In this study we will discuss the major applications of sentiment analysis and sentiment analysis on different languages.

2.1 Applications of Sentiment Analysis

Many researchers worked in this area to find out the emotions of general public. Researchers have taken reviews of products, comments on issues specified in Facebook and Twitter, and also opinions through articles presented in different blogs and public newspapers, etc.

In the paper [1], author specified a methodology to find out the age group of people who were tweeted in twitter. Many people do not like to reveal their personal information in the social media but they would like to share their opinion. So they simply hide their details from the public. Hear Author collected 70,000 sentences through Twitter API, and divided into two parts. One is through comments given and other one is through user general profile. In this work author found Teenagers and Adult age groups by the way they use punctuation marks, length of the text, and also usage of short forms, etc. This classification has been done by using DCNN algorithm and proved as a best classifier for this application.

Authors Kai Yang et al. [2] constructed a domain specific sentiment dictionary which can handle domain specific conflicts and a hybrid model to get high accuracy in sentiment classification. The general dictionary which is used to differentiate sentiment polarities is not suitable to all applications. So author constructed a domain-specific dictionary. Author also proposed a Hybrid model to eliminate the deficiencies of single models like SVM and GBDT.

Sentiment analysis has been done on the training programs conducted by an organization. It is extended to educational field. In the paper [3] author considered the feedback forms submitted by participants. On those text comments they applied sentiment analysis and found the sentiment score of each program. They compared their result with the actual rating of the program given by the participant and visualized the results.

Sentiment Analysis has extended its wings towards the cloud. Author Alkalbani et al. proposed [4] a method to do sentiment analysis on SaaS reviews. Every client takes the service from the cloud and provides the feedback after completion of the service. This feedback has been taken and applied sentiment analysis to find the polarity of each feedback. On that they applied SVM to classify the Good and Bad cloud service providers in SaaS. This sentiment analysis also helps to identify the hacking activists on the web [5]. In this work, author collected twitter data and analyzed by applying sentiment analysis and regression technique.

2.2 Sentiment Analysis on Different Languages

Many researchers worked on sentiment analysis by applying it on different languages. In the paper [6], author collected Facebook posts which are written in Arabic and generated a corpora. This corpora is used by classifiers to do analysis on these posts. In paper [7], Authors Y. M. Aye and S. S. Aung have designed a lexicon-based sentiment analysis in the application of feedback given on foods and restaurants in Myanmar text. Nepali is a free word order language. Author worked [8] on Nepali language and applied three classifiers on it to find the best suitable classifier on Nepali language. Multinomial Naive Bayes classifier performs better on this language. Another Author worked on Portuguese language [9]. In this work they compared the three POS Taggers and resulted Tree Tagger as the best among FreeLing, CitiusTagger and TreeTagger. Sentiment analysis also applied on the predominant language like Urdu [10]. In this method author considered the comments in Urdu from different web sites and generated corpus and calculated the polarity.

3 Methodology

This work consists of sequential phases which are shown in Fig. 1.

3.1 Data Collection

In this phase, collected data is from twitter using Next Analyst. Next Analyst helps the researchers to extract useful tweets from twitter. Initially we need to login to developer twitter site. Develop an Application by giving details like Application name, Purpose, etc. It generates an application id and key. By using these credentials we could download the tweets on required topic directly into an Excel Sheet.

Fig. 1 Work flow

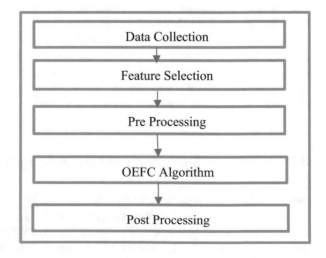

3.2 Preprocessing

Twitter data consists of lots of noise. It needs to be preprocessed to get quality and task relevant data. The extracted twitter dataset consists of numerous features. But for our task we need specific features. These specific features are selected from the dataset using feature selection. These selected features are tweets along with language and their favorite count. Our proposed algorithm (OEFC) could not handle all the languages. So we need tweets only in English. Based on that criteria we selected instances which are in English only. To prepare data for the task, we need to remove special symbols from the tweets, stop words, white spaces, punctuation marks, digits, and links, convert all the upper case letters into lowercase and apply stemming. After cleaning the data, the data should be ready for task.

Our proposed algorithm needs Positive and Negative word datasets. The set of Positive and Negative words are available on online. These individual datasets should represent task specific Positive and Negative words. So, we need to add task specific words into these list. After adding all these words the final version of positive words set is named as Positive and negative words set is called as Negative.

3.3 Proposed Algorithm

OEFC algorithm consists of the following steps:

Input: Preprocessed tweets dataset, Positive words list stored as Positive and Negative words list stored as Negative.

Start

Step 1: Create the number of columns based on the number of words in every tweet.

Step 2: For all tweets
Find the number of positive terms and Negative terms by comparing every term in the each Row with Positive and Negative Vectors.

Step 3: For all the tweets
Find the Polarity of every tweet based on the highest polarity count. For example, If Number of Positive terms is greater than Number of Negative terms in a tweet then the polarity of that tweet is Positive.

If Number of Positive terms is less than Number of Negative terms in a tweet then the polarity of that tweet is Negative.

Step 4: Consider the Favorite count of every tweet and generate two vectors named
as Positivity and Negativity.

$$\text{Find such that}: \text{Positivity} = 1 * \text{Favorite count};$$
$$\text{Negativity} = 1 * \text{Favorite count};$$

Step 5: Find the Average

Find the sum of Positivity vector value and store in pos.

Find the sum of Negativity vector value and store in neg.

Step 6: Find the final opinion

$$\text{if pos} > \text{neg then Opinion} = \text{Positive};$$
$$\text{if neg} > \text{pos then Opinion} = \text{Negative};$$

Stop.

3.4 Post Processing

Any opinion extracted should be shown in a proper way for better analysis. For this purpose we need to post process the result. For better analysis we are preparing a word cloud to show the occurrence of words in the task. Opinion could be visualized with a graph.

4 Experimental Work

We implemented this work in R Studio. Installed all these set of packages which are shown in Table 1.

Table 1 List of packages

S.no	Packages	S.no	Packages	S.no	Packages
1	Xlsx	5	devtools	9	dplyr
2	data.table	6	sentimentR	10	tidyr
3	Tm	7	stringi	11	purrr
4	twitteR	8	stringr	12	wordcloud

4.1 Data Collection

Data has been collected using Next Analyst. Using this add into the Excel, extracted recent 400 tweets from the twitter with 45 features. We imported that excel sheet into a "mydata" data frame in RStudio. Even though we extracted 45 features, the task relevant features are only three. Those are full_text, user_lang, and user_favourites_count. The tweets are available in full_text feature. We are going to use only English dictionary. So, we extracted only English language tweets by using user_lang field. Consider the tweets only when user_lang = ="en". When we observed these tweets, some of them were relevant to the GST file returns and problems related to returns. Our objective is to extract the opinion of public on GST and we are not interested to consider the GST filing web site. We removed such tweets and considered only task relevant tweets. Finally our dataset consists of 190 tweets with three features. These are stored in "mydata3" data frame. To apply our algorithm on the dataset we need two more vectors. Those are positive [11] and negative [12] list of words stored as positive and negative vectors. The number of positive words in the positive vector is 2022 and the number of negative words in the negative vector is 4794.

4.2 Preprocessing

The Tweets which are listed under full_text feature are consisting of noisy data. The tweets are having Retweets, @, and also links along with tweet. We removed all these from the full_text. In the next phase of preprocessing, removed the punctuation marks, Numbers/Digits, tab spaces, blank space at the beginning and ending, and stop words. Converted the tweets from upper case to lower case. Now our dataset is ready for the task and stored this full_text feature in another new data frame "doc".

4.3 Opinion Extraction with Favorites Count (OEFC) Algorithm

In the first step we applied string split function to these tweets which are stored in "doc" to split every word as a separate column by row wise by using string_split_fixed function. It has given 21 columns with 190 rows in "m2" which is another data frame. 21 indicates the maximum number of words in a row in the dataset. The data in "m2" data frame is applied to the second step of our algorithm along with positive and negative vectors. Sample data of m2 could be seen in Fig. 2. The word cloud of preprocessed data is shown in Fig. 3.

SNO	V1	V2	V3	V4	V5	V6	V7	V8	V9	V10	V11	V12	V13	V14	V15	V16	V17	V18	V19	V20	V21
1	gstn	leaks																			
2	whyonce	think	sir	gst	payment	basis	receipt	basis	exclusiveservice	sectorsbelow9k	turnover										
3	queries	raised	gst	helpdesk	unanswered	since	long														
4	rttrader	kills	self	sales	fail	post	gst	gst													
5	broken	wilsonleather	collegiate	gst	football	ntfns	made	usasebay													
6	broken	wilsonleather	collegiate	gst	football	ntfns	made	usasebay													
7	rtgst	may	will	successful	remove	sectiontax	deposit	monthlygst	returns	quaerly	gst	w8€									
8	rtcome	realitythe	gst	pcool	still	working	properly														
9		us	thiswe	unable	earn	single	penny	implementation	gst	nw	ur	asking	penalty	whats							
10	gst	system	working	fine																	
11	gst	return	filing	really	headache	god	help	us													
12	indian	jails	train	prisoners	file	gst	return	punishment	rather	breaking	stones	will	tougher	taskunicodedcdude							
13	govt	searching	tax	payer	economically	backwardclass	even	can	pay	higher	gst	luxury	car								
14	rtayalam	workinguploads	invoice	reflecting	retruns	release	dont	miss	lea€												
15	indian	jails	train	prisoners	file	gst	return	punishment	rather	breaking	stones	will	tougher	taskunicodedcdud							
16	gst	site	working	fabulous	now	think	gst	site	insomniacs												
17	rtjobs	demo	scam	farmer	suicides	hasty	gst	rolloutgdp	gloom	amp	since	oppn	cant	take	insideparliament	till	nov	tu€			
18	gst	depaments	penalised	mercy	tax	payers	full	pardon	tax	collectors	paining	us									
19	rtonly	tax	evaders	amppeople	dealing	informal	black	sector	complaining	paying	gst	honest	tax	payers	happy	gstsi€					
20	rtsorry	sir	several	hours	uploading	gst	invoices	gstr	reflect	done											
21	rtwith	gst	place	tax	gdp	ratio	nowwill	go	toso	stop	worrying	fiscal	deficit	huge	spending€						
22	rtsystem	workinguploaded	invoice	reflecting	retruns	please	dont	miss	lea€												
23	rtgrowth	prospects	deceleration	demonitization	amp	gst															
24	rton	asked	problems	gst	super	replywe	indians	want	switzerland	facilities	want	pay	tax								
25	rtwhoever	says	gst	pcool	working	fine	will	face	mirror	gstnfailed											
26	will	gst	empower	homebuyers	strike	better	dealwill														

Fig. 2 Sample data after preprocessing and tokenization

Fig. 3 Word cloud of the preprocessed data

In the second step, compare every term from the first tweet row to last tweet row with positive and negative vectors and store the match with value 1. i.e., If the word in the m2 matches with positive vector word then make value as 1 in the "pos.-matches" data frame otherwise 0. In the same way if the word in the m2 matches with negative vector word then make value as 1 in the "neg.matches" data frame otherwise 0. Insert these pos.matches and neg.matches into mydata2 data frame as mydata2$pos1 and mydata2$neg1columns.

In the third step, find the polarity by comparing the pos1 with neg1. Find the difference between pos1 and neg1 and store it in polarity feature. If pos1 in a tweet row counts more than tweet row count of neg1 then consider as Positive. If neg1 in a tweet row counts more than tweet row count of pos1 then consider as Negative. If both are same consider it as Neutral. Our concentration is on positive and negative polarities only. To represent this scenario Find the value of polarity in terms of integers. Create a column named value. Based on the polarity the value field stores 0, 1, and 2. If you find polarity with negative number then that corresponding value

	full_text	user_favourites_count	posf	negf	polarity	value
1	: GSTN leaks	61567	0	1	-1	0
2	@askGST_GoI @GSTCouncils: why/once think sir GS...	217	0	0	0	2
3	queries raised on GST helpdesk are unanswered sin...	430	0	0	0	2
4	RT @news24tvchannel: Trader kills self as sales fall ...	73	0	2	-2	0
5	"Broken In" #Wilson 1003 #Leather #Collegiate GST ...	281	0	1	-1	0
6	"Broken In" #Wilson 1003 #Leather #Collegiate GST ...	118	0	1	-1	0
7	RT @ASHUTOSHCA2000: @adhia03 gst may will be s...	2977	1	0	1	1
8	RT @jishnucb72: @gstindia Come to reality @gstindi...	49	1	0	1	1
9	@Tejasparekh32 @PMOIndia @narendramodi @adhia...	40	0	2	-2	0
10	@askGSTech GST system is working fine https://t.co...	12	1	0	1	1
11	Gst return filing is really headache, god help us	0	0	1	-1	0
12	@askGSTech @askGST_GoI @askGST_GoI Indian jails...	1020	1	1	0	2

Fig. 4 Sample data with positive and negative count tweets wise

Fig. 5 General sentiment analysis algorithm

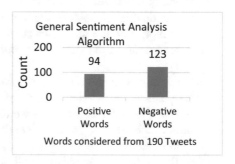

field is 0. If you find polarity with positive number then that corresponding value field is 1. If the polarity field is 0 then make value field 0. This is shown in Fig. 4.

In the fourth step, consider the user_favourites_count feature in mydata2. Create the vectors named "pos" and "neg". Assign the value of user_favourites_count into that "pos" vector if and only if the value of polarity is positive to that tweet. Assign the value of user_favourites_count into that "neg" vector if and only if the value of polarity is Negative to that tweet. In the fifth step, find the sum of pos vector which has given the value of 1,36,502. In the same way find the sum of neg vector which has given the value of 2,71,838.

In the final step, we could say that public is really against to the GST based on the values of pos and neg. The neg value is greater than pos.

When we considered 190 tweets the general sentiment analysis algorithm has given positive and negative word count which is shown in Fig. 5. Figure 6 shows the OEFC algorithm result which extracted the opinion of 4,08,340 twitter users.

Fig. 6 Performance of OEFC
algorithm

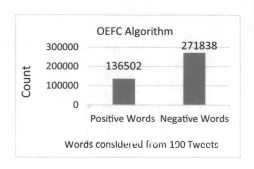

5 Discussions

- In this work, we considered user_favourites_count feature which really tells the opinion of other users even though they will not write any textual comments. If an individual found any comment interesting and matched to their opinion then they simply like it which is calculated as a user_favourites_count. In this we considered only 190 twitter users. But extracted the opinion of 4,08,340 twitter users.
- Instead of considering the tweets together as a single list, we considered tweet wise polarity which actually helps to extract more accurate result.
  For example, a Tweet with two words which is positive and a Tweet with ten words which are negative then overall polarity should be Neutral but as per our previous methods we will get negative because total positive terms two and negative terms ten. So polarity will be Negative which is incorrect.
- We ignored the Neutral opinion on GST, because here in our tweets if the number of positive words in a tweet is equal to number of negative words then we have given polarity as Neutral. Even though the count of positive words is Zero and Negative words is Zero the result becomes Zero. So we ignored Neutral tweets from our analysis.
- We considered only Twitter data because Extracting data from Twitter is more comfortable than facebook. If we want to extract comments from facebook we need to specify the facebook page related to GST. I found a GST filing page but that does not have comments on GST decision. All those comments are on filing only. When we use this Twitter we could extract any tweets related to the keyword GST.

6 Conclusion

In this work, we extracted the opinion of twitter users on GST using 190 tweets which are extracted from 400 tweets collected using Next Analyst. We just applied our proposed algorithm on the 190 user tweets but considered the opinion of more than 4lacks twitter users. Finally this analysis has given the Negative opinion on GST. We compared the result with existing methods and proved this algorithm as efficient. This work could be adopted to any application.

References

1. Guimarães, R.G., Rosa, R.L., De Gaetano, D., Rodríguez, D.Z., Bressan, G.: Age Groups Classification in Social Network Using Deep Learning. IEEE Access. **5**, 10805–10816 (2017)
2. Yang, K., Cai, Y., Huang, D., Li, J., Zhou, Z., Lei, X.: An effective hybrid model for opinion mining and sentiment analysis. In: 2017 IEEE international conference on big data and smart computing. (BigComp), pp. 465–466. Jeju (2017)
3. Ravi, K., Siddeshwar, V., Ravi, V., Mohan, L.: Sentiment analysis applied to educational sector. In: 2015 IEEE international conference on computational intelligence and computing research (ICCIC), pp. 1–6. Madurai (2015)
4. Alkalbani, A.M., Ghamry, A.M., Hussain, F.K., Hussain, O.K.: Sentiment analysis and classification for software as a service reviews. In: 2016 IEEE 30th international conference on advanced information networking and applications (AINA), pp. 53–58. Crans-Montana (2016)
5. Hernández, A., et al.: Security attack prediction based on user sentiment analysis of Twitter data. In: 2016 IEEE international conference on industrial technology (ICIT), pp. 610–617. Taipei (2016)
6. Itani, M., Roast, C., Al-Khayatt, S.: Corpora for sentiment analysis of Arabic text in social media. In: 2017 8th international conference on information and communication systems (ICICS), pp. 64–69. Irbid (2017)
7. Aye, Y.M., Aung, S.S.: Sentiment analysis for reviews of restaurants in Myanmar text. In: 2017 18th IEEE/ACIS international conference on software engineering, artificial intelligence, networking and parallel/distributed computing (SNPD), pp. 321–326. Kanazawa, Japan (2017)
8. Thapa, L.B.R., Bal, B.K.: Classifying sentiments in Nepali subjective texts. In: 2016 7th International conference on information, intelligence, systems & applications (IISA), pp. 1–6. Chalkidiki (2016)
9. Freitas, L.A.D., Vieira, R.: Exploring resources for sentiment analysis in Portuguese language. In: 2015 Brazilian conference on intelligent systems (BRACIS), pp. 152–156. Natal (2015)
10. Rehman, Z.U., Bajwa, I.S.: Lexicon-based sentiment analysis for Urdu language. In: 2016 sixth international conference on innovative computing technology (INTECH), pp. 497–501. Dublin (2016)
11. http://ptrckprry.com/course/ssd/data/positive-words.txt
12. http://ptrckprry.com/course/ssd/data/negative-words.txt

Graph-Based Sentiment Analysis Model for E-Commerce Websites' Data

Monali Bordoloi and Saroj Kumar Biswas

Abstract E-Commerce has evolved tremendously in the past few years. To enhance the existing business position, the commercial sites need to understand the underlying sentiment of the customers. To do so, efficient sentiment analysis technique is highly desirable in order to deeply understand the underlying meaning and sentiment of the customers. This paper proposes an effective sentiment analysis model that makes use of graph-based keyword extraction using degree centrality measure and domain dedicated polarity assignment techniques for the advanced analysis of mobile handset reviews collected from different electronic commercial sites. The proposed model outperforms some of the existing models.

Keywords Sentiment analysis · Keyword extraction · Graph-based approach · POS tagging

1 Introduction

With the evolution of user-friendly and cheaper devices, the use of Internet has accelerated a lot in the recent years. Different electronic commercial portals have come up with the idea of letting people order almost anything from anywhere. However, with the availability of different brands for a single item, people often face difficulty while making the correct choice. Though the shopping sites have the provision of presenting the comments and star rankings provided by the previous customers to guide the probable customers, the number of existing reviews/ comments can be so large that it can be a hectic task to go through all those thousands of reviews. To ease such a complex situation, sentiment analysis proves

M. Bordoloi (✉) · S. K. Biswas
Department of Computer Science and Engineering,
National Institute of Technology, 788010 Silchar, Assam, India
e-mail: monali.bordoloi@gmail.com

S. K. Biswas
e-mail: bissarojkum@yahoo.com

© Springer Nature Singapore Pte Ltd. 2019
P. K. Mallick et al. (eds.), *Cognitive Informatics and Soft Computing*,
Advances in Intelligent Systems and Computing 768,
https://doi.org/10.1007/978-981-13-0617-4_45

to be a great solution. Sentiment analysis can provide a better platform to sum up the overall impact of a particular product on the reviewers while eliminating the need to go through all the thousands of reviews before buying that product. Sentiment analysis which is also known as "opinion mining" is the art of judging the polarity of a sentence. With the help of sentiment analysis techniques, processing millions of documents and producing an overall "opinion" about the product is easy.

Sentiment analysis can be performed using different techniques. Machine learning tools like Naïve Bayes (NB), Support Vector Machines (SVM), Maximum Entropy (MaxEnt), etc., have been widely used in different sentiment analysis models. Sentiment analysis using keyword extraction followed by polarity assignment by using Sentiment Dictionaries is another such technique which is in wide use. However, these techniques face certain limitations. Most of the machine learning techniques is really time consuming and do not consider much about the domain of the document in which the word occurs. Keyword Extraction methods use generalized dictionaries which give inappropriate results. The experimentation for this paper was first based on two different models using different machine learning tools. For the first model SVM and NB were used for sentiment analysis [1]. For the second model MaxEnt was used for Sentiment Analysis [2]. But because of the above-mentioned limitations results were not satisfactory. The use of only the useful features leads to the development of a better model. By using graph concept along with degree centrality measure the most important vertices (features or words) can be retrieved [3]. Considering the semantics of the words even better keywords can be extracted [4]. Adjective and adverbs in a sentence better depict the underlying sentiment of the whole sentence. So, implementation of Parts of Speech (POS) tagging was done to get only those words which are adjectives and adverbs. The next task was to assign polarity to the extracted keywords, i.e., whether they depict positivity or negativity. Many generalized dictionaries are available for this task. But they give improper results; sentiment of a word may change according to its domain. So, development of a domain dedicated polarity assignment method is done which has been described in details in later part of the paper. Thus, using keyword extraction and polarity assignment techniques an advanced sentiment analysis model called GB-SAME (Graph-Based Sentiment Analysis Model for E-Commerce) has been proposed that can provide a better platform to the sellers of the products or the online shopping sites themselves to sum up the thousands and thousands of reviews regarding each and every product and provide the customers with an overall polarity of the different opinions, which will help the new customers to acquire a generalized opinion or sentiment purely based on the reviews of previous customers about the product they are interested in before buying that particular product from the site.

In the next section we provide an overview of background and literature survey, followed by the methodology of the proposed model, an illustrative example, experimental results obtained and at last the conclusion.

2 Background and Literature Survey

The transparency maintained by the online shopping sites regarding the sales and feedbacks provided by the customers, helps the probable customers to reach an optimal decision. The need of finding an optimal decision by acquiring more details boosts the need to generate efficient sentiment analysis techniques. The literature provides numerous algorithms and techniques to solve the main aim of sentiment analysis, i.e., to categorize text as positive, negative, or neutral or even into higher levels of categorization such as highly positive, slightly positive, highly negative, and slightly negative.

Rana and Singh [1] proposed a model which was basically focused for sentiment analysis of movie reviews in order to draw a comparison between the two of the popular machine learning tools namely SVM and NB. Yan and Huang [2] used MaxEnt method to perform the sentiment analysis of Tibetan sentence by using the probability difference between positive and negative. Sharma and Dey [5] established a commendable comparison of seven different existing machine learning techniques in association with different feature selection methods that can be used affectively in sentiment analysis of online movie reviews. The work of Sharma and Dey [5] is mostly similar to the model by Tan and Zhang [6], where sentiment analysis of different domains namely education, movie and house, are performed using different feature selection and machine learning techniques, however, Tan and Zhang [6] used Chinese text POS tool ICTCLAS (Institute of Computing Technology, Chinese Lexical Analysis System) to parse and tag Chinese review documents before performing feature selection and classification. Boiy and Moens [7] used a cascaded approach of three different machine learning techniques namely SVM, Multinomial Naïve Bayes (MNB) and MaxEnt to find the sentiment associated with multilingual text representing different blog, review, and forum texts using unigram feature vectors.

With the ever increasing amount of data and their diversity, the number of features or words in the models using n-grams keeps on increasing. In this context, keyword extraction plays a vital role which can be used to extract the set of words or phrases that best describe the underlying meaning of a document. Using graphs for keyword extraction provides a better scenario to find the important set of words instead of considering all the words. Vector Space Model (VSM) and Graph-based model are two popular keyword extraction approaches for text representation. For keyword extraction a graph of the given dataset is created by treating each word as a node in the graph. If two words co-occur in the dataset an edge is created between the two words to show relation. There exist many centrality measures such as degree centrality, closeness centrality, and betweenness centrality which can be applied to graphs to determine the most important words in the document, which can be used as the keywords for the document. In the work by Nagarajan et al. [3], important or stronger keywords are extracted from the graph using the concept of degree centrality measure. Yadav et al. [4] proposed a model to extract keywords using different graph theoretic approaches such as Degree, Eccentricity, Closeness,

Centrality, etc., by giving more importance to the semantics of the words for the keyword extraction. Bronselaer and Pasi [8] proposed a novel work where they deliberately used POS tagging for graph-based representation of textual documents.

3 Proposed GB-SAME Model

The proposed model called GB-SAME is represented using a flowchart as shown in Fig. 1. The raw data, i.e., the reviews are collected along with the ratings from different commercial sites and assigned a class, i.e., whether positive or negative to represent a proper pattern. Product reviews having rating 3 and above are assigned class 1 (indicating positive) while others are assigned class 0 (indicating negative). The model consists of six main stages as explained in details below:

3.1 Preprocessing

Each and every word and character is not always important for analyzing the sentiments. The patterns are preprocessed as follows:

i. URL Removal: We removed the associated URLs of different forms like "www. or http. or https." or "abc.def.com or abc.def.in or abc.def.net or abc@def.net or abd.de@fgh.co.uk" from the original reviews.
ii. Emoticons removal: The different emoticons like ☺, :p, ☹, :*, etc., are removed from the original reviews.
iii. Conversion to lower case: All the reviews are converted to lower case say, for example, the original review: "The Lenovo K3 has a GOOD camera quality" is converted to "the Lenovo K3 has a good camera quality".
iv. Removal of stopwords: The stop words such as about, almost, be, etc., that do not provide any sentiment in the review are removed. A standard list of stopwords is created and used. A sample of those stopwords is shown in Table 1.
v. Omission of punctuations, numbers, and control characters.

3.2 Feature Vector Extraction

In this step each and every preprocessed patterns are tokenized. All the token words for a particular pattern are stored in an array. A feature list containing unique token words from feature vectors of all reviews is created.

Fig. 1 The proposed
GB-SAME model

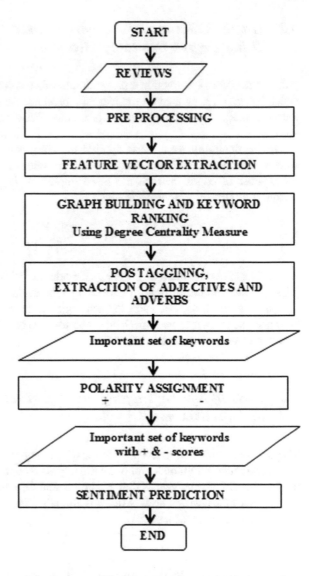

Table 1 Sample of standard
list of stopwords

a, able, about, above, according, accordingly, across, actually,
after, afterwards, again, became, because, become, becomes,
becoming, been, before, beforehand, behind, being, believe,
below, beside, besides, best

3.3 Graph Building and Keyword Ranking Using Degree Centrality Measure

After feature vector extraction a set of unique token words is obtained for the whole dataset. A graph is created by the textual representation of these token words. Each token is made the node. An edge is created between a pair of tokens if they co-occur.

Degree centrality measure is applied to this graph to find the important keywords. This measure is based on the number of edges incident on a node. The more the number of edges incident the more important the node is in the graph. The degree centrality of a node V_i is given by

$$CD(V_i) = \frac{|N(V_i)|}{|V| - 1},$$
(1)

where $CD(V_i)$ is the degree centrality of node V_i, $|V|$ is the number of nodes and $|N(V_i)|$ is the number of nodes connected to the node V_i.

This measure assigns a score to each node on the basis of which the words can be ranked according to its importance. The top five hundred words with the highest scores will be treated as the keywords.

3.4 Parts of Speech Tagging and Extraction of Adjectives and Adverbs

The keywords obtained in the previous step are tagged for their POS using the POS tagger available in Python3. Out of all the keywords only adjectives and adverbs are used for sentiment analysis as they best represent the sentiment of a pattern. This is the new set of keywords now. After this the adjectives and adverbs are chosen for the polarity assignment task.

3.5 Domain-Dedicated Polarity Assignment

In this step one keyword is considered at a time and its pair is taken with every other keyword in the list of keywords. The number of times the pair occurs in positive reviews and negative reviews in the training dataset is counted.

Let for each word W_i,

$CP(P_{i,j})$ = number of times the pair (i,j) occurs in positive reviews in training set
$CN(P_{i,j})$ = number of times the pair (i,j) occurs in negative reviews in training set

After this we take all the reviews in which the keyword W_i occurs singly, i.e., the particular review has no other keyword other that keyword W_i.

$CP(W_i)$ = number of times the word W_i occurs singly in positive reviews in training set

$CN(W_i)$ = number of times the word W_i occurs singly in negative reviews in training set

Now for each W_i its positive and negative score is calculated as follows:

$$\text{Positive\_score}(W_i) = \sum_j \frac{CP(P_{i,j})}{CP(P_{i,j}) + CN(P_{i,j})} + \frac{CP(W_i)}{CP(W_i) + CN(W_i)} \quad (2)$$

$$\text{Negative\_score}(W_i) = \sum_j \frac{CN(P_{i,j})}{CP(P_{i,j}) + CN(P_{i,j})} + \frac{CN(W_i)}{CP(W_i) + CN(W_i)} \quad (3)$$

Normalization of the polarity scores

The scale of the positive and negative scores may be different. So a normalization process is used to curb the variation. The range of scores is brought between $(0, 1)$.

Normalization formula is adopted from the paper by Biswas et al. [9]:

$$y = \frac{(\text{ymax} - \text{ymin}) * (x - \text{xmin})}{\text{xmax} - \text{xmin}} + \text{ymin}, \quad (4)$$

where x is data to be normalized, i.e., polarity score, xmax is maximum value of all polarity scores, xmin is minimum value of all polarity scores, y is normalized polarity score, ymax is desired maximum polarity score, and ymin is desired minimum polarity score.

3.6 Sentiment Prediction of Patterns

For the sentiment prediction of the patterns, a pattern having review R is taken and its feature vector is extracted as in step 2. For each token word W_i in the feature vector of R it is checked whether the token is one of the keywords from the set of important keywords extracted in step 4. If true then a score to W_i will be assigned based on the polarity score for that word using the following rule:

If W_i is a keyword,

$$\text{Score}(W_i) = \text{Positive\_score}(W_i) - \text{Negative\_score}(W_i) \quad (5)$$

else

$$\text{Score}(W_i) = 0 \quad (6)$$

Now score is assigned to R by combining the score of its each word

$$Score(R) = \sum_i Score(W_i) \qquad (7)$$

Finally the polarity of a review is assigned as 1 for positive and 0 for negative according to following rule:

If (Score (R) >=0) then (Polarity (R) = 1) Else (Polarity (R) = 0).

4 Experimental Results

The data were collected for the mobile phone brand Lenovo K3 from Amazon and Flipkart. The description of the data used is shown in Table 2. Reviews were divided into training and testing sets and experiment was performed using Python 3. 90% of the data set is used for training the sentiment model while 10% is used for the testing.

Two existing models are considered as a baseline for the comparison and validation of our proposed model as stated below

1. Baseline Model 1: The model by Rana and Singh [1]

    i. Using NB and
    ii. Using SVM

2. Baseline Model 2: The model by Yan and Huang [2] using MaxEnt.

Performance Measure

Confusion matrix is used for the calculation of accuracy. Table 3 presents the confusion matrix for a two-class classifier. The accuracy is the proportion of the total number of predictions that are correct. It is determined using the Eq. 8.

$$Accuracy = \frac{TP + TN}{TP + TN + FP + FN} \qquad (8)$$

Table 2 Dataset description

Dataset	Size	Feature type	Number of class
Lenovo K3	1500	Mixed	2

Table 3 Confusion matrix

		Predicted	
		Positive	Negative
Intelligence	Positive	TP	FN
	Negative	FP	TN

Table 4 Accuracy table for the different datasets

Datasets	Proposed model	Base model 1 using Naive Bayes (1a)	Base model 1 using SVM (1b)	Base model 2 using maximum entropy (2)
Lenovo K3	82.67	81.33	78.67	72

Table 4 depicts the accuracy of the proposed model for the dataset used in comparison to the two existing models. It is observed from Table 4 that the accuracy for the dataset Lenevo K3 is far better than the existing models. The accuracy of the proposed model is better than that of NB, SVM and MaxEnt. The performance of MaxEnt is observed as worst in comparison to the other two classifiers.

5 Conclusion

An effective sentiment analysis model using graph-based method has been proposed in this paper which intends to determine the overall underlying sentiment or opinion presented towards a popular product, while investigating thousands of reviews about that particular product from different electronic commercial sites and reducing all the noise and informality that comes along with a set of text data (review). This paper examines the use of keyword extraction method using graph in sentiment analysis. Also, instead of using all the features from the dataset, selecting only the important keywords solves the problem of dealing with irrelevant or redundant data. A domain-dedicated polarity assignment technique has been achieved in this paper to assign the relevant polarity to the important set of keywords extracted using degree centrality measure.

The proposed model uses dataset comprising of reviews by customers on a popular and trending mobile phone and outperforms the two existing models to a great extent. The proposed model shows comparatively better performance than the three popular machine learning techniques used in this domain namely SVM, NB, and MaxEnt.

References

1. Rana, S., Singh, A.: Comparative analysis of sentiment orientation using SVM and Naïve Bayes techniques. In: 2nd international conference on next generation computing technologies, IEEE (2016)
2. Yan, X., Huang, T.: Tibetan sentence sentiment analysis based on the maximum entropy model. In: 10th International conference on broadband and wireless computing, communication and applications, IEEE (2015)
3. Nagarajan, R., Anu, S., Nair, H.: Keyword extraction using graph based approach. Int. J. Adv. Res. Comput. Sci. Softw. Eng. **10**(10) (2016)

4. Yadav, C.S., Sharan, A., Joshi, M.L.: Semantic graph based approach for text Mining. In: International conference on issues and challenges in intelligent computing technique, IEEE (2014)
5. Sharma, A., Dey, S.: A comparative study of feature selection and machine learning techniques for sentiment analysis. RACS'12, Oct 23–26, 2012, San Antonio, TX, USA. pp. 1–7, ACM (2012) (978-1-4503-1492)
6. Tang, S., Zhang, J.: An empirical study of sentiment analysis for chinese documents. Expert Syst. Appl. **34**, 2622–2629 Elsevier (2008)
7. Boiy, E., Moens, M.F.: A machine learning approach to sentiment analysis in multilingual Web texts. Inf. Retrieval **12**, 526–558, https://doi.org/10.1007/s10791-008-9070-z. Springer (2009)
8. Bronselaer, A., Pasi, G.: An approach to graph-based analysis of textual documents. In: 8th Conference of the European Society for Fuzzy Logic and Technology (EUSFLAT 2013), Advances in Intelligent Systems Research, Atlantis Press (2013)
9. Biswas, S.K., Marbaniang, L., Purkayastha, B., Chakraborty, M., Heisnam, R.S., Bordoloi, M.: Rainfall forecasting by relevant attributes using artificial neural networks—a comparative study. Int. J. Big Data Intell. **3**(2), 111–121, Inderscience (2015)

Isolated Word Recognition Based on Different Statistical Analysis and Feature Selection Technique

Saswati Debnath and Pinki Roy

Abstract Isolated word recognition serve as an important aspect of speech recognition problem. This paper contributes a solution of speaker-independent isolated word recognition based on different statistical analysis and feature selection method. In this work different parametric and nonparametric statistical algorithm such as analysis of variance (ANOVA) and Kruskal–Wallis are used to rank the features and incremental feature selection (IFS) to find the efficient features set. The objective of applying statistical analysis algorithm and feature selection technique on the cepstral feature is to improve the word recognition performance using efficient and optimal number of feature set. The experimental analysis is carried out using two machine learning techniques such as Artificial Neural Network (ANN) and Support vector machine (SVM) classifier. Performance of both the classifier has been evaluated and described in this paper. From the experimental analysis it has been observed that statistical analysis with feature selection technique provides better result for the two classifier as compared to original all cepstral features.

Keywords Isolated word recognition · Cepstral features (MFCC)
Statistical analysis · Feature selection · Machine learning

1 Introduction

One of the important aspects to increase the performance of the speech recognition system is feature evaluation mechanism. There is some drawback for that in practical application; recognition accuracy does not reach high. In this paper we contribute the solution of speaker independent isolated word recognition problem

S. Debnath (✉) · P. Roy
Department of Computer Science and Engineering, NIT, Silchar, Assam, India
e-mail: dnsaswati@gmail.com

P. Roy
e-mail: pinkiroy2405@gmail.com

© Springer Nature Singapore Pte Ltd. 2019 463
P. K. Mallick et al. (eds.), *Cognitive Informatics and Soft Computing*,
Advances in Intelligent Systems and Computing 768,
https://doi.org/10.1007/978-981-13-0617-4_46

and address the feature selection method with statistical approach. Acoustic feature extraction has been a subject of increasing interest in recent years of speech recognition. It is also important work to recognize efficient and optimum number of features, because all the features are not equally significant for recognition. Therefore to select valuable features here we introduce statistical analysis of features which also reduce the dimension of feature set. First we extract acoustic features of a speech signal and after that select the important features using statistical analysis and incremental feature selection technique. In this work we have extract the MFCC [1] feature of speech signal and each word is modeled by calculating mean values of the cepstral features. After that in feature selection, analysis of variance (ANOVA) and Kruskal–Wallis test are used to rank the features along with incremental feature selection (IFS) to determine the optimal features set. Statistical algorithms are used to calculate the Chi-square value of an individual feature so that important feature can be extract. Highest Chi-square or F value has highest priority and according to that we rank the features. Feature selection method improves the recognition accuracy while reducing the dimension of the feature set. All the experiments are carried out using two classifier such as ANN and SVM.

Comparative analysis has been done on performance obtained from without feature selection technique, using statistical algorithm on features and feature selection technique to show that statistical analysis with feature selection technique improves the performance of isolated word recognition. The paper is organized as follows: In Sect. 2 we give the literature review of some paper on isolated word and digit recognition. Section 3 gives the proposed methodology of isolated word recognition. Section 4 gives database description and the experimental analysis and discussion is given in Sect. 5. Finally we conclude our paper with future work in Sect. 6.

2 Literature Review

In the paper [4] speaker-independent isolated word recognition has been done using MFCC and HMM. System has developed for kannada language. They used phone level and syllable level HMM for recognition. Isolated word recognition using Liquid State Machine (LSM) has been introduced by in paper [5]. In [3] author contributes feature selection technique in speech recognition and speaker recognition. They have used ANOVA, LDA (linear discriminative analysis), SVM-REE (support vector machine recursive feature elimination) to reduce the dimension of the feature set and increase recognition rate.

Mishra et al. [6] proposed isolated digit recognition using HMM and MFCC for hindi digit. They have used HTK and MATLAB for evaluating their proposed model. In this paper [7] author described real-time speaker-dependent isolated word speech recognition system for human computer interaction. MFCC and dynamic

programming algorithm have been used for recognizing the words. In [8] introduced speaker-independent continuous speech recognition and isolated word
recognition using VQ and HMM. They have proposed PLP feature extraction with
combine VQ + HMM technique and observed that system performs well with
hybrid technique.

Paper [9] described isolated English word recognition using MFCC and dynamic
time warping (DTW). In [10] author presented speech recognition system for isolated English digit recognition using improved feature for DTW (IFDTW) with
13th weighted MFCC coefficients. Author also proposed solar-based technique for
reducing the complexity of recognition and faster implementation of IFDTW.

This paper [11] proposed Speech recognition system for isolated and connected
word recognition of Hindi language using HMM and MFCC.

3 Proposed Methodology of Isolated Word Recognition

Proposed methodology consists of different steps of database collection, feature
extraction, feature selection, classification using different classifier. The architecture
of proposed model is shown in Fig. 1. The phases of isolated word recognition are
as follows:

1. **Feature extraction**: 19th order of Mel frequency cepstral coefficients (MFCCs)
   are used as an acoustic feature set.
2. **Feature selection**: Statistical algorithms such as analysis of variance (ANOVA)
   and Kruskal–Wallis with incremental feature selection (IFS) are used to rank the
   feature. Ranking is based on their F-statistics or Chi-square value. This is our
   main contribution to this work.

$$F = \frac{\text{variation between sample means}}{\text{variation within the samples}}$$

3. **Classifier**: In order to analyze performance of the system the experiment has
   been carried out using two different classifiers-ANN and SVM.

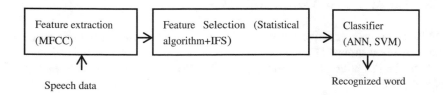

Fig. 1 Block diagram of proposed model

3.1 MFCC Feature Extraction

Acoustic feature extraction has been an important method in recent years of speech recognition and speaker recognition. Mel frequency cepstrum coefficients (MFCC), is the most popular acoustic feature extraction was first introduced in 1980 by David and Mermelstein 1980 [1]. Mel Frequency Cepstral Coefficients (MFCC) [4, 12–14] is a short term spectral feature. At the first we analyze each speech signal into short time frame. 50% overlap is applied to the frames. After framing the signal a hamming window function is applied with every frame of speech signal. Discrete Fourier Transform (DFT) is used to extract spectral information for a discrete frequency bands. Fast Fourier Transform (FFT) is used; it is the most commonly used algorithm to compute DFT. After that calculate Mel filter bank energy in a mel scale. For a given frequency f we can use the following formula to compute the Mels in Hz: Mel scale [13] is defined as:

$$m_f = 1125 \ln\left(1 + \frac{f}{700}\right), \tag{1}$$

where f is the actual frequency in Hz.

In the final step take the logarithm of Mel filter bank energies. The result is called the Mel frequency cepstrum coefficients (MFCCs) [14]. Here we extract 19th order of MFCC features.

3.2 Feature Selection

Selecting the effective feature set that can increase the classification accuracy is the main aim of feature selection technique, while reducing the irrelevant and redundant feature [16–23]. In this work for each recognition model two algorithms is used, one is to rank the feature and other one is to select the feature subset. Analysis of variance (ANOVA) and Kruskal–Wallis are used to rank the feature with incremental feature selection (IFS) technique which selects the relevant features.

3.2.1 The Analysis of Variance (ANOVA)

ANOVA is a very effective parametric statistical method to find the difference in means between groups [17].

According to principle of ANOVA [17, 22, 23] the score (F) of dataset can be defined [17]

$$F(\xi) = \frac{sB^2(\xi)}{sW^2(\xi)}, \tag{2}$$

where $sB^2(\xi)$ is the sample variance between groups (also called Mean Square Between, MSB) and $sW^2(\xi)$ is sample variance within groups (also called Mean Square Within, MSW) [17]. Decision is made using F-statistics [17]. One-way ANOVA is used in this work.

3.2.2 Kruskal–Wallis Test

Kruskal–Wallis is a rank-based nonparametric test. It is used to find the significant difference between two or more groups of an independent variable. It is a Chi-square distribution. The test is performed to get the Chi-square value of each feature and rank those feature in a decreasing order.

3.2.3 Incremental Feature Selection (IFS)

IFS method evaluates performance of the system using different combination of features. The features subset from feature with the highest F value or Chi-square value to lowest is in the ranked feature set. A new feature subset is produced when the feature with the second highest F value is added.

One-way ANOVA with IFS and Kruskal–Wallis with IFS have been done in this experiment for selecting the optimal number of features set.

3.3 Classification

3.3.1 Classification Based on ANN

Artificial neural networks (ANN) [24] are network of simple processing elements called neuron and inspired by observations made in the study of biological systems. ANN is used for pattern classification because of its capability of nonlinear, non-parametric relationships between input data and output.

3.3.2 Classification Based on SVM

SVM [25] is a supervised machine learning that estimates decision surfaces directly rather than modeling a probability distribution across the training data [25]. SVM classifies the nonlinear output vector to a higher dimensional feature space. In this experiment fourfold cross validation has been used. We are using different kernel function of SVM classifier such as "radial basis function", "polynomial", and "linear" and select the best kernel function of the classifier. The model has been selected based on feature selection and best kernel function.

4 Database Collection

A speech database has been recorded in a lab environment. Data from 10 adult male speakers of age group 24–30 years have been recorded where each speaker has repeated each word 20 times. Sampling rate is 16 kHz and the sampling bit resolution is 16 bits/sample. So for one word 200 data sample was recorded. Database has been uploaded in the website [26].

5 Experimental Results and Discussion

5.1 Performance of System Using All MFCC Features

Nineteenth-dimensional cepstral feature has been extracted from speech signal. First we have done the experimental analysis on original 19th-dimensional MFCC matrix and classify using two classifier, i.e., ANN and SVM. Fourfold cross validation has been carried out for each of the classifier. The performance of the system can be calculated by total number of words correctly recognized during testing phase.

Tables 1 and 2 show the system accuracy using ANN and SVM classifier with all cepstral features.

Table 1 Performance of ANN using original all cepstral features

Exp. no.	No. of hidden layer	No. of nodes	Iteration	System accuracy (%)
1	2	30, 20	100	90.10
2	2	40, 30	100	91.02
3	2	50, 40	100	92.32
4	2	60, 50	100	*92.93*
5	2	70, 60	100	92.82

Table 2 Performance of different kernel function of SVM using original cepstral features

Exp. no.	Kernel function	System accuracy (%)
1	Radial basis function (rbf)	*88.67*
2	Linear	68.2207
3	Polynomial	84.76

5.2 Performance of the System Using ANOVA with IFS

It is very important to discover whether all features are significant to develop a system or not. The performance of the individual feature can be evaluated and also observe the performance when concatenating with other features. Incremental feature selection (IFS) technique gradually concatenates the features. In this experiment to select the features and concatenate the features statistical analysis has been used for ranking the feature. ANOVA test is carried out to rank the feature based on the F statistics and after that IFS is used to select and adding the significant features. F statistics calculate the statistical information such as mean and variance. Feature with highest F value ranked one and the features with second highest value ranked two and so on.

Fourfold cross validation has been done for each classifier to get efficient recognition accuracy. Table 3 gives the performance of isolated word recognition using statistical analysis and feature selection technique for ANN classifier and it shows that for 10 number of feature it gives significant result. Performance of SVM classifier for different kernel function has been shown in Table 4. SVM classifier with kernel function "rbf" gives the highest accuracy of 95.03% using eight cepstral features.

Table 3 Performance of ANN classifier after statistical analysis (ANOVA) and feature selection (IFS)

Exp. no.	No. of hidden layer	No. of hidden units	Iterations	No. of features	Sequence of features according to rank	System accuracy (%)
1	2	30, 20	100	13	1, 2, 5, 3, 10, 4, 11, 13, 16, 7, 12, 9, 6	92.06
2	2	40, 30	100	10	1, 2, 5, 3, 10, 4, 11, 13, 16, 7	94.78
3	2	50, 40	100	14	1, 2, 5, 3, 10, 4, 11, 13, 16, 7, 12, 9, 6, 15	90.88
4	2	60, 50	100	10	1, 2, 5, 3, 10, 4, 11, 13, 16, 7	*93.96*
5	2	70, 60	100	10	1, 2, 5, 3, 10, 4, 11, 13, 16, 7	90.34

Table 4 Performance of SVM classifier with statistical analysis ANOVA and IFS

Exp. no.	Kernel function	No. of features	Sequence of features according to rank	System accuracy (%)
1	rbf	8	1, 2, 5, 3, 10, 4, 11, 13	*95.03*
2	Linear	19	1, 2, 5, 3, 10, 4, 11, 13, 16, 7, 12, 9, 6, 15, 19, 17, 14, 8, 18	82.34
3	Polynomial	14	1, 2, 5, 3, 10, 4, 11, 13, 16, 7, 12, 9, 6, 15	90.65

5.3 Performance of Feature Selection Technique Using Kruskal–Wallis with IFS

Kruskal–Wallis is used to rank the mfcc features. All the 19th coefficients are ranked after the evaluation of chi-square value or F value. After Kruskal–Wallis test IFS has been done to select optimal feature set for every classifier.

Here Tables 5 and 6 give the recognition rate of Kruskal–Wallis with IFS technique using ANN and SVM classifier. ANN classifier gives 92.18% accuracy for 12 features whereas SVM gives 95.20% recognition rate using 10 numbers of features.

5.4 Analysis of Result

In this experiment different parametric and nonparametric statistical analysis and feature selection techniques are proposed and implemented for isolated word recognition. Experimental analysis shows both the result without using feature selection technique and using feature selection technique. After applying the statistical analysis with feature selection technique such as ANOVA with IFS and Kruskal–Wallis with IFS, it has been observed that for both the classifier, system give better recognition accuracy. So performance of the system depends on how efficiently we select the feature. In this experiment using English digit dataset SVM

Table 5 Performance of ANN classifier after statistical analysis of Kruskal–Wallis and feature selection (IFS)

Exp. no.	No. of hidden layer	No. of hidden units	Iterations	No. of features	Sequence of features according to rank	System accuracy (%)
1	2	30, 20	100	12	1, 2, 5, 10, 3, 4, 11, 13, 9, 12, 6, 7	92.18
2	2	40, 30	100	13	1, 2, 5, 10, 3, 4, 11, 13, 9, 12, 6, 7, 16	91.68
3	2	50, 40	100	13	1, 2, 5, 10, 3, 4, 11, 13, 9, 12, 6, 7, 16	91.85
4	2	60, 50	100	19	1, 2, 5, 10, 3, 4, 11, 13, 9, 12, 6, 7, 16, 19, 15, 14, 17, 8, 18	90.75
5	2	70, 60	100	15	1, 2, 5, 10, 3, 4, 11, 13, 9, 12, 6, 7, 16, 19	90.56

Table 6 Performance of SVM classifier with Kruskal-wallis and IFS

Exp. no.	Kernel function	No. of features	Sequence of features according to rank	System accuracy (%)
1	rbf	10	1, 2, 5, 10, 3, 4, 11, 13, 9, 12	95.20
2	Linear	18	1, 2, 5, 10, 3, 4, 11, 13, 9, 12, 6, 7, 16, 19, 15, 14, 17, 8	81.25
3	Polynomial	13	1, 2, 5, 10,3,4,11, 13,9,12,6,7,16	92.78

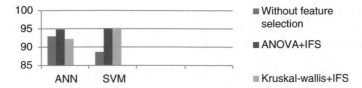

Fig. 2 Comparison graph of performance between existing cepstral feature and Statistical analysis + IFS

Table 7 Sequential rank of the features according to their F value obtained through statistical analysis of ANOVA test and Kruskal-wallis test

ANOVA				Kruskal-wallis			
Rank	Features	Rank	Features	Rank	Features	Rank	Features
R1	1	R11	12	R1	1	R11	6
R2	2	R12	9	R2	2	R12	7
R3	5	R13	6	R3	5	R13	16
R4	3	R14	15	R4	10	R14	19
R5	10	R15	19	R5	3	R15	15
R6	4	R16	17	R6	4	R16	14
R7	11	R17	14	R7	11	R17	17
R8	13	R18	8	R8	13	R18	8
R9	16	R19	18	R9	9	R19	18
R10	7			R10	12		

classifier perform best with feature selection technique. ANOVA with IFS system gives 95.03% recognition accuracy using eight cepstral features. Similarly ANN performs better when select efficient feature using statistical test and IFS feature selection technique. Comparison of word recognition accuracy between feature selection technique and without feature selection technique for individual classifier is given below in Fig. 2.

Table 7 gives the sequential rank of features according to their chi-square value using ANOVA and Kruskal–Wallis.

6 Conclusion

Acoustic feature of speech signal MFCC is used which extract cepstral features of speech for isolated word recognition. We have introduced statistical analysis and feature selection technique to improve word recognition accuracy. ANOVA and Kruskal–Wallis with IFS feature selection select the subset of cepstral feature which significantly reduce redundant feature and rank the efficient feature according to their F value. All the experiment is done using machine learning technique ANN and SVM. Therefore from experimental analysis it has been observed that instead of taking all the cepstral features randomly it is good to rank the efficient feature and according to that ranking select the best feature subset.

Acknowledgements The Authors gratefully acknowledge Dr. Pradip K. Das, Department of Computer Science and Engineering, Indian Institute of Technology, Guwahati (IITG) and also acknowledge his students worked under his guidance for providing database support for this work. Dr. Pradip K. Das (http://www.iitg.ac.in/pkdas/), professor of IIT, Guwahati has the research interest of Digital Signal Processing, Speech Processing, Man-Machine Intelligence Systems and this work has been supported by him.

References

1. Davis, S.B., Mermelstein, P.: Comparison of parametric representation for monosyllabic word recognition in continuously spoken sentences. IEEE Trans. Acoust. Speech Sig. Process. **28**(4), 357–365 (1980)
2. Rabiner, L., Juang, B.H.: Fundamentals of speech recognition. Prentice Hall, Upper Saddle River (2012)
3. Promotor: Prof. Dr. ir. D. Van Compernolle Co-Promotor: Prof. Dr. ir. H. Van hamme. Wu, T.: Feature Selection in Speech and Speaker Recognition (2009)
4. Thalengala, A., Shama, K.: Study of sub-word acoustical models for Kannada isolated word recognition system. Int. J. Speech Technol. **19**, 817–826 (2016). https://doi.org/10.1007/s10772-016-9374-0
5. Verstraeten, D., Schrauwen, B., Stroobandt, D., Van Campenhout, J.: Isolated word recognition with the Liquid State Machine:a case study. Inf. Process. Lett. **95**, 521–528 (2005)
6. Mishra, A.N., Biswas, A., Chandra, M.: Isolated Hindi Digit Recognition: a comparative study. Int. J. Electron. Commun. Eng. (IJECE) **3**(1), 229–238 (2010)
7. Nandyala, S.P.: Real time isolated word speech recognition system for human computer interaction. Int. J. Comput. Appl. **12**(2) Nov (2010)
8. Revathi, A., Venkataramani, Y.: Speaker independent continuous speech & isolated digit recognition using VQ & Hmm. In: IEEE, pp. 198–202 (2011)
9. Limkara, M., Raob, R., Sagvekarc, V.: Isolated Digit Recognition Using MFCC and DTW. IJAEEE **1**(1), 59–64 (2012)
10. Chapaneri, S.V., Jayaswal, D.J.: Efficient speech recognition system for isolated digits. IJCSET **4**(3), 228–236 (2013)
11. Choudhary, A., Chauhan, R., Gupta Gautam, S.: Automatic speech recognition system for isolated and connected words of Hindi language by using hidden markov model toolkit (HTK). In: Association of computer electronics and electrical engineers (ACEEE) (2013)
12. Soni, B., Debnath, S., Das, P.K.: Text-dependent speaker verification using classical LBG, adaptive LBG and FCM vector quantization. Int. J. Speech Technol. **19**(3), 525–536 (2016)
13. Gold, B., Morgan, N.: Speech and audio signal processing. John Wiley and Sons, New York, NY (2000)
14. Becchetti, C., Ricotti, L.P.: Speech recognition. John Wiley and Sons, England (1999)
15. Davis, S.B., Mermelstein, P.: Comparison of parametric representation for monosyllabic word recognition in continuously spoken sentences. IEEE Trans. Acoust. Speech Signal Process. **28**(4), 357–365 (1980)
16. Chandrashekar, G., Sahin, F.: A survey on feature selection methods. Comput. Electr. Eng. **40**(1), 16–28 (2014)

17. Ding, H., Feng, P.M., Chen, W., Lin, H.: Identification of bacteriophage virion proteins by the ANOVA feature selection and analysis. Mol. Bio. Syst. **10**(8), 2229–2235 (2014)
18. Chan, Y., Walmsley, R.P.: Learning and understanding the Kruskal-Wallis one-way analysis-of-variance-by-ranks test for differences among three or more independent groups. Phys. Ther. **77**(12), 1755–1761 (1997)
19. Niu, B., Huang, G., Zheng, L., Wang, X., Chen, F., Zhang, Y., Huang, T.: Prediction of substrate-enzyme-product interaction basedon molecular descriptors and physicochemical properties. J. Proteomics **75**, 1654–1665 (2012)
20. Settouti, N., Bechar, M.E.A., Chikh, M.A.: Statistical comparisons of the top 10 algorithms in data mining for classification task. Int. J. Interact. Multimed. Artif. Intel. **4**(1), 46–51 (2016)
21. Kumari, P., Vaish, A.: Feature-level fusion of mental task's brain signal for an efficient identification system. Neural Comput. Appl. **27**(3), 659–669 (2016)
22. Ding, H., Guo, S.H., Deng, E.Z., Yuan, L.F., Guo, F.B., Huang, J., Rao, N.N., Chen, W., Lin, H.: Chemom. Intell. Lab. Syst. **124**, 9–13 (2013)
23. Lin, H., Chen, W., Ding, H.: PLoS ONE **8**, e75726 (2013)
24. Pujari, J.D., Yakkundimath, R., Byadgi, A.S.: SVM and ANN based classification of plant diseases using feature reduction technique. Int. J. Interact. Multimed. Artif. Intel. **3**(7), 6–14 (2016)
25. Ganapathiraju, A., Jonathan, E., Hamakerand, J., Picone, J.: Applications of support vector machines to speech recognition. IEEE Trans. Signal Process. **52**(8) August (2004)
26. http://www.iitg.ernet.in/pkdas/digits.rar

Keyword Extraction from Tweets Using Weighted Graph

Saroj Kumar Biswas

Abstract One of the most important tasks of sentiment analysis of twitter contents is automatic keyword extraction. Vector Space Model (VSM) is one of the most well-known keyword extraction techniques; however it has some limitation such as scalability and sparsity. Graph-based keyword extraction approach is used to overcome those limitations. This paper proposes an unsupervised graph-based keyword extraction method, called Keyword from Weighted Graph (KWG) which uses Node Edge (NE) rank centrality measure to calculate the importance of nodes closeness centrality measure to break the ties among the nodes. The proposed method is validated with two datasets: Uri Attack, and American Election. From the experimental results it is observed that the performances of the proposed method outperform the eigen vector centrality and the textrank centrality measures. The performances are shown in terms of precision, recall, and *F*-measure.

Keywords Extraction · Sentiment analysis · Graph-based model
Centrality measure · Vector space model

1 Introduction

Keyword extraction is the way of finding key terms from a text that can appropriately represent the subject of the text [1]. But, assigning keywords to documents/ texts manually is a very costly, time-consuming and tedious task. Consequently automatic keyword extraction from text of social networking site has attracted the interest of researchers over the last years as it is helpful in several applications, such as automatic indexing, automatic summarization, automatic classification, automatic clustering, and automatic topic detection.

S. K. Biswas (✉)
Department of Computer Science and Engineering,
National Institute of Technology, Silchar, India
e-mail: bissarojkum@yahoo.com

© Springer Nature Singapore Pte Ltd. 2019
P. K. Mallick et al. (eds.), *Cognitive Informatics and Soft Computing*,
Advances in Intelligent Systems and Computing 768,
https://doi.org/10.1007/978-981-13-0617-4_47

Social networks are formed by social interactions like co-authoring, advising, supervising, sharing interests, making phone calls, etc. One of the most popular social networking sites is Twitter where people put their opinions about various topics like politics, brands, products, and celebrities', etc. [2]. This growth needs study and analysis of contents of tweets in hope of summarizing the huge collection of posts, which is called sentiment analysis. Hence sentiment analysis has been an important topic for data mining [3]. Sentiment classification is intensively helpful and useful in business intelligence applications, recommender systems, and political and administrative decisions. One of the most important tasks in sentiment analysis is keyword extraction because if keywords are properly extracted, subject of the text can be studied and analyzed comprehensively and good decision can be made on a text.

When texts are represented using the well-known VDM [4], it creates sparse matrices which to be dealt with computationally. Besides, while target application involves twitter contents, compared with traditional text collections, this problem becomes even worse as twitter contains. Diversity, informality, grammatical errors, buzzwords, slangs, etc., in its contents. An effective technique is required to extract useful keywords [5, 6]. Graph-based approach to extract keywords is an appropriate one in such situation. Therefore this paper proposes a graph-based keyword extraction method, called Keyword from Weighted Graph (KWG) where NE rank centrality measure is used to calculate relevance of the nodes/keywords and closeness centrality measure is to break the ties among the keywords in a dataset built from twitter messages. NE centrality works well with disconnected graph as term frequency and degree of nodes are used as parameters to compute this centrality. The proposed method is also compared with eigen vector centrality and textrank centrality measures as relevance of the keywords in the same framework.

2 Literature Survey

Some of the recent keyword extraction methods are reported in this section. Hulth [7] proposed an approach that incorporates linguistic knowledge along with the statistical features of the text. Nguyen et al. [8] presented a key phrase extraction algorithm for scientific papers in which they use linguistic features like position of phrases in the document which shows a significant improvement in performance. Witten et al. [9] introduced an algorithm called Key phrase Extraction Algorithm (KEA) which extracts candidate key phrases using lexical methods. Zhang et al. [10] proposed a method that uses global context information and local context information along with SVM to extract keywords. Medelyan et al. [11] introduced KEA++ which aims at improving the results given by KEA. Abilhoa et al. [12] proposed a keyword extraction method from tweet collections that represents texts as graphs and applies centrality measures-degree, closeness and eccentricity, for finding the relevant vertices (keywords). Nagarajan et al. [13] presented a graph-based keyword extraction algorithm which identifies the relationship between the words of the documents by edges. Song et al. [14] proposed a method which considers three major factors that make it different from other keyword extraction methods.

3 Proposed KWG Model

The flow chart of the proposed KWG model is shown in Fig. 1. The whole process of the proposed model is segregated in three phases: Preprocessing, Textual Graph Representation and Keyword Extraction. The details of the phases are given below.

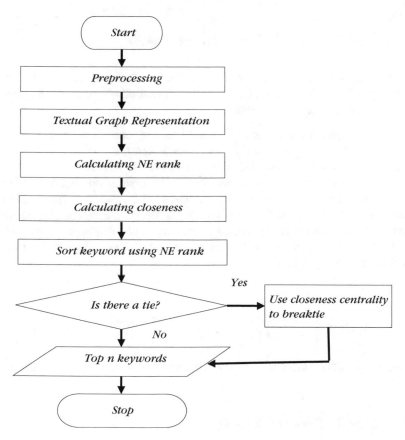

Fig. 1 Flow chart of KWG model

3.1 Phase 1: Preprocessing

Tokenization and removal of stop words are done in this phase and the steps are given below in detail

- **Tokenization**: Each term in a tweet belonging to a dataset is treated as a token. Tokens are the basic constituents of a tweet/document. Let T be the set of tweets which is represented as $T = \{T_1, T_2, T_3 \ldots T_i|$ i is the number of tweets$\}$. Then each tweet in T is preprocessed and its terms are treated as tokens. Let t be the set of tokens represented as $t = \{t_1, t_2, t_3 \ldots t_k\}$ t includes tokens from all the tweets of T where the number of tokens in the set T is k.
- **Stop word removal**: A standard list of stop words is created and these stop words are then removed from the set, t.

3.2 Phase 2: Textual Graph Representation

In textual graph the entity set comprises of the set of tokens. Let $G = (V, E)$ is a graph where V is the set of vertices and E is the set of edges. Then the steps for building a textual graph are as follows.

- **Vertex assignment**: One vertex is created for one token. The set of vertices are created from the set of tokens in the vertex assignment.
- **Edging**: The edges are established by pairs of tokens read in the same sequence in which they appear in the original tweets/documents. So an edge E_{ij} is generated for each token "i" and its immediate successor j. Edge weights are assigned which are equal to the co-occurrence of the nodes. An adjacency matrix for the textual graph is made which represents weights of edges. Let A is the adjacency matrix for the textual given graph. Then, the weights in A are calculated by Eq. 1 given below.

$$A(i,j) = \begin{cases} c, \text{where } c \text{ is the co-occurence frequency of } i \text{ and } j \\ \text{Inf,if there is no direct edge between } i \text{ and } j \end{cases}, \quad (1)$$

where, c represents weight of the edge generated between node i and node j.

3.3 Phase 3: Keyword Extraction

The proposed model includes the following steps for keyword extraction.

- **Calculate the NE rank**: NE rank method uses the weight of nodes with Textrank method which results in a node edge weighting approach called Node Edge Rank [15]: The rank/relevance of node v_i using NE rank is calculated by Eq. 2 given below.

$$R(v_i) = (1 - d) \cdot W(v_i) + d \cdot W(v_i) \cdot \sum_{j:v_j \to v_i} \frac{w_{ji}}{\sum_{k:v_j} w_{jk}} R(v_j) \qquad (2)$$

Here, d is the damping factor which denotes the probability of jumping from a node to the next node. It is usually set to 0.85. Similarly, $(1 - d)$ denotes the probability of jumping to a new node. $W(v_i)$ is the weight of the current node. w_{ji} is the weight of the edge from the previous vertex v_j to the current vertex v_i and $\sum_{k:v_j} w_{jk}$ is the summation of all edge weights in the previous node v_j. $R(v_j)$ denotes the rank/relevance of node v_j. The proposed model considers normalized TF in lieu of TF-IDF as weight of nodes as it gives better result. d is the damping factor which is the probability of jumping from a node to the next node and is set to 0.85. $(1 - d)$ is the probability of jumping to a new node.
- **Calculate the closeness centrality**: The closeness centrality, $CC(v)$ of a vertex v is the inverse of farness, i.e., the sum of the shortest distances between the vertex v and all the other vertices in a graph. Suppose d_{vu} is the shortest path between vertices u and v. The closeness centrality of a vertex v is calculated by Eq. 3 given below.

$$CC(v) = \frac{1}{\sum_{v \neq u} d_{vu}} \qquad (3)$$

- **Sort the keywords**: Do the following for all the nodes.

Let i and j are two terms/nodes such that j occurs immediately after i in the list/text. Then:	
a	If NE(i) = NE(j)
	If degree(i) < degree(j) Swap (i, j)
b	Else if NE(i) < NE(j) Swap (i, j)
c	Otherwise No action

Finally, n best ranked vertices are selected as keywords.

4 Results and Discussions

The experiments are conducted with two datasets: Uri Attack and American Election. Uri Attack and American Election datasets are collected from twitter. Each of the datasets consists of 500 tweets. Performance of the proposed model is compared with that of Textrank and eigen vector centrality measures for two datasets. Three edge weighting possibilities are taken which are same weight assignment (W_1), weight as co-occurrence frequency (W_f) and weight as inverse co-occurrence frequency $(W_{1/f})$. Mihalcea et al. [16] have used textrank centrality measure as relevance of the keywords.

From the datasets keywords are defined manually because there is no correct set of keywords for a given dataset not even humans may agree on the keywords they extract from a given dataset. Therefore to do the experiments, three persons are invited to suggest an unspecified number of keywords from the datasets. The intersection and union of the sets identified by three people for each dataset are determined. Table 1 contains keywords extracted by three persons for two datasets. Intersection sets are represented by bold face in Table 1.

Precision (Pr), Recall (Re), and F-measure performance measures are used as evaluation metrics for keyword extraction and are given in Eq. (4) to (6).

$$Pr = \frac{|\{Relevant\} \cap \{Retrieved\}|}{|\{Retrieved\}|} \tag{4}$$

$$Re = \frac{|\{Inter\_Relevant\} \cap \{Retrieved\}|}{|\{Inter\_Relevant\}|} \tag{5}$$

Table 1 Keywords extracted by three persons for four datasets

Reader	Dataset	Extracted keywords
1	Uri Attack	**Martyred, uri, attack, terror,** pm, **army,** soldiers, condemns, **surgical, strike,** terrorist, **India, Pakistan,** surgicalstrike, pak, Kashmir, **uriattack**
	American Election	America, campaign, supporters, Russia, riot, land, media **American,** Donald, **presedential,** People, **election, trump, Hilary,**2016
Reader 2	Uri Attack	**uri, attack, army, surgical, strike, uriattack, india, terror, Martyred,** people, **Pakistan,** fawad, ban, artist, jawans,, pm
	American Election	**American,** 2016, **presedential, election, trump, Hilary,** media, Clinton, Obama, shock, refugee, flag, won, result, supporters
Reader 3	Uri Attack	**uri, attack, army, surgical, strike, uriattack, india, terror,** surgicalstrike, **Martyred,** Kejri, Aap, Ban, Modi, terrorist, issue, Rahul, blame, **pakistan**
	American Election	**American, presedential, election, trump, Hilary,** people, muslims, Donald, president, US, democracy, like,lost, canada, women

$$F - \text{measure} = 2 \times \frac{\text{Pr} \times \text{Re}}{(\text{Pr} + \text{Re})} \tag{6}$$

To compute Pr, *Relevant* denotes the number of retrieved keywords which appear in at least one of the human lists/sets. To compute Re, *Inter_Relevant* denotes the number of retrieved keywords which appear in the intersection set of the three human lists.

Tables 2 and 3 show the keywords extracted; and performances by Textrank and eigen vector centrality measures, and the proposed model for Uri Attach and American Election respectively. It is observed from Table 2 that in most of the cases the proposed method performs better than others in Uri Attach dataset however textrank, [16] performs little better than the method in *F*-measure with W_1 and W_f. Table 3 shows that the proposed method performs far better than all other models in all the cases of American Election dataset. The significant improvements are observed in both the datasets by the proposed model because NE centrality works well with disconnected graph as term frequency and degree of node are used as parameters to compute this centrality. It can also be observed that the

Table 2 Performances in Uri Attack dataset

Centrality measure/method	Weight assignment	Pr	Re	F-measure
Eigen centrality	W_1	30	30	30
	W_f	30	30	30
	$W_{1/f}$	60	30	40
Textrank centrality [16]	W_1	90	70	84.7
	W_f	90	70	84.7
	$W_{1/f}$	90	70	78.75
Proposed method (NE centrality)	W_1	90	70	78.75
	W_f	90	70	78.75
	$W_{1/f}$	90	70	78.75

Table 3 Performances in American election dataset

Centrality measure/method	Weight assignment	Pr	Re	F-measure
Eigen centrality	W_1	50	60	54.54
	W_f	60	80	68.57
	$W_{1/f}$	60	60	60
Textrank centrality [16]	W_1	60	80	68.57
	W_f	60	80	68.57
	$W_{1/f}$	60	80	68.57
Proposed method (NE centrality)	W_1	70	100	82.35
	W_f	70	100	82.35
	$W_{1/f}$	70	100	82.35

performances by the proposed method with W_1, W_f and $W_{1/f}$ are equal. This is possibly due to the small size of the datasets. From all the experimental results, it is observed that the proposed method is superior to others. Textrank [16] also shows significant performance but finally it can be concluded that overall performance of the proposed method is far better than the Textrank and eigen vector centrality measures.

5 Conclusions

This paper proposes a graph-based automatic keyword extraction model, called KWG. The proposed model consists of 3 phases: preprocessing, textual graph representation, and keyword extraction. In preprocessing, tokenization of tweets and stop word removal is done. In textual graph representation, the set of vertices are created from the set of tokens and edges are established by pairs of tokens read in the same sequence in which they appear in the original tweets. Finally, NE rank centrality measure is used to calculate the importance of keywords and closeness centrality measure is used to break the ties among the keywords for keyword extraction.

To assess performance of the proposed model, experiments are conducted with two datasets and performances are evaluated with precision, recall and F-measure performance measures. The proposed model is compared with existing graph-based keyword extraction techniques. From the experimental results it is observed that performances of the proposed model outperform the existing graph-based models. The proposed method can be used as a keyword extraction method by taking into account its performance.

Other centrality measures can be used individually or in combination in future to get better results. Other edge weighting and graph building mechanisms can also be used/proposed to improve the results.

References

1. Beliga, S., Mestrovic, A., Martincic-Ipsic, S.: An overview of graph-based keyword extraction methods and approaches. JIOS **39**(1), 1–20 (2015)
2. Savita, D.B., Gore, D.: Sentiment Analysis on Twitter Data Using Support Vector Machine. Int. J. Comput. Sci. Trends Technol. (IJCST) **4**(3), 365–370 (2016)
3. Hemalatha, I., Saradhi Varma, G.P., Govardhan, A.: Sentiment Analysis Tool using Machine Learning Algorithms. Int. J. Emerg. Trends Technol. Comput. Sci. (IJETTCS) **2**(2), 105–109 (2013)
4. Salton, G., Yang, C.S., Yu, C.T.: A theory of term importance in automatic text analysis. J. Am. Soc. Inf. Sci. Technol. **26**(1), 33–44 (1975)
5. Russell, M.A.: Mining the social web: data mining Facebook, Twitter, LinkedIn, Google+, GitHub, and More. O'Reilly Media Inc. (2013)

6. Ediger, D., Jiang, K., Riedy, J., Bader, D.A., Corley, C., Farber, R., Reynolds, W.N.: Massive social network analysis: mining twitter for social good. In: 2010 39th international conference on parallel processing (ICPP), pp. 583–593, IEEE (2010)
7. Hulth, A.: Improved automatic keyword extraction given more linguistic knowledge. In Proceeding EMNLP '03 proceedings of the 2003 conference on Empirical methods in natural language processing, pp. 216–223 (2003)
8. Nguyen, T.D., Kan, M.Y.: Keyphrase Extraction in scientific publications. In: Proceeding ICADL'07 proceedings of the 10th international conference on Asian digital libraries: looking back 10 years and forging new frontiers, pp. 317–326 (2007)
9. Witten, I.H., Paynter, G.W., Frank, E., Gutwin, C., Nevill-Manning, C.G.: KEA: Practical automatic keyphrase extraction. In: Proceedings DL '99 proceedings of the fourth ACM conference on Digital libraries, pp. 254–255 (1999)
10. Zhang, K., Xu, H., Tang, J., Li, J.: Keyword extraction using support vector machine. In: Proceeding WAIM '06 proceedings of the 7th international conference on Advances in Web-Age information management, pp. 85–96 (2006)
11. Medelyan, O., Witten, I.H.: Thesaurus based automatic keyphrase indexing. In: Proceeding JCDL '06 proceedings of the 6th ACM/IEEE-CS joint conference on digital libraries, pp. 296–297 (2006)
12. Abilhoa, W.D., de Castro, L.N.: A keyword extraction method from twitter messages represented as graphs. Appl. Math. Comput. 240, 308–325 (2014)
13. Nagarajan, R., Nair, S.A.H., Aruna, P., Puviarasan, N.: Keyword extraction using graph based approach. Int. J. Adv. Res. Comput. Sci. Softw. Eng. 6(10), 25–29 (2016)
14. Song, H.J., Go, J., Park, S.B., Park, S.Y., Kim, K.Y.: A just-in-time keyword extraction from meeting transcripts using temporal and participant information. J. Intell. Inf. Syst. (Springer) 48(1), 117–140 (2017)
15. Bellaachia, A., Al-Dhelaan, M.: NE-Rank: a novel graph-based key phrase extraction in twitter. In: Proceedings WI-IAT'12 proceedings of the 2012 IEEE/WIC/ACM international joint conferences on web intelligence and intelligent agent technology, vol. 1, pp. 372–379 (2012)
16. Mihalcea, R., Tarau, P.: TextRank: bringing order into texts. In: Lin, D. Wu, D. (eds.) Proceedings of the 2004 conference on empirical methods in natural language processing, Barcelona, pp. 404–411, Association for Computational Linguistics, Spain (2004)

Analysis of Active and Reactive Power and Voltage Control in the Power Structure with the Help of Unified Power Flow Controller

Sahil Rotra, Inderpreet Kaur and Kamal Kant Sharma

Abstract The power system composed of power generation, transmission, distribution, and the utilization. In this paper we are considering Flexible A.C transmitting structure (Facts) devices known Unified power flow controller (UPFC). The UPFC is used for to increase controllability and also capable to expand power transport capacity of the networks. UPFC have the features to control true power, Reactive power, Line impedances, phase angle, power quality and to adjust the transmitting real-time voltage command in the transmission line. This appliance provides significant power quality flow in the power system; this characterizes even increased important and sensitive that UPFC can be put into the transmitting line with their boundaries. The device gives unique features to command the true and reactive power move and the voltage command. This paper concerning 5-bus transmission systems in which we placed UPFC to improve the true and reactive power, Line impedances, phase angle, and voltage control. By performing with and without using UPFC, when with the use UPFC device they help to command the true power and reactive power move and the voltage command by charge the controller at the forward end by simulation tools and when without not using UPFC, they cannot help to improve or command the power flow and voltage command. The power structure model was made on simulation version 2010b. By composition power model we see the results with and without using UPFC. On the conclusion we consider that the UPFC is an ideal device to improve the true or real and reactive power move and voltage command in power structure.

Keywords Flexible A.C transmitting structure (Facts) · Unified power flow\move controller (UPFC) · True (active) power (P) and reactive power(Q) Power structure simulation · Mat lab version 2010b

S. Rotra (✉) · I. Kaur · K. K. Sharma
EED, Chandigarh University, Mohali, India
e-mail: inder_preet74@yahoo.com

S. Rotra
e-mail: samrotra@gmail.com

K. K. Sharma
e-mail: Sharmakamal2002@gmail.com

© Springer Nature Singapore Pte Ltd. 2019
P. K. Mallick et al. (eds.), *Cognitive Informatics and Soft Computing*,
Advances in Intelligent Systems and Computing 768,
https://doi.org/10.1007/978-981-13-0617-4_48

1 Introduction

In the past few years, significant requirement have placed in the transmitting and distribution networks. In the present power structure, almost all the system is interconnected power structure which is very complicated in nature; there is a need to improve the utilization in the power system, while still we have to preserve the structure security and reliable in the power system networks. As power flows in the networks the system will become more complicated, less assured, less reliable, big power flow with no command, excess reactive power flows in the networks, while some transmitting band are impose up with boundary load and some transmitting band may have become overburden, which may have great result on the voltage control, power flows, and also effect on system stability and security in the networks [1]. In the late year 1980s, the Electric power research institute (EPRI), launch a newly technique to improve the power flow, line impedances, phase angle, and the voltage control in the electric transmission system called the FACTS (Flexible A.C Transmission system) technology. The technology helps us to find the solution in the networks. The term FACTS devices are used to command the P and Q power flow and voltage profile and also enhance the power transfer capabilities in the networks. The facts is a family of appliance which can be put in series, shunt and in some times both series and shunt. In common the Facts devices can be split into two parts converters facts controller devices and non-converters facts controller devices [2]. Non converters facts controllers counting, Thyristor controlled series capacitor (TCSC) and Static VAR compensators (SVC) have the advantages to Absorbed or generated the Q power lacking the use of reactor and ac capacitor. Converters based facts controller devices counting UPFC, STATCOM, SSSC, and IPFC which have the advantages too independently to command the P, Q power flow and also increase the transfer capabilities in the transmission lines [3].

Power flow between two buses line loss transmitting line is specified by

$$P = \frac{E_i \cdot V_j}{x_{ij}} \sin(\delta_{ij}),\tag{1}$$

where, E_i and δ_i are the ith bus voltage significance and phase angle, and where V_j and δ_j are the jth bus voltage significance and phase angle, X is the line Reactance's. From first calculation, the power flow in the transmitting band is a consequence of Line impedances, sending and the collect voltages and the phase angles connecting the voltages. The command of P and Q power, voltages, phase angles in transmitting lines is possible by the use of one or a compound of the power move positioning. The bus voltages, phase angle, and the line impedances in the power structure can be adjust and flexible with FACTS technology such as static synchronous series compensator (SSSC), static VAR compensator (SVC) and the Unified power flow controller (UPFC), etc. [1].

The most flexible and the secure appliance in the facts devices is the UPFC. The UPFC is a amalgamation of three characterizes, i.e., line impedances, phase angle, and the voltages magnitude. The UPFC appliance is an amalgamation of two FACTS devices STATCOM (static synchronous compensator) and the SSSC (static series synchronous compensator). SSSC is basically is used to control the phase angle and the real-time voltage control in series with line, spell STATCOM is shunt converter used to provide Q power. In this paper we used the FACTS devices called UPFC to command the P and Q power flow and the true power control in transmitting line, the UPFC is 5-bus transmission systems. The power system model was made on simulation version 2010b. By composition power model, we see the results with and without using UPFC [4].

2 Power Flow and Voltage Profile in A.C Transmission Structure

In the power system, there is a power flows in the transmission system called as active power also known as real power, true power (P) and reactive power (Q) and the apparent power (A) in the system. (P) Power is that power which we actual consumes in the transmission system measured in the Kilowatt (K·W). Q Power is that power that magnetic equipment (transformers, motors, and relay) needs to produce the flux (Q) power is measured in KVAR. And the third power is the (A) power is the sum of (P) and (Q) power, i.e., the summation of K·W + KVAR = KVA. Now the term introduced in the power system known as power factor. Power factor is explained as the angle between V and I are known as power factor [4].

$$\text{power factor} = \frac{K \cdot W}{KVA}$$

$$\text{Therefore, } Cos\theta = \frac{K \cdot W}{KVA}$$

$$\text{and, } Sin\theta = \frac{KVAR}{KVA}$$

From above equation, it is clear that the power factor is equal to the P power divided by the amount of P power and Q power and that power is known as the apparent power in the power structure. The more Q power (KVAR) in the structure, the lower in the ratio of real power (K·W) therefore the power factor becomes lower. The less the reactive power (KVAR), the highest the ratio of real power (K·W). In fact if the Q power near zero, the power factor near 1.0. In order to have well-organized system the power factor should have close to 1.0. Almost all of the loads in power structure are the inductive loads (Transformers, induction motors, generators, high-intensity discharge lightning), which are the cause of Reactive power. Q power is produced by inductive loads which increase the amount of apparent power (KVA) in the giving out system. The greater in apparent power and the Q power result in larger angle (δ) [5] (Fig. 1).

Fig. 1 Power triangle in AC
transmission system [5]

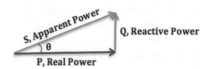

In an A.C power system, the generation and the load must be balance at all the time. If the electrical generation is less than the load there is frequency drop in the transmitting band and thereby the load goes down equal to the generation minus the transmission losses. The active power flows from the surplus areas to deficit areas and the power flow through all parallel paths. One of the most important problem in power system is the (Q) power command and the voltage profile command in the networks, throughout the everyday process the power system may experienced both over voltage and under voltage violation that can be overcome by the command V/VAR command [3]. The main problem in the power system is that when the reactive power flows from the inductive reactance of the power lines there is a voltage drop in the line [1]. As many FACTS devices in the power system which we used in series or in shunt or in both series and shunt operations to improve the control of reactive and active powers FACTS has a lot to do with reactive power compensation. Following are the different FACTS devices to upgrade the voltage strength and control in the power structure are as D-STATCOM, STATCOM, SERIES CAPACITORS, SHUNT CAPACITORS, SVC (Synchronous VAR Compensator), SHUNT REACTORS, SYNCHRONOUS CONDENSER.

3 Future Extent of the Work

The goal of this paper work is with the Utilize of FACT devices called UPFC which installed in the transmission lines to improve the followings which are as under [3].

- P and Q power flow control in transmitting system.
- Voltage control and stability in the transmitting lines.
- Phase angle, line impedances control.
- Sending end bus voltage control in the transmission structure.
- D.C bus voltage magnitude.

4 Unified Power Flow Controller

The UPFC is a FACT devices was given by Gyugui in 1991. The UPFC was used for the actual hour command and very efficient recompense technique in the A.C power transmission system, providing multi-functionality reliability need to solve the difficulty covering the power delivering industries. Basically UPFC are able to command all the variable in the transmitting lines they help to control the voltage stability also improve the P and Q power, line impedances, and phase angle in the transmission lines that is why they named as "Unified". Alternatively UPFC can separately command the P and Q power flow in the transmitting lines [6].

4.1 Basic Principle Operation of UPFC

An easy two appliance power arrangement is shown in Fig. 2 in which V_s is the transmitting end voltage and V_r is the accepting end voltage and X is the line impedances. The UPFC is a generalized synchronous voltage source (SVC) constitutes power structure frequency by voltage phasor V_{pq} with controllable magnitude V_{pq} and phase.

Angle (δ) is in series with transmission lines. The figure below acutely includes angle regulation and voltage, the synchronous voltage source exchanges both P and Q power in the transmitting lines, the synchronous voltage source are able to create only Q power swap the P power must be give to it or take in it [5, 6] (Fig. 3).

The above figure, the UPFC is composed of two switching voltage source converters. These rear to rear converters, i.e., converter 1 and converter 2 run by usual dc link providing by dc storage capacitor, this positioning purpose is an perfect ac to ac power converters in which the P power are free to move or flow in the orientation between the station of two converters and each one converter can

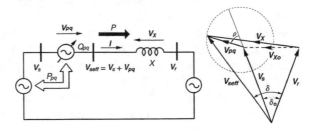

Fig. 2 Description of UPFC in a two appliance power structure [5]

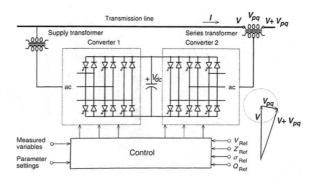

Fig. 3 Execution of UPFC by two back to back voltage source converters [5]

absorb or make the (*Q*) power at its own production End. Converter 2 supply the *P* power and the (*Q*) power command flow of UPFC by introduce the voltage V_{pq} with magnitude V_{pq} and phase angle which is in series with line via a transformer. The main purpose of converter 1 is to ingest the *P* power by converter 2 at the usual dc link. Converter 1 can also generates or absorbs control *Q* power and provides parallel *Q* power compensation technique for the transmission line. There is no *Q* power move between the dc links [4, 5].

5 Simulation with and Without Using UPFC Device

A UPFC is applying to *P* and *Q* power for 500 kV/230 kV transmitting line. The system consists of 5-buses namely (b1, b2, b3, b4, b5) attached by the three transmitting lines namely (l1, l2, l3) and two power plant 500 kV/230 kV transformer named as TR1 and TR2. The total capacity produce by the plant is 1500 mv which is arrange to 500 kV, 1500 mva and 500 MW load add to b3. Each power plant should focus on their speed regulation and power system stabilizer (PSS). The power plant 2 given to the 500 kV equivalent in the advancement of 2400 mva transformers add to buses b4 and b5. The UPFC parameters are given in the below model, UPFC is placed at right to the line L2 is utilized the control of *P* and *Q* power flow in 500 kV bus b3, and the power bus is b-UPFC. The UPFC apparatus composed of series converter and shunt converter IGBT type, that the sequence converter is rated as 100 mva with highest voltage prevention of 0.1 pu and the parallel converter is evaluate as 10 mva, verify the parallel converter is in voltage regulation manner and sequence converter is in *P* and *Q* powers command. The UPFC references the *P* power and the *Q* power are set in bar labeled P(pu) and Q(pu). The blink at breaker is closed and accrue powers at bus b3 is 587 mw and −27 mvar. At time = 10 s., P(pu) is enlarge by one pu (100 MW), from 5.87 pu to 6.87 pu, Qref (pu) is retain consistent on −0.27 pu. The model is made on MATLAB 2010b SIMULINK in sim power all the parameters are present there. The output of the UPFC Model gives the system stability and also enhances to increase the power transfer ability [1, 5] (Figs. 4, 5, 6, 7, 8, 9 and 10).

Fig. 4 Voltage waves in b1 b2 b3 b4 b5 with UPFC

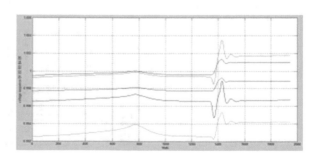

Fig. 5 Active power flow waveforms in b1 b2 b3 b4 b5 with UPFC

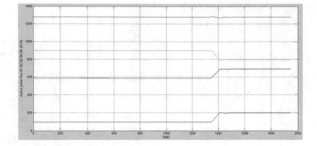

Fig. 6 Q (reactive) power flow waveforms in different buses with UPFC

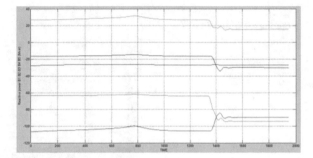

Fig. 7 UPFC controllable region

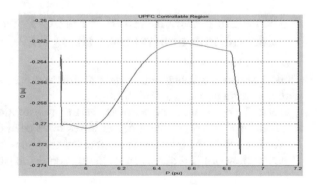

Fig. 8 Voltage wave forms without UPFC

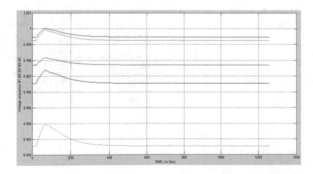

Fig. 9 Active power flow
waveforms without UPFC

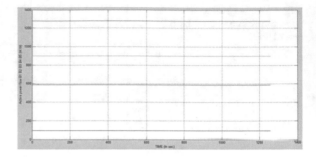

Fig. 10 Reactive power flow
wave forms without UPFC

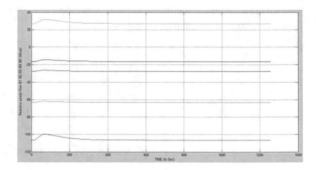

6 Simulation Conclusion of P and Q Power and Voltage Control with and Without Using UPFC

Table 1 shows the voltage magnitude at different buses of the power grid system with and without use of UPFC [2].

Table 2 shows the active power in the transmitting lines, performances of simulink of P powers of the different buses have been demonstrated as under [2].

Table 3 shows the Q power in the transmitting lines, performance of simulink of Q power flow of the different buses have been demonstrated as under [2].

Table 1 Voltage of
magnitude

BUS	With UPFC	Without UPFC
Bus1	0.9967	0.9965
Bus2	1.002	0.9993
Bus3	1.001	0.9995
Bus4	0.9942	0.9925
Bus5	0.9989	0.9977

Table 2 Active (P) powers flow

BUS	With UPFC	Without UPFC
Bus1	196.6	95.16
Bus2	689.7	5888.8
Bus3	687	587
Bus4	796	898.7
Bus5	1277	1279

Table 3 Reactive (Q) power flow

BUS	With UPFC	Without UPFC
Bus1	−30.06	−16.34
Bus2	−94.06	−63.27
Bus3	−27	−27.79
Bus4	15.57	26.89
Bus5	−89.32	−106.4

7   Conclusion

In this paper by analyzing the facilities of Unified power move controller in the transmitting lines, it is main in the transmission lines to control of the voltage stability, line impedances, true (P) power, Q power, line parameters and the phase angle. In the transmitting structure it is evitable to maintain the reactive power move for better voltage regulation and better performance in the system, the UPFC have the capability to command the power move, i.e., P and Q power and also enhance to grow the power shift capabilities in the transmission lines. This research regarding with 5-bus transmitting system apply the UPFC appliance to command the power and to enhance the structure to be stable. When there is no UPFC is installed in the transmitting line the power flow and the voltage profile cannot be improved and when placed UPFC devices in the transmitting line the P and Q powers, voltage balanced and the system stability will be improved the results shown above with and without the use of UPFC device in MATLAB simulation. From simulation results we consider that the UPFC is an ideal device which helps to grow the power shift capabilities and also help for controlling of P power and Q power move in the system.

References

1. Kuralkar, V.: Active and reactive power control using UPFC. Int. J. Eng. Res. Gen. Sci. **2**(6), ISSN 2091–2730, Oct–Nov (2014)
2. Komoni, V., Krasniqi, I., Kabashi, G., Alidemaj, A.: Control active and reactive power flow with UPFC connected in transmission line. In: 8th Mediterranean conference on power generation, transmission, distribution and energy conversion. Medpower (2012)

3. Akwukwaegbu, I.O., Okwe Gerald, Ibe.: Concepts of reactive power control and voltage stability methods in power system network. IOSR J. Comput. Eng. (IOSR-JCE) 11(2), 15–25 e-ISSN: 2278-0661, P-ISSN: 2278-8727 (May–June 2013)
4. Inderpreet, K., Kamal, S., et al.: Power system statbiliy for the islanding operation of microgrids. Indian J. Sci. Technol. 9(38), 1–5 (2016). https://doi.org/10.17485/ijst/2016/v9i38/101474
5. Hingorani, N.G., Gyugi, L.: Understanding FACTS: concepts and technology of flexible AC transmission system. IEEE Press, New York, NY (2000)
6. Yadav, M., Soni, A.: Improvement of power flow and voltage stability using unified power flow controller (UPFC)'. In: International conference on electrical, electronics, and optimization techniques (ICEEOT) (2016)

Clustering and Predicting Driving Violations Using Web-Enabled Big Data Techniques

Lakshmi Prayaga, Krishna Devulapalli, Srinath Devulapalli
and Keerthi Devulapalli

Abstract When quintillion bytes of data on a multitude of topics is being generated each day creating Big data in size and in scope, the need for analyzing such voluminous data, extract meaning from it and providing a visualization is also increasing. Driving violations is one of the topics that have been recorded over multiple years. Several studies have been conducted to predict driver behavior using simulations and other tools such as built-in sensors in the vehicles. This research activity focuses on the design of an interactive Big data web application to analyze a given dataset using techniques such as cluster analysis and predict driving violations based on available demographics. The rest of the paper describes the suite of technologies for Big data analytics that facilitated this development and the implications of this study.

Keywords Big data · Interactive data analytics · Cluster analysis
Clustering counties · Types of driving violations

L. Prayaga (✉)
Department of Information Technology, University of West Florida,
Pensacola, USA
e-mail: Lprayaga@uwf.edu

K. Devulapalli
Director-Grade Scientist and Head of University Computer Center,
IICT, Hyderabad, India
e-mail: krishnad2@gmail.com

S. Devulapalli
Dell, Austin, TX, USA
e-mail: sriniris@gmail.com

K. Devulapalli
Osthus, Melbourne, FL, USA
e-mail: keerthi0589@gmail.com

© Springer Nature Singapore Pte Ltd. 2019
P. K. Mallick et al. (eds.), *Cognitive Informatics and Soft Computing*,
Advances in Intelligent Systems and Computing 768,
https://doi.org/10.1007/978-981-13-0617-4_49

1 Introduction

Analyzing and predicting driver behavior and traffic violations have been an area of interest to insurance companies, city administrators, psychological cancellers and others for many years. These studies provide important information and affect business decisions in calculating actuaries, insurance amounts, prescribing medications, etc. Several tools including in vehicle electronic devices and data recorders, [1, 2] and simulations have also been used to predict and modify driver behavior. However, simulators have some disadvantages including that they are not able to account fully for the real-world scenarios and may not accurately reflect parameters such as human error, external distractions and physical characteristics [3–5]. This study aims to achieve two goals: a. Design a web interface to provide data analytics and visualizations on a publicly available data set that describes the demographics and traffic violations which were not a primary focus of other studies, and b. Provide the schematics for using EC2 from AWS that integrates a set of technologies to design a web interface using techniques for Big data analytics for the dataset used in this study. We also note that IBM, Microsoft and several other major companies have web-based interfaces for data analytics. However, they are expensive for individuals and small businesses to use in their research projects, or make educated decisions. Our approach in this project used open source tools which are available for individuals and small businesses free of cost.

2 Dataset Used

For this study, the dataset on driver demographics and traffic violations available publicly at [6] was used. The dataset resource contains data for all 50 states in the United States, however we chose to use data for the state of Florida for this study. This dataset contains 5,421,446 million (uncleaned), 3,689,126 (cleaned) records and twenty attributes. Our study focused on driver demographics and the violations committed by the drivers.

Our research questions were to observe

a. Specific types of driving violations attributed to race and or gender.
b. If we could cluster counties exhibiting similar types of driving violations.

This dataset is also drilled down to the county level in the state to give a more focused view on the types of violations that occurred most often in each county. Such an analysis might offer insight to city officials and insurance companies to target their message to drivers, encourage them to utilize safe driving practices to avoid common mistakes. Additionally, driver behavior analysis is also helpful in designing appropriate training materials that might help drivers avoid accidents, better engineering of automobiles and design of automated driver assistance packages [7].

3 Workflow of Technologies Used

The web-based Analytics System of driving violations was launched on a t2. Micro instance of the Amazon Elastic Compute Cloud (EC2) under Amazon Linux (CentOS) Operating System. Initially R Studio Server was installed by making use of Louis Aslett's [8] R Studio Server Amazon Machine Image (AMI). This AMI automatically creates the R Studio Server instance. The Amazon instance was then accessed locally through PUTTY by using proper authentication information. The latest version of Apache Spark (version 2.2.0) was downloaded and installed on the Amazon website. R Studio Server was configured to use Spark and SparkR in its application. Utilizing R Studio Server, a Shiny App was developed to perform various operations relating to the web application like data collection, data cleaning, data exploration, data summary and aggregation, and data visualization. The Shiny App was also used to perform cluster analysis and display the results on the same web page. The Shiny App can display all the results through various Menu options in the same web page. Data visualization like bar charts, dendrograms, etc., were also included in the Menu options. Figure 1 describes the workflow of the web application using Amazon EC2.

4 Implementation and Output of the Analysis

This section provides details on the implementation and the output of the techniques used to perform the data analytics. Details on the web interface, a description of each of the options and the results are presented in Table 1.

Figure 2 displays the details of the data utilized in the study and Fig. 3 gives the display of some sample utilized for the analysis.

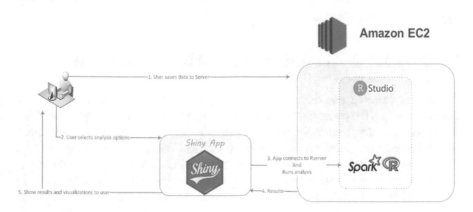

Fig. 1 Workflow of the web application

Table 1 Description of the web application and the menu options

Menu item	Description
Data description	Gives the details of the data such as the number of records, and records utilized for final analysis after data cleaning
Explore data	Gives preview of the data used. Sample data will be displayed
Cross tabs	
(i) Race versus gender (ii) Race versus violation	Violations—gender-wise for each race Cross tabulation Violations—violation-wise counts for each race cross tabulation
Cluster analysis	Cluster analysis model results
Plots	
(i) Race versus gender (ii) Dendrogram	Bar plot of Gender-wise violations for each race Cluster analysis dendrogram plot

Fig. 2 Data description

5 Cross Tabulation Results—Race-Wise and Gender-Wise

To answer research question 1, cross tabulations were evaluated race-wise and gender-wise and the percentage distribution of the violations for each gender within every race are presented in the following Fig. 4. One can observe from this table, that the percentage violations in males are more than Females in each race. For e.g., the violation percentages are 36.5% in females and 63.5% in males for the Asian Community. Similar results were observed in other races demonstrating that the traffic violations in males are much higher compared to females.

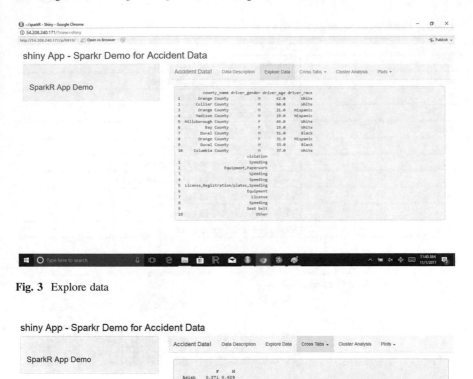

Fig. 3 Explore data

shiny App - Sparkr Demo for Accident Data

Fig. 4 Cross tabulation Gender versus violation

6 Cross Tabulation Results—Race-Wise and Violation-Wise

The original 187 traffic violations are reduced to 11 violations by combining all the related violations into 11 subgroups. These 11 subgroups are abbreviated as DUI, Equ, Lic, Lig, Oth, Pap, Reg, Saf, Sea, Spe and Tru. The details of these 11 violations are given in Table 2.

Cross Tabulations have been worked out Race-wise and Violation-wise and the results are presented in Fig. 5. Each row in this table gives the percentage violation figures of the race for various types of violations. Thus for the Black Race, the violation percentages for DUI, EQU, Lic etc. are 0.00145, 0.14754, 0.11331 and so on. Similarly one can view for other races.

The percentage distribution of violations for each race (row-wise figures) is given in Fig. 5. In each race, it is observed that the violations due to Spe (Speeding) are the highest. These percentage violations are 2.318, 35.776, 34.960, 47.910, and 40.964 among the races Asian, Black, Hispanic, Other and White. Violations due to

Table 2 Grouping of types of violation

Violation	Description
DUI	Violations relating to DUI
Equ	Violations relating to equipment
Lic	Violations relating to license
Lig	Violations relating to light
Oth	Violations relating to others
Pap	Violations relating to paper
Reg	Violations relating to registration
Saf	Violations relating to safety movement
Sea	Violations relating to seat belt
Spe	Violations relating to speed
Tru	Violations relating to truck

shiny App - Sparkr Demo for Accident Data

SparkR App Demo

Accident Data!   Data Description   Explore Data   Cross Tabs ▾   Cluster Analysis   Plots ▾

	DUI	Equ	Lic	Lig	Oth	Pap	Reg	Saf	Sea	Spe	Tru
Asian	0.00000	0.07285	0.08609	0.01325	0.02649	0.07285	0.05960	0.10596	0.01325	0.52318	0.02649
Black	0.00145	0.14754	0.11331	0.00386	0.02700	0.09836	0.04966	0.09016	0.05882	0.35776	0.05207
Hispanic	0.00552	0.12389	0.09122	0.00339	0.03564	0.07934	0.04328	0.12770	0.04703	0.34560	0.07340
Other	0.00322	0.09968	0.06431	0.00322	0.03537	0.08682	0.04180	0.10611	0.04502	0.47910	0.03537
White	0.00620	0.13283	0.06443	0.00477	0.03357	0.09322	0.05679	0.09609	0.05791	0.40564	0.04454

Fig. 5 Cross tabulations of race versus violation

Fig. 6 Percentage of accidents by race and gender

Equ(equipment), Pap(paper) and Saf(Safety) occupy next position. Figure 6 is bar chart displaying results for the percentage of accidents distribution by Race and Gender.

7 Cluster Analysis

A cluster analysis was performed to answer research question 2, if we could cluster counties exhibiting similar types of driving violations. The results are displayed in Figs. 7 and 8. Figure 7 displays the results for a two cluster analysis and Fig. 8 displays the results in a dendrogram.

A dendrogram is a tree type of diagram, showing relationships between similar sets of data. This dendrogram shows how the Counties having similar violations are clustered. Counties close to the same height are similar to each other and counties with different heights are dissimilar—the greater the difference in height, the more is the dissimilarity. At height 1.0, we observe, there are three clusters, whereas at height 1.5, there are two clusters. When we consider a two cluster solutions, the counties Miame-Dade County, Palm Beach County, Broward county and Orange County form into one cluster, while all the other counties form the second cluster. A close inspection of these four counties of first cluster violations with the violations of other counties violations from cross table results of Table 2 suggest that the violations in these four counties are much higher than the other counties. The population densities of these four counties from literature indicate that the population densities of these four counties are much higher than the other counties thus contributing to one of the reasons for higher violations of these four counties.

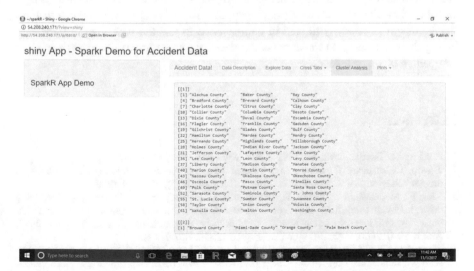

Fig. 7 Cluster analysis results for 2 clusters

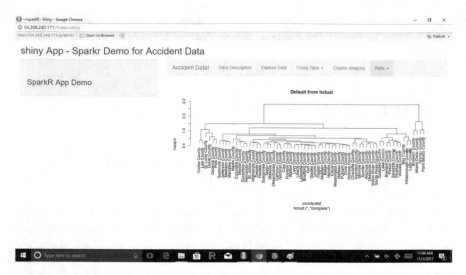

Fig. 8 Dendogram of the cluster analysis

8 Limitations and Future Development

While we believe that this study is limited to one state, we plan to conduct the study over all fifty states and provide focused results for individual states and at the same time also observe any commonalities. We also note that the techniques described here provide an interesting method to design a web interface with open-source technologies for Big data applications. However, it is limited in scope to users who are trained statisticians or data scientists. Our future work will focus on designing a web interface that will allow citizen data scientists who are professionals in their fields but not necessarily trained in data science to use the web interface for analyzing their datasets and make educated decisions to improve the growth of their companies or other areas.

References

1. Road Safety International.: Teen black box, www.roadsafety.com. (2008)
2. Toledo, T., Musicant, O., Lotan, T.: In-vehicle data recorders for monitoring and feedback on drivers behavior. Trans. Res. Part C: Emerg. Technol. **16**(3), 320–331 (2008)
3. Sato, T., Akamatsu, M., Shibata, T., Matsumoto, S., Hatakeyama, N., Hayama, K.: Predicting driver behavior using field experiment data and driving simulator experiment data: assessing impact of elimination of stop regulation at railway crossings. Int. J. Veh. Technol. **2013**, Article ID 912860, 9 (2013). https://doi.org/10.1155/2013/912860
4. Wang, W., Xi, J., Chen, H.: Modeling and recognizing driver behavior based on driving data: a survey. Math. Prob. Eng. **2014**, Article ID 245641, 20 (2014). https://doi.org/10.1155/2014/245641, retrieved Oct 28th, 2017, from: https://openpolicing.stanford.edu/data/

5. U.S Department of Transportation.: https://vtechworks.lib.vt.edu/bitstream/handle/10919/55090/DriverInattention.pdf?sequence=1
6. Pierson, E., Simoiu, C., Overgoor, J., Corbett-Davies, S., Ramachandran, V., Phillips, C., Goel, S.: A large-scale analysis of racial disparities in police stops across the United States (2017)
7. Bifulco, G.N., Galante, F., Pariota, L., Russo Spena, M., Del Gais, P.: Data collection for traffic and drivers' behaviour studies: a large-scale survey. In: Procedia—social and behavioral sciences. vol. 111, 2014, pp. 721–730, ISSN 1877-0428, https://doi.org/10.1016/j.sbspro.2014.01.106, http://www.sciencedirect.com/science/article/pii/S1877042814001074
8. Aslette, L.: Rstudio server amazon machine image(AMI), retrieved on 27th Oct 2017 from: http://www.louisaslett.com/RStudio_AMI/

S-RLNC Multi-generation Mixing Assisted MAC Optimization for Multimedia Multicast over LTE-A

Veeresh Patil, Sanjeev Gupta and C. Keshavamurthy

Abstract The high pace emergence in communication systems and associated demands has triggered academia-industries to achieve more efficient solution for Quality of Service (QoS) delivery for which recently introduced Long-Term Evolution (LTE) or LTE-Advanced (LTE-A) has been found as a promising solution. However, enabling QoS and Quality of Experience (QoE) delivery for multimedia data over LTE has always been a challenging task. QoS demands require reliable data transmission with minimum signaling overheads, computational complexity, minimum latency, for which classical Hybrid Automatic Repeat Request (HARQ) based LTE-MAC is not sufficient. To alleviate these issues, in this paper a novel and robust Multiple Generation Mixing (MGM) assisted Systematic Random Linear Network Coding (S-RLNC) model is developed to be used at the top of LTE-MAC protocol stack for multimedia data transmission over LTE-A system. Our proposed model incorporated interleaving and coding approach along with MGM to ensure secure, resource efficient and reliable multiple data delivery over LTE systems. One of the key novelties of our proposed system is the use of MGM concept that significantly reduces the signaling overheads, redundant packet needed for each generation that consequently optimizes bandwidth utilization. This feature augments our proposed system to ensure reliable and resource-efficient data transmission over LTE systems. The simulation results reveal that our proposed S-RLNC-MGM based MAC can ensure QoS/QoE delivery over LTE-A systems for multimedia data communication.

V. Patil · S. Gupta (✉)
Department of Electronics and Communication Engineering,
AISECT University, Bhopal, India
e-mail: sanjeevgupta73@yahoo.com

V. Patil
e-mail: veeresh91@yahoo.co.in

C. Keshavamurthy
Department of Electronics and Communication Engineering,
ACS College of Engineering, Bangalore, India
e-mail: keshava_uvce@yahoo.com

© Springer Nature Singapore Pte Ltd. 2019
P. K. Mallick et al. (eds.), *Cognitive Informatics and Soft Computing*,
Advances in Intelligent Systems and Computing 768,
https://doi.org/10.1007/978-981-13-0617-4_50

Keywords LTE · Multimedia broadcast · Multicast transmission
Random linear network coding · MAC optimization · Multi-generation mixing

1 Introduction

In last few years, wireless communication has gained an immense space across
human presence. Communication system being an inevitable need of modern day
human society encompassing civil, defense, scientific or industrial purposes has
motivated academia-industries to achieve better wireless communication solution.
The emerging demand of mobile communication has also raised the need of QoS/
QoE provision. The typical LTE-A systems are designed to provide high data rate
communication. However, the exponential and consistent rise in QoS/QoE demands
has triggered industries to further enhance LTE-A protocol. To meet QoS demands,
3rd Generation Project (3GPP) standards incorporated different technologies to
enable Multimedia Broadcast and Multicast Services (MBMS) for LTE-A systems
[1]. To achieve QoS delivery optimizing MAC layer of the protocol stack can be of
paramount significance [2]. Now, considering the transmission nature of LTE-A
systems where the HARQ are used as FEC to perform successful data transmission,
it is obvious that HARQ introduces significant signaling overhead in acknowl-
edging transmitter about successful packet delivery. To alleviate these issues, NC
technique can be applied [3]. NC scheme can be a better alternative to alleviate
limitations of HARQ that as a result can augment classical Forward Error
Correction (FEC) process [4] for delivering enhanced MBMS transmission [5]. It
can be significant for multimedia data transmission which requires timely data
delivery with minimum latency and resource consumption. Recently, a variant of
NC named Random Network Coding (RNC) has been applied in the latest generation of smart-phones at the application layer to ensure reliable multimedia delivery
over wireless links [6]. RNC functions on the basis of random linear coding [7]
concept that can be used as a rate-less coding scheme for unicast/multicast transmission [8] and as a NC scheme to enhance throughput in cooperative communication environment [9]. A more sophisticated model called Unequal Error
Protection (UEP) model [10] was developed by amalgamating sparse RNC [11] so
as to lower computational complexity, error-resilience and content-aware data
transmission. With the efficacy of RNC scheme to alleviate the gap of upper-layer
media compression/packetization and the lower-layer wireless packet transmission,
in this paper a robust S-RLNC is developed to be applied at the top of existing
HARQ [12] to assist reliable and QoS oriented multimedia data delivery over
LTE-A. In this paper, our proposed S-RLNC model based MAC protocol intends to
facilitate a sophisticate and efficient RAN-wide NC MAC sub-layer that could
enhance the packet delivery across a multi-hop RAN topologies, particularly
LTE-A Heterogeneous Networks (HetNets) [13]. Considering bandwidth and signaling overhead efficacy it our proposed model applies MGM concept that transmits
single redundant packet for the samples generated after multiple generation of

S-RLNC. This as a result avoids need of per generation-redundant packet and hence optimizes both signaling overhead as well as bandwidth utilization. S-RLNC model is applied in between the transport and network layer, functional at the top of MAC standard. In addition, our method intends to assist multimedia content delivery awareness within the lower-layer LTE RAN protocols. S-RLNC model has been examined in terms of throughput, delay, packet loss, number of redundant packets for successful multimedia packet delivery.

2 Our Contribution

In this research paper a robust systematic RLNC-based MAC optimization model is developed. The overall proposed multimedia multicast model intends to enhance Packet Delivery Ratio (PDR), reliable transmission, minimum latency and resource efficient transmission while ensuring minimum computational overheads or signaling overhead. Undeniably, RLNC method is found to assist reliable data transmission over wireless network. Unlike traditional NC schemes, our proposed system incorporates systematic RLNC approach with pre-coding and interleaving mechanism that assure reliable delivery while ensuring minimum data drop and retransmission probability. In addition, our method incorporates MGM technique that facilitates both high data rate transmission and reliable communication over network because of its ability to reduce high redundant packet requirement. Noticeably, in proposed MGM model single redundant packet is transmitted over the mix of the samples (data) from multiple generations. Additionally, the Iterative Buffer Flush (IBF) concept plays vital role in reducing unwanted bandwidth occupancy.

A brief of the proposed system is discussed in sub sections.

2.1 Pre-coding and Interleaving Based S-RLNC for MBMS

The efficacy of S-RLNC algorithm for multicast transmission was assessed in our previous work [14]. Our proposed S-RLNC method employs RLNC for groups of packets called generations that undeniably augments the usability of NC scheme for real-time applications. It strengthens our proposed method to deal with real-time packet erasures, delays and topology changes conditions. Considering multimedia data transmission where jitter and latency adversely affects the QoS and QoE, our proposed method is developed in such manner that it supports minimum but significant network-coded redundant packets as FEC to ensure reliability and delay-resilience. These redundant NC packets perform FEC better than classical ARQ based schemes to combat packet erasures [15]. Considering the issues in our previous work [14], in this paper a robust Pre-Coding and Interleaving based S-RLNC model is developed that augments its efficiency to alleviate packet losses.

2.1.1 S-RLNC

A schematic of the proposed PCI-based S-RLNC transmission model is given in Fig. 1. As depicted, our proposed model encompasses three consecutive phases, packet encoding at the source node, process at the intermediate node and packet decoding at the sink node. These processes are given as follows:

Packet Encoding at the Source Node

Our proposed pre-coding method combined source packets linearly so as to generate a set of linear packet combinations. Consider $[X_M]_{n \times s}(M = 1, 2, \ldots m)$, where $m \geq n$, be $n \times m$ source packets where every packet is a $1 \times s$ matrix containing symbols from a GF of size 2^F, where F states the order of GF.

Pre-coding give rise to a $m \times m$ matrix containing the linear combinations, $[Y_M]_{m \times s} (M = 1, 2, \ldots, m)$, by performing (1).

$$[Y_M]_{m \times s} = w_{m \times n} \cdot [X_M]_{n \times s} \tag{1}$$

In (1), states a $m \times n$ fraction of the Symbol Combination Matrix (SCM). In proposed system, to enhance the robustness against packet loss, the pre-coded packet combinations are processed for interleaving with each other that eventually generates $\left[\acute{Y}_\delta\right]_{m \times p} (\delta = 1, 2, \ldots, m)$ in such manner that,

$$\left[\acute{Y}_\delta\right]_{m \times s} = \gamma\left([Y_M]_{m \times s}\right) \tag{2}$$

where γ is obtained through (3).

$$\acute{Y}_\delta(M, w) = Y_M(\delta, w) \tag{3}$$

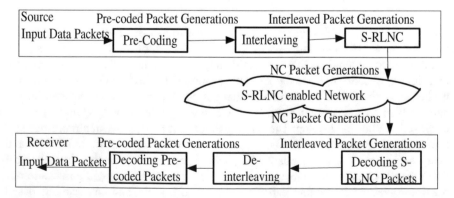

Fig. 1 Block diagram of pre-coding and interleaving assisted S-RLNC for MBMS

In (3), $w (1 \leq w \leq s)$ signifies the position of a symbol in a packet.

S-RLNC has been employed to $\left[\acute{Y}_\delta\right]_{m \times s} (J = 1, 2, \ldots, m)$ that helps in generating or retrieving $[Z_\delta]_{1 \times s} (\delta = 1, 2, \ldots, m)$, where $1 \geq m$. Here, the generations are obtained by means of $[Z_\delta]_{l \times s} (\delta = 1, 2, \ldots, m)$, where individual component of C_δ refers a symbol belonging to 2^F. Thus, the output generated would be the linear packet combinations given by $[Z_\delta]_{1 \times s} (\delta = 1, 2, \ldots, m)$. Mathematically,

$$[Z_\delta]_{l \times s} = [C_\delta]_{l \times m} \times [\acute{Y}_\delta]_{m \times s} \tag{4}$$

As depicted in (4), $m \times 1$ number of generated packet combinations, given as $(Z_1, \ldots, Z_m)$ are retrieved and identified to be in the same interleaved group. Noticeably, the individual set of 1 linear packet combinations (i.e., $[C_\delta]_{1 \times m}[Z_\delta]_{1 \times s}$) originates from a single generation.

Intermediate Nodes

In our proposed method, at intermediate node S-RLNC performs linear combination of the packets from same generation. Noticeably, all data packets belonging to the same generation have the same sequence number or rank and interleaved group. Therefore, at this node the S-RLNC generation, sequence number and interleaved group are appended to the output packet combinations.

Packet Decoding at Sink Nodes

Consider the packets retrieved by the sink node is in the form of $\left[\left[\widehat{C}_\delta\right]_{\eta\delta \times m} \left[\widehat{Z}_\delta\right]_{\eta\delta \times s}\right]$, where η_δ signifies the number of linear packet combinations pertaining to δth S-RLNC generation, obtained at the sink. Let, $[C_J]_{\eta\delta \times m} (J = 1, 2, \ldots, m)$ be the coefficient matrix and $\left[\widehat{Z}_J\right]_{\eta\delta \times m} (J = 1, 2, \ldots, m)$ be the linear combination matrix obtained at the sink node. With the retrieved variables, the sink node selects m linearly independent combinations from each generation (5).

$$\left[\left[\widehat{C}_\delta\right]_{m \times m} \left[\widehat{Z}_\delta\right]_{m \times s}\right] = k\left(\left[\left[\widehat{C}_\delta\right]_{\eta\delta \times m} \left[\widehat{Z}_\delta\right]_{\eta\delta \times s}\right]\right) \tag{5}$$

In (5), the parameter κ selects m linearly independent combinations from δth the generation, provided conditions; $\eta_\delta \geq m$ and the sink node have received m combinations out of η_δ linear packet combinations. Now, S-RLNC exploits the selected packet combinations and coefficients retrieved from δth generation for generating an interleaved generation such that,

$$\left[\widehat{Y}_\delta\right]_{m\times s} = \left[\widehat{C}_\delta\right]_{m\times m}^{-1} \cdot \left[\widehat{Z}_\delta\right]_{m\times s} \tag{6}$$

In our model, the m interleaved generations pertaining to an interleaved group are ordered by means of a rank value, which is then followed by de-interleaving so as to generate pre-coded generations. Mathematically, (7)

$$\left[\widehat{Y}_M\right]_{m\times s} = \overline{Y}\left[\widehat{Y}_\delta\right]_{m\times s} \tag{7}$$

Here, $\overline{Y}$ is processed in such manner that,

$$\widehat{Y}_M(\delta, w) = \widehat{Y}_\delta(M, w) \tag{8}$$

In our proposed method, the n combinations from each m sets of de-interleaved combinations pertaining to a pre-coded generation are taken into consideration along with their allied coefficients. Let, the selected coefficients and combinations be $[\omega_M]_{n\times n}\left[\widehat{Y}_M\right]_{m\times s}$, then the original packets are decoded (9).

$$\left[\widehat{X}_M\right]_{n\times s} = [\omega_M]_{n\times n}^{-1}\left[\widehat{Y}_M\right]_{n\times s} \tag{9}$$

2.1.2 MGM Based S-RLNC for Multimedia Data Transmission

To enable packet loss resilient transmission, our proposed S-RLNC model employs MGM mechanism. S-RLNC-MGM multiple generations are collected and is transformed into combination sets, which is then followed by combining data packets pertaining to a specific set so as to enhance Packet Loss Resiliency (PLR) [16]. To enable S-RLNC-MGM based multicast, let the size of a combination set be z generations. The schematic of the multiple generations is given in Fig. 2. Here, we assign a location index for each generation which is appended in the combination matrix. If the initial generation in the combination is one and the final generation in the combination set is z, then the position index would be $(1 \leq d \leq z)$. As depicted in Fig. 2, the source node linearly combines the initial d.n numeral of source packets to generate a total of d.m linear combinations, where $m \geq n$. To ensure the successful delivery of the data packet combinations to the sink under the probability of packet loss, the z. $(m-n)$ additionally transmits packet combination. Consider, $[X_d]_{n\times s}$ be the n source packets pertaining to the dth generation in the matrix, where the individual source packet refers a $1 \times s$ matrix of the symbols obtained from the Galois Field of size 2^F, where F refers GF's order. In S-RLNC model, at the source node, $A[C]_{m\times(d.n)}$ matrix with rank d.n is obtained by means of the variables retrieved from the similar GF (2^F).

Fig. 2 S-RLNC with
Multi-Generation Mixing
(MGM)

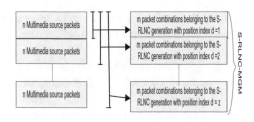

Now, from the dth combination matrix, a total of output packet mixtures are obtained using (10).

$$[Y]_{m \times s} = [C]_{m \times (d.n)} \cdot [X]_{(d.n) \times s} \qquad (10)$$

The generated m combinations allied with d · zth S-RLNC generations are then transmitted from the source node. Sink nodes collects packet combinations from a suitable combination set. Sink nodes decode the $d.n$ S-RLNC-MGM packets while the rank of the received matrix from the initial d generations turns out to be $d . n$. Consider, $d . n$ unique packets be $\left[\left[\widehat{C} \right]_{(d.n) \times (d.n)} \left[\widehat{Y} \right]_{(d.n) \times s} \right]$, where $\left[\widehat{C} \right]_{(d.n) \times (d.n)}$ signifies the coefficients pertinent to generate the received packets $\left[\widehat{Y} \right]_{(d.n) \times s}$, in S-RLNC model the packets are decoded using (11).

$$\left[\widehat{X} \right]_{(d.n) \times s} = \left[\widehat{C} \right]_{(d.n) \times (d.n)}^{-1} \cdot \left[\widehat{Y} \right]_{(d.n) \times s} \qquad (11)$$

Noticeably, in case of packet loss, $(m - n)$ numbers of redundant packets are generated from d generations which are used to perform decoding. As novel solution, in our proposed model, the redundant packets are generated for multiple generation fixed packet combination that avoids the need of redundant packet per generation demands. It significantly enhances bandwidth optimization as well as signaling overheads. Furthermore, in our proposed model a novel function called IBF has been applied that once the packet is decoded successfully at the receiver, it flushes the buffers and hence enables sufficient resource for further data transmission. It enhances resource efficacy of the proposed model, which is of vital significance for multimedia data transmission over LTE-A networks. Taking into consideration of the use of our proposed research work for multimedia data communication in LTE-A networks, we have derived our proposed S-RLNC-MGM model to cope up with content as well as network awareness to deliver QoS assured multicast transmission. A brief of the derived network and content awareness model is given as follows:

2.1.3   Network and Content Awareness of S-RLNC-MGM Model for Multicast Transmission over LTE

In our proposed model, to enhance the likelihood of decoding of each S-RLNC generation, it is intended to distribute redundant packet in such manner that it would assist deciding all data packets while ensuring minimum redundant packets demand. A snippet of decoding multimedia data packets at the receiver (in LTE, UEs) is given as follows:

Decoding Likelihood Optimization (DLO) Each Generation

S-RLNC-MGM model incorporates an optimization model to perform efficient redundant packet allocation across network that intends to increase the probability of decoding at receiver. We consider source nodes, intermediate nodes and the sink nodes. In proposed model, source node collects data packets from the application layer and prepares it as initial PDCP packet, which is then prepared as a sets of $\sum_{d=1}^{z} n_d$ NC-PDCP packets where z signifies the total number of generations and n_d refers the total source packets allied with the generation having position index d. Here, each source packet contains the factors selected from GF (2^F). Here, the source packets combined linearly by means of the variables selected from the similar GF to generate $\sum_{d=1}^{z} m_d$ S-RLNC packets allied with multiple generations $g_1, g_2, ..., g_z$, where m_d signifies the total number of S-RLNC coded packets belonging to g_d generation. In our model, the m_d number of S-RLNC-MGM data packets retrieved from the g_dth generation (here, $0 < d \leq z$), are generated by means of combining the initial $\sum_{v=1}^{d} n_v$ source packets linearly. Thus, the optimal number of S-RLNC-MGM coded packets (m_d) generated for g_d for retrieving the optimal probability of packet decoding is estimated using following method.

Consider, the probability of a S-RLNC-MGM coded packet reaching to the UEs be α_{avg}. Mathematically,

$$\alpha_{avg} = \frac{\sum_{v=1}^{\varphi} \alpha_v}{\varphi} \tag{12}$$

Noticeably, the generation g_d and its previous generations, $g_1, ..., g_{d-1}$, are capable of being decoded if $\sum_{v=1}^{d} n_v$ number of linearly dependent packet combinations are received from the different generations $g_1, ..., g_d$. It enables receiver to decode each generations $g_1, ..., g_d$ if generation g'_d (where $0 \leq d' < d$), is decoded and $\sum_{v=d'+1}^{d} n_v$ linearly independent S-RLNC-MGM coded packets are received from the generations $g'_{d+1}, ..., g_d$. Consider, the probability of decoding g_d after receiving packets from $g_1, ..., g_d$ be ς_d, then it is presumed that an adequately large size of GF is considered for computation. In such cases, with $d = 1$, ς_1 is equivalent to the probability of receiving at least n_1 S-RLNC-MGM coded packets out of the transmitted m_1 coded packets from generation g_1. Hence,

$$P_{\text{dec}}(n, m, \alpha_{\text{avg}}) = \sum_{v=n}^{m} \binom{m}{v} \alpha_{\text{avg}}^{v} (1 - \alpha_{\text{avg}})^{m-v} \tag{13}$$

Now, deriving by (13)

$$\varsigma_1 = P(n_1, m_1, \alpha_{\text{avg}}) \tag{14}$$

In the similar way, ς_2 equals the likelihood of receiving at least n_2 packet combinations from g_2 if g_1 is decoded. Additionally, g_1 and g_2 are also capable of decoding if the sum of $n_1 + n_2$ packet mixtures are received from the respective generations. Consequently, ς_2 is estimated using (15).

$$\varsigma_2 = \varsigma_1 \cdot P(n_1, m_1, \alpha_{\text{avg}}) + (1 - \varsigma_1) \cdot P\left(\sum_{v=1}^{2} n_v, \sum_{v=1}^{2} m_v, \alpha_{\text{avg}}\right) \tag{15}$$

$$
\begin{aligned}
\varsigma_z = \ &\varsigma_{z-1} \cdot P(n_z, m_z, \alpha_{\text{avg}}) \\
&+ (1 - \varsigma_{z-1}) \cdot \varsigma_{z-2} \cdot P\left(\sum_{v=z-1}^{z} n_v, \sum_{v=z-1}^{z} m_v, \alpha_{\text{avg}}\right) \\
&+ \cdots \\
&+ (1 - \varsigma_1) \cdot (1 - \varsigma_2) \ldots (1 \\
&- \varsigma_3) \cdot P\left(\sum_{v=1}^{z} n_v, \sum_{v=1}^{z} m_v, \alpha_{\text{avg}}\right)
\end{aligned} \tag{16}
$$

Now, considering above Eqs. (14)–(16), ς_d is derived into a generalized form

$$
\varsigma_d = \begin{cases}
P(n_1, m_1, \alpha_{\text{avg}}) & ; \quad d = 1 \\
\left\{ \begin{array}{l} \varsigma_{d-1} \cdot P(n_d, m_d, \alpha_{\text{avg}}) + \\ \left[\displaystyle\sum_{v=1}^{d-1} \varsigma_{d-v-1} \cdot \prod_{w=d-v}^{d-1} \left((1 - \varsigma_k) \cdot P\left(\begin{array}{l} \displaystyle\sum_{w=d-1}^{d} n_w, \\ \displaystyle\sum_{w=d-1}^{d} m_w, \alpha_{\text{avg}} \end{array} \right) \right) \right] \end{array} \right\} & ; \quad d \geq 1
\end{cases}
$$

where, $\varsigma_0 = 1$. Furthermore, it must be noted that g_d is capable of getting decoded only when $\sum_{v=d}^{\tilde{d}} n_v$ linearly independent packets are received from generations $g_d, \ldots, g_{\tilde{d}}$ where $d \leq \tilde{d} \leq z$. Consider the probability of decoding g_d after reaching packet combination be ρ_d, then with $d = z$,

$$\rho_z = \varsigma_z \tag{17}$$

Similarly, with $d = (z - 1)$,

$$\rho_{z-1} = \varsigma_{z-1} + (1 - \varsigma_{z-1}) \cdot \varsigma_z \tag{18}$$

when $d = 1$,

$$\rho_1 = \varsigma_1 + (1 - \varsigma_1) \cdot \varsigma_2 + \cdots + (1 - \varsigma_1) \ldots (1 - \varsigma_{z-1}) \varsigma_z \tag{19}$$

Now, using (17)–(19), t; a generalized form is obtained (20).

$$\rho_d = \begin{cases} \varsigma_d & d = z \\ \varsigma_d + \sum\limits_{v=d+1}^{z} \left[\varsigma_v \cdot \prod\limits_{w=d}^{v-1} (1 - \varsigma_w) \right] & d < z \end{cases} \tag{20}$$

The average decoding probability for S-RLNC generation, $\bar{\rho}$, is estimated using (21).

$$\bar{\rho} = \frac{\sum_{d=1}^{z} \rho_d}{z} \tag{21}$$

The major significance of $m_1, m_2, \ldots, m_z$ is that it applies the decoding probability with $\mathrm{Max}(\bar{\rho})$ in such way that

$$\sum_{d=1}^{z} (m_d) = R \tag{22}$$

where, R signifies the summation of the packet combinations.

2.1.4 Received Video Quality

In our proposed S-RLNC model an enhancement is made by means of the precedence of scalable packet layers that enables optimal redundancy distribution. Here, by applying a scalable multimedia data with z scalability layers, which are generated in such manner that every successive layer is encoded at a superior quality by managing the quantization factor, spatial resolution or temporal resolution. Thus, each successive scalability layer received by UEs will increase the quality of data received. Almost all multimedia data layers connected to the dth scalability layer in a Group of Packets (GOP) are assigned to the generation g_d (Fig. 3). Similar to the decoding probability optimization, the enhancement is made by considering average probability of a coded packet reaching UEs or receiver α_{avg}.

Fig. 3 Multimedia data packet assignment to S-RLNC generations in MGM approach

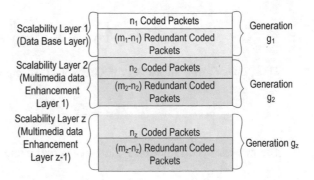

The probability of retrieving each multimedia layer merely equal to the dth scalability layer, σ_i, is equivalent to the likelihood of decoding generations $g_1,\ldots,g_i$ and not decoding the generation. Therefore, g_{i+1} is obtained using (23).

$$\sigma_d = \left[\prod_{v=1}^{d} \rho_v\right] \cdot \left[1 - \rho_{d+1}\right] \tag{23}$$

Let q_d be the multimedia quality when each layer equal to the dth scalability layer is received and then the received data quality of the users $\varnothing$ is obtained as (24).

$$\phi = \sum_{d=1}^{z} \sigma_d \cdot q_d \tag{24}$$

The optimal value of m_1, m_2, ..., m_z signifies the received data quality at the users is estimated using an enhancement model Max($\varnothing$). Mathematically (25),

$$\sum_{d=1}^{z} (m_d) = R \tag{25}$$

The aforementioned model ensures data recovery at the receiver through efficient redundant packets. Despite of deciding the base layer, a UE cannot playback the multimedia data constantly and therefore augmentation of the likelihood of decoding at the base layer of the multimedia data at every UE would enhance the likelihood of unremitting multimedia view, while QoS assurance. To deal with this, a redundant packet distribution optimization is performed that ensures the sufficient and optimal number of redundant packets is distributed across network that ensures multimedia data decoding while preserving original quality. We have applied a scalable multimedia multicast model, where the received data quality can be estimated at each UE by replacing with the likelihood of a coded data reaching each UE μ, α_μ rather than α_{avg}. Here, each user can be assigned redundant packet using a rank value of π_μ in such way that

$$\pi_\mu = \begin{cases} \varnothing_\mu & \rho 1_\mu > \theta \\ 0 & \rho 1_\mu \leq \theta \end{cases} \tag{26}$$

Thus, considering all users in the network (φ), the mean rank is obtained as

$$\bar{\pi} = \frac{\sum_{\mu=1}^{\varphi} \pi_v}{\varphi} \tag{27}$$

The average quality per user can be obtained using (28).

$$\sum_{d=1}^{z} (m_d = R) \tag{28}$$

2.2 S-RLNC-MGM-Assisted MAC for LTE/LTE-A

Figure 4 presents the classical protocol stack functional for the downlink IP packet flow in LTE protocol. As depicted by applying the Packet Data Conversion Protocol (PDCP), IP data packets reach the base station known as eNB. Now, the PDCP/IP packets are transmitted to the Radio Link Control (RLC) layer once processing the header compression and ciphering, which is then followed by segmentation and (or) concatenation of the packets to fit the MAC frame size. It is then followed by the conversion into specific data packets to be fit with the physical layer (PHY) Transport Block (TB) size for further transmission. Dependable MAC frame delivery above eNB/UE connection is supported using MAC-HARQ approach for unicast services. In major traditional approaches, in case HARQ retransmissions fails, RLC-layer Acknowledged Mode (AM) is employed that uses additional RLC-level ARQ to assure RLC packet delivery [17]. In our proposed model S-RLNC has been applied at the top of the classical MAC RNC, as depicted in Fig. 4.

In process, the PHY TB is mapped to a unit of PHY time-frequency resources on the wireless link known as Resource Block Pair (RBP). Typically, unit PHY RBP equals 1 ms of time period also called Transmission Time Interval (TTI). The classical LTE-MAC with MAC-HARQ imposes significant computational or (and) signaling overheads during FEC [12]. The predominant solution originates from the RLC layer to deliver RLC PDCP packet above the eNB/UE line. This solution is the normalized method that transmits IP packet through parallel, autonomous as well as time-interleaved delivery of associated disjoint segments, which is supported through the MAC-HARQ protocol to assist reliable multimedia data delivery (Fig. 4). In this mechanism the IP packet is accepted once every of its RLC/MAC segments are received at the receiver or the UE [17]. Unlike traditional MAC-HARQ, we have applied our proposed S-RLNC model at the top of classical MAC protocol. Unlike classical LTE-MAC, IP packet is not segmented in our

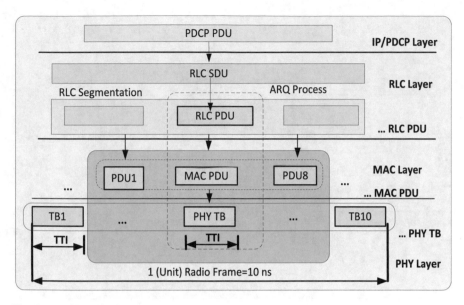

Fig. 4 MAC-HARQ solution

proposed S-RLNC-MAC model and RLC layer compresses the IP packet directly into the RLC packet that maintains the reliability of IP packets containing multimedia data components [18]. As depicted in Fig. 5, in case of PDCP packets, RLC layer concatenates several packets into a single RLC packet. Further, in MAC layer, RLC acknowledgement is processed with the proposed S-RLNC-MGM model where the data packet is split into K same size source symbols from which a stream of S-RNLC processed symbol is generated. In our proposed model, S-RLNC-MGM with GF of size 8 has been taken into consideration to perform multicast transmission over LTE-MAC. Here, MAC S-RLNC generated suitable equal-length data packets which are collected into a MAC segment to fit the future PHY TBS. In our model, MAC segment is accumulated in the PHY TB. The received PHY TB at UE with additional redundant packets is decoded using S-RLNC-MGM-MAC sub-layer. Once receiving linearly independent encoded packets from the stream of MAC segments, the UE MAC gives back an individual ACK deciding the RLC packet transmission. We have applied real-time multimedia data packets, where realistic LTE-A EUTRAN data delivery model is developed by broadcasting image data for unicast as well as multicast multimedia transmission services. The data transmission is considered where IP data streams transmit video substance to the UEs. Here, every IP image data is assigned a set of time-frequency resources at the eNB/UE radio-interface, defined using a set of PHY RBPs above a time-sequence of TTI periods.

Considering MBMS transmissions, S-RLNC-MDM exploits network conditions which have been simulated with Gilbert Elliot model, and accordingly performs resource allocation. Even exploiting channel state conditions (CSI), eNB can use

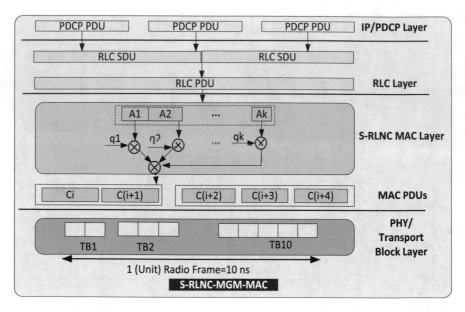

Fig. 5 S-RLNC MGM-MAC solution

and schedule certain suitable coding as well as modulation system over PHY TBs for upcoming TTIs. On the contrary, the currently applied 3GPP standard of multimedia broadcast, the eNB applies a fixed PHY broadcast model targeting total cellular reporting regardless of the single UE channel conditions.

In S-RLNC-MGM-based MAC, we considered a LTE/LTE-A method containing a macro-cellular network with six macro-cells positioned in three tiers in the region of a central eNB. The QoS of the proposed model is assessed with regard to PDR, loss ratio, delay, peak signal to noise ratio (PSNR). The outcomes obtained are discussed as follows.

3 Results and Discussion

In this research, the emphasis was made on developing a robust and efficient multimedia multicasting model for LTE-A system. Considering the need of QoS/QoE experience in LTE broadcast demands, here we focused on enhancing NC scheme by incorporating systematic NC scheme where only a fraction of input data is enquired to be encrypted. This enables our proposed system to be computationally efficient. The development of interleaving and coding was applied to strengthen transmission reliability. Unlike classical NC scheme, we developed a MGM model to strengthen or optimize redundant packet distribution across the UEs in LTE network. Thus, the overall proposed NC scheme was derived as

S-RLNC-MGM model to be further applied for LTE-MAC as a substitute or at the top of traditional LTE-MAC-HARQ. This as a result enhances overall signaling overheads by reducing iterative ARQ to the transmitter. We applied an IBF model that once receiving data packet at the receiver flushes transmitter buffer and hence preserves bandwidth significantly. We derived MGM model that transmits redundant packet at once after mixing encoded packets from multiple generation. It reduces computational complexity and buffer occupancy. To assess the error-proneness of the proposed S-RLNC-MGM model, we assessed our proposed system with different loss conditions, which was incorporated by varying network loss pattern. We used COST231 (an enhanced Okumura Hata loss Model (OHM) which functions in the frequency range of 150–1500 MHz) to examine PLR under varying loss conditions. Noticeably, COST231 loss model is applicable for the frequency range of 1500–2000 MHz, which is sufficient for LTE systems. This simulation case is applicable for the UE's speed of 30 km/h. Our proposed system was tested with six UEs distributed in a three layers (source, intermediate and sink) distributed across LTE network. We tested performance with the Galois Field of size 4, 8 and 16, and the results for each test conditions (GF = 4/8/16) was examined. In S-RLNC-MGM, we applied image data as multimedia input. As sample, we applied standard images of different size such as Lena, Photographer, and Baboon to test efficacy in terms of PSNR, MSE. The simulation was done for different datasets with different GF size and packet rate. The overall simulation model is developed using MATLAB 2015a tool.

At first S-RLNC-MGM model has been examined for throughput PDR and packet loss performance with different GF size and the number of generation. Since S-RLNC-MGM applies redundant packet transmission for multiple generations at a time rather than redundant packets-per generation. It enhances buffer occupancy while maintaining optimal data decoding at the receiver. Considering above results to further examine the performance of the proposed S-RLNC-MGM model for multimedia multicast over LTE, in our research we applied the following test conditions. Here, the images of different sizes (Windows-22 kb, Lena-100 kb, Baboon-175 kb) have been applied for performance evaluation (Figs. 6, 7, 8, 9, 10, 11, 12 and 13; Table 1).

Fig. 6 Throughput versus galois field size

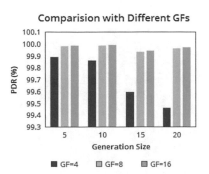

Fig. 7 Packet loss versus
generations (GF = 4)

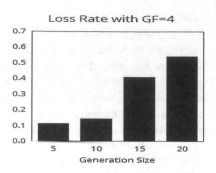

Fig. 8 Packet loss versus
generations (GF = 8)

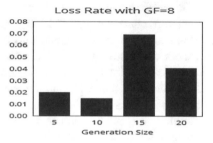

Fig. 9 Packet loss versus
generations (GF = 16)

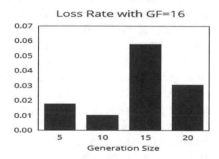

Fig. 10 Throughput versus
redundant packets

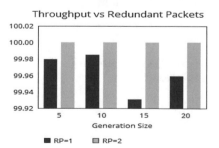

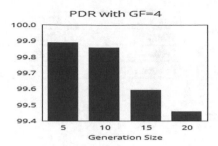

Fig. 11 Throughput versus no. of generations (GF = 4)

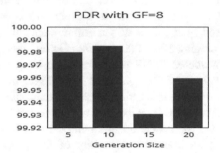

Fig. 12 Throughput versus no. of generations (GF = 8)

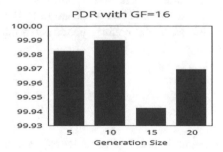

Fig. 13 Throughput versus no. of generations (GF = 16)

Table 1 Simulation conditions

Parameter	Value
Galois field	8
Number of generations	10
Redundant packet	2
Network loss condition (probability)	0.005
Loss model	Gilbert Elliot
Total number of cells	6
Layer of multicast	3
Packet size	1000
Data types	Multimedia (image) data

Table 2 Comparative
analysis of multimedia data
transmission

Image data	MSE	PSNR
Window	21.20	39.51
Leena	27.67	33.74
Baboon	132.01	26.95

3.1 Comparative Analysis

Our proposed system was assessed with image multimedia data and hence avoided
any channel coder such as H.264/AVC or SVC. Here, we encoded the base layer
and two enhancement layers, which were taken into consideration in such way that
the individual scalable layer possesses a distinct discrepancy in bit-rates and
quality. The considered image data was accommodated in S-RLNC-MGM with
different symbol size ranging 500–2000 from the Galois Field of varying size (4, 8,
and 16). We have considered PSNR of the received image at the receiver as the
quality index. The maximum PSNR obtained in the PNC was approximate 35 dB,
while in our work the simulation with image data exhibited maximum PSNR of
39.51 dB. The results in Table 2 reveal that with increase in data size or payload
PSNR reduces; however the reduction is significantly low to make any decisive
quality degradation. It signifies the suitability of the proposed S-RLNC-MGM
model towards multimedia data transmission or multicast over LTE-A protocol
standard. Recently, authors [19] developed a Practical Network Coding
(PNC) scheme for content and network aware multimedia multicast. Though, our
work exploits multicast nature of the PNC proposed in [19], the implementation of
our proposed S-RLNC-MGM model enables better performance in terms of multimedia data transmission.

4 Conclusion

This research paper presented an enhanced S-RLNC based transmission model,
which has been armored with Multi-Generation-Mixing (MGM) concept and
Iterative Buffer Flush (IBF) that cumulatively reduces signaling overheads and
resource consumption. In addition, the use of MGM with S-RLNC strengthened
proposed multimedia multicast model to ensure optimal redundant packet transmission by reducing huge redundant packet demands for decoding that eventually
augments computational efficacy as well as resource utilization. Furthermore,
realizing the practical network conditions, the proposed S-RLNC-MGM model was
examined with varying network loss conditions, where it was found robust in terms
of successful data delivery and decoding at the receiver. Considering the limitations
of the existing LTE HARQ-MAC protocol, particularly in terms of signaling
overhead, Round Trip Time (RTT) and latency, in this research paper, the proposed
S-RLNC scheme has been applied on the top of LTE-MAC has that augments

existing system to deliver reliable multimedia data with almost 100% success rate with low latency and minimum error. Considering quality preserving nature, the simulation with multimedia image data has revealed minimum mean square error and higher signal to noise ratio, which signifies its suitability for real-time multimedia data transmission over LTE. In future, the proposed S-RLNC-MGM model can be applied with LTE standard simulation models with video data under dynamic network conditions. The usefulness of H264/SVC encoding can also be assessed to augment data rate transmission in LTE.

References

1. Gruber, M., Zeller, D.: Multimedia broadcast multicast service: new transmission schemes and related challenges. IEEE Commun. Mag. **49**(12), 176–181 (2011)
2. Vukobratovic, D., Khirallah, C., Stankovic, V., Thompson, J.: Random network coding for multimedia delivery over LTE-advanced. In: 2012 IEEE International Conference Multimedia and Expo (ICME), pp. 200–205 (2012)
3. Hamdoun, H., Loskot, P.: Implementing network coding in LTE and LTE-A. In: The First International Workshop on Smart Wireless Communications. Luton, UK (2012)
4. Ahlswede, R., Cai, N., Li, S.R., Yeung, R.W.: Network information flow. IEEE Trans. Inf. Theor. (2000)
5. 3GPP TR 36.913 v8.0.1 (Release 8): Requirements for further advancement for (E-UTRA) (2009)
6. Shojania, H., Li, B.: Random network coding on the iPhone: fact or fiction? In: ACM NOSSDAV 2009. Williamsburg. USA (2009)
7. Chou, P.A., Wu, Y., Jain, K.: Practical network coding. In: Allerton Conference 2003, Monticello, IL, USA (2003)
8. Liva, G., Paolini, E., Chiani, M.: Performance versus overhead for fountain codes over GFq. IEEE Comm. Lett. **14**(2), 178–180 (2010)
9. Chou, P.A., Wu, Y.: Network coding for the internet and wireless networks. IEEE Signal Proc. Mag. **24**(5), 77–85 (2007)
10. Vukobratovi´c, D., Stankovi´c, V.: Unequal error protection random linear coding strategies for erasure channels. IEEE Trans. Commun. **60**(5), 1243–1252 (2012)
11. Schotsch, B., Lupoaie, R., Vary, P.: The performance of low-density random linear fountain codes over higher order galois fields under maximum likelihood decoding. In: Allerton 2011, USA (2011)
12. Khirallah, C., Vukobratovi´c, D., Thompson, J.: Performance evaluation and energy efficiency of random network coding in LTE advanced. IEEE Trans. Wireless Commun. **11**(12), 4275–4285 (2012)
13. Khandekar, A., Bhushan, N., Tingfang, J., Vanghi, V.: LTE-advanced: heterogeneous networks. In: European Wireless EW 2010, pp. 978–982 (2010)
14. Veeresh, P., Sanjeev, G., Keshavamurthy, C.: An enhanced network coding based MAC optimization model for QoS oriented multicast transmission over LTE networks. Int. J. Comput. Sci. Inf. Secur. **14**(12) (2016)
15. Jaggi, S., Langberg, M., Katti, S., Ho, T., Katabi, D., Medard, M.: Resilient network coding in the presence of Byzantine adversaries. In: IEEE INFOCOM 2007. 26th IEEE International Conference on Computer Communications, pp. 616– 624 (2007)
16. Halloush, M., Radha, H.: Network coding with multi-generation mixing: a generalized framework for practical network coding. IEEE Trans. Wireless Commun. **10**(2), 466–473 (2011)

17. Holma, H., Toskala, A.: LTE for UMTS: Evolution to LTE-Advanced, 2nd edn. Wiley, Hoboken, New Jersey (2011)
18. Wiegand, T., Sullivan, G.J., Bjontegaard, G., Luthra, A.: Overview of the H. 264/AVC video coding standard. IEEE Trans. Circ. Syst. Video Tech. **13**(7), 560–576 (2003)
19. de Alwis, C., Arachchi, H.K., Fernando, A., Pourazad, M.: Content and network-aware multicast over wireless networks. In: 10th International Conference on Heterogeneous Networking for Quality, Reliability, Security and Robustness, pp. 122–128. Rhodes (2014)

A Novel Method for Epileptic EEG Classification Using DWT, MGA, and ANFIS: A Real Time Application to Cardiac Patients with Epilepsy

Mohanty Madhusmita, Basu Mousumi, Pattanayak Deba Narayan and Mohapatra Sumant Kumar

Abstract The automatic diagnosis of heart patients with epilepsy by reviewing the EEG recording is highly necessary. It aims to enhance the significant statistical parameters. In this paper a composite method is proposed for seizure classification of cardiac patients. Firstly DWT is employed to analyze the EEG data and obtain the time and frequency domain features. Second, the extracted features were inputted to the ANFIS network to classify the seizure EEG and seizure free EEG signals. Third to improve the statistical performances a modified genetic algorithm (MGA) is used to optimize the classifiers. Sensitivity (SEN), Specificity (SPE), Accuracy (ACC), metric G-mean and Average detection Ratio (ADR) is used to evaluate the performance of this method. The SEN of 99.73%, SPE of 99.12%, ACC of 99.35%, G-mean 99.42% and ADR of 99.43% are yielded on the real patient specific EEG database. The comparison with other detection methods shows the superior performance of this method, which indicates its potential for detecting seizure events in clinical practice of heart patients.

Keywords EEG · Epilepsy · DWT · MGA · ANFIS

M. Madhusmita (✉)
Department of Electronics and Communication Engineering,
Gandhi Engineering College, Madanpur, Odisha, India
e-mail: madhusmitap2003@yahoo.co.in

B. Mousumi
Department of Power Engineering, Jadavpur University,
Saltlake Campus, Kolkata, West Bengal, India
e-mail: mousumibasu@yahoo.com

P. D. Narayan
Department of Electrical and Electronics Engineering,
Trident Academy of Technology, Bhubaneswar, Odisha, India
e-mail: debapattanaik33@yahoo.com

M. S. Kumar
Department of Electronics and Telecom Engineering,
Trident Academy of Technology, Bhubaneswar, Odisha, India
e-mail: sumsusmeera@gmail.com

P. K. Mallick et al. (eds.), *Cognitive Informatics and Soft Computing*,
Advances in Intelligent Systems and Computing 768,
https://doi.org/10.1007/978-981-13-0617-4_51

525

1 Introduction

Deficiency of EEG signal transmission due to pumping problem in heart creates epileptically abnormalities in human brain of the cardiac patient. Due to the above cause, error occurs for blood circulation to the human brain. The epileptic seizures are generated due to abnormality of neural activity in the brain. Normal EEG signal is used to analyze both (brain and heart) activity as a threshold level for whole classification process. 0.6–0.8% of the world's population are affected by epilepsy that shows the common neurological disorder in brain [1]. Electroencephalogram (EEG) signal provides an effective observation of the brain's electrical activity. Liu et al. [2] extracted the features of EEG signals into different time and frequency domain DWT. According to the approach of different authors, features in terms of relative sub-band energy, frequency identification were calculated at the standard scales for seizure EEG detection. Patidar et al. [3] proposed a tunable Q-wavelet transform for feature extraction on EEG signal. The EEG signal is highly affected by alcohols and it is analyzed by an integrated alcoholic index [3]. The epileptic seizure prediction is accomplished by using EEG signal [4]. Das et al. [5] approached a classification of EEG signal using inverse Gaussian parameter using dual-tree complex wavelet transform domain simple classification of focal and non-focal EEG signals using entropy based features in the EMDDWT domain is proposed [6]. In [7] an automated detection system using Empirical wavelet transform is used and the neural computing is utilized for whole process. In [8, 9] SVM is used for proper classification. Aarabi et al. [10], Zhang et al. [11] and Qi et al. [12] are emphasized on improved computerized model for proper diagnosis of raw EEG and affected EEG signals classification.

2 Materials and Methods

For the proposed work, multichannel biomedical EEG signal is used to evaluate the method. This case conducted in NIMHANS, Bangalore, India. Normal EEG signal is taken from physionet data base (https://physionet.org/physiobank/database/eegdb/eeg_healthy.dat). The seizure EEG signal recorded from cardiac patients with epilepsy (6 patients) in the NIMHANS center by using 10–20 electrode international standard. Details of the patients are mentioned in Table 1.

2.1 DWT

The DWT has a large number of applications in biomedical engineering, technology and mathematics. DWT decomposition accommodate with a suitable correlation between time and frequency domain. This formulation of DWT depends on the

Table 1 Detailed clinical information's of the patients with epilepsy

PI	Age/ Gender	No. of seizure events/Time (s)	No. of channels	Total seizure time (s)	Total seizure free time (s)	Seizure origin
1	17/M	6(47–112)	21	328	32,208	Right frontal lobe
2	12/F	4(21–93)	23	924	18,088	Left and middle temporal lobe
3	08/M	3(53–118)	21	454	25,124	Frontal lobe
4	16/F	7(91–178)	21	323	19,978	Absence seizure
5	6/M	5(6–18)	23	302	19,772	Right central origin
6	4/M	7(154–212)	23	203	14,787	Left front and middle temporal lobe

relationship to develop frequency sub-bands with discrete samplings of a mother wavelet function. The output of DWT after each decomposition level contains two parts one is detail coefficients part and approximation coefficients part. For the simplification of classification in our work only low frequency parts, i.e., approximately co-efficient parts are taken for the succeeding stage of decomposition in both for numerical and functional analysis.

The DWT of inputted data are calculated by passing it across a cascading arrangement of filters. Sample values taken for consideration are applied across the low pass filter. Filtered output having with impulse response "g" derived from the convolution values of the two factors. Basic formula for DWT decomposer is expressed as

$$y[k] = (x * g)[k] = \sum_{n=-\infty}^{\infty} x[n]g[k-n] \tag{1}$$

The low pass and high pass filters are shown in Eqs. (2) and (3)

$$Y_{\text{low}}[k] = \sum_{n=-\infty}^{\infty} x[n]g[2k-n] \tag{2}$$

$$Y_{\text{high}}[k] = \sum_{n=-\infty}^{\infty} x[n]h[2k-n] \tag{3}$$

Figure (1) shows three levels of decomposition, but in our application four level decompositions is used. In the present model we have employed sub-band analysis of the wavelet co-efficient.

Fig. 1 Block diagram of
three-level DWT

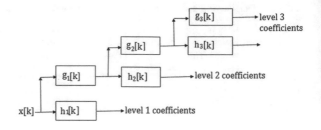

2.2 Modified Genetic Algorithm (MGA)

The proposed modified genetic algorithm (MGA) has the high accuracy prediction ability with high convergence speed. It has only one solution at one time but in case of GA, there is collection of solutions. The stepwise process of MGA is described as

Step 1. The population size, maximum generation size, cross over rate (R_C), mutation rate (R_m) and the increment or decrement rate (improvement rates, R_i) are determined.

Step 2. Create an initial solution as current solution (C_s) and temporary solution (T_S).

Step 3. Maintain the population in a systematic way

$$e_g = \begin{cases} 1 & s_g < s'_g \\ 0 & s_g = s'_g \\ -1 & s_g > s'_g \end{cases} \quad g = 1, 2, \ldots, n \tag{4}$$

Where

$$E = \{e_1, e_2, \ldots, e_n\}, S = \{s_1, s_2, \ldots, s_n\}, s' = \{s'_1, s'_2, \ldots, s'_n\}$$

Step 4. The fitness values of each experience are to be properly maintained or sorted in a descending order.

Step 5. Select two chromosomes of two parents.

Step 6. In this step, a new offspring is generated from step 5 depending upon their cross over rate.

Step 7. In the step (new offspring) a genetic diversity is maintained by which mutation rate is determined.

Step 8. Based on step 6 and step 7, a new solution is generated as

$$S'_g = \begin{cases} S_g + R_m \times |S_g|, & e_g = 1 \\ S_g, e_g = 0 & g = 1, \ldots, n \\ S_g - R_m \times |S_g|, & e_g = -1 \end{cases} \tag{5}$$

Where

$$E = \{e_1, e_2, \ldots, e_n\},$$
$$S_g = \{s_1, s_2, \ldots, s_n\},$$
$$S'_g = \{s'_1, s'_2, \ldots s'_n\}$$

Step. 9.   Repeat the steps from 4 to 9 consecutively if the stopping criterion is not satisfied to get the current solution.

3   Experimental Results and Discussion

The values of true positive are detected as the number of seizure epoch identified by the experts and by the algorithm (MGA). True negative value is expressed as the non-seizure epochs identified both by experts and algorithms. In our method nine rule ANFIS is used as classifier as shown in Fig. 2. The Proposed method is shown in Fig. 3.

These outcomes are expressed as below.

$$\text{Sensitivity (SEN)} = \frac{\text{TP}}{\text{TP} + \text{FN}} \times 100 \tag{6}$$

$$\text{Specificity (SPE)} = \frac{\text{TN}}{\text{TN} + \text{FP}} \times 100 \tag{7}$$

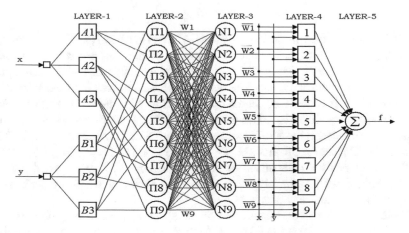

Fig. 2   Block diagram of nine-rule ANFIS

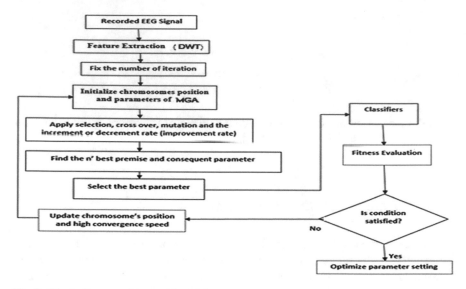

Fig. 3 Block diagram of proposed model

$$\text{Accuracy (ACC)} = \frac{\text{TN} + \text{TP}}{\text{TN} + \text{TP} + \text{FN} + \text{FP}} \times 100 \tag{8}$$

The expression for G-mean is

$$= \sqrt{\text{SEN} * \text{SPE}} \tag{9}$$

Average detection ratio (ADR) is expressed as

$$\text{ADR} = \frac{\text{SEN} + \text{SPE}}{2} \times 100 \tag{10}$$

EEG data are experimented using MATLAB version R2010a. Table 1 mentions the detail specification (such as age, sex, number of channels and seizure origin) of six numbers of patients. Data of the 5th number patient taken as test signal to update the parameters of the classifiers. The resulted value compares both the seizure and non-seizure labels of the EEG signal by comparing with the normal EEG signal.

At window size 2000, based on the method DWT + ANFIS + MGA calculated values are of sensitivity 99.73%, specificity of 99.12%, accuracy of 99.35%, G-mean 99.42% and ADR 99.43% (TP = 743, FP = 11, TN = 1244, FN = 02).

Table 2 shows the comparison of measured accuracy. The proposed methods ANFIS trained with MGA algorithm shows the accuracy which outperforms as compared to existing methods.

Table 3 shows a comparison analysis of specificity, sensitivity G-mean and ADR of the proposed method (DWT + ANFIS + MGA) with existing methods. It shows the outperformance in all respects for better clinical applications.

Table 2 Comparison of measured accuracy of proposed methods with existing methods

Authors	Reference	Year	Features/method	Analysis/ classification	Accuracy (%)
Patider et al.	[3]	2016	TQWT	LS-SVM	97.02
Ataee et al.	[4]	2016	Wavelet features	Neural network	94.00
Das et al.	[5]	2016	Dual-tree	SVM	96.28
Das and Bhuiyan	[6]	2016	EMDDWT, log-entropy	KNN	89.4
Bhattacharyya et al.	[7]	2016	RPS, CTM	LS-SVM	90
Proposed method			DWT	ANFIS-MGA	99.35

Table 3 Comparison analysis of specificity, sensitivity G-mean and ADR of the proposed method with existing methods

Reference	Methods	Sensitivity (%)	Specificity (%)	G-Mean (%)	ADR (%)
[8]	WT and SVM	94.46	95.26	94.86	94.86
[9]	Support Vector Machine	92.4	98.6	95.26	95.32
[10]	Fuzzy rule	98.72	68.27	82.1	83.50
[11]	Fractal dimensions and gradient boosting classifier	91.27	90.59	90.93	90.94
[12]	WPT-WELM	97.73	90.33	93.96	94.3
Proposed method	DWT-ANFIS-MGA	99.73	99.12	99.42	99.43

Figure 4 represents two different significant signals taken from patient number 5 as listed in Table 1. Figure 4a shows the normal EEG signal taken from the database and Fig. 4b represents the recorded seizure EEG signal from the cardiac patient. Figure 5 shows the output of the DWT decomposer of EEG signal. Figure 6 shows Comparison curve of EEG Signal between original signal and DWT + ANFIS Output. Figure 7 shows the comparison curve of EEG Signal between original signal and DWT + ANFIS + MGA output. Figure 8 shows the ROC curve of four methods (DWT-ANFIS-MGA, WOFE-ANFIS-MGA, DWT-ANFIS, and WOFE-ANFIS). The out performance of different calculated methods are clearly mentioned in ROC diagram where the curve "A" of the proposed method using ANFIS classifier has best performance as compared to others.

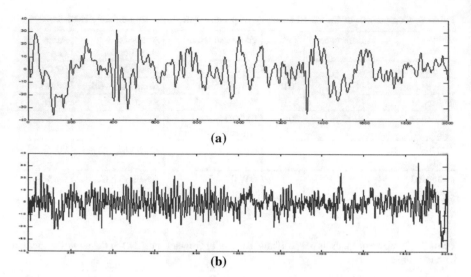

(a)

(b)

Fig. 4 **a** Normal EEG signal from healthy subject **b** recorded seizure EEG signal as examples PI5

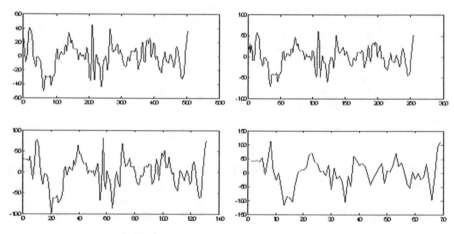

Fig. 5 Four-level DWT output of EEG signal

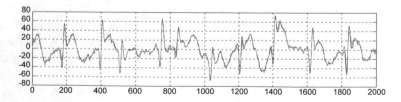

Fig. 6 Comparison curve of EEG signal between original signal and DWT + ANFIS output

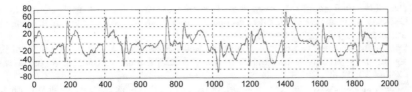

Fig. 7 Comparison curve of EEG signal between original signal and DWT + ANFIS + MGA output

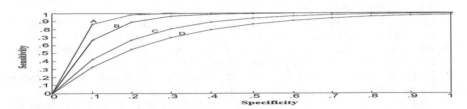

Fig. 8 ROC curve of the proposed model **a** DWT-ANFIS-MGA **b** WOFE-ANFIS-MGA **c** DWT-ANFIS **d** WOFE-ANFIS. *WOFE* Without Feature Extraction

4 Conclusion

In this study a novel classified learning machine is proposed to detect seizure activities in cardiac patients with epilepsy. The time and frequency domain features estimated on the component signals from the discrete wavelet decomposition to characterize the behavior of the EEG signals. The MGA optimized ANFIS network is employed for the EEG signal classification. The proposed method outperforms in all respects compared with the existing methods. It is planned in future to evaluate the performance of the method using much larger amounts of true continuous EEG data of cardiac patients with epilepsy.

References

1. Adelui, H., Zhou, Z., Dadmehr, N.: Analysis of EEG records in an epileptic patient using wavelet transform. J. Neurosci. Methods **123**, 69–87 (2003)
2. Liu, Y., Zhou, W., Yuan, Q., Chen, S.: Automatic seizure detection using wavelet transform and SVM in long-term intracranial EEG. IEEE Trans. Neural. Syst. Rehabil. Eng. **20**, 749–755 (2012)
3. Patidar, S., Pachori, R.B., Upadhyay, A., Rajendra Acharya, U.: An integrated alcoholic index using tunable-Q wavelet transform based features extracted from EEG signals for diagnosis of alcoholism. Appl. Soft Comput. **4**, 112–118 (2016)
4. Ataee, P., Avanaki, A.N., Shariatpanahi, H.F., Khoee, S.M.: Ranking features of wavelet-decomposed EEG based on significance in epileptic seizure-prediction. In: IEEE 14th European Signal Processing Conference, pp. 1–4 (2006)

5. Das, A.B., Bhuiyan, M.I.H., Alam, S.S.: Classification of EEG signals using normal inverse Gaussian parameters in the dual-tree complex wavelet transform domain for seizure detection. Signal Image Video **10**, 259–266 (2016)
6. Das, A.B., Bhuiyan, M.I.H.: Discrimination and classification of focal and non-focal EEG signals using entropy-based features in the EMDDWT domain. Biomed. Signal Process. Control **19**, 11–21 (2016)
7. Bhattacharyya, A., Sharma, M., Pachori, R.B., Sircar, P., Acharya, U.R.: A novel approach for automated detection of focal EEG signals using empirical wavelet transform. Neural Comput. Appl. **8**, 1–11 (2016)
8. Liu, Y.X., Zhou, W.D., Yuan, Q., Chen, S.S.: Automatic seizure detection using wavelet transform and SVM in long-term intracranial EEG. IEEE Trans. Neural. Syst. Rehabil. Eng. **20**(6), 749–755 (2014)
9. Chandaka, S., Chatterjee, A., Munshi, S.: Cross-correlation aided support vector machine classifier for classification of EEG signals. Expert Syst. **36**, 1329–1336 (2009)
10. Aarabi, A., Fazel-Rezai, R., Aghakhani, Y.: A fuzzy rule-based system for epileptic seizure detection in intracranial EEG. Clin. Neurophysiol. **120**, 1648–1657 (2009)
11. Zhang, Y., Zhou, W., Yuan, S., Yuan, Q.: Seizure detection method based on fractal dimension and gradient boosting. Epilepsy Behav. **43**, 30–38 (2015)
12. Yuan, Q., Zhou, W., Zhang, L., Zhang, F., Fangzho, X., Leng, Y., Wei, D., Chen, M.: Epileptic seizure detection based on imbalanced classification and wavelet packet transform. Seizure **50**, 99–108 (2017)

Exploring Hybrid Recommender Systems for Personalized Travel Applications

R. Logesh and V. Subramaniyaswamy

Abstract The recent research in the recommender systems domain has attracted many researchers due to its increasing demands in the real world. To bridge the real-world issues of the users with the problems of the researchers in the digital world, we present hybrid recommendation techniques in e-Tourism domain. In this paper, we have explained the research problems in the e-Tourism applications and presented the possible solution to achieve better personalized recommendations. We have developed a Personalized Context-Aware Hybrid Travel Recommender System (PCAHTRS) by incorporating user's contextual information. The proposed PCAHTRS is evaluated on the real-time large-scale datasets of Yelp and TripAdvisor. The experimental results depict the improved performance of the proposed approach over traditional approaches. We have concluded the paper with future work guidelines to help researchers to achieve fruitful solutions for recommendation problems.

Keywords Recommender systems · Collaborative filtering · Context-aware recommendations · Clustering · E-Toursim · Hybrid systems

1 Introduction

The recent development in the web has created a huge requirement for the assistive tools such as recommender systems to support common man in their daily life. Recommender systems are the best decision support tools that had been successfully employed in various domains such as e-commerce, movies, TV, and e-Tourism. As the e-Tourism is very much integrated with user's activities, it is found to be more adventurous and interesting. Generation of personalized recommendations is not an easier task. Especially, generating recommendations to the

R. Logesh (✉) · V. Subramaniyaswamy
School of Computing, SASTRA Deemed University, Thanjavur 613401, India
e-mail: logeshr@outlook.com

V. Subramaniyaswamy
e-mail: vsubramaniyaswamy@gmail.com

© Springer Nature Singapore Pte Ltd. 2019
P. K. Mallick et al. (eds.), *Cognitive Informatics and Soft Computing*,
Advances in Intelligent Systems and Computing 768,
https://doi.org/10.1007/978-981-13-0617-4_52

user in the travel is much complex task. Converting the virtual recommendations according to the user's current physical world context is the personalization problem and it has yet to be addressed. Analysis of information of user and items filtering into demographic, collaborative, and content-based systems are the ways according to which the recommender systems are traditionally classified [1–4].

2 Recommendation Generation Approaches

The extent to which similarity exits between the items suggested and the users is calculated by the Content-based systems. The process involves the comparison between the preferences of the user's and the item features. Hence the alternatives and the users are represented commonly. The extent to which the user profiles and alternatives are matched is represented by an overall score of performance. High performance score indicates high performance with respect to the alternative considered. User's histories are also considered sometimes. Accuracy in the knowledge of user preference is required here in order to filter the relevant and appropriate items. "Cold start" problem that is, when a new user is encountered the suggestions are poor due to inadequacies in the knowledge of the user initially [5–8].

Collaborative systems consider user groups that have similar likings and preferences to make the recommendations. The user ratings of items are used to determine how similar the user's preferences are. When a set of users is determined such that the current user has similar preferences with that set, then the suggestions are made to the current user based on the preferences of the determined set. Demographic-based systems use the demographic information of users such as the country, age and level of studies in order to provide the suggestions. The classification of the stereo-typical classes here is unique, which is different from other recommender systems in which tourist behavior variation with respect to different locations are considered [9–13]. Each of the defined stereotype class describes the tourist behavior in terms of what and how the tourist would behave in different scenarios [14–19]. Hybrid recommendation method can be used to provide efficient recommendations to the tourists.

In order to overcome the drawbacks of these models, hybrid models involving a combination of more than one model is generally used in practice. Figure 1 represents the generic framework of a hybrid recommender system.

3 Travel Recommender Systems

The enormous amount of information is available over the web, thus the process of making a decision on the large-scale data becomes slow, complicated, and time consuming. Recommender system alleviates the information overload problem and generates suggestions based on user's preferences, interests, and locations. Collaborative filtering plays a vital role in recommender systems, which

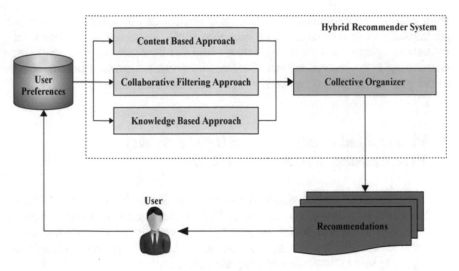

Fig. 1 Generic framework of a hybrid recommender system

recommends item based on ratings or behavior of other users whose rating profile is close to that of the target user [20]. Personalized recommendation significantly increases the likelihood of user compared with generic solutions.

The general approaches used in collaborative filtering are memory-based, model-based and hybrid approach. In the memory-based collaborative filtering, relationship between user and item is computed and user-item rating matrix builds. In model-based approaches, a model is built based on the user's profile or history using probabilistic methods. The hybrid approach integrates the advantages of both the memory-based and model-based approach to overcome the limitations and to improve the prediction performance [21].

Based on user ratings/feedbacks, hybrid collaborative filtering recommendation algorithm clusters similar users or items into a group to identify a neighborhood [22]. Then by measuring the similarity from item-based cluster and user-based cluster, personalized recommendation can be generated effectively based on the user's point of interest(POI) similar to that of the nearest neighborhood [23]. By integrating clustering technology with the collaborating filtering approaches, one can overcome the overgeneralization and scalability problem [24].

In addition opinion mining technology is also employed into the recommender system to refine the user's POI and tourism related decisions [25]. Opinion mining determines the attitude of the user with respect to any item or the overall contextual information and builds a profile. Contextual information can be implicit or explicit and can be inferred from user behavior [26]. By integrating all the factors and employed into the recommender system, hybrid collaborative filtering algorithm predicts and generates the list of tourism destination based on user's preferences.

For effective context-dependent travel recommendation, contextual factors such as time, location or social company and user's own sensibility to specific contextual factors should be considered [27]. Changing context can disappoint the user and significantly affect the travel related decisions [28]. The context-aware travel recommender system exploits the context of both the user and the item while recommending POI. The major issue of context-aware travel recommender system is to compute effective recommendation even with less preference data [29].

4 Personalized Context-Aware Hybrid Travel Recommender System (PCAHTRS)

The main objective of the proposed model is to design a hybrid collaborative filtering travel recommender system that offers personalized tourist spots based on his ratings and preferences. The proposed system combines the contextual information with the opinion mining and generates the top-N recommendations by employing hybrid collaborating filtering algorithm.

Our proposed context-aware hybrid travel recommender model comprises of three major phases. The clear organization of the proposed recommendation model is illustrated in Fig. 2. In the first phase, the recommender model is constructed by clustering the user's ratings on POI and then the similarity matrices are computed for each cluster.

In the second phase, we extract the text reviews for the major tourism factors from the database. Then by using opinion mining, we build the user preference

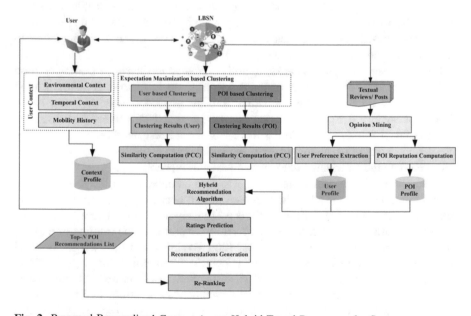

Fig. 2 Proposed Personalized Context-Aware Hybrid Travel Recommender System

profile and POI opinion reputations. Finally, the hybrid recommendation algorithm employs user-based similarity, POI-based similarity, implicit user profile and POI opinion reputation similarity to predict the ratings for a tourist spot. Based on the predicted ratings, a set of recommendation is generated.

And in the third phase, Contextual information such as current location, time and user's mobility history are considered to model a context profile based on POI. Then the context profile is used to re-rank the recommendations and generates Top-N tourist POI recommendations.

4.1 Clustering and Similarity Computation

In the first phase, user's ratings on tourist spots are clustered using an Expectation Maximization (EM) algorithm. To predict the effective, personalized recommendations, the target user is assigned to be one of the determined clusters. Then we perform Pearson Correlation Coefficient (PCC) on each cluster to estimate similarity between the users. Similarly, the same procedure is carried out for the POI based recommendations.

The EM algorithm works in an iterative manner to determine the maximum likelihood until the algorithm converges on a fixed point. Expectation and Maximization are two steps, used repeatedly to calculate the maximum likelihood of incomplete dataset. In E-step, unobserved or missing variable is estimated from the given observed parameters and in the M-step, new maximum likelihood is calculated from the completed data.

PCC is used to estimate correlation between two users or items. PCC measures the strength of the similarity between two users. The coefficient value can range between -1 (negatively correlated) and $+1$ (positively correlated). By applying PCC, users with similar taste are effectively revealed.

4.2 Opinion Mining

From the text review dataset, we extract the crucial information such as (i) features of POI, (ii) user's positive and negative opinion on each feature, (iii) POI reputation and (iv) user preferences. Then by employing opinion mining techniques, implicit user profile and POI reputations are built. At the end of opinion mining, each user and POI are assigned a value from the text review dataset. A beta probability density function is used to build POI opinion reputation rating matrix by all the reviewing users. To determine the user's preference, we use Inverse Document Frequency method to measure the weight of the feature. Because the feature word appears in many text reviews cannot effectively distinguish the user's preference rather than in fewer reviews.

In the final step, hybrid recommendation algorithm employs computed implicit user profile, POI reputations, and also the user-based similarity and POI based similarity computed in the first phase. By taking all into an account, hybrid recommendation algorithm predicts the ratings for a tourist place and generates the set of POI travel recommendations.

4.3 Context-Aware Recommendations

In this phase, recommendation system incorporates contextual information to build a context profile in order to provide satisfactory personalized tours for the target user. Contextual information such as user's current location, time, weather, POI, history of visited tours and its rating should be considered.

Moreover, new user with less prior ratings may not know all the needs and cause the recommender system to suffer from the cold-start problem. To overcome this problem, case-based reasoning method is employed to retrieve relevant POI that is similar to that of user's query case. Now the recommender system re-computes the ratings for all possible POIs and generates the new Top-N travel recommendations to the target user-based on his POI and some contextual information.

5 Experiments and Discussions

The major challenge in the development of recommender model is scalability and sparsity. In the proposed PCAHTRS, clustering and similarity prediction methods are used to overcome these issues In addition, opinion mining and contextual information is also employed to overcome the cold-start problem, and thereby increases the accuracy of the recommendation. To evaluate the model, a large-scale real-time datasets of Yelp and TripAdvisor is used. The results provided by recommender model showed that the use of contextual information with the aid of clustering, similarity computation and opinion mining is effective in improving the performance of the recommender system.

With regard to the scalability, the proposed model enhanced the scalability of the recommender model that use clustering and the similarity prediction technique is substantially higher than other methods. With regard to the sparsity, the hybrid collaborative filtering algorithm outperformed the baseline approaches. In addition, the model uses opinion mining technique for better prediction accuracy compared to the other models. The proposed model takes the opinion of the user as an input to determine the relationship between the POI and their related ratings. The improvement in the accuracy of the recommendation is because of the recommendation algorithm uses additional contextual information.

To evaluate the PCAHTRS recommender model, three standard evaluation metrics such as RMSE, Coverage and F-Measure is used to estimate prediction accuracy and user satisfaction. Figures 3, 4 and 5 presents the comparisons of the RMSE, Coverage and F-Measure over existing recommendation approaches. The recommendation model needs enough user feedbacks/ratings to generate valid recommendations; else this may lead to inaccurate recommendations. The prediction accuracy denotes the degree to which the generated recommendation suits the user's POI. And user satisfaction is evaluated by the degree to which the recommended tourist place impresses the user.

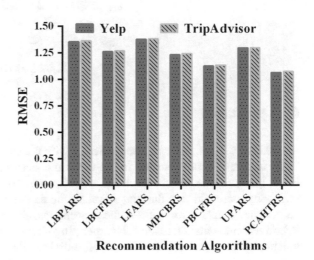

Fig. 3 Comparison of RMSE over different recommendation approaches

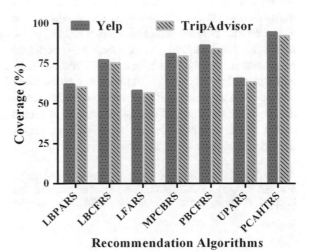

Fig. 4 Comparison of Coverage over different recommendation approaches

Fig. 5 Comparison of
F-Measure over different
recommendation approaches

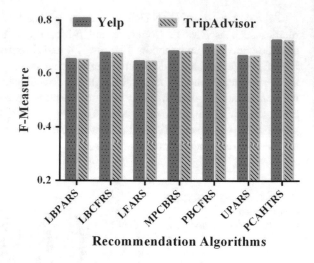

6 Conclusions

As the final section, we conclude the article with the analysis of the discussed works
to enhance the future research in the recommender systems domain. Every traditional recommendation approach suffers from various problems such as sparsity,
scalability, and cold start during generating the recommendations to the target user.
As a better solution for address these problems, the researcher can move to the
hybrid recommendation models. Incorporating user's contextual information is very
much helpful in solving the issues of individuality. Usage of the implicit and
explicit preferences of users extended with the semantic models addresses the
problem of uncertainty in the recommendation process. The present study proposed
the personalized context-aware hybrid travel recommender system based on the
user contextual information and opinion mining technique to improve the prediction
accuracy. We analyzed the accuracy and performance of the proposed recommendation model with the help of real-world datasets of Yelp and TripAdvisor. The
experimental results confirmed the enhanced quality of recommendations by means
of prediction accuracy and performance. As the success of the recommender system
lies in the satisfaction of the target user to whom the recommendations are generated, the researchers should be focused in the development of user satisfiable
recommender systems in future.

Acknowledgements Authors thank the Science and Engineering Research Board for their
financial support (YSS/2014/000718/ES). Authors also express their gratitude to SASTRA
Deemed University for the infrastructure facilities and support provided to conduct the research.

References

1. Batet, M., Moreno, A., Sánchez, D., Isern, D., Valls, A.: Turist@: Agent-based personalised recommendation of touristic activities. Expert Syst. Appl. **39**, 7319–7329 (2012)
2. Huang, Y., Bian, L.: A Bayesian network and analytic hierarchy process based personalized recommendations for tourist attractions over the Internet. Expert Syst. Appl. **36**, 933–943 (2009)
3. Niaraki, A.S., Kim, K.: Ontology based personalized route planning system using a multi-criteria decision making approach. Expert Syst. Appl. **36**, 2250–2259 (2009)
4. Noguera, J.M., Barranco, M.J., Segura, R.J., Martínez, L.: A mobile 3D-GIS hybrid recommender system for tourism. Inf. Sci. **215**, 37–52 (2012)
5. Agalya, D., Subramaniyaswamy, V.: Group-Aware recommendation using random forest classification for sparsity problem. Indian J. Sci Technol. 9(48), art. no. 107960 (2016)
6. Saipraba, N., Subramaniyaswamy, V.: Enhancing stability of recommender system: an ensemble based information retrieval approach. Indian J. Sci. Technol. 9(48), art. no. 107979 (2016)
7. Logesh, R., Subramaniyaswamy, V.: A reliable point of interest recommendation based on trust relevancy between users. Wireless Pers. Commun. **97**(2), 2751–2780 (2017)
8. Logesh, R., Subramaniyaswamy, V., Vijayakumar, V., Gao, X.Z., Indragandhi, V.: A hybrid quantum-induced swarm intelligence clustering for the urban trip recommendation in smart city. Future Gener. Comput. Syst. 1–38 (2017)
9. Logesh, R., Subramaniyaswamy, V.: Learning recency and inferring associations in location based social network for emotion induced Point-Of-Interest recommendation. J. Inf. Sci. Eng. **33**(6), 1629–1647 (2017)
10. Logesh, R., Subramaniyaswamy, V.: A collaborative location based travel recommendation system through enhanced rating prediction for the group of users. Comput. Intell. Neurosci. 1–28, art. no. 1291358 (2016)
11. Logesh, R., Subramaniyaswamy, V., Malathi, D., Senthilselvan, N., Sasikumar, A., Saravanan, P., Manikandan, G.: Dynamic particle swarm optimization for personalized recommender system based on electroencephalography feedback. Biomed. Res. **28**(13), 5646–5650 (2017)
12. Logesh, R., Subramaniyaswamy, V., Vijayakumar, V.: A personalized travel recommender system through utilizing social network profile and accurate GPS data. Electron. Gov. Int. J. (2017)
13. Subramaniyaswamy, V., Logesh, R.: Adaptive KNN based recommender system through mining of user preferences. Wireless Pers. Commun. **97**(2), 2229–2247 (2017)
14. Subramaniyaswamy, V., Logesh, R., Abejith, M., Sunil, U., Umamakeswari, A.: Sentiment analysis of tweets for estimating criticality and security of events. J. Organ. End User Comput. **29**(4), 1–20 (2017)
15. Subramaniyaswamy, V., Vijayakumar, V., Indragandhi, V., Logesh, R.: Data mining-based tag recommendation system: an overview. Wiley Interdisc. Rev. Data Min. Knowl. Disc. **5**(3), 87–112 (2015)
16. Subramaniyaswamy, V., Logesh, R., Chandrashekhar, M., Challa, A., Vijayakumar, V.: A personalised movie recommendation system based on collaborative filtering. Int. J. High Perform. Comput. Networking **10**(1–2), 54–63 (2017)
17. Arunkumar, S., Subramaniyaswamy, V., Devika, R., Logesh, R.: Generating visually meaningful encrypted image using image splitting technique. Int. J. Mech. Eng. Technol. **8**(8), 361–368 (2017)
18. Saravanan, P., Arunkumar, S., Subramaniyaswamy, V., Logesh, R.: Enhanced web caching using bloom filter for local area networks. Int. J. Mech. Eng. Technol. **8**(8), 211–217 (2017)
19. Senthilselvan, N., Udaya, Sree N., Medini, T., Subhakari, Mounika G., Subramaniyaswamy, V., Sivaramakrishnan, N., Logesh, R.: Keyword-aware recommender system based on user demographic attributes. Int. J. Mech. Eng. Technol. **8**(8), 1466–1476 (2017)

20. Nilashi, M., Ibrahim, O., Bagherifard, K.: A recommender system based on collaborative filtering using ontology and dimensionality reduction techniques. Expert Syst. Appl. **92**, 507–520 (2018)
21. Nilashi, M., Bagherifard, K., Ibrahim, O., Alizadeh, H., Lasisi, A., Roozegar, N.: Collaborative filtering recommender systems. Res. J. Appl. Sci. Eng. Technol. **5**, 4168–4182 (2013)
22. Pham, M.C., Cao, Y., Klamma, R., Jarke, M.: A clustering approach for collaborative filtering recommendation using social network analysis. J. Univers. Comput. Sci. **17**(4), 583–604 (2011)
23. Gong, S.J.: A collaborative filtering recommendation algorithm based on user clustering and item clustering. J. Softw. 5(7) (2010)
24. Kushwaha, N., Vyas, O. P.: SemMovieRec: extraction of semantic features of DBpedia for recommender system. In: Proceedings of the 7th ACM India Computing Conference. p.13 (2014)
25. Zheng, X., Luo, Y., Xu, Z., Yu, O., Lu, L.: Tourism destination recommender system for the Cold start problem. KSII T. Internet Info. Syst. 10(7) (2016)
26. Hariri, N., Zheng, Y., Mobasher, B., Burke, R.: Context-aware recommendation based on review mining. Gen. Co-Chairs (2011)
27. Adomavicius, G., Tuzhilin, A.: Context-Aware Recommender Systems. In: Ricci, F., Rokach, L., Shapira, B. (eds.) Recommender Systems Handbook. Springer, Boston, MA (2015)
28. Braunhofer, M., Ricci, F.: Selective contextual information acquisition in travel recommender systems. Inf Technol Tourism. (2017)
29. Bahramian, Z., Abbaspour, R.A., Claramunt, C.: A cold start context-aware recommender system for tour planning using artificial neural network and case based reasoning. Mob. Inf. Syst. (2017)

Aggregated Rank in First-Fit-Decreasing for Green Cloud Computing

Saikishor Jangiti, E. Sri Ram and V. S. Shankar Sriram

Abstract Virtual Machine (VM) placement in a cloud data center is a Vector Bin-Packing (VBP) problem to minimize the number of PMs used for hosting the given VM requests. First-Fit-Decreasing (FFD) variants are widely used for VM placement. In this paper, a novel FFD variant, Aggregated Rank in FFD (FFD-AR) is proposed for VM placement. Simulation experiments were carried out using two datasets: a dataset inspired by Amazon EC2 instances and another is a synthetic dataset. The packing efficiency of the proposed FFD-AR results is better as compared to all the other baseline FFD variants. We believe the proposed FFD-AR can be applied to wide applications of VBP like production planning and logistics.

Keywords Cloud computing · VM placement · VM consolidation
Infrastructure as a service · Vector Bin-packing

1 Introduction

Hassle-free provision and management of computing infrastructure is the most important factor in the growth of cloud computing. According to Gartner [1], "*By the year 2020, a corporate not using cloud computing will be rare*". Cloud Data Centers (CDCs) receive millions of infrastructure requests from clients across the globe. Each request is allotted with a Virtual Machine (VM) as per the resource demands by sharing the actual resources of a Physical Machine (PM) to multiple requests. One important focus of Cloud Service Provider (CSP) high availability (99.999%) of services and it leads to high energy consumption in CDCs. CSP needs to provide services at low cost without compromising quality to attract customers

S. Jangiti (✉) · V. S. Shankar Sriram
Centre for Information Super Highway (CISH),
SASTRA Deemed University, Thanjavur, Tamilnadu, India
e-mail: sai@cse.sastra.edu

E. Sri Ram
School of Computing, SASTRA Deemed University, Thanjavur, Tamilnadu, India

© Springer Nature Singapore Pte Ltd. 2019
P. K. Mallick et al. (eds.), *Cognitive Informatics and Soft Computing*,
Advances in Intelligent Systems and Computing 768,
https://doi.org/10.1007/978-981-13-0617-4_53

and compete in the industry. A recent survey by the Department of Energy, USA has shown that VM consolidation used in a data center can save up to 520 billion kWh of energy by 2020. VM consolidation minimizes electricity costs and makes CSP achieve the objective of low-cost resource provision without compromising the quality and trust [2].

The Information Technology (IT) field is undergoing an astonishing transformation, with the advent of technologies like cloud computing, virtualisation, IoT, data analytics, machine learning, etc. One of these key technologies of the day, the cloud computing allows on-demand access to the remote IT infrastructure and services over the internet in a pay per use model. Infrastructure as a Service (IaaS), Platform as a Service (PaaS) and Software as a Service (SaaS) are the well-known cloud offerings among "anything as a Service" (XaaS) models provided by a CSP. This paper focuses on reducing the energy consumption in a cloud data center offering IaaS. IaaS offers computing infrastructure in the form of Virtual Machines (VMs). Each VM is allotted with the required parts of the actual Physical Machine (PM) resources. The allotment of VMs onto a minimum number of PMs is known as VM placement. VM placement is an NP-Hard problem called Vector Bin-Packing (VBP) problem, also known as multi-capacity or multi-dimensional bin-packing problem.

Many heuristics came up for solving the VBP problem. Figure 1 presents a typical VBP problem with five vector items each with four resource requirements, also a solution packed into two bins. After the advent of virtualisation and cloud, the VBP problem has again started demanding a better consideration. Here in VM placement, both VMs and PMs have multiple resources and is a typical

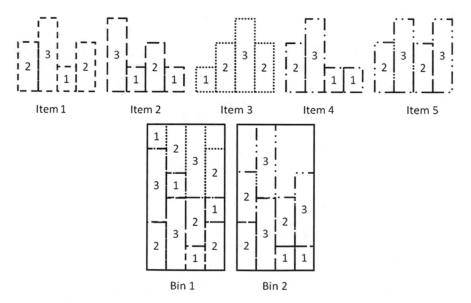

Fig. 1 VBP of 5 items into two bins, both bin and item have four capacities

VBP. A resource capacity of all the VM requests should be less than the corresponding PM resource capacity and is ensured by validating all the VM requests as follows:

$$\text{VM}_j \cdot R_i \leq \text{PM} \cdot R_i \quad \forall j \in \{1, 2, 3, \ldots, n\}, \ \forall i \in \{1, 2, 3, \ldots, r\}$$

A cloud data center receives a large number of VM requests with different resource requirements and the cloud data center model considered for experimentation of VM placement simulation is as shown in Fig. 2. First-Fit Decreasing (FFD) is a 1D bin-packing heuristic, and it produces an approximation solution for 1D bin-packing. Next-Fit Decreasing and Best-Fit Decreasing are two other approaches for filling 1D bins. VM placement is a VBP, and there is no exact or approximate algorithm for VBP. Even though FFD can do the 1D packing, by introducing a vector to scalar conversion, the FFD pseudo code is given in Fig. 3.

Several variants of FFD like FFDProd, FFDSum, FFDAvgSum, FFDExpSum are in use to solve VBP. Panigrahy et al. [3] studied these FFD variants and designed new geometric heuristics, Norm-Based Greedy (NBG) and Dot Product (DP) for VM Placement. DP and NBG dominated the other heuristics in solution quality. Microsoft Cloud Platform system center (Virtual Machine Manager) also uses DP and NBG as an internal function for VM consolidation [4]. Wei Zhu et al. [5] recently used the DP method based on PM residual resource capacities and proposed a vector packing, Heuristic Virtual Resource Allocation Algorithm (HVRAA).

DP and NBG dominated all the FFD variants by producing energy-efficient packing results, but the employment of *continuous sorting* to determine the largest item among remaining items upon placing each item is delaying the packing speed,

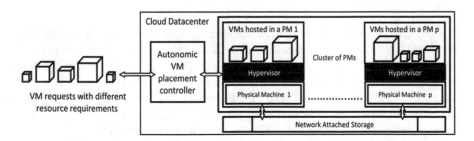

Fig. 2 Cloud data center model considered for VM Placement experiments

```
First Fit Decreasing (FFD)
1. Sort items in decreasing order of size
2. For each item i = 1 to n do
3.    Place item i in the first bin where it will fit.
4. end for
```

Fig. 3 FFD placement strategy [3]

Table 1 Time complexity analysis

VBP heuristic	Time complexity
FFDProd [3]	$O\ (n \log n)$
FFDSum [3]	$O\ (n \log n)$
FFDAvgsum [3]	$O\ (n \log n)$
FFDExpSum [3]	$O\ (n \log n)$
Dot product [3]	$O\ (n^2 \log n)$
Norm-based greedy [3]	$O\ (n^2 \log n)$
HVRAA [5]	$O\ (n^2 \log n)$
Proposed FFD-AR	$O\ (r\ n \log n)$
n—number of VMs, r—number of resources	

and also it is intractable when items are too large. The time complexities of all the FFD variants are listed in Table 1. By observing the time complexities and knowing that usually r < n, the proposed FFD-AR will do a faster packing than DP, NBG and HVRAA. Not only better speed FFD-AR also dominated these three VBP algorithms in better packing efficiency, which results energy efficiency in the cloud data center.

The organization of the remaining portion of this paper as follows: The VBP problem is defined formally in Sect. 2. Section 3 details the proposed FFD-AR heuristic approach. Section 4 compares the simulation results of FFD-AR with other baseline heuristics. Finally, Sect. 5 provides the conclusions of the study and future work.

2 Problem Formulation

In CDCs, while dealing with VM placement as a VBP problem, it is necessary to consider, the multiple resources in a PM like CPU, memory, disk storage, and network bandwidth. The energy consumption of PMs holds a major part of the total energy consumption in a CDC. Another point to be highlighted here is a PM consumes 60% of its maximum power consumption even when idle, although PM's energy consumption is proportional to its CPU utilization. Hence, our focus is directly minimizing the number of PMs used to host the given VMs.

In this VM placement research work, it is considered that each of the n VMs and all the PMs consist of four resources. Let matrix P of size $b \times n$ be the outcome of hosting n VMs using b PMs, where $P_{ki} = 1$, if VM_i is hosted in PM_k and $P_{ki} = 0,$, otherwise. Let $f(PM_k)$ be a function such that $f(PM_k) = 1$, if PM_k is hosting minimum one VM and $f(PM_k) = 0$, otherwise. The VM placement problem of assigning the given VMs onto a minimum number of PMs and the constraints involved in the placement can be formulated as:

$$\text{minimize } \sum_{k=1}^{b} f(\text{PM}_k) \tag{1}$$

$$\text{subject to } \sum_{i=1}^{n} P_{ki} \cdot R_j \leq PM \cdot R_j \; \forall j \in \{1, 2, \ldots, r\}, \forall k \in \{1, 2, \ldots, b\} \tag{2}$$

$$\sum_{k=1}^{b} P_{ki} \leq 1 \quad \forall i \in \{1, 2, 3, \ldots, n\} \tag{3}$$

Equation (1) defines the objective of the optimisation problem, i.e., to minimize the PMs used while satisfying the constraints in Eqs. 2 and 3. The constraint in Eq. 2 imposes a limit on maximum utilization of all the r resources in a particular PM. The constraint verified by Eq. 3 is to ensure that a VM is placed in not more than one bin.

3 First-Fit Decreasing—Aggregate Rank

First-Fit Decreasing (FFD) variants are widely used to solve the VM placement problem to minimize the number of PMs used to host the given VM requests. In this paper, a novel FFD variant, Aggregated Rank in FFD (FFD-AR) is proposed. The FFD algorithm in Fig. 3 initially sorts all the 1D items in descending order of their item-size and places all the items into bins using First-Fit strategy (Step 3). Multi-capacity items are having multiple sizes represented as a vector of sizes. Any FFD variant needs to combine the vector of sizes into a scalar-value to determine the current largest item and to make use of FFD strategy. For example, sum or product of all the values in the vector results in a scalar. While all the FFD variants used for VBP are commonly using FFD strategy, the main difference in these variants is the model used for the vector to scalar conversion. The FFD variant called Dot Product is exhibiting better packing efficiency among the FFD variants, however it is suffering from a limitation—repeated sorting [6]. The proposed FFD-AR is free from the above limitation.

The vector to scalar conversion models employed by other FFD variants is described in Table 2. The proposed FFD variant, FFD-AR assigns a rank to VM by aggregating all the ranks of various resource demands to determine the largest item. The item with the largest rank is the largest item. Algorithm 1 details the VM placement using Aggregated Rank in FFD.

Table 2 Vector to scalar conversion methods of FFD variants

FFD-variant	Vector to scalar conversion method
FFDProd [3]	Product of resource demands of a VM
FFDSum [3]	Weighted sum of resource demands of a VM
FFDAvgsum [3]	Weighted sum using average demand by all VMs as a weight for the same demand
FFDExpSum [3]	Exponential for average demands as weights
Dot product [3]	Sum of products of remaining capacities of a PM to average resource requirement
Norm-based greedy [3]	Difference between a VM resource demand and remaining resource capacity in PM as weights
HVRAA [5]	Dot product based

Algorithm 1: First Fit Decreasing with Aggregate Rank (FFD-AR)

Input: pmList, vmList **Output:** VM_placement_map
1 sort the vmList based on aggregate rank
2 **for each** vm in vmList **do**
3   **for each** pm in pmList **do**
4   **if** (isAllocable(vm,pm)) **then**
5   VM_placement_map.place(pm,vm)
6   **break**
7 **return** VM_placement_map

3.1 Aggregated Rank and Its Working

Let us consider six VMs each with a requirement vector of four resources as follows.

$$VM_0 = \{512, 624, 256, 342\}$$

$$VM_1 = \{1024, 356, 212, 256\}$$

$$VM_2 = \{800, 912, 256, 624\}$$

$$VM_3 = \{356, 256, 128, 512\}$$

$$VM_4 = \{200, 400, 624, 812\}$$

Table 3 Ranks of VMs concerning each requirement

Rank	Order of VMs concerning each requirement			
	Requirement 1	Requirement 2	Requirement 3	Requirement 4
1	VM_4	VM_3	VM_3	VM_1
2	VM_3	VM_1	VM_1	VM_0
3	VM_0	VM_4	VM_0	VM_3
4	VM_2	VM_0	VM_2	VM_5
5	VM_1	VM_5	VM_5	VM_2
6	VM_5	VM_4	VM_4	VM_4

Table 4 Aggregated Ranks of VMs

Aggregated rank	VM
7	VM_3
10	VM_1
12	VM_0
16	VM_4
19	VM_2
20	VM_5

$$VM_5 = \{1024, 624, 256, 512\}$$

The VM_0 strictly needs 512 "units" of the first resource and 624 units of the second resource, etc., all together in a single PM. Table 3 shows the sorted VMs concerning each requirement. Table 4 presents the aggregated rank of each VM. The aggregated rank of a VM is the sum of ranks of the particular VM across all the resource requirements. The aggregated rank of VM_3 is $2 + 1 + 1 + 3 = 7$.

Thus, according to aggregated ranks, the decreasing order of items for use in FFD is $VM_5 \geq VM_2 \geq VM_4 \geq VM_0 \geq VM_1 \geq VM_3$. The next largest item is VM_5.

4 Results and Discussion

4.1 Experimental Setup

The VM placement simulation experiments were carried out in Intel® Core ™ i7 processor @ 2.5 GHz system with 4 GB RAM running Windows 10 operating system. Both the existing VBP algorithm HVRAA and the proposed FFD-AR are implemented in C++ environment.

Table 5 Standard VM
instances from Real Dataset –
EC2 [7], their resource ratio
and sizes

S. no	Instant name	Cores (size)	RAM (GB)
1	t2 . micro	1	1
2	t2 . small	1	2
3	t2 . medium	2	4
4	m3 . medium	1	3.75
5	m3 . large	2	7.5
6	m3 . xlarge	4	15
7	m3 . 2xlarge	8	30
8	c3 . large	2	3.75
9	c3 . xlarge	4	7.5
10	c3 . 2xlarge	8	15
11	c3 . 4xlarge	16	30
12	c3 . 8xlarge	32	60
13	r3 . large	2	15.25
14	r3 . xlarge	4	30.5
15	r3 . 2xlarge	8	61
16	r3 . 4xlarge	16	122
17	r3 . 8xlarge	32	244

4.2   Datasets Description

A standard dataset for VM placement called **Real Dataset—EC2** [7] is used,
Amazon EC2 IaaS instances inspire all the VM instances in this dataset. Real
Dataset-EC2 consists of compute-optimized and memory-optimized instances five
each, and the rest seven are general purpose instances as shown in Table 5. Another
synthetic dataset has been generated with varying number of VM requests and sizes
to test the packing efficiency and speed of the proposed algorithm.

4.3   Related Work

Even though the engineering applications of bin-packing problem date back to
40 years, VM placement shed more light on VBP in the recent time. Wu et al. [8]
solved the VM placement problem using a genetic algorithm focusing on minimizing PMs and network bandwidth. Furlong et al. [9] proposed and compared the
performance of six variants of FFD namely PM Load, Percentage Utilization,
Absolute PM Capacity, FFD Sorted PM and FFD Residual Load algorithms.
Metaheuristics like ant colony system algorithm [10], particle swarm optimisation
[11], and cuckoo search optimisation [12] were also applied for VBP. However, in
the literature of above metaheuristics, as well as heuristics proposed by Furlong
et al., the VBP of VM requests is tested with not more two resources. Stillwell et al.

[13] proposed and evaluated the packing efficiency of few more FFD variants and also empirically proved that FFD variants outperforms a genetic algorithm for VBP. Panigrahy et al. [3] studied all the baseline FFD variants suitable for VBP of VM requests with more than two resource demands (any number of resources in general) and suggested two new FFD variants using residual resources in PM as weights to overcome the limitations in baseline variants. The FFD variant using Dot Product outperformed all the other FFD variants used for performance comparison.

4.4 Simulation Experiments and Performance Comparison

In this paper, a novel FFD-AR heuristic for the VM placement, an optimisation problem is implemented as a VBP and its packing efficiency as well as speed is evaluated against another well-known heuristic, HVRAA [5]. HVRAA uses a Dot product based FFD approach for VM placement. Dot Product based FFD is also used by Microsoft's Virtual Machine Manager [4].

To test the packing efficiency and processing speed of the proposed FFD-AR compared to the existing VBP heuristics, FFD-AR is compared with HVRAA using 3000 VM requests both from Real Dataset EC2 and the synthetic dataset. Figure 4 shows that FFD-AR is carrying out an efficient packing of VM requests compared to HVRAA by utilizing less PMs in all the rounds. The same set of VM requests is supplied for both the VBP heuristics starting from a sample size 100 up to 3000. Figure 5 shows that FFD-AR is processing faster compared to HVRAA. Also, FFD-AR processing time is not affected as much HVRAA processing time is affected according to the number of VM requests supplied. Table 6 is also listing the details about the total number of PMs used by HVRAA and proposed FFD-AR along with the processing time taken.

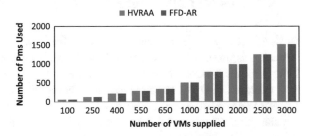

Fig. 4 A comparison of PMs used by HVRAA and proposed FFD-AR

Fig. 5 A comparison of processing time taken by HVRAA and proposed FFD-AR

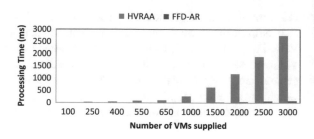

Table 6 PMs used and processing times of HVRAA and proposed FFD-AR

S. no	No. of VMs	PM size in		PMs used by		Processing time	
		Resource 1	Resource 2	HVRAA	FFD—AR	HVRAA	FFD—AR
1	100	50	50	58	58	6	3
2	250	150	175	126	126	35	7
3	400	4000	400	217	216	48	4
4	550	450	435	285	284	89	6
5	650	720	602	337	337	109	8
6	1000	900	800	508	508	268	15
7	1500	500	500	788	787	635	28
8	2000	100	200	994	994	1185	45
9	2500	1500	1500	1256	1255	1883	70
10	3000	1000	1000	1525	1524	2746	78
Total				6094	6089	7004	264

5 Conclusion and Future Work

FFD-AR, a novel VM placement heuristic for VBP of VM requests with multiple resources is proposed in this paper. Simulation experiments conducted showed that FFD-AR uses fewer PMs compared to other VBP heuristics to host the same set of VM requests. FFD-AR supports the green computing in a cloud data center offering Infrastructure as a service. The number of PMs used directly affects the power consumption of the cloud data center. The use of FFD-AR reduces operational expenses. Also, the speed and scalability of the VM Placement are improved. In future, we will focus on further improvements in the optimal VM placement and deal with heterogeneous PMs.

Acknowledgements The authors thank the Department of Science and Technology for their financial support (SR/FST/ETI-349/2013) under Fund for Improvement of S&T Infrastructure in Universities and Higher Educational Institutions.

References

1. Gartner Says By 2020, a Corporate "No-Cloud" Policy Will Be as Rare as a "No-Internet" Policy Is Today, http://www.gartner.com/newsroom/id/3354117
2. Somu, N., Kirthivasan, K., VS, S.S.: A computational model for ranking cloud service providers using hyper graph based techniques. Futur. Gener. Comput. Syst. **68**, 14–30 (2017)
3. Panigrahy, R., Talwar, K., Uyeda, L., Wieder, U.: Heuristics for vector bin packing. Res. Microsoft., 1–14 (2011)
4. Lee, S., Prabhakaran, R.P.V., Ramasubramanian, V., Uyeda, K.T.L., Wieder, U.: Validating heuristics for virtual machines consolidation. Res. Microsoft. 81–97 (2010)
5. Zhu, W., Zhuang, Y., Zhang, L.: A three-dimensional virtual resource scheduling method for energy saving in cloud computing. Futur. Gener. Comput. Syst. **69**, 66–74 (2017)
6. Jangiti, S., Shankar Sriram, V.S.: Scalable and direct vector bin-packing heuristic based on residual resource ratios for virtual machine placement in cloud data centers. Comput. Electr. Eng. **68**, 44–61 (2018). https://doi.org/10.1016/j.compeleceng.2018.03.029. ISSN 0045-7906
7. Zheng, Q., Li, R., Li, X., Shah, N., Zhang, J., Tian, F., Chao, K.M., Li, J.: Virtual machine consolidated placement based on multi-objective biogeography-based optimization. Futur. Gener. Comput. Syst. **54**, 95–122 (2016)
8. Wu, G., Tang, M., Tian, Y.C., Li, W.: Energy-efficient virtual machine placement in data centers by genetic algorithm. In: Lecture Notes in Computer Science (including subseries Lecture Notes in Artificial Intelligence and Lecture Notes in Bioinformatics), pp. 315–323 (2012)
9. Furlong, J., Shi, L., Wang, R.: Empirical evaluation of vector bin packing algorithms for energy efficient data centers—Vinicius (consolidação). In: 2013 IEEE Symposium on Computers and Communications (ISCC) (2013)
10. Gao, Y., Guan, H., Qi, Z., Hou, Y., Liu, L.: A multi-objective ant colony system algorithm for virtual machine placement in cloud computing. J. Comput. Syst. Sci. **79**, 1230–1242 (2013)
11. Xiong, A., Xu, C.: Energy efficient multiresource allocation of virtual machine based on PSO in cloud data center. Math. Probl. Eng. **2014**, 1–8 (2014)
12. Sait, S.M., Bala, A., El-Maleh, A.H.: Cuckoo search based resource optimization of datacenters. Appl. Intell. **44**, 489–506 (2016)
13. Stillwell, M., Schanzenbach, D., Vivien, F., Casanova, H.: Resource allocation algorithms for virtualized service hosting platforms. J. Parallel Distrib. Comput. **70**, 962–974 (2010)

Real-Time Vehicle-Type Categorization and Character Extraction from the License Plates

Sneha Pavaskar and Suneeta Budihal

Abstract Modern-day Intelligent Transportation System (ITS) needs heavy attention due to presence of existing infrastructure of roadways that is worsening with growing traffic. Thus, monitoring and controlling the traffic becomes a tedious task, hence automatic control is required rather than manual controlling. The vehicles are classified into the following types: Bike, Car, Auto-rickshaw, and HMV. Detection of vehicles is the main key task for the classification and also keeping a count of it. The proposed framework includes the vehicle-type classification by considering two features namely, contour formation for detection of vehicles and other is the concept of convex hull, which helps in classifying the vehicles. Text extraction from the vehicle license plates is another necessary task for ITS. KNN algorithm is used to create the xml les that is utilized to identify the characters and accordingly display it on the image and displays its effectiveness. This technique holds good for both single-lined and double-lined license plate reading efficiently, the concept of Tesseract-OCR is also used for character recognition and then the correctness is being compared for their effectiveness.

Keywords Convex Hull · Heavy motor vehicles (HMV) · Intelligent transport systems (ITS) · K-Nearest neighbors (KNN) · Tesseract-OCR (Optical character Recognition)

1 Introduction

According to the current survey, most of the places around the world are equipped with surveillance cameras, which are capable of defining different intelligent alert systems. Trending interest of people in Image Processing can be widely classified

S. Pavaskar · S. Budihal (✉)
School of ECE, KLE Technological University, Hubballi, India
e-mail: Suneeta_vb@bvb.edu

S. Pavaskar
e-mail: pavaskarsneha@gmail.com

into major divisions as enhancement and improvisation in properties of the images that could help for easier human understanding, and other look out for the division would be processing the acquired images that could be compressed for data handling, storage, transmission and making the data compatible with external perception of the machine.

Video analytics being active topic for research becomes a demanding issue for the development and production of wide range in Intelligent Transport Systems. Video analytics becomes more crucial for its exactness in operations when the background clutter, placement of the video capture devices, orientation and poses of vehicles are denser than usual environments. This instance could be compared with traffic on highway and traffic at the important junctions of the city. The recognition and classification of the vehicles depends on the shape of the vehicle, size of the vehicle, different orientation of the vehicle and many more other external features for their detection, tracking and recognition during the analysis of particular incidents. We can observe variations in the density of the traffic when considered on a daily basis, thus accurate analysis is required after understanding the nature of the location that is to be monitored.

Camera-based vehicle recognition is categorized into detection, classification and identification. The important stage in the whole process is extracting the license plate from the video sequence or the image. This could be the most difficult part due to changing weather condition and the quality of the video frame capture that might be blurry, foggy, illumination, etc. The degraded frames of vehicle images require extensive computational processing.

2   Related Work

Normally, the related papers with active traffic video surveillance is centered around tallying vehicles utilizing fundamental image processing systems to acquire insights about the resource utilization. However, there are many works that point to give complex evaluations of vehicle elements and their features, measurements to order them as light motor or heavy motor vehicles (LMV/HMV). In [1] the exemplars of each category in parameterized 3D models is used with 2D projection of camera viewpoint. Thus the projections decide the vehicle class. In [2] the knowledge of 3D projections based on different approximations for classification. The 3D structure defining vehicle lengths and heights is projected on the roads and its matching, classifies the vehicle.

In [3] the method of deep learning and deep neural networks is used, where deep learning is used for feature extraction aspects and shallow learning is used for feature dimension extraction leading to vehicle classification. In [4] weakly supervised Convolutional Neural Network (CNN) with training relied on image-level labels for vehicle detection and recognition system. In [5] the vehicle detection and other morphological operations are carried out with the usage of Gaussian mixture models and vehicles are classified based on vehicle features and centroid measurement, KNN as a classifier for distinguishing the vehicles based on virtually divided detection zones in the video.

In [6] the license plates are segmented using the sliding concentric windows which are later probabilistic neural networks for the alphanumeric detection and veri ed. To detect and recognize the vehicle-plates in almost every environments. In [7] fuzzy logic disciplines tries to extract license plates from the image, which are then conceptualized by neural subjects aiming to identify the numbers. In [8] there is compilation of different technologies that have been used for the license plate recognition, basically the usage of statistical/hybrid models and other classifiers for feature extractions by quantization's and other algorithms, referred for character recognition. In [9] template matching with OCR is used for license plate recognition and also using the Maximum Average Correlation Height (MACH) filter, Log r-theta Mapping techniques to recognize the type of vehicle.

3 Proposed Methodology

The proposed system focuses to solve the two main troubles prevalent in educational Intelligent Transport Systems, namely keeping a track of license plate numbers and type of vehicles. The system consists of two major components.

Source for video capture.

The application that is developed for the stated problem.

The block diagram for the proposed system is as shown in Fig. 1. According to the flow diagram, the processing of such data is carried out over the sequence of images or on a video. Then the constant background is nullified, so that one can define the unusual changes in the image after its preprocessing. Then define the algorithms that categorizes the vehicle-type and other part extracts the characters

Fig. 1 Proposed system-flow diagram

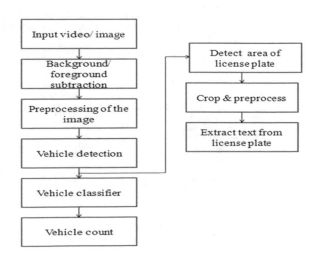

from the license plates. The classification process holds good for single lane traffic and the license plate detection is carried out with the images obtained.

3.1 Vehicle-Type Classification

Majority of these frameworks utilize limited identification zones for tallying vehicles. Once the location zones are set apart on the picture, the pixel estimates in every identification zone is checked for a change after some time. Joining this straightforward procedure with a few heuristics gives precise vehicle numbers in ideal conditions.

The vehicle categorization is carried out in accordance to the algorithm stated in Fig. 2, which totally depends on the computer vision techniques. To begin with the process, firstly we define the bounding rectangles that are supposed to be bound around the vehicles which are defined on the aspect ratio that is calculated from the bounding height, bounding width, bounding area, and diagonal size. Equations (1)

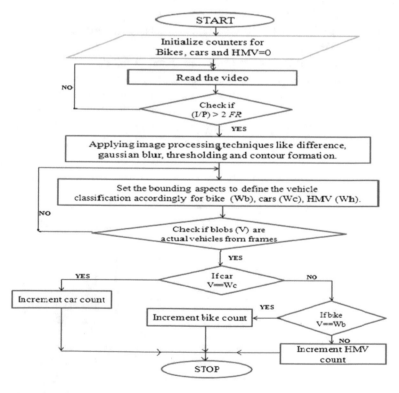

Fig. 2 Flow-chart for vehicle classification

and (2) describes the basic step to determine the blob size and the contour formed around the vehicle.

$$\text{Aspect Ratio} = \frac{\text{Height}}{\text{Width}} \qquad (1)$$

$$\text{Area Ratio} = \frac{\text{Area}}{\text{Region of interest}} \qquad (2)$$

When the video is being played, then always two frames are considered for the process of classification. To check the frames for not being duplicated such verification is necessary. Thus the considered frames have to undergo frame difference measurement.

Frame Difference Measurement being the simple and easy method to work with, and also helps in subtracting the background and retain the required information from the video images. It looks at each frame of the video to the past edge, and just passes the pixels that have changed. So it highlights moving articles and makes stationary things vanish.

Now the convex hull, which are the connecting points of the contours formed at the boundaries and is the size of it. The convex hull created is now parameterized for its values that will undergo comparison with the actual aspects set for the vehicle categorization. The distinguishing parameters would be the bounding rectangle area, height, width, diagonal size, aspect ratios that are set within the maximum and minimum range and is counter-checked by the contour area and actual size of the current contour detect.

This step being the major one determines the actual categorization of the vehicles. As blob around the vehicle gets closer to the virtual line drawn then the threshold of the blobs set and the contour-blob formed is compared. Thus, vehicle gets detected and recognized, then the counter increments as per its type and the image of vehicle is saved.

3.2 Character Extraction from License Plates

Normally, LPR framework contains four preparing steps. The initial step is to get the picture or video from the camera. The second step is LP identification from the picture. The third step is to extricate the characters from the LP and the last step is to perceive the separated characters utilizing diverse classifiers. These four stages can be accomplished by the blend of various systems of picture preparing and pattern recognition. The process-flow is as explained in Fig. 3.

The alphabet-based aspect ratio is compared with the set value and tries to check with the knn training set. The KNN training is carried out by preparing two sets of xml les namely the classifications xml and the images xml. These two xml's determine the training of the alphabet and the numerical by bounding the rectangle

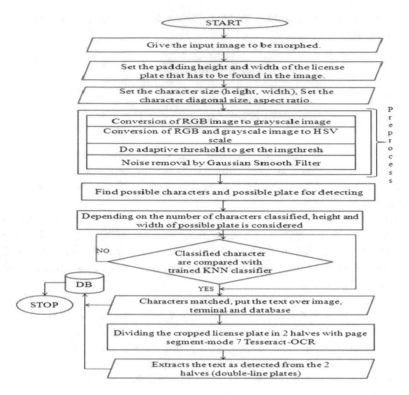

Fig. 3 Flow-chart for character extraction from the license plates

over it. That carries out all the basic image processing techniques and recognizes the characters by bounding the letters with green box around the letter. This indicates that the character is being identified and its features are being extracted and stored in the xml les. Hence the actual character from the license plates, then it sets the characters aspect bounds and then compares with the xmls created by its training. As the possible plate is detected based on the maximum number of characters detected in a row, thus the KNN gets applied to it.

Tesseract being the pre-trained library of characters which is present in various languages. The extracted license plate is set within the rectangle and is considered as an image, and the characters from the rectangle are displayed.

4   Results and Analysis

In this section, we first describe the setup of our implementation environment and the traffic surveillance vehicle dataset. Classification experiments and location prediction performance are then evaluated.

4.1 Implementation Setup and Datasets of Traffic Surveillance

To implement our vehicle recognition system, a computer with 32 GB memory and 4 GB RAM is employed. The program runs on a 64-bit Ubuntu 16.04 LTS operating system with QT-5.7, OpenCV-3.2 with other added packages like OpenCV-Contrib, Tesseract-OCR. SQLite3 for database connection is installed. The captured images have resolution of 1860 1024 pixels. The length of considered videos is of about 15 min.

4.1.1 Vehicle Classification

The dataset was recorded by traffic surveillance cameras installed at two different places, i.e., at KLE bridge, Belgaum and the other at the under bridge as shown in Figs. 4 and 5 respectively. The vehicles aspect ratio of the contours formed, decides the type of vehicle. The vehicle-type determined is further counted to determine its flow over the particular pathway. This data could be further used for other Intelligent Traffic Systems.

Fig. 4 Video clip of an underbridge

Fig. 5 Outcome of vehicle classification from a live capture of KLE still

4.1.2 Character Extraction from License Plates

The images considered are of size 1080 × 1024. The whole image is scanned for vehicle plate and is extracted for feature extraction and feature tracking speed for the vehicles. The preprocessing and the results are discussed for single-lined license plates in Fig. 6 and for double-lined license plate in Fig. 7 is given. Table 1 describes the correctness of the system to detect the characters, accordingly based on the trained xml.

Fig. 6 Output of single-lined license plate with characters extracted

Fig. 7 Output of double-lined license plate with characters extracted

Table 1 Percentage correctness of license plate extraction

Sl. no.	Detected license plates numbers			
	Original	Extracted	KNN	Tesseract-OCR
1.	GM 2169-13	GM 2169-13	100	100
2.	GM 1733-12	GM 1733-J2	87.5	90
3.	GW 9177 Y	GW 9177 V	85.7	90
4.	GR 3982-09	GR 39BZ-09	75	85

5　Conclusion

We have used easy and relatively effective convex hull and aspect ratio concept for vehicle counting and classification. It helps to differentiate between Bike, Car, and HMV (Heavy Motor Vehicles). Contours pointing around the boundary of the vehicle in the first frame, then detection of the vehicles are initiated and convex hull forms rectangle on the vehicles, considering the parameter like Bounding area, aspect ratio, diagonal size, height and width of the frame. Classification of the vehicles is successfully completed, thus the results with counting and classification of the vehicles gives about 85% accuracy.

The training carried out with KNN for about 15 different font styles of alphabet and numerals is converted to xml les, namely the classification.xml and images.xml which has exactly recognized and produced the same characters. The process of learning both single-lined and double-lined license plates was checked and verified with both KNN and Tesseract-OCR for its correctness and thus successful extraction of text from the license plate is stored into the database. Table 1 describes the effectiveness of the techniques like KNN and Tesseract-OCR which decides the efficient method for character extraction from the snapshots captured.

References

1. Gupte, S., Masoud, O., Martin, R.F.K., Papanikolopoulos, N.P.: Detection and classification of vehicles. IEEE Trans. Intell. Transp. Syst. **3**(1), 37–47 (2002)
2. Baker, K.D., Sullivan, G.D.: Performance assessment of model-based tracking. In: Proceedings of IEEE Workshop Applications of Computer Vision, Palm Springs, CA, p. 2835 (1992)
3. Zhang, Z., Xu, C., Feng, W.: Road vehicle detection and classification based on deep neural network, acknowledged by National Natural Science Foundation of China. In: The Fundamental Research Funds for the Central Universities, IEEE, pp. 675–678 (In press, 2016)
4. Jiang, C., Zhang, B.: Weakly supervised vehicle detection and classification by convolutional neural network. In: 9th International Congress on Image and Signal Processing, Bio Medical Engineering and Informatics, pp 570–575 (2016)
5. Seenouvong, N., Watchareeruetai, U., Nuthong, C.: Vehicle detection and classification system based on virtual detection zone. In: 13th International Joint Conference on Computer Science and Software Engineering (JCSSE) (In press, 2016)
6. Chang, S.L., Chen, L.S., Chung, Y.C., Chen, S.W.: Automatic license plate recognition. IEEE Trans. Intell. Transp. Syst. **5**(1), 42–53 (2004)
7. Anagnostopoulos, C.N.E., Anagnostopoulos, I.E., Psoroulas, I.D., Loumos, V., Kayafas, E.: License plate recognition from still images and video sequences: a survey. IEEE Trans. Intell. Transp. Syst. **9**(3), 377–391 (2008)
8. Du, S., Ibrahim, M., Shehata, M., Badawy, W.: Automatic license plate recognition (ALPR): A state-of-the-art review. IEEE Trans. Circ. Syst. Video Technol. **23**(2), 311–325 (2013)
9. Puranic, A., Deepak, K.T., Umadevi, V.: Vehicle number plate recognition system: a literature review and implementation using template matching. Int. J. Comput. Appl. **134**(1), 12–16 (2016)

Ranking Decision Rules Using Rough Set Theory—A Comparative Analysis

M. K. Sabu

Abstract Rule induction is one of the fundamental tasks in knowledge discovery system. The aim of rule mining is to discover hidden and interesting patterns from data contained in large datasets. Rough Set Theory provides efficient mathematical models suitable for decision rule induction especially in the case of inconsistent datasets. To develop measures for extracting significant rules automatically from a large number of rules generated by a data mining system is a challenging problem in rule generation. In this paper, decision rules produced by a conventional rule mining system are ranked based on a rule evaluation measure called Degree of Rule Significance. Based on this measure a rule ranking algorithm is designed to rank the generated decision rules. The ranking given by the algorithm is evaluated with the help of two rough set based rule evaluation approaches. Experimental results show the effectiveness of Degree of Rule Significance to identify important rules.

Keywords Rule induction · Rough set theory · Rule evaluation
Reduct rules · Discernibility matrix

1 Introduction

Rough Set Theory (RST), proposed by Z. Pawlak is a mathematical approach to handle vagueness present in data acquired from experience. Basic concepts of this theory are classifications and categories. In a dataset the collection of attributes is divided into two—a set of conditional attributes and another set of decision attributes. Usually, we consider a dataset having only one decision attribute. In RST, the partition of a dataset with respect to a given set of conditional attributes is created with the help of an equivalence relation known as indiscernibility relation. This indiscernibility relation is the mathematical basis of RST. The classification of

M. K. Sabu (✉)
Department of Computer Applications, Cochin University of Science
and Technology, Cochin, Kerala, India
e-mail: sabumk@cusat.ac.in

© Springer Nature Singapore Pte Ltd. 2019
P. K. Mallick et al. (eds.), *Cognitive Informatics and Soft Computing*,
Advances in Intelligent Systems and Computing 768,
https://doi.org/10.1007/978-981-13-0617-4_55

a dataset based on the similarity of the considered decision attribute is known as categories or concepts. If each of these concepts is representable as a union of some of the elementary sets in the equivalence class structure produced by the indiscernibility relation, then we say that the knowledge represented by the decision attribute is definable with the knowledge given by the conditional attributes. But this is not the case always because some concepts cannot be expressed exactly by employing the knowledge represented by the conditional attributes. In other words, the concept may include and exclude objects contained in the same equivalence class induced by the indiscernibility relation. Such concepts are known as vague (or imprecise) concepts. In RST, a vague concept can be approximated using the equivalence class structure produced by the indiscernibility relation by constructing its lower and upper approximations [1]. The lower approximation or positive region is the union of equivalence classes where each equivalence class is a subset of the selected vague concept. Generally, the lower approximation of a concept is the set of objects of the dataset which can positively be classified as its members [1]. Similarly, the upper approximation of a concept is defined as the union of all equivalence classes that cannot be positively be classified as members of its complement.

Rule induction is a basic data mining task because regularities hidden in data are frequently expressed in terms of rules. It always serves as a tool to extract valuable and interesting information for decision making from large volumes of data. The rule generation phase is generally followed by a rule evaluation phase to select important rules which are the outcome of an efficient rule induction process. The main difficulty with conventional rule induction algorithms is that too many rules are generated even from a medium-sized dataset and hence it is challenging for the users to analyze these rules to identify significant ones. To overcome this limitation, the rule induction process should include a mechanism to select interesting and important rules automatically from the rule set generated by the process. In this paper, using a Rough Set based rule evaluation measure, *Degree of Rule Significance* [2], an algorithmic frame work is presented to rank and extract significant rules generated by a conventional rule mining algorithm. The algorithm is applied on a real-life dataset and the ranked rules are evaluated based on two Rough Set based strategies.

The rest of the paper is organized as follows: A brief review of conventional rule evaluation approaches is presented in Sect. 2. Section 3 discusses two popular rough set based rule evaluation approaches. The concept of a new rough set based rule evaluation measure is presented in Sect. 4 followed by the experimental design to rank the decision rules generated by a conventional rule mining algorithm is discussed. The experimental analysis based on a real-life learning disability dataset is described in Sect. 6. In Sect. 7, significance of the results obtained as a result of the experiment is evaluated based on two rough set based strategies and Sect. 8 concludes the paper.

2 Rule Evaluation Approaches

Rule induction always serves as a tool to assist humans in extracting knowledge from huge volumes of data and thereby facilitating the decision making process. In general, an input dataset may contain redundant and inconsistent information and the rules induced from such an information system may not be equally important and interesting. Determining or identifying what is interesting is a major challenge in the area of data mining especially in rule induction. Similarly when the size of the dataset increases, the number of rules mined from the dataset also increases and consequently the analysis of these rules becomes unwieldy for the users of the data mining system. An efficient rule mining algorithm should extract only interesting rules from the large amount of data contained in the dataset. This is achieved by introducing rule interestingness measures to evaluate various rules generated by the rule induction process. Rule interestingness measures actually differentiate the "valid, potentially useful and understandable" rules from those that are not [3]. Many interestingness measures have been suggested to evaluate the significance of a rule [4–6]. These measures are broadly classified into subjective interestingness, objective interestingness, and impartial interestingness [3, 7, 8].

Subjective interestingness relies on domain expert's prior opinions towards the particular application [9]. These are the most suitable measures to evaluate rules because 'what is interesting' is ultimately subjective. Sometimes subjective interestingness measures become infeasible because the evaluators have to search the entire set of rules to select the significant rules. Conventional rule evaluation focus mostly on objective measures [10]. These measures are independent of any predefined opinions or actual knowledge represented by the domain of the data. Objective interestingness of a rule is measured by considering the statistical properties of the data, the attribute value pairs involved in the rule and the number of cases satisfied by that particular rule. The rule interestingness measures support and confidence used in *apriori* association rule mining algorithm [11], are best examples for objective measures. An interestingness measure which is independent of the domain, task and user is known as impartial interestingness measure [7].

3 Rough Set Based Rule Evaluation Approaches

Use of RST in rule mining, especially the concept of reduct and core, is very helpful in deriving important and interesting rules from a dataset. Decision rules generated from reducts of a dataset can fully characterize the entire dataset [12]. By taking the advantages of RST, Jiye Li proposed two popular rule evaluation approaches namely rule importance measure and rules as attribute measure.

3.1 Rule Importance Measure

Rule importance measure [4] is a popular Rough Set based rule evaluation measure used to identify meaningful, important and interesting rules from huge amount of rules generated by a conventional rule mining system. This rule evaluation measure is defined by considering various reducts of the considered dataset. A reduct of a dataset represents a subset of conditional attributes which are sufficient to classify the data contained in the dataset. The common attribute contained in all the reducts gives the core of the dataset which includes the most important attributes. The rule set generated from a reduct always represents important and interesting rules that can be extracted from the dataset. Since, reduct is not unique and each reduct is a representative of the original dataset, multiple rule sets are generated by these reducts. Jiye Li [4] argue that the frequency of each rule in all such generated reduct rule sets decides the importance of the respective rules and hence these frequencies determine a ranking for various rules. Based on this frequency value, Li proposed the Rule Importance measure for rule evaluation. For the ith reduct rule R_i the Rule Importance measure [4, 12] is defined as:

$$\text{Rule Importance } (R_i) = \frac{\text{Frequence of rule } R_i \text{ in all generated reducts}}{\text{Total number of reducts}} \quad (1)$$

Since, computing all reducts of a high-dimensional dataset is NP hard, the rule importance measure computation is time consuming. Another limitation of Rule Importance measure is that when there is only one reduct for a dataset, this measure outputs all the generated rules as rules with importance 100%. This implies that for dataset having only one reduct, the Rule Importance measure does not make any difference between the generated rules. Hence, rule evaluation using this measure is infeasible in some situations where the given dataset is having only a single reduct or datasets with more number of attributes.

3.2 Rules as Attribute Measure

Rules as Attribute measure is a Rough Set based rule evaluation measure proposed by Li [4]. Using this measure, it is possible to identify the important and interesting rules from a large number of rules generated by the rule mining algorithm. In this approach, to identify the significant rules, as in Rule Importance measure, the concept of reduct is used. To explain the process of evaluation, consider a decision table $T = (U, A, d)$, where U is the non-empty finite set of objects $\{x_1, x_2, \ldots, x_m\}$ called the universe, $A = \{a_1, a_2, \ldots, a_n\}$ is a non-empty finite set of independent variables called conditional attributes, and d is the dependent variable called the decision attribute. From T, using a conventional rule mining process various decision rules are generated first. Then a new decision table T' is constructed from

T by considering the generated rules as conditional attributes. In T', the decision attribute d and its values remain the same as with the original decision table. Various entries of the new decision table T' represents the information obtained as a result of applying each rule to various objects of the decision table [13]. If both the antecedent and the consequent of a rule R_j appear together in an object x_i of the decision table, then we say x_i follows rule R_j [4]. In this case a value 1 is assigned to the corresponding cell (i, j) of T'; otherwise the value 0 is assigned. After filling all the entries of T', reducts are generated from T'. Since the conditional attributes of the new decision table are the rules generated from the original decision table, the attributes representing the reducts are the rules itself. These reduct rules represents the important representative rules fully characterizing the original decision table. The core attributes of the new decision table T' contains the most important rules. Hence, this RST based user-independent and domain-independent rule evaluation method leads to the extraction of significant rules automatically from the given dataset. As pointed out in the Rule Importance measure, the main limitation of this approach is the reduct computation time, because computing all the reducts of a dataset is NP hard.

4 Rule Evaluation Measure—Degree of Rule Significance

A rule evaluation measure becomes efficient only when it is objective and at the same time it is domain-dependent. In an earlier paper [2], the author proposed such an evaluation measure by taking the advantages of both the rough set based rule evaluation approaches discussed in previous sections. The proposed rule evaluation measure, *Degree of Rule Significance* (DRS) [2], is a modification of the Rule Importance Measure proposed by Jiye Li and Nick Cercone [9]. For the modification, the concept of rules as attribute measure is used. In this approach, in order to eliminate the reduct computation, the idea of discernibility matrix in RST [4, 6, 14] is used in a new dimension as follows:

Consider the decision table $T = \{U, A, d\}$. From this decision table construct a modified discernibility matrix D of size $p \times n$ where p is the total number of pairs (x, y) of the objects in U such that $d(x) \neq d(y)$ and n is the number of conditional attributes in T. Various entries D_{ij} of the matrix D is defined as

$$D_{ij} = \begin{cases} 1 & a_j(x) \neq a_j(y), \quad i = 1, 2, 3, \ldots, p, \text{ and } j = 1, 2, 3, \ldots, n \\ 0 & \text{otherwise} \end{cases} \quad (2)$$

where (x, y) be the ith pair of objects satisfying $d(x) \neq d(y)$ in U and j represents the index of the conditional attribute. The sum of all the entries in a column of this matrix indicates discrimination frequency of the corresponding conditional attribute in discriminating the objects in various pairs. The larger the sum is more example pairs the attribute can discriminate and hence the significance of that attribute is high compared to the other attributes. Therefore these frequency values give a ranking for the conditional attributes to select the significant attributes automatically.

To define DRS [2], by considering rules generated from T using a conventional rule mining algorithm as conditional attributes, a new decision table T' is constructed as explained in Sect. 3.2. From this newly constructed decision table T', a modified discernibility matrix D is constructed using Eq. 2. Since, the entire conditional attributes of the modified decision table T' are rules, the sum of the entries of the columns of D gives the significance (frequency) of these rules in discriminating various objects in the decision table. Hence it is possible to consider these frequency values as a measure to quantify the importance of each rule in the domain based on the discrimination power of these rules. The DRS of each rule is calculated based on the corresponding discrimination frequency value. It is actually a normalized value of the discrimination frequency. To perform the normalization, each frequency value is divided with the total number of object pairs available in T'. The *Degree of Rule Significance* [2] is defined as follows:

Definition: If f_i represents the discrimination frequency of the ith rule R_i to discriminate various object pairs (x, y) in T' with $d(x) \neq d(y)$ and c represents the total number of object pairs (x, y) satisfying the condition $d(x) \neq d(y)$ in T', then the *Degree of Rule Significance* of ith rule, $\delta_i = f_i/c$, where $i = 1, 2, \ldots, n$.

With the help of DRS, it is possible to rank all the rules generated by the rule mining algorithm. This ranking provides a domain dependent method to identify significant rules automatically from the large number of rules generated. The advantage of this objective and user independent rule evaluation measure is that it can bring together the domain related knowledge (i.e., discrimination information) and statistical properties of the data in rule evaluations.

5 Experimental Design

Let $R = \{R_1, R_2, \ldots, R_n\}$ be the set of decision rules generated by a conventional rule mining algorithm from the decision table $T = \{U, A, d\}$. Then a modified decision table T' of size $m \times (n + 1)$ is constructed by considering these derived rules $\{R_1, R_2, \ldots, R_n\}$ as conditional attributes and the decision d from the original decision table as the decision attribute. The row s of this new table represent the objects of the original decision table and the different entries $T'[i, j]$ of this new decision table are defined as

$$T'_{ij} = \begin{cases} 1 & \text{if object } x_i \text{ follows rule } R_j, \\ 0 & \text{otherwise} \end{cases} \tag{3}$$

where $i = 1, 2, 3, \ldots, m; j = 1, 2, \ldots, n$. The entry $T'[i, n + 1]$ represents the same decision attribute value of the ith object in the original decision table. From T' by constructing the Boolean matrix D, DRS of each rule is calculated and rules are ranked as per their significance. The overall process of ranking the generated rules is formulated as a modified rules ranking algorithm RR. The algorithm is presented below and in this algorithm, the original decision table T is the input and rules ranked as per their significance in the data domain is the output.

Algorithm: RR

Step 1: Input the original decision table T.

Step 2: Generate the decision rules from the decision table T.

Step 3: Considering rules generated in Step 2 as conditional attributes, construct a modified decision table T' using Eq. 3.

Step 4: Sort the rows of the decision table T' in ascending order of the decision attribute values.

Step 5: Generate a Boolean matrix D from the decision table obtained in Step 4 using Eq. 2.

Step 6: Find the discrimination frequency of each rule as the sum of the corresponding column of the matrix D.

Step 7: Calculate DRS of each rule by dividing the column sum obtained in Step 6 with the total number of rows of D.

Step 8: Select various rules from the decision table D and arrange according to the descending order of their DRS calculated in Step 7.

Step 9: Output the sorted list of rules.

The algorithm RR will produce all the rules generated from the original decision table but ranked as per their significance in the domain. With the help of a suitable threshold value for DRS, important rules can be selected automatically from among the large number of rules generated by the conventional rule mining algorithm.

6 Experimental Analysis and Results

In order to apply the proposed methodology, a real dataset consisting of the signs and symptoms of the Learning Disabilities in school age children is selected. It is collected from various sources which include a child care clinic providing assistance for handling learning disability in children and three different schools conducting such Learning Disability (LD) assessment studies. Learning disabilities may affect the children in various phases of learning such as acquisition, organization, understanding, retention and/or use of information. About 10% of children enrolled in schools are affected with this LD. If appropriate supports are provided at the right time, children with learning disabilities can experience success. The collected dataset is helpful to identify the presence of LD while diagnosing its presence in a new case.

The dataset contains 249 student records with 16 conditional attributes as signs and symptoms of LD and the existence of LD in a child as decision attribute. All these attributes takes either t or f as its values depending on the presence of the

Table 1 Key used for representing the symptoms of LD

Key/ Abbreviations	Symptoms	Key/ Abbreviations	Symptoms
DR	Difficulty with reading	LM	Lack of motivation
DS	Difficulty with spelling	DSS	Difficulty with study skills
DH	Difficulty with handwriting	DNS	Does not like school
DWE	Difficulty with written expression	DLL	Difficulty in learning a language
DBA	Difficulty with basic arithmetic skills	DLS	Difficulty in learning a subject
DHA	Difficulty with higher arithmetic skills	STL	Is slow to learn
DA	Difficulty with attention	RG	Repeated a grade
ED	Easily distracted	LD	Learning disability
DM	Difficulty with memory		

considered symptoms and the corresponding decision. Table 1 gives key used for representing the considered signs and symptoms and its abbreviations [2].

To implement the algorithm, decision rules are generated from the LD dataset using the *apriori* association rule mining algorithm. From the complete set of rules generated by the algorithm, a subset of first 125 rules is selected for the experiment. By considering these 125 rules, a new decision table $T'_{249 \times 126}$ is constructed. In T' the selected 125 rules are the conditional attributes and the decision LD is the decision attribute [4]. Various entries of the matrix T' is generated by gathering the information regarding whether different objects of the dataset follows a particular rule. For example, as per the collected data, rule R_1 (DR = t) & (DWE = t) & (RG = f) ==> (LD = t) cannot be applied to the first object, because the antecedent (DR = t) & (DWE = t) & (RG = f) not satisfied in this object. Therefore, we assign $T'[1, 1] = 0$. But when R_1 is applied to second object, both the antecedent (DS = t) & (DWE = t) & (RG = f) and the consequent (LD = t) appear together in the rule. Hence, $T'[2, 1] = 1$. In this way all the entries are filled in the matrix T'.

In T', to simplify the processing, the value 'f' of the decision attribute LD is set to 0 and value 't' to 1. There is no inconsistent information present in the new decision table T'. In the next step of the algorithm, a Boolean matrix D consisting of 14850 rows is generated using Eq. 2. The Boolean matrix generated by the algorithm (not all the entries) is shown in Table 2. The sum of all the entries in each column of T' gives the discrimination frequency of the respective rules. The *Degree of Rule Significance* each rule is then calculated by dividing each discrimination frequency value with the total number of rows of D. The set of ranked rules together with the corresponding *Degree of Rule Significance* is presented in Table 3.

In Table 3, rules R_{100} and R_{117} are ranked in the first position, because these two rules have the same value for the *Degree of Rule Significance* which is calculated as 12969/14850 (=0.8733). This is because, the frequency with which the rule can

Table 2 Boolean matrix generated by RR algorithm

Row No.	R_1	R_2	R_3	R_4	R_5	...	R_{123}	R_{124}	R_{125}
1	0	0	0	0	0	...	0	0	0
2	1	1	1	1	1	...	1	0	0
3	1	1	1	1	1	...	1	0	0
4	0	0	0	0	0	...	0	0	0
5	1	1	1	1	1	...	1	0	1
⋮	⋮	⋮	⋮	⋮	⋮	...	⋮	⋮	⋮
14849	1	1	1	1	1	...	1	1	0
14850	1	1	1	1	1	...	1	1	0
Sum	10494	10494	9504	9504	9405	...	12474	4950	3168
DRS	0.7067	0.7067	0.6400	0.6400	0.63333	...	0.8400	0.3333	0.2133

Table 3 Rules ranked as per their degree of rule significance

Rank	Rule no.	Rule	Degree of significance
1	100	DR = t & LM = t & DNS = f ==> LD = t	0.8733
2	117	DR = t & DWE = t ==> LD = t	0.8733
3	123	DS = t & DWE = t ==> LD = t	0.84
4	11	DS = f & DSS = f ==> LD = f	0.8182
5	16	DR = f & DS = f ==> LD = f	0.798
6	50	DSS = t & DNS = f & RG = f ==> LD = t	0.78
7	49	DSS = t & DNS = f & STL = t ==> LD = t	0.78
8	53	DS = t & STL = t ==> LD = t	0.7667
9	24	DS = f & LM = f ==> LD = f	0.7475
10	23	DS = f & DM = f ==> LD = f	0.7374
11	25	DS = f & DBA = f ==> LD = f	0.7374
12	2	DS = t & DWE = t & RG = f ==> LD = t	0.7067
13	1	DR = t & DWE = t & RG = f => LD = t	0.7067
14	65	DWE = t & DA = f & DSS = t ==> LD = t	0.6867
15	68	DS = t & DA = t & RG = f ==> LD = t	0.6667

discriminate various object pairs (x, y) with $d(x) \neq d(y)$ is 12969 and the total number of object pairs satisfying the same condition is 14850. The next rule in ranking is R_{123} with D*egree of Rule Significance* 0.84. In this way rules can be ranked. By setting a suitable threshold value for *Degree of Rule Significance*, significant rules can be selected automatically from among the large number of rules generated by the rule mining algorithm. The ranking given in Table 3 is mainly based on the discernibility information collected by considering various cases present in the dataset. In other words, this ranking is subjective to the problem domain because it is based on the knowledge available in the dataset. The result of this experiment shows that DRS is a best choice for selecting significant rules from the large number of rules generated by conventional rule mining algorithms.

7   Evaluation

To evaluate the significance of the results generated by the proposed method, two different Rough Set based strategies are adopted. In the first strategy, reduct/core attributes are generated from the original dataset and verified the involvement of these attributes in the ranked rules. Since reduct/core is a set of representative and important attributes, the rules containing these attributes are considered as significant and can fully describe the decision attribute.

In the second approach, reducts are generated from the decision table constructed by considering rules as attributes. The reducts of this new decision table actually represent the reduct rules of the original decision table and hence they are significant. Similarly, the core attributes of this new decision table represents the most important rules to describe the concepts of the original decision table. The presence of these reduct/core rules in top ranked rules with high *Degree of Rule Significance* shows the efficiency of the proposed method.

7.1   Evaluation of the Results Using Approach—I

The Heuristic search algorithm of ROSE2 generates six reducts. These reducts are {DR, DS, DH, DWE, ED, DLL, RG}, {DR, DS, DWE, DHA, ED, DLL, RG}, {DR, DS, DWE, DA, ED, DLL, RG}, {DR, DWE, DA, ED, DM, DLL, RG}, {DR, DS, DWE, DBA, DHA, ED, RG}, and {DR, DS, DWE, DBA, DA, ED, RG}. The intersection of these six reducts, that is, the core attributes are {DR, ED, RG, DWE}. All the 125 ranked rules are analyzed for the existence of the derived reduct/core attributes. For the experiment a subset of 15 top ranked rules are randomly selected. These selected rules are shown in Table 3. In these 15 ranked rules all the rules except one are reduct rules because reduct/core attributes are involved in these rules. Also it is observed that the last ranked rule, i.e., Rule 125, does not contain any of the reduct/core attributes. This shows that the proposed ranking process is an effective technique to discover the important rules automatically from among the large number of rules generated by the rule mining algorithm.

7.2   Evaluation of the Results Using Approach—II

From the decision table generated from the LD dataset by considering rules as attributes, the Heuristic search algorithm of ROSE2 generates 10 reduct rule sets as shown in Table 4.

From the ranked set of rules the selection of 50% of the top ranked rules gives a subset of 63 significant rules. These rules are shown in Table 5.

Table 4 Reduct rules generated from LD Dataset

No.	Reducts	No.	Reducts
1	$\{R_{11}, R_{18}, R_{23}, R_{28}, R_{32}, R_{82}, R_{100}\}$	6	$\{R_{11}, R_{18}, R_{24}, R_{32}, R_{42}, R_{82}, R_{100}\}$
2	$\{R_{11}, R_{18}, R_{23}, R_{32}, R_{42}, R_{82}, R_{100}\}$	7	$\{R_{11}, R_{20}, R_{24}, R_{28}, R_{32}, R_{82}, R_{100}\}$
3	$\{R_{11}, R_{20}, R_{23}, R_{28}, R_{32}, R_{82}, R_{100}\}$	8	$\{R_{11}, R_{20}, R_{24}, R_{32}, R_{42}, R_{82}, R_{100}\}$
4	$\{R_{11}, R_{20}, R_{23}, R_{32}, R_{42}, R_{82}, R_{100}\}$	9	$\{R_{11}, R_{18}, R_{23}, R_{28}, R_{34}, R_{82}, R_{100}\}$
5	$\{R_{11}, R_{18}, R_{24}, R_{28}, R_{32}, R_{82}, R_{100}\}$	10	$\{R_{11}, R_{18}, R_{23}, R_{34}, R_{42}, R_{82}, R_{100}\}$

Table 5 50% top ranked rules produced by RR algorithm from LD dataset

Rank	Rule no.	DRS	Rank	Rule no.	DRS	Rank	Rule no.	DRS
1	100	0.8733	22	54	0.6162	43	115	0.5267
2	117	0.8733	23	78	0.6067	44	99	0.5267
3	123	0.84	24	85	0.5867	45	15	0.5267
4	11	0.8182	25	84	0.5867	46	17	0.52
5	16	0.798	26	60	0.5859	47	121	0.5133
6	50	0.78	27	87	0.58	48	20	0.5067
7	49	0.78	28	6	0.5667	49	21	0.5067
8	53	0.7667	29	103	0.5667	50	18	0.5067
9	24	0.7475	30	104	0.5667	51	19	0.5067
10	23	0.7374	31	90	0.5667	52	122	0.5
11	25	0.7374	32	91	0.5667	53	22	0.5
12	2	0.7067	33	7	0.5533	54	26	0.4733
13	1	0.7067	34	108	0.5467	55	27	0.4733
14	65	0.6867	35	10	0.54	56	28	0.46
15	68	0.6667	36	96	0.54	57	29	0.4533
16	3	0.64	37	111	0.54	58	30	0.4533
17	4	0.64	38	8	0.54	59	31	0.44
18	72	0.64	39	9	0.54	60	32	0.44
19	5	0.6333	40	13	0.5333	61	33	0.44
20	76	0.62	41	14	0.5333	62	34	0.44
21	75	0.62	42	12	0.5333	63	39	0.4333

From Table 4, various reduct rules are R_{11}, R_{18}, R_{20}, R_{23}, R_{24}, R_{28}, R_{32}, R_{42}, R_{82}, and R_{100}. All these reduct rules except R_{42} and R_{82} are present in the selected set of rules. For example, consider rule R_{100}, the rule comes first in the ranking process with a *Degree of Rule Significance* 0.8733. This rule represents a core rule. Similarly, rule R_{11} ranked in the fourth position is also a core rule. The rule ranked in the last position is R_{125}. The degree of significance of this rule is 0.2133. It is not involved in any of the reducts shown in Table 4. This shows that, rule R_{125} is not that much significant in this domain to take a decision regarding the existence of LD. In RR algorithm, ranking is based on the discrimination frequency of each rule obtained from the information when this rule is applied on various objects. This shows the effectiveness of the proposed *Degree of Rule Significance* in identifying the significant rules. By adding more number of rules in the ranking process the domain of discrimination can be expanded and hence it is possible to generate better results.

8 Conclusion

Rule mining is a process to discover interesting relationships (associations) among data items in large datasets. A challenging problem in rule generation is the assessment of derived rules based on the quality. Quality of a decision rule is measured in terms of its predictive accuracy, comprehensiveness and interestingness. As a post processing operation to rule mining, efficient and effective methods are required to extract interesting, relevant, and novel rules automatically. This is achieved by introducing rule interestingness measures to evaluate various rules generated by the rule induction process. Two important RST-based rule evaluation measures viz. Rule Importance Measure and Rules as Attribute Measure are discussed in this paper. By taking the advantage of these two measures, an objective and domain dependent measure called *Degree of Rule Significance* is considered as a new rule evaluation measure. Based on this measure a modified Rule Ranking (RR) algorithm is proposed to rank and extract significant rules produced by a conventional rule mining system. In RR algorithm the idea of decision relative discernibility matrix in RST is used to eliminate the expensive reduct computation.

Experiments on the Learning Disability dataset demonstrate the effectiveness of this new measure in ranking the generated decision rules. Two different approaches are adopted to prove the efficiency of the rule evaluation measure DRS. In the first approach, reduct/core attributes are generated from the original dataset. The involvement of these attributes in the top ranked rules shows the effectiveness of the proposed measure in selecting the significant rules. In the second approach, reduct/core rules are generated by constructing a new decision table by considering the generated decision rules as conditional attributes. The presence of these reduct/core rules in the set of top ranked rules with high *Degree of Rule Significance* proves that the proposed method provides an automatic and effective way for ranking decision rules.

References

1. Pawlak, Z.: Rough Sets: Theoretical Aspects of Reasoning About Data. Kluwer Academic Publishers, Norwell, MA, USA (1992)
2. Sabu, M.K., Raju, G.: A Rough Set Based Approach for Ranking Decision Rules, pp. 671–682. Springer, Berlin, Heidelberg (2011)

3. Sahar S.: Interestingness via what is not interesting. In: Proceedings of the Fifth ACM SIGKDD International Conferences on Knowledge Discovery and Data Mining, pp. 332–336 (1999)
4. Li, J.: Rough set based rule evaluations and their applications, Ph.D. Thesis (2007)
5. Bruha, I.: Quality of decision rules: definitions and classification scheme for multiple rules. In: Machine learning and Statistics, pp. 107–131 (1997)
6. Tan, P.N., Kumar, V.: Interestingness measures for association patterns: a perspective. In: Proceedings of Workshop on Postprocessing in Machine Learning and Data Mining (2000)
7. Sahar, S.: Interestingness preprocessing. In: Proceedings of the IEEE International Conference on Data Mining, pp. 489–496 (2001)
8. Sahar, S.: Exploring interestingness through clustering : a framework. In: Proceedings of the IEEE International Conference on Data Mining, pp. 677–680 (2002a)
9. Sahar, S.: On Incorporating subjective interestingness into the mining process. In: Proceedings of the IEEE International Conference on Data Mining, pp. 681–684 (2002b)
10. Hilderman, R.J., Hamilton, H.J., Oa, C.S.S.: Principles for Mining Summaries Using Objective Measures of Interestingness, pp. 72–81 (2000)
11. Silberschatz, A., Member, S., Tuzhilin, A.: What makes patterns interesting in knowledge discovery systems. IEEE Trans. Knowl. Data Eng. **8**(6), 970–974 (1996)
12. Li, J., Cercone, N.: Discovering and ranking important rules. IEEE Int. Conf. Granular Comput. **2**, 506–511 (2005)
13. Li, J., Cercone, N.: A method of discovering important rules using rules as attributes. Int. J. Intell. Syst. **25**, 180–206 (2010)
14. Tan, S., Wang, Y., Cheng, X.: An efficient feature ranking measure for text categorization. In: Proceedings of the 2008 ACM Symposium on Applied Computing, SAC' 08, pp. 407–413 (2008)

Performance Comparison of Clustering Algorithms Based Image Segmentation on Mobile Devices

Hemantkumar R. Turkar and Nileshsingh V. Thakur

Abstract In general, clustering concept is used for segmentation of images. In literature, it is found that different clustering approaches are proposed for the purpose of image segmentation. This paper presents comparative analysis of clustering algorithms namely, k-Means (KM), Moving k-Means (MKM), and Enhanced Moving k-Means (EMKM) for image segmentation on mobile devices. Experimentations are carried out on natural images with RGB and HSV color spaces which are used in mobile devices. Performance of KM, MKM, EMKM algorithms is evaluated using qualitative and quantitative parameters, particularly, using Mean Square Error. The obtained results show that the EMKM algorithm is the most suitable technique for image segmentation.

Keywords Clustering algorithm · RGB · HSV · Image segmentation

1 Introduction

The process of grouped samples [1] is called as clustering technique. Clustering technique widely used in data mining, image analysis, and statistical data analysis, etc. In an image processing domain, segmentation is the basic step for image classification and description. Recently, the clustering algorithms are widely used for image segmentation [2–11] in different image processing domain. As of now, to produce the quality segmentation, different clustering algorithms are proposed. The k-Means clustering algorithm (KM) [12] is an iterative technique and the most

H. R. Turkar (✉)
Department of Computer Science and Engineering, Rajiv Gandhi College
of Engineering and Research, Nagpur, Maharashtra, India
e-mail: hemantturkar@rediffmail.com

N. V. Thakur
Department of Computer Science and Engineering, Nagpur Institute
of Technology, Nagpur, Maharashtra, India
e-mail: thakurnisvis@rediffmail.com

© Springer Nature Singapore Pte Ltd. 2019
P. K. Mallick et al. (eds.), *Cognitive Informatics and Soft Computing*,
Advances in Intelligent Systems and Computing 768,
https://doi.org/10.1007/978-981-13-0617-4_56

efficient algorithm because of its simplicity and easy to implement among all the existing clustering algorithms. Initially, the pixels are grouped based on their texture, shape, and color features and then pixel groups are merged into the different number of regions. Although it is a simple technique but it contains some drawbacks [13, 14], i.e., depends on k-value initialization, missing of a small cluster. Mashor [15] proposed the advancement over existing k-means algorithm, i.e., Moving k-Means clustering (MKM). This algorithm finds the fitness of each center and if certain criteria cannot satisfy center then it will be shifted to the active center [16] during the clustering process. Enhanced Moving k-means Algorithm (EMKM) [17] finds clustered into non-overlapping regions. The other content of the paper is described as follows. The color spaces and experimental results on RGB and HSV color spaces are discussed in Sect. 2 and Sect. 3, respectively. Paper ends with conclusion and future scope in Sect. 4 followed by references.

2 RGB and HSV Color Space

Color can be quantified using many methods. Colors are present or represented in visual experiences. Color spaces indicate colors from the three colors—Red, Green, and Blue [18]. RGB (Red, Green, and Blue) color space is used by scanners and digital cameras and it is an additive representation of all the colors [19, 20]. Qualitative analysis is widely used in the probabilistic statement to carry out weakness and effectiveness of the algorithm on visual perception. The color space hue, saturation, value (HSV) is derived from RGB intensity values. In HSV, a pixel is represented by its H-hue, S-saturation, and V-value coordinates. Hue is a quantified representation of color; any specific color is a hue. Saturation determines the domination of particular channel; it is the purity or shade of a color [20, 21].

3 Experimental Results on RGB and HSV Color Space

Various experimentations are carried out on mobile devices with different values of k for RGB and HSV color space. Three images, namely, apple logo; Beach; and Red flower shown in Fig. 1 are used for comparative analysis. Comparative analysis of KM, MKM, EMKM algorithms is carried out using Mean Square Error (MSE).

Figures 2, 3, 4 5, 6 and 7 shows the segmentation results of RGB and HSV images respectively for KM, MKM, and EMKM algorithms. All the images are segmented by considering the cluster value $k = 3$, 4, and 5 clusters.

Quantitative Analysis is carried out in terms of Mean Square Error (MSE) to evaluate the performance of algorithms. MSE is calculated by using an Eq. (1), where V_i is the pixel which is part of jth cluster and N is an image.

Fig. 1 Original images: **a** Apple logo, **b** beach and **c** red flower

Fig. 2 Segmentation of RGB images by selecting cluster value $k = 3$

$$\text{MSE} = \frac{1}{N} \sum_{j=1}^{k} \sum_{i \in c_j} \left\| v_i - c_j \right\|^2 \tag{1}$$

The lesser difference between the output and original image conveys that the data in the considered region is located near its center. Comparison of implemented algorithms is carried out in terms of MSE. MSE is calculated for KM, MKM and EMKM for RGB and HSV images with the cluster count of k, where $k = 3, 4$ and 5 respectively. Tables 1 and 2 shows the summary of MSE evaluation for RGB and HSV color spaces respectively.

Graphical analysis of the obtained results of MSE evaluation of KM, MKM, and EMKM algorithms with $k = 3$ for RGB and HSV color spaces is shown in Figs. 8

Fig. 3 Segmentation of RGB images by selecting cluster value $k = 4$

Fig. 4 Segmentation of RGB images by selecting cluster value $k = 5$

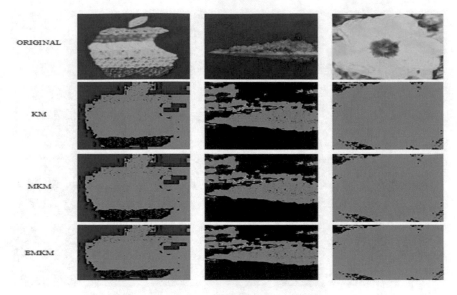

Fig. 5 Segmentation of HSV images by selecting cluster value $k = 3$

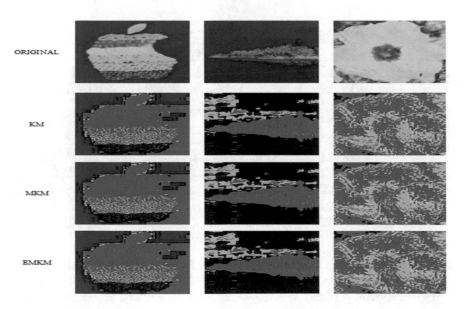

Fig. 6 Segmentation of HSV images by selecting cluster value $k = 4$

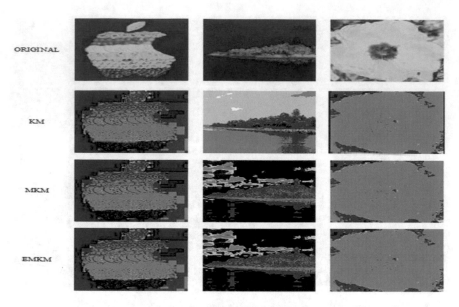

Fig. 7 Segmentation of HSV images by selecting cluster value $k = 5$

Table 1 Quantitative evaluations: MSE on RGB color space

RGB color space		Mean Square Error (MSE)		
No. of cluster		$k = 3$	$k = 4$	$k = 5$
KM	Apple logo	7168.87	20394.11	19452.40
	Beach	8363.37	5265.377	15448.00
	Red flower	47728.39	68459.36	447728
MKM	Apple logo	6523.64	19718.67	19241.24
	Beach	8044.28	4950.32	14602.65
	Red flower	47491.00	67963.56	407828
EMKM	Apple logo	5755.68	19344.71	18259.02
	Beach	7307.00	3995.83	14579.00
	Red flower	46266.99	67090.56	389875

and 9. Obtained results show that the EMKM algorithm has less MSE value for every image and for every type of cluster in RGB and HSV color space. So, the EMKM algorithm performs better in comparison with KM and MKM clustering algorithms.

Table 2 Quantitative evaluations: MSE on HSV color space

HSV color space		Mean Square Error (MSE)		
No. of cluster		$k = 3$	$k = 4$	$k = 5$
KM	Apple logo	7880.91	32629.31	12648.99
	Beach	17715.61	2085.10	15448.00
	Red flower	49697.19	19964.35	44767.87
MKM	Apple logo	7816.30	32541.64	12215.10
	Beach	17132.21	1988.74	13844.60
	Red flower	48905.45	19066.19	44100.46
EMKM	Apple logo	6734.99	31608.94	11221.31
	Beach	16604.67	1509.47	12522.55
	Red flower	48483.65	18753.02	43725.61

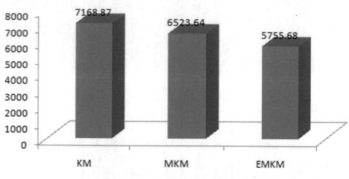

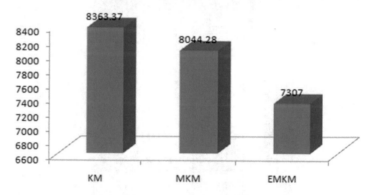

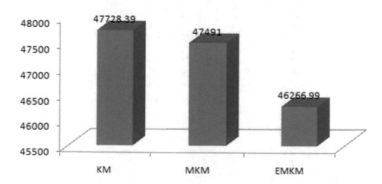

Fig. 8 MSE evaluation of KM, MKM, and EMKM algorithms with $k = 3$ for RGB color space

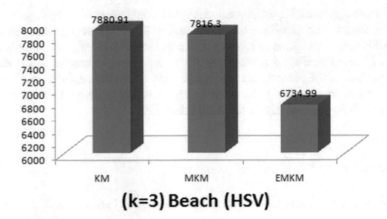

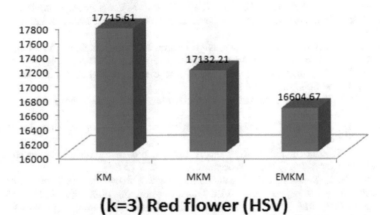

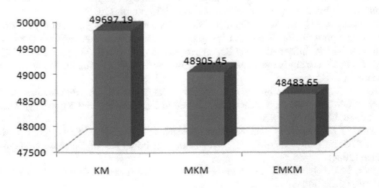

Fig. 9 MSE evaluation of KM, MKM, and EMKM algorithms with $k = 3$ for HSV color space

4 Conclusion and Future Scope

An EMKM algorithm is an extended version of conventional MKM algorithm and KM algorithm. The experimental results justify that EMKM algorithm generates less MSE value. The comparative result shows that EMKM algorithm finds better results of segmentation with lesser sensitivity and cluster variance to the defined center values in the color spaces on mobile devices. In future, there is a scope of improvement in modifying the existing clustering algorithm on different color spaces for image segmentation on mobile devices.

References

1. Gonzalez, R.C., Woods, R.E.: Digital Image Processing, 2nd edn. Prentice hall, New Jersey (2012)
2. Damahe, L.B., Janwe, N.J., Krishna, R.K., Thakur, N.V.: Segmentation, threshold and classification in microscopic images: an overview. Proc. Int. Conf. Data Manag. ICDM **2010**, 203–211 (2010)
3. Damahe, L.B., Krishna, R.K., Janwe, N.J., Thakur, N.V.: Segmentation based approach to detect parasites and RBCs in blood cell images. Int. J. Comput. Sci. Appl. **4**(2), 71–81 (2011)
4. Khaire, P.A., Thakur, N.V.: An overview of image segmentation algorithms. Int. J. Image Process. Vision Sci. **1**(2), 62–68 (2012)
5. Khaire, P.A., Thakur, N.V.: A fuzzy set approach for edge detection. Int. J. Image Process. **6** (6), 403–412 (2012)
6. Ranit, S.B., Thakur, N.V.: Image segmentation using various approaches. Int. J. Image Process. Vision Sci. **2**(2), 3–4 (2014)
7. Parihar, V.R., Thakur, N.V.: Graph theory based approach for image segmentation using wavelet transform. Int. J. Image Process. **8**(5), 255–277 (2014)
8. Mat Isa, N.A., Salamah, S.A., Ngah, U.K.: Adaptive fuzzy moving K-means clustering algorithm for image segmentation. IEEE Trans. Consum. Electron. **55**(4), 2145–2153 (2009)
9. Sulaiman, S.N., Mat Isa, N.A.: Adaptive fuzzy-K-means clustering algorithm for image segmentation. IEEE Trans. Consum. Electron. **56**(4), 2661–2668 (2010)
10. Kulkarni, S.S., Thakur, N.V.: An approach for image segmentation based on region connectivity. Int. J. Adv. Electr. Electron. Eng. **4**(3), 55–68 (2015)
11. Mat-Isa, N.A., Mashor, M.Y., Othman, N.H.: Comparison of segmentation performance of clustering algorithms for Pap smear images. In: Proceedings of International Conference on Robotics, Vision, Information and Signal processing (ROVISP2003), pp. 118–125 (2003)
12. Turi, R.H.: Clustering-based colour image segmentation. Ph.D. Thesis, Monash University, Australia (2001)
13. Ordonez, C.: Clustering binary data streams with K-means. In: Proceedings of the 8th ACM SIGMOD Workshop on Research Issues in Data Mining and Knowledge Discovery, San Diego, California (2003)
14. Agarwal, P.K., Mustafa, N.H.: K-means projective clustering. In: Proceedings of the 23rd ACM SIGMOD-SIGACT-SIGART Symposium on Principles of Database Systems, Paris, France (2004)
15. Mashor, M.Y.: Hybrid training algorithm for RBF network. Int. J. Comput. Internet Manage. **8**(2), 50–65 (2000)
16. Naik, Devarshi, Shah, Pinal: A review on image segmentation clustering algorithms. Int. J. Comput. Sci. Inf. Technol. **5**(3), 3289–3293 (2014)

17. Siddiqui, F.U., Mat Isa, N.A.: Enhanced moving K-means (EMKM) algorithm for image segmentation. IEEE Trans. Consum. Electron. **57**(2), 833–841 (2011)
18. Cantrell, K., Erenas, M.M., Orbe-Paya, I., Capitan-Vallvey, L.F.: Use of the hue parameter of hue, saturation, value color space as a quantitative analytical parameter for bitonal optical sensors. Anal. Chem. **82**(2), 531–542 (2010)
19. Kekre, H.B., Sonawane, K.: Bin pixel count, mean and total of intensities extracted from partitioned equalized histogram for CBIR. Int. J. Eng. Sci. Technol. **4**(3), 1233–1243 (2008)
20. Solunke, R.P., Patil, A.A.: Image processing for mango ripening stage detection: RGB and HSV method. In: Proceedings of Third International Conference on Image Processing (2015)
21. Wang, X., Jia, K., Sun, Z.: An efficient method of shadow elimination based on image region information in HSV color space. In: Proceedings of International Conference on Computational Intelligence and Communication Networks (2015)

An Enhanced Approach to Memetic Algorithm Used for Character Recognition

Rashmi Welekar and Nileshsingh V. Thakur

Abstract Character recognition is a best case to apply logics from Memetic Algorithms (MA) for image processing. In cases, like finger print matching, cent percent accuracy is expected but the character recognition on other hand can auto correct some errors. Time of processing is not the first criteria in figure print analysis but accuracy is a must, whereas while extracting characters from image speed of processing becomes more important parameter. This aspect of character recognition provides wide scope of implementing MA. The typing on QWERTY keyboard is the best example of brain using MA and dividing the character search in two parts with 13 characters for left hand and 13 for right. We never need to cross hands for typing next character as the design of keyboard is ensures that in most of the cases consecutive characters appear in specific sequence and brain keeps itself already prepared to hit next key but waits for confirmation. As we move dipper into string the search starts reducing which further enhances the predictive capacity of brain for expected next character. This can be seen as local search and cultural evolution which is performed by brain with each successive character in a string in [1, 2]. Hence character recognition is more of string recognition than treating every character in isolation. This paper explains the above theory with results and also presents an enhanced MA for character recognition.

Keywords Memetic algorithm · Character recognition · Local search
Enhanced mcmetic algorithm

R. Welekar (✉)
Department of Computer Science and Engineering, Shri Ramdeobaba College
of Engineering and Management, Nagpur, Maharashtra, India
e-mail: rashmi.welekar@gmail.com

N. V. Thakur
Department of Computer Science and Engineering, Nagpur Institute
of Technology, Nagpur, Maharashtra, India
e-mail: thakurnisvis@rediffmail.com

© Springer Nature Singapore Pte Ltd. 2019
P. K. Mallick et al. (eds.), *Cognitive Informatics and Soft Computing*,
Advances in Intelligent Systems and Computing 768,
https://doi.org/10.1007/978-981-13-0617-4_57

1 Introduction

Image processing and character recognition are two very old but still very fresh, new, and crucial topics. These topics form the very foundation of modern human communication.

A large data is available in form of different scripts on stone, copper plates, in form of fossil, and even degraded papers. All such text is available in form of images and we need to extract all information without loss of data.

On other side, in modern world, we share many documents in form of images. Hence with progress in technology and exponentially rising processing power of computer systems there is a big scope to introduce novel and faster yet efficient method of image processing for character recognition.

There are many conventional methods and logics which are static in nature. But self-learning and self-evolving algorithms like genetic algorithms with support of artificial intelligence and neural networks are much efficient but time-consuming and lengthy processes. Now we find another dimension to this area of research which is based on local evolution of every element in isolation and self-maturing process before genetic mutation. The memetic algorithms (MA) are similar to genetic algorithms but with a feature of improving cultural and behavioral characteristics of individual genomes locally and in isolation.

In classical genetics, every element is a single static sample. These samples are mixed only during processing or crossover and the result of genetic mutation generate next level of another set of static samples. But in memetic logic, every individual sample element can improve its own characteristics and prepare itself for better crossover. This local and isolated improvisation in every element makes MA more better choice for character recognition.

2 Prior Work

In our previous work published in SEMCCO 2014 [5], we have suggested, and tested certain hybrid functions and methodology where MA was implemented with background of other Evolutionary Algorithms [3, 4]. The MA implemented in previous paper different is from conventional Classical Evolutionary (CA) Algorithm. In above mentioned work, the suggested approach is to force each child to go through a local search (LS) at each generation a step, which makes the algorithm "memetic". Two children are generated after one point crossover. For the first time the fittest child is inserted into the next generation. From the next iteration the LS is applied. The minimum edit distance of the new child is computed and it is compared with the previous child. If we obtain reduced edit distance, then only the new child is inserted into the next generation. This technique enhances the quality of the next generation meme. Our ultimate solution is to find the string with minimum value of minimum edit distance. When the population has a large variety of fitness values, the LS works toward optimization and less fit child tend to be rejected.

We have been working on other parameters for improvement of the current implementation of MA. In our previous paper we have presented results based on population size and Time taken for result generation for MA by considering two different sets of training samples per class for character recognition [5].

The above statistical in Table 1 and Fig. 1 with analysis shows that for a minor correction in error rate, there is a considerable rise in recognition time. This time is bound to increase as sample size grows and the real time data with thousands of different handwriting styles is loaded in database [6, 7]. Hence we have worked on a concept of MA where more emphasis is given on LS and building strings of every character in shortest possible length. The results of improved MA are as follows.

The enhancement in recognition time is 9.09% according to Fig. 2. This shows that the MA approach with LS gives better results [8–10]. From the very concept of

Table 1 Error rate and time for result generation

Set 1	Set 2
Population Size For memetic algorithm:	Population Size For memetic algorithm:
Number of authors: 4	Number of authors: 6
Total samples: 104	Total samples: 156
Number of training samples per class: 6	Number of training samples per class: 6
error rate: 3.84%	error rate: 3.84%
Average time for result generation: 2.0 s	Average time for result generation: 2.2 s

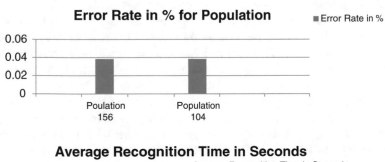

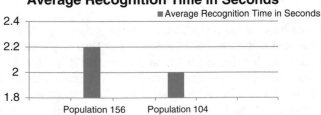

Fig. 1 % of Error rate and time for result generation

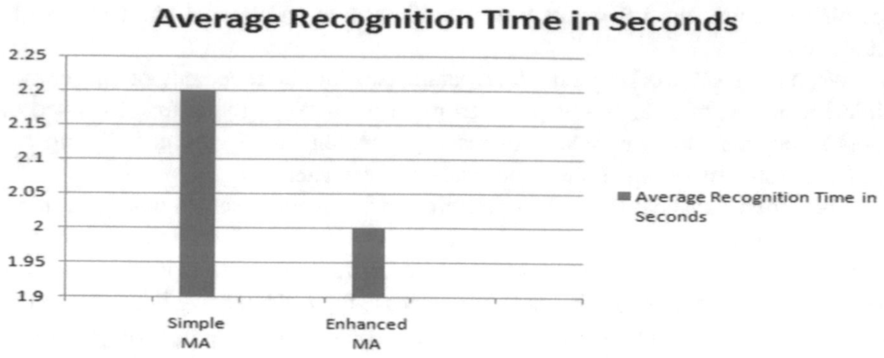

Average Recognition Time in Seconds

Fig. 2 Results for enhanced MA

MA we have decided to improve LS technique and make the process self-learning in nodal level.

3   The Improvised Approach with Local Search in Memetic Algorithm for Character Recognition

The case of character recognition falls under the category of approximate computing. In this case an approximate result is sufficient for many accurate estimates and predictions. Even if we can hit the probability of accuracy to 0.6, it can be turned up into 100% accuracy in character recognition using features of AI, clubbed with MA as a foundation of character recognition process [10, 11].

The main focus in this paper is on proposing the new ways to improve local learning mechanism of recognition algorithm using MA. Now in this paper, we are taking the process one step forward with introducing certain new parameters which can be used to improve LS with local learning ability of algorithm. As the memetic theory suggests, it is better to improve every genom locally and as we say, in isolation before it takes part in crossover mutation and produce next generation genom. The whole process can be seen as using as much variables with ++x notations or analogy than x++ so that every new iteration takes into consideration, a locally, self improved genome for mutation. This is bound to give better crossover results.

With proposed logic of improvised MA, we suggest an algorithm which takes into account the previously searched character and then applies the recognition logic to a subset rather than full English character set [12–15]. Thus the local node develops culturally with localized learning and then with more learned and evolved logic goes into the next iteration of character search. The procedure can be repeated till white space is encountered. This makes character recognition more purpose oriented and expedites the string identification process rather than identifying every

Fig. 3 Reduction in
character search

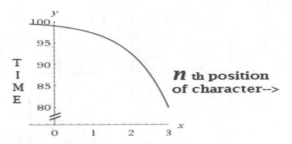

Proposed Reduction in subsequent
character search due to
cultural evolution and self learning

individual character as a subset from full set of 52 patterns or characters including capital and small. As we increase the population size, it leads to increase in character set by fold of 26 or 52. But if we go by proposed algorithm based on design concept of QUERTY keyboard, we can reduce the set of characters to be searched which makes sufficient room to increase population and add more styles of selected patterns rather than having too many unwanted characters in a search set [15[1], 16]. This is what a cultural learning of node means and this learning becomes more and more predictive as we proceed further in string. For example once we are in last four character search of strings like station, relation, automation etc. our MA becomes more predictive and just searches for specific characters rather than searching the whole set of images. On last character we need to just figure the existence of letter "n". This makes the MA as a confirmation test than search algorithm [17–19].

With this local and cultural evolution we expect the recognition time to reduce exponentially as we move deeper and deeper in string from nth position to $n + 1$ position which is shown in Fig. 3 and Tabulated in Table 2.

The simulation has been done with a dataset of four authors and then six authors. Twenty-six character samples have been taken from authors. Their offline images have been converted to strings. All of them have been stored in database. The test strings have been taken from an author whose data is not present in database. This test string is tested on database and the results for all characters have been obtained in similar manner.

To make the process of self learning procedure evolutionary we are considering the following case of character recognition which is carried out in our brain [20].

Our new approach is based on the same memetic tracking and tracings of learning curve to make fittest logic for patterns recognition or character recognition [21, 22]. Consider the following paragraph:

[1]The research was funded by National Science Foundation grants BCS 0957074 and BCS 1257272.

R. Welekar and N. V. Thakur

Table 2 Comparison of cultural evolution

No. of Authors	Without cultural evolution				Expected results with cultural evolution			
	Sample set [population]	Rise in sample set (%)	Recognition time in sec	Rise in recognition time	Sample set [population]	Recognition time	Rise n recognition time	Net saving in time
4	104	Ref Val (%)	2.0	Ref Val (%)	52	0.75	Ref Val (%)	1.25
6	156	50.0	2.2	10.00	78	0.84	12.0	1.36
10	260	66.6	2.8	27.27	130	1.02	21.40	1.78
50	1300	400	16.3	482.14	650	4.1	301.9	12.2

Acocdnnig to an enlgnsihunviesitrystduy, the oredr of ltretrs can be jmbueld and oen is stlil able to raed the txetwiohtutdclftfuiiy.

The reading and understanding of string is possible brain uses "lexical" and "semantycal" ways for character recognition and word processing, instead of the phonological way. The lexical and semantycal ways consist of every word you ever heard and their meaning. The phonogical way is used only during learning new words. The lexical and semantycal way forms the memetic evolution and the phonogical way forms the LS [23]. With better volume in dataset, the brain is more powered to crack the expected character and just tries to confirm it using LS in phonogical way. If we see a ccrtain collection of letters or words, our brains jump to conclusions about what comes next [24]. Now consider the second example:

S1M1L4RLY, Y0UR M1ND 15 R34D1NG 7H15 4U70M471C4LLY W17H0U7 3V3 N 7H1NK1NG 4B0U7 17.

In case of having numbers in place of some letters, a 2007 study by cognitive scientists in Spain found that brain does not search for digits. This proves that the letter-like appearance of the digits, as well as their context, has a stronger influence on our brains than their actual status as digits. The researchers think some sort of top-down feedback mechanism normalizes the visual input, allowing us to read the string easily [25].

Taking the clue from above biological facts of processing powers of human brain, we are suggesting a memetic approach for character recognition. Instead of going the conventional way using standard procedure of preprocessing, feature extraction, and post processing, the suggested algorithm improves with each iteration and improves its predicting power using memetic approach.

The error percentage will reduce and processing time will be shortened as crossover, mutation continuously improve processing power and predictive decision making faster, thereby reducing total computational complexity. From the above findings by Simei Gomes Wysoski, LubicaBenuskova, and Nikola Kasabov [25, 26], there is a large scope to conclude that the character recognition is more of memetic image processing for our brain. As the brain goes through writing and reading process and increases in its vocabulary, it starts to relate every shape, may it be numbers or any absurd designs to characters. The improvement is local as it has to no connection with IQ level or the schooling and teaching process. Every genome starts relating every shape with some type of character or symbol that can initiate communication [27]. These areas where brain executes LS and improves the minimum edit distance locally. Further study shows that the brain tries to hit the recognition probability to 0.6 and then the AI takes over and hence there is no need to develop an algorithm which will focus on 100% error free character recognition, but the emphasis should be on improving the local learning with cultural parameters which can enhance the local learning of genome in isolation which can be termed as memevolution [28, 29].

Hence instead of focusing character recognition in isolation we are attempting to hit string recognition with each probability of 0.6–0.7 for correctness of each character available in string. This is where local improvisation can be added. Just like common pairs of letter keys on typewriter are mechanically separated to avoid jamming effect which used to halt typing, we can separate the letter search in reduced selective set to make search faster. This technique has to be implemented locally before going into search of $n + 1$th character in string when we are processing nth character [10–12]. The logic of the QWERTY layout was based on letter usage in English rather than letter position in the alphabet.

4 The Proposed Algorithm in a Nutshell

Repeat till white space
  With each nth character in string

1. Select a more improvised and reduced character subset based on local evolution.
2. Process the character with best edit distance in subset.
3. Match the outcome with available database of strings.
   If no match found repeat from step 2.
4. Perform confirmation test and add the element to String Chain.

5 Conclusion and Future Scope

Using the LS in selected subset of characters and by performing predictive confirmation tests instead of recognition for approximately 50% of characters in a string; we can save considerable time in text extraction from any image. As shown in Table 2, with population size of 600 we can save 12 s per character search. The proposed work aims at 100% accuracy but with a combined approach of MA and predictive theory of neurology. The algorithm is expected to increase scanning rate. The cultural evolution in string processing where results of nth recognized character will be passed on to next gene to make further search simpler and easier with predictive approach. The application of MA is expected to boost efficiency, reduce memory consumption enabling better utilization of available resources of processing systems. The future scope of this approach can be to implement predictive analysis extended to implement applications like forensic reports for DNA string matching where the algorithm can predict the remaining string after limited number of searches.

References

1. Smith, J., Krasongor, N.: A tutorial for competent memetic algorithms model taxonomy and design issues. IEEE Trans. Evol. Comput. **9**(5), ISSN 1089-778X. Bristol (2005)
2. Moscato, P.: On evolution, search, optimization—GAs and martial arts: toward Memetic Algorithms. Technical Report. California Institute of Technology, Pasadena, CA. Caltech Concurrent Computational Programming Rep. 826 (1989)

3. He, L., Mort, N.: Hybrid genetic algorithms for telecommunications network back-up routing. BT Technol. J. **18**(4), 42–56 (2000)
4. Morris, G.M., Goodsell, D.S., Halliday, R.S., Huey, R., Hart, W.E., Belew, R.K., Olson, A.J.: Automated docking using a Lamarkian genetic algorithm and an empirical binding free energy function. J. Comput. Chem. **14**, 1639–1662 (1998)
5. Welekar, R., Dr. Thakur, N.V: Memetic algorithm used in character recognition. In: International Conference on Swarm, Evolutionary, and Memetic Computing (SEMCCO), pp. 636–646. Bhuvaneshwar (2014)
6. Malik, L., Deshpande, P.S, Bhagat, S.: Character recognition using relationship between connected segments & Neural Network. WSEAS Trans. Comput. **5**(1) (2006)
7. Flowers, D.L., Jones, K., Noble, K., VanMeter, J., Zeffiro, T.A., Wood, F.B, Eden, G.F.: Attention to single letters activates left extrastriate cortex. Neuro Image **21**, 829–839 (2004)
8. Han, S., Northoff, G.: Culture-sensitive neural substrates of human cognition: a transcultural neuroimaging approach. Nat. Rev. Neurosci. **9**, 646–654 (2008)
9. Kasabov, N.: Evolving spiking neural networks and neurogenetic systems for spatio-and spectrotemporal data modelling and pattern recognition. In: Liu, J., Alippi, C., Bouchon-Meunier, B., Greenwood, G.W., Abbass, H.A. (eds.) Advances in Computational Intelligence, pp. 234–260. Springer, Berlin (2012)
10. Wysoski, S., Benuskova, L., Kasabov, N.: On-line learning with structural adaptation in a network of spiking neurons for visual pattern recognition. In: International Conference on Artificial Neural Networks (ICANN), pp. 61–70 (2006)
11. Casasanto, D., Jasmin, K., Brookshire, G., Gijssels, T.: The QWERTY effect: how typing shapes word meanings and baby names. In: Proceedings of the 36th Annual Conference of the Cognitive Science Society. Cognitive Science Society, Austin, TX (2014)
12. MacKenzie, I.S.: Modeling text input for single-switch scanning. In: International Conference on Computers for Handicapped Persons (ICCHP), pp. 423–430. Springer, Berlin (2012)
13. MacKenzie, I.S.: The one-key challenge: searching for a fast one-key text entry method. In: Proceedings of the 11th International ACM SIGACCESS Conference on Computers and Accessibility, pp. 91–98. ACM, New York (2009)
14. Christopher Latham Sholes Biography (1819–1890). http://www.madehow.com/inventorbios/13/Christopher-Latham-Sholes.html#ixzz4wfpWkGOV
15. Snyder, K.M., Ashitaka, Y., Shimada, H. et al.: What skilled typists don't know about the QWERTY keyboard. Atten. Percept. Psychophys. **76**, 162 (2014). https://doi.org/10.3758/s13414-013-0548-4
16. Jasmin, K., Casasanto, D.: The QWERTY effect: how typing shapes the meanings of words. Psychon. Bull. Rev. **19**(3), 499–504 (2012). https://www.ncbi.nlm.nih.gov/pubmed/22391999
17. Waddington, C.T., MacKenzie, I.S., Read, J.C., Horton, M.: Comparing a scanning ambiguous keyboard to the on-screen QWERTY keyboard. In: Proceedings of the 31st International British Computer Society Human-Computer Interaction Conference—HCI 2013. British Computer Society, London (2017). http://dx.doi.org/10.14236/ewic/HCI2017.103
18. Tak-Shing, T.C, Geraint, A. W.: Memetic network of musical agents. Department of Computing, City University, London
19. McNamara, A.: Can we measure memes? Department of Psychology, University of Surrey, Surrey, UK (2011)
20. OECD: Understanding the brain -towards a new learning science, OECD, Paris. OECD/CERI International Conference (2008)
21. Higgs, P.G.: The mimetic transition: a simulation study of the evolution of learning by imitation. In: Proceedings of the Royal Society of London B: Biological Sciences, vol. 267, pp. 1355–1361. Iacoboni, M (2005)
22. Kasabov, N., Feigin, V., Hou, Z.G., Chen, Y., Liang, L., et al.: Evolving spiking neural networks for personalized modelling, classification and prediction of spatio-temporal patterns with a case study on stroke. Neurocomputing **134**, 269–279 (2014)

23. Rizzolatti, G., Fadiga, L., Gallese, V., Fogassi, L.: Premotor cortex and the recognition of motor actions. Brain Res. Cogn. Brain Res. **3**, 131–141 (1996)
24. Gautier, I., Tarr, M.J.: Becoming a "Greeble" expert: exploring mechanisms for face recognition. Vision Res. **37**, 1673–1682. PMID 9231232 (1997)
25. Saleh, A.Y., Shamsuddin, S.M., Hamed, H.N.A.: Memetic harmony search algorithm based on multi-objective differential evolution of evolving spiking neural networks. Int. J Swarm Intel. Evol. Comput. **5**(1) (2016)
26. Hamed, N., Kasabov, Z., Shamsuddin, S.M.: String pattern recognition using evolving spiking neural networks and quantum inspired particle swarm optimization. In: International Conference on Neural Information Processing (ICONIP) 2009, LNCS, vol. 5864, pp. 611–619. Springer, Berlin (2009)
27. Kasabov, N.: Integrative probabilistic evolving spiking neural networks utilising quantum inspired evolutionary algorithm: a computational framework. In: Köppen, M., Kasabov, N., Coghill, G. (eds.) International Conference on Neural Information Processing (ICONIP) 2008, LNCS, vol. 5506, pp. 3–13. Springer, Heidelberg (2009)
28. Grossberg, S., Merrill, J.W.L.: The hippocampus and cerebellum inadaptively timed learning, recognition, and movement. J. Cogn. Neurosci. **8**(2), 257–277 (1996)
29. Scliar-Cabral, L.: Neuroscience applied to learning alphabetic principles: new proposal. Emeritus- Universidade Federal de Santa Catarina. http://dx.doi.org/10.5007/2175-8026. 2012n63p187

A Review on Sentiment Analysis of Opinion Mining

Sireesha Jasti and Tummala Sita Mahalakshmi

Abstract In the recent era of Internet, social network plays very important role and occupies majority of share in data sharing between various groups. The data in social sites contain multidimensional data posted by different types of people. The posting contain people observations, thoughts, opinions, decisions and the rationale behind those decisions. Based on these postings or tweets one can analyse the sentiment about that specific product, service, event or any other participating by sharing their opinions, activity thoughts and ideas. In this paper, efficient algorithms are discussed for sentiment analysis of the tweets. The opinion on a specific topic mainly depends on the people, also the accuracy of opinions mining depends on the polarity strength. In this paper various Machine learning algorithms and various pre-processing techniques that make the data ready for opinion mining are discussed.

Keywords Opinion mining · Sentiment analysis · Social media
Internet · Feature extraction · Tweets · Filtering · Pre-processing

1 Introduction

In the present digital world news travel at much faster rate compared to print and television. Traditional media has been leveraging digital channels to be competitive. Recent statistics have shown that maximum number of content consumers have been using various digital platforms. Netizen are not only quick in consuming the content but are also open in expressing their views.

S. Jasti (✉) · T. S. Mahalakshmi
GITAM University, Visakhapatnam, India
e-mail: sireeshjasti@gmail.com

T. S. Mahalakshmi
e-mail: Sita.tummala@gitam.edu

S. Jasti
Department of CSE, Malla Reddy Engineering College (A), Secunderabad, India

© Springer Nature Singapore Pte Ltd. 2019
P. K. Mallick et al. (eds.), *Cognitive Informatics and Soft Computing*,
Advances in Intelligent Systems and Computing 768,
https://doi.org/10.1007/978-981-13-0617-4_58

Social networks play vital role in providing a platform for netizen to express their views on the news and the circumstances surrounding the event. It is also evident that social networks played a key role in influencing the opinions, views, sentiments of a larger population resulting in significant and highly impactful changes be it political, cultural, etc.

Social Networks are a huge source of information which contains unstructured content in the form of tweets, postings and discussions. Analysing these data sources and deriving insightful views is a challenging exercise. However one should not undermine the importance and the wealth of information these data sources provide us in understanding the target audience and serving them better.

Generally, the posts, tweets, discussions on social platforms are either positive or negative or at times neutral in nature. Classifying posts as positive, negative or neutral is highly important to derive better insights. The classification requires trained data sets and effective algorithms. And the phenomena of analysing various tweets posted online are called data mining.

In some cases it is also important to consider the scenarios where the posts might not have any subjective material. Hence it is tedious to derive genuine and valid opinions from such type of non-subjective information. However any article will contain information related to good or bad. This helps in classifying the content into good news and bad news. And hence sentiment analysis can result in terms of number fleece towards the news and number of supporters towards the news.

In the present paper different existing opinion mining techniques are discussed along with their merits and demerits. Along with the literature about the opinion mining and the framework of the opinion mining is explored. In opinion mining it is very important to select the correct post which helps in framing news. The major challenge in opinion mining is deriving insights from unstructured data. Hence it is important to convert unstructured data into structured data [1]. In static data, once the data is trained further changes will not be done. There are lots of chances of having structured data in statistical data mining techniques. Handling and analysis this structured database does not create any problem. In the dynamic data mining technique there are lot more chances of having unstructured data. Hence efficient data mining techniques are required to handle unstructured data of tweets on social sites. But handling and analysing structured data in both statistical and dynamic data mining is an easy task.

Generally the web is a rich source of variety of data. This data can be categorised into passive and active. The passive data contains all characteristics of attributes, events. The active data contains all attributes like sentiments, emotions, and feelings which are assessed by events and entities [2]. The opinion mining has spread its applications in wide areas like decease mining in health diagnosis, mining of online customer reviews on products, movies, and news etc. Opinion mining is also used in several other domains like search engines, Human Computer Interface, policy decision-making, marketing and quality control [1]. The importance of opinion mining also extended to machine learning, artificial intelligence, and in prediction systems. In most of the applications the major challenges arise in terms of scalability, precision in extracting data, validity of data, natural language processing and quality. There are efficient methods and algorithms that are applied to achieve some of the above goals up to some extent.

2 Opinion Mining

Opinion mining is a technique which is used to extract the meaningful information from the structured or unstructured data collected from various sources like web, text documents or posts in social network sites. In the current work online posts/ tweets are used as experimental data sets. These tweets are analysed to conclude and interpret the current events like political, finance, sports, movie news, etc. The tweets are characterised based on their internal meaning like user feelings, sentiments, understandings, expectations and attitude. Some of these characteristics of the tweets plays major role in accomplishing meaning full opinion about the current affairs. Based on the literature it is proven that in most of the cases the opinion mining depends on major areas which are discussed as follows [2].

i. Sentiment analysis: The primary task of sentiment analysis is classification of polarity discussed in the first section of this paper. The overall content is analysed in binary format with the help of machine learning algorithms and is derived as either positive or negative. The binary format includes thumb indicated reviews. That is thumb up and thumb down will decide feedback [3]. Based on the feedback the percentage of polarity strength will be derived. This makes the review more effective and reliable [4].

ii. Agreement detection: It is another type of sentiment analysis functioned based on the binary format. Initially the documents are given sentiment based labels according to the strength of the polarity. In agreement detection similar or different labels are provided to the pair of documents based on their sentiment. After the detection of polarity difference of the documents, it assigns the grade of the positivity to the polarity [3].

iii. Feature-based sentiment analysis: In this the content is analysed based on attitude, events, and reviews. This process requires more data mining and filtering than other methods.

iv. Opinion Summarization: Is the process of collecting different types of tweets from different persons at different time to finalise and create meaningful and human understandable statements. The summaries are prepared based on sentiments and user reviews. Generally the sentiment analysis of tweets depends on their relative information. Achieving relative information is very difficult in sentiment analysis as it depends on information polarity and polarity strength. If the polarity of the summarised data is positive, negative or neutral, then the polarity strength may become strongly positive, moderately positive and weakly positive.

In the associated field of current topics the sentiment analysis and attitude (opinion) excavating is being considered under the same umbrella as they belong to same kind of turf. As both the fields concentrating on recognition of emotions and detection of polarity of the state of statements. Hence, natural language processing must be used to repossess and filter the information from World Wide Web. But the NLP is complex since it requires through analysis of semantic, lexical, unstructured language rules.

2.1 Challenges in Opinion Mining

1. Most of the existing methods of opinion and sentimental analysis systems facing saviour problems due to lexical and semantic analysis.
2. It is a major challenge to separate subjective information from the document. If the document contain fewer subjects and has few opinions it is difficult to frame opinion based on lack of subject field.
3. The polarity consideration of a word or words become confusing because of their use of context may be different at different situations. For example, the phrase "accidentally" has different meaning in the sentence "He fell on wet floor accidentally" and in the sentence "I met my school friend accidentally in Visakhapatnam". In the former case the phrase shows positive polarity and in the later case it shows negative polarity.
4. Small changes of the text may not change full meaning in traditional text processing. But in opinion mining shows difference.
5. In most of the cases peoples have contradiction in the polarity of the statements they are posting in the same context.
6. This is important to consider duplicate tweets on web posting. Filtering plays very important role in identifying repeated or same tweets.
7. Classification based on polarity is also one of the interesting challenges to take right decision in opinion mining. This can be overcome with efficient algorithms.

3 Evolutions of Opinion Mining

Social Networks are rich sources of tweets by different people from different sectors. The tweets may contain information from different sectors like movies, online product reviews, gossips and political news. Hence complex procedure is required to analyse multidimensional data. Hence effective algorithms are required to analyse these procedures. In this section the flow of opinion retrieval flow is discussed. The opinion retrieval is done through various steps like pre-processing, classification, detecting polarity measurement and finally extraction of opinion is shown in Fig. 1.

Fig. 1 Data flow of a basic
opinion retrieval system

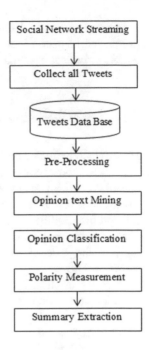

3.1 Social Networking Streaming

It is one of the contexts mostly discussed in the field of Data Mining and the schemes used in data mining technique used to retrieve the information and discover the data from huge web documents and services. In the particular context there are three different types of information can be processed during data discovery and retrieving the web data mining. First one is web user mining deals with extracting the useful knowledge from logging the information produced by web servers. The web content is the second category of mining produces various retrieved information acquired from the web. The reviewed information attained from the web is stored in server log files, to know the interest of users. The last but not the least, the third one Web Structure mining tries to ascertain useful information after the structured hyperlinks [5].

3.2 Collect All Tweets

How people behave on the social media platform will be known by collecting the data from the twitter. Now it is a great exercise in data science. Posts collected by the users are stored in database for further investigation. Pull various data structures from the twitter like tweets, user profiles, user friends and followers, symbols, etc. There are three methods to collect the data. First one SEARCH API allows to search old tweets and the REST API is the second API which let to gather the profiles of the user, friends and groups and the streaming API is the third API which is able to collect dynamic tweets in real time scenarios.

3.3 Pre-processing

Major challenge after collecting the data is to tame it in terms of emotions, noise, relevance, folksonomies and slangs. As part of this, following attributes are to be considered with respect to a tweet. They are text, created-at, favourite count, re tweet count favourite, re tweeted, language, id, place that is geo location, author's full profile, entities like URL, @ mentions, hash tags, and symbols, user identifiers-if a specific user receives a reply from and status identifier id—which is a reply of the specific status by the tweet. The structure of trained data model is shown in Fig. 2.

A series of short sentences are pre-processed (that is punctuation marks, tokenization and word normalisation, etc.) to provide only significant information [6]. Pre-processing makes the text uniform, through which the classification can make easier by representing multiple words in the same way. In pre-processing we have different methods like cleaning, emotion, negation, dictionary stemming, etc.

Sometime tweets are not useful and understandable by the system. In this case it is very important to convert the collected information in system understandable level. The data may contain username, passwords, special characters and duplicate posts. The hyperlinks are removed from the database. In order to filter out in the

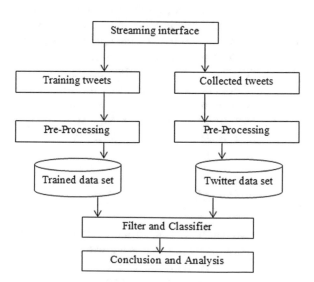

Fig. 2 Architecture of trained data model

data pre-processing suggest three basic methods to train the model based on the features extracted here.

The k-nearest neighbour, Support Vector Machine, BaysNet, Logistic Regression, Random Forest and Naïve Bayes are the different machine learning methods used for sentiment classification. They classify sentiment in terms of positive, or negative. Where, there exists no subjective material for deciding the opinion, such states are labelled as neutral. Hence extensive styles of features are required in order to recognise the opinion and their polarity [7]. Extensive algorithms are used for opinion text categorisation [8, 9]. The process of obtaining accurate sentiment analysis contain several eliminations like filter out unnecessary symbols which does not show effect on opinion classification, and reduce noise from original text. This phenomenon is called text pre-processing. The pre-processed text is used for preparing trained data set. The steps involved in pre-processing are discussed in Sect. 4. The k-nearest neighbour and random forest produces more accurate prediction results than naïve Bayes and BaysNet algorithm.

3.4 Opinion Text Mining

For efficient opinion mining one has to observe four effective parameters, they are polarity, opinion score, quality, trust. Different mining techniques like lexical method, linguistic approach, classification techniques and Machine learning techniques are some of the current methods used in opinion detection.

3.5 Opinion Classification

The people opinion can be collected based upon three levels. The first level is, the whole document is categorised into various states like positive, negative, and neutral. The second level analyses the individual sentence to classify them as positive, negative, or neutral. In the third level extract the features and analyse them for positive, negative and neutral features.

3.6 Polarity Measurement

The classified sentences are categorised based on positive, negative or neutral polarity. The polarity of a document or a statement plays very important role in analysis of sentiment which helps in giving clear and understandable conclusion.

3.7   Summary Extraction

In this section provides concise view of large number of opinions. It finally produces a meaningful opinion based on their tweets.

4   Text Pre-processing

The following steps are performed to make the text ready for the trained data model [10–12] as shown in Fig. 3.

4.1   Tokenization

It is the first step of pre-processing, where the data stream is partitioned into bag-of-words and tokens. Bag-of-words is a process uses a document with individual words in Euclidean space. Each word in a statement treated as feature and they are extracted here for sentiment analysis [13].

4.2   Removing Stop-Words

Ignore all the stop-words (includes the, is, as, or, and, also, which, etc.) such as separators, articles, prepositions, and frequently used functional words. As the stop-words do not have any effect on final conclusion, they can be removed from bag-of words.

Fig. 3 Process steps tangled
in pre-processing

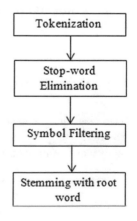

4.3 Symbols

There are certain symbols which are not at all required in opinion analysis. Symbols like @, #, logos, user names, passwords and twitter symbols can be removed from the tweets.

4.4 Stemming

Stemming is used to replace all associate words with their root words. For instance required, requirement, requiring, and all other related words are considered as their base word as require. Hence it reduces the size of bag-of-words.

5 Trained Model

Feature extraction plays very important role in extracting the features of the opinion data. In this section selected methods are used to extract use full words from the tweets and remove unwanted text or noises. This helps in obtaining accurate conclusion.

5.1 Unigram

It is one of the simple methods used to select a feature from use full words. It also helps to reduce unwanted words causes noise. At a time the unigram fetches one word from the text or data which is mined from web.

5.2 N-Gram Features

It is defined as developing a set of sequential words. For instance if $N = 2$, it represents a bigram, where two words are accessed at a time. Unigram has better performance on movie reviews for classification of polarity [14]. But bigram can be used where sequence of words decides the sentiment polarity strength.

5.3 External Lexicon

The performance of sentiment analysis further improved by external lexicon. The external lexicon includes all predefined words from tweets with respective to polarity.

6 Conclusion

The sentiment analysis can be done based upon the polarity strength of the tweets. Major machine learning algorithms can be implemented in the process of classification of sentimental information. Feature extraction, and Feature filtering plays important role in obtaining sequence of words from the tweets. The opinion mining derives the final conclusion that is summary of the tweets posted on the social web sites. The opinion mining has wide variety of applications in multidimensional areas. Filtering features such as unigrams, bigrams are discussed in process of feature extraction.

References

1. Khan, K., Baharudin, B., Khan, A., Ullah, A.: Mining Opinion Components from Unstructured Reviews: A Review, pp. 1319–1578 (2014/2012)
2. Bhatia S., et al.: Strategies for mining opinions: a survey. In: IEEE 2nd International Conference on Computing for Sustainable Global Development (INDIACom) (2015)
3. Cambria, Erik, et al.: New Avenues in Opinion Mining and Sentiment Analysis. Published by the IEEE Computer Society, IEEE Intelligent systems (2013)
4. Rao K.M., et al.: An efficient method for parameter estimation of software reliability growth model using artificial bee colony optimization. Lect. Notes Comput. Sci. (LNCS-Springer series), **8947**, 765–776 (2015)
5. Kumari, N., et.al.: Sentiment analysis on E-commerce application by using opinion mining. In: 6th International Conference—Cloud System and Big Data Engineering (2016)
6. Angiani, G., Ferrari, L., Fontanini, T., Fornacciari, P., Iotti, E., Magliani, F., Manicard, S.: A Comparison Between Preprocessing Techniques for Sentiment Analysis in Twitter, vol. 6, pp. 417–422 (2006)
7. Pang, B., Lee, L.: Opinion mining and sentiment analysis. Found Trends Inf Retrieval **2**, 1–135 (2008)
8. Pak, A., Paroubek, P.: Twitter as a corpus for sentiment analysis and opinion mining. In: LREC, 2010, pp. 1320–1326
9. Genkin, A., Lewis, D.D., Madigan, D.: Large-scale Bayesian logistic regression for text categorization. Technometrics **49**, 291–304 (2007)
10. Salton, G., McGill, M.J.: Introduction to Modern Information Retrieval (1986)
11. Zhang, L.: Sentiment Analysis on Twitter with Stock Price and Significant Keyword Correlation (2013)
12. Jivani, A.G.: A comparative study of stemming algorithms. Int. J. Comp. Tech. Appl **2**, 1930–1938 (2011)
13. Vijayarani, S., Janani, R.: Text Mining: Open Source Tokenization Tools—An Analysis, vol. 3 (2016)
14. Pang, B., Lee, L., Vaithyanathan, S.: Thumbs Up?: Sentiment Classification Using Machine Learning Techniques. In: Proceedings of the ACL-02 Conference on EMPIRICAL Methods in Natural Language Processing, vol. 10, pp. 79–86 (2002)
15. Domingos, P., Pazzani, M.: On the optimality of the simple Bayesian classifier under zero-one loss. Mach. Learn. **29**, 103–130 (1997)

Performance Analysis of Extreme Learning Machine Variants with Varying Intermediate Nodes and Different Activation Functions

Harshit Kumar Lohani, S. Dhanalakshmi and V. Hemalatha

Abstract Feedforward Neural Networks are the type of Artificial Neural networks, which follow a unidirectional path. The input nodes are associated with the intermediate layers and the intermediate layers are associated with the output layer. There are no connections which feedback to the input or the intermediate layer and thus are different from the recurrent neural networks. Extreme Learning Machine (ELM) is an algorithm that has no feedback path and the data flows in a single direction, i.e., from input to output. ELM is an emerging algorithm and is widely used for but not limited to classification, clustering, regression, sparse approximation, feature learning, and compression with a single layer or multi-layers of intermediate nodes. The best-preferred standpoint of ELM is that there is no requirement for the intermediate layer factors to be tuned. The intermediate layer is randomly generated and is never updated thereafter.

Keywords ELM · Clustering · Regression · Sparse approximation
Neural networks

1 Introduction

Feedforward Neural Networks were the first Artificial Neural Networks to be devised. In Feedforward neural networks, there is an input layer, followed by an intermediate layer (can be more than one or even absent in some cases) and finally

H. K. Lohani · S. Dhanalakshmi (✉) · V. Hemalatha
Faculty of Electronics and Communication Engineering,
Kattankulathur Campus, SRM University, Chennai, India
e-mail: dhanalakshmi.s@ktr.srmuniv.ac.in

H. K. Lohani
e-mail: 555harshit@gmail.com

V. Hemalatha
e-mail: hemalatha.v@ktr.srmuniv.ac.in

© Springer Nature Singapore Pte Ltd. 2019
P. K. Mallick et al. (eds.), *Cognitive Informatics and Soft Computing*,
Advances in Intelligent Systems and Computing 768,
https://doi.org/10.1007/978-981-13-0617-4_59

613

the output layer. The way is associated from the data-in layer to the intermediate data set and the intermediate data set is associated with the data-out level. Thus the data flow in a single direction. There are no feedback paths from the output of the intermediate layer or input layer by which the system can learn from the output. The input layer nodes are associated with the intermediate layer nodes, intermediate layer nodes to additionally intermediate layers nodes (if there should arise an occurrence of various intermediate layer systems) and the last intermediate layer nodes to the output layer nodes by methods of weights. These weights are responsible for the learning process of the network. The higher the weight value, the stronger the connection is. The weights can be increased or decreased depending on the results obtained from training the network. Also, there is another important element responsible for training is the bias connected to the intermediate layers. Bias element is attached to the input and each intermediate layer and is not swayed by the values in the previous layer. In other words, these neurons do not have any incoming connections. Bias units still have outgoing connections and they can contribute to the output of the ANN [1, 2].

The paper highlights some of the most widely used variants of the ELM in Sect. 2. The decision in activation function and the quantity of intermediate nodes in each layer and the quantity of intermediate layers assumes a fundamental part in the execution of the system and thus Sect. 3 shows a comparison of performance using different activation functions and with different intermediate layer nodes.

2 ELM Variants

ELM is different from other feedforward neural networks because of its characteristics. While the primitive feedforward neural networks needed numerous iterations to learn and provide a satisfactory result, the ELM was devised to learn in a single iteration. The primitive ones needed the weights and the bias to be manually set initially and was updated after each iteration [3]. This increased the complexity as well as required a lot of computation power and time. In the original ELM proposed by G.-B. Huang, there was a single layer of the intermediate layer. The factors associated with the intermediate layer such as the weights and the bias are generated randomly once and there is no requirement for these factors to be updated afterward. The learning process is thus completed in a single iteration. As per to the experiments performed by G.-B. Huang, the ELM proved to be 1000 times faster and provided better accuracy in most of the cases [4].

The following subsections show a few important variants of ELMs and their algorithms.

2.1 Single Intermediate Layer ELM

A single intermediate layer ELM Architecture is as presented in Fig. 1a. The "x" denotes the intermediate layer, "h" the intermediate layer and "t" denotes the target layer. The "w" denotes the weights associated with the connections, connecting the data-in layer and the intermediate layer and "B" denotes the weights related to the links connecting the intermediate layer and data-out layer. The bias connected at the intermediate layer is denoted by "b". The calculation for the ELM with a solitary intermediate layer is as follows [5, 6].

<u>**Algorithm:**</u>
Initialization

Step 1: *Introduce the intermediate layer factors as well as the numeral of intermediate nodes for the given in going information. The numeral of in going information nodes has not been exactly or else equal to the numeral of intermediate nodes.*
The input layer data are represented by x, the weights from the input layer to the intermediate layer are denoted by w, bias is denoted by b.

Learning

Step 2: *Figure out the intermediate layer h by performing*

$$h = (x * w) + b$$

Step 3: *The obtained intermediate layer is activated using activation function to achieve H*
H = g(h), where the function g() depends upon the selected activation function.

Step 4: *The intermediate layer and the data-out layer weights are computed by the accompanying*
*B = H' * t, t(target) associated with the inputs.*
and H' = $H^T(HH^T)^{-1}$ for fat matrix

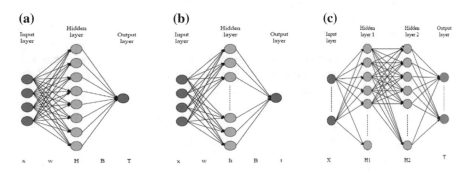

Fig. 1 a Single intermediate layer ELM. **b** I & E ELM. **c** Two intermediate layered ELM

and $H' = (HH^T)^{-1} H^T$ for skinny matrix
where H^T is the transpose of H.

Step 5: *Target is calculated by*
$$T = H * B$$

2.2 Incremental and Enhanced ELM

ELM has proved to have a superior performance than the rest of the feedforward neural networks, and this attracted a lot of researchers. From the research made on ELM, it was found that in some cases, a few intermediate nodes played a minor role as compared to the amount if complexity created by them [7]. So in order to further optimize the algorithm and avoid the above-mentioned situation, G.-B. Huang proposed an algorithm which was based on incremental feedforward networks based on generalized approximation with random generated intermediate nodes and was named as Incremental ELM [8]. This calculation was additionally ad-libbed by Lei Chen and was alluded as Incremental and Enhanced ELM (I & E ELM). In I & E ELM, in each of its learning steps, various intermediate nodes are whimsically produced. In the midst of them, the intermediate node with principal residual error diminishing is being supplemented to the predominant system and the data-out layer weights of the system are ascertained in an indistinguishable straightforward way from done in the novel Incremental ELM [9].

Algorithm:
Initialization

Step 1: *Let L_{max} be the most added number of intermediate nodes to the system and residual E = t where t = $\{t_1, t_2, t_3, ..., t_n\}^T$.*

Step 2: *Also, set the maximum number of trails k for passing on random intermediate nodes at each step and expected to learn accuracy ε.*

Learning

Step 3: *Whenever $L < L_{max}$ and $||E|| > \varepsilon$*

- *Increment quantity of Intermediate Nodes L by 1*
  $L = L+1$
- *For i in 1:k*

  - *Allocate the factors $(a_{(i)}, b_{(i)})$ for the new intermediate node L*
  - *Compute the data-out layer encumbrance $\beta_{(i)}$ for the new intermediate node*

$$\beta(i) = \frac{E \cdot H_{(i)}^T}{H_{(i)} \cdot H_{(i)}^T}$$

- *Analyze the residual error*

$$E_{(i)} = E - \beta_{(i)} \cdot H_{(i)}$$

- *End for*

- *Let i* = {i|min1 ≤ i ≤ k ||E(i)||}*
- *Set E = $E_{(i)}$, $a_L = a_{(L*)}$, $b_L = b_{(L*)}$, and $\beta_L = \beta_{(L*)}$.*

Step 4: *End While*

2.3   Two Intermediate Layered ELM

A feedforward system with single intermediate layer assigns all the weights in a random fashion [10].

Algorithm:
Initialization

Step 1: *Given is an arrangement of N sets (xi, ti) and 2L intermediate neurons, activation function g(x) while the weights are associated arbitrarily.*

Learning

Step 2: *The first intermediate layer h1 has to be found out by*

$$h1 = (X * Wh) + B1$$

where
X = Input layer
Wh = Weight associated with the 1st intermediate layer
B1 = Bias associated with the 1st intermediate layer

Step 3: *Activate the obtained intermediate layer h1*

$$H1 = g(h1)$$

where g() represents the activation function expression

Step 4: *The result for the second intermediate layer to be sorted out by*

$$H1 = T * \beta'$$

where $\beta' = (\beta^T \beta)^{-1} \beta^T$ if $\beta^T \beta$ is singular
else $\beta' = \beta^T (\beta^T \beta)^{-1}$

Step 5: *Calculate the augmented Whe by*

$$Whe = g^{-1}(H1)He'$$

where He', the universal inverse of He' = $[1\ H]^T$
where 1 = Single column matrix of N elements
and g^{-1} is the inverse of the activation function

Step 6: Calculate the output of the second intermediate layer H2 by

$$H2 = g(Whe * He)$$

Step 7: Result for second intermediate layer is given by

$$\beta new = H2' * T$$

where H2', the universal inverse of H2.
Step 8: Finally

$$f(x) = H2 * \beta new$$

3   Activation Functions

In case of computational networks, the activation functions express the output of a node for a particular input or set of inputs. A typical processor circuit is like a digital system of activation functions that can be either OFF or ON depending on the input factors. While if there should be an occurrence of naturally roused neural systems, the activation function is a portrayal of the rate of action potential terminating of the cell. The following are a few frequently used activation function with the ELM Algorithm [11].

3.1   Logistic or Sigmoidal Function

The Logistic or Sigmoidal function is given by [12, 13]

$$F(x) = \frac{1}{1 + e^{-x}}$$

where

x value to be activated

$F(x)$ activated value for the variable x and the value of x can range from $-\infty$ to ∞.

3.2 *Inverse Tangent Function*

There are six trigonometric functions but none of them are one to one when it comes to the reverse of the functions. The most widely used inverse function is inverse tangent or inverse tan or arctan function [14]. It is given by

$$y = \arctan(x)$$

which means $x = \tan(y)$.

3.3 *Leaky RELU*

RELU stands for Rectified Linear Unit. Leaky RELU is defined by Qian et al. [15]

$$f(x) = \begin{cases} x & \text{if } x > 0 \\ 0.01 * x & \text{otherwise} \end{cases},$$

where x can be any real number.

3.4 *Parametric RELU*

The unit which makes use of a rectifier is called as a RELU. Parametric RELU or PRELU is given by Jiang et al. [16]

$$f(x) = \begin{cases} x & \text{if } x > 0 \\ a * x & \text{otherwise} \end{cases},$$

where x can be any real value and $a \leq 1$.

3.5 Softmax Activation Function

The softmax activation function is given by [17, 18]

$$f(x) = \frac{1}{1 + |x|}$$

3.6 Hyperbolic Tangent Function

The hyperbolic Tangent function or the tanh function is the solution to the equation $F = 1 - f^2$ and the function is given by

$$y = \tan h(x),$$

where x can be any real value and the function gives an output in the range of $(-1,1)$.

4 Performance Analysis

The Figs. 2 and 3 shows the graphs for the activation functions mentioned above.

The decision of functions required for activating and the number of intermediate nodes for calculations assume an incredible part in the exactness and the execution time and the accompanying charts demonstrate a correlation of the execution time and precision of the calculation with the above-depicted activation function versus the number of intermediate nodes utilized in preparing the calculation (Fig. 3).

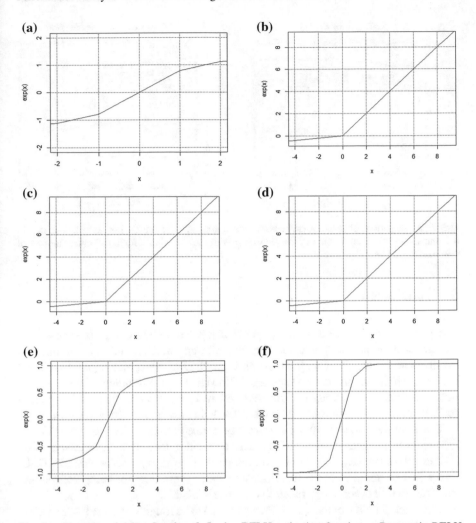

Fig. 2 **a** Arctan activation function. **b** Leaky RELU activation function. **c** Parametric RELU activation function. **d** Sigmoidal activation function. **e** Softmax activation function. **f** Tanh activation function

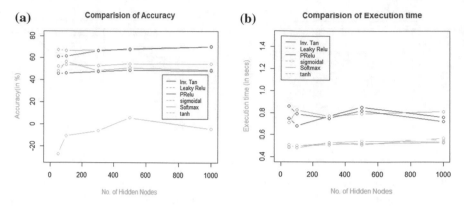

Fig. 3 **a** Comparison of accuracy of the activation functions versus number of intermediate nodes. **b** Comparison of execution time of the algorithm with the activation functions versus number of intermediate nodes

5   Conclusion

The primary charts give a point by point perspective of how the exactness of a similar calculation can fluctuate with the chosen function and the number of intermediate nodes. The Leaky RELU and the PRELU performed equally well and have the maximum accuracy. They are followed by the Sigmoidal, Softmax, and Arctan activation functions perform equivalent to each other. The tanh function performed very badly and is not suitable for being applied to the algorithm.

The second graph is all about the execution time consumed by the algorithm for training with the activation functions. The sigmoidal, tanh and softmax activation functions consumed the same execution time roughly while the PRELU, Leaky RELU, and the Arctan functions took quite more time for being trained and the difference between these functions is clearly visible.

The execution time of the algorithm was found to be roughly around 1 s with 1000 intermediate nodes when performed in R Studio Version 1.0.153 being installed on an Intel® core™ i5-3230M CPU clocking at 2.60 GHz based personal computer. This execution time is little when contrasted with the time taken by other feed forward neural systems. So the prime concern now is the accuracy of the system. Considering the accuracy provided by the activation function considered for the experiment, Leaky RELU and PRELU were considered the most appropriate ones.

The ELM is another calculation and is showing signs of improvement consistently with the various research into being performed on it. There is a few problem too, like the operation on the singular matrices being produced in the intermediate stages of the training algorithm and the optimum number of intermediate layers and intermediate nodes to be considered for training the algorithm which has not been addressed properly and needs to be considered for the further development of the algorithm.

References

1. Huang, G., et al.: Trends in extreme learning machines: a review. Int. Neural Netw. Soc. Eur. Neural Netw. Soc. Jpn. Neural Netw. Soc. **16**, 32–48 (2015)
2. Wan, Y., et al.: Twin extreme learning machines for pattern classification. Neurocomputing **260**, 235–244, 18 Oct 2017
3. Huang, G.-B.: Extreme learning machine: theory and applications. Neurocomputing **70**, 489–501 (2006)
4. Huang, G.-B.: Extreme learning machine for regression and multiclass classification. IEEE Trans. Syst. Man Cybern. **42** (2011)
5. Wang, H., Brandon Westover, M., Bai, Z.: Sparse extreme learning machine for classification. IEEE Trans. Cybern. (2014)
6. Lan, Y.: Constructive intermediate nodes selection of extreme learning machine for regression. Neurocomputing, 16–18, 3191–3199 (2010)
7. Chen, L., Huang, G.-B.: Enhanced random search based incremental extreme learning machine. Neurocomputing **71** (2008)
8. Chen, L., Huang, G.-B.: Convex incremental extreme learning machine. Neurocomputing **70**, 3056–3062 (2007)
9. Huang, G.-B.: Incremental extreme learning machine with fully complex intermediate nodes. Neurocomputing (2007)
10. Huang, G.-B.: On-line sequential extreme learning machine. In: The IASTED International Conference on Computational Intelligence, Canada (2005)
11. Qu, B.Y., et al.: Two-intermediate-layer extreme learning machine for regression. Neurocomputing **175**, 826–834 (2016)
12. Zhu, Q., Siew, C., Huang, G.B.: Extreme learning machine: a new learning scheme of feedforward neural networks. In: Proceedings of the 2004 IEEE International Joint Conference on Neural Networks (2004)
13. Mishra, A., et al.: Bi-modal derivative adaptive activation function sigmoidal feedforward artificial neural networks. Appl. Soft Comput. **61**, 983–994 (2017)
14. Sonoda, S., Murata, N.: Neural network with unbounded activation functions is universal approximator. Appl. Comput. Harmonic Anal. **43**(2), 233–268 (2017)
15. Qian, S., et al.: Adaptive activation functions in convolutional neural networks. Neurocomputing, 6 July 2017
16. Jiang, X., et al.: Deep neural networks with elastic rectified linear units for object recognition. Neurocomputing, 23 Sept 2017
17. Cao, J., et al.: Randomly translational activation inspired by the input distributions of ReLU. Neurocomputing, 20 Sept 2017
18. Sun, W., Su, F.: A novel companion objective function for regularization of deep convolutional neural networks. Image Vis. Comput. **60**, 58–63 (2017)

Role of Filter Sizes in Effective Image Classification Using Convolutional Neural Network

Vaibhav Sharma and E. Elamaran

Abstract Over the past few years, Deep Neural Networks have provided us the best results on a variety of problems, such as pattern recognition, computer vision, and Speech recognition and image classification. Convolutional neural networks are one of the deep learning models which are mostly used in image classification and are the base of many other deep neural network models. Convolution neural network uses convolution and pooling layers for feature abstraction. Unlike a regular Neural Network, the layers of a Convolutional neural Network have neurons arranged in three dimensions: width, height, depth, and filter sizes of different dimensions are used for feature reduction. But the problem with the convolution neural network is that it is difficult to train and can led to overfitting. There are many factors to look for while designing Convolutional Neural Networks one of them is filter size. Dimensions of filter sizes play a very important role for effective training in the convolutional neural network, So, here in this paper we compared the results of 3×3, 5×5 and 7×7 filter sizes and checked training accuracy, test accuracy, training loss, test loss as constraints.

Keywords ConvNet · Deep learning · Activation function · Max pooling
Filter size · Classification · Convolutional neural network

1 Introduction

In recent years, deep convolution neural networks [1, 2] have led to a series of modulation for image classification. Deep networks integrate different levels of features or filters [1, 2] in multilayer fashion known as convolution layers which can be enriched by increasing the number of stacks (depth).

V. Sharma (✉) · E. Elamaran
SRM University, Kattankulathur, India
e-mail: vbsmandi@gmail.com

E. Elamaran
e-mail: elamaran.e@ktr.srmuniv.ac.in

© Springer Nature Singapore Pte Ltd. 2019
P. K. Mallick et al. (eds.), *Cognitive Informatics and Soft Computing*,
Advances in Intelligent Systems and Computing 768,
https://doi.org/10.1007/978-981-13-0617-4_60

Most of the Deep networks inherit characteristics of convolutional neural network (ConvNet). ConvNet is very similar to regular neural networks. They are made up of neurons comprising of learnable weights and biases. Each neuron receives some inputs, performs a dot product with weights add biases to it and optionally follows it with a nonlinearity. It can be seen like this, we have the network comprising of raw image pixels on one end and class score at the other. The whole network gives a single differentiable score function from the raw image pixels on one end to class scores at the other. And they still have a loss function (e.g., SVM/Softmax) [3, 4] on the last (fully connected) layer.

ConvNet are widely used for classification of images. ConvNet architectures make the assumption that the raw image pixels are inputs, which allows us to encode certain properties into the architecture. This effectively down sample the amount of features and vastly reduce the dimension layer by layer in the network.

Regular neural networks (Perceptrons) [5, 6] receive an input (a single vector or a 1D array of multiple nodes), and pass it through multiple hidden layers. Here each node is fully connected to all the nodes in the previous layer, and node in a single layer work completely independent of other nodes in that layer. The fully connected layer is called the "output layer". We classify this layer to give valid a class score. Regular Neural Nets do not scale well to full images [7]. In Convolutional Neural Networks, input consists of images; the layers of a ConvNet have neurons arranged in 3 dimensions: width, height, depth. For example, the input images we are using are taken from the MNIST dataset [8]. These images are considered as an array of input volume of dimensions $28 \times 28 \times 1$ (width, height, depth respectively, as the images are grayscale so depth is taken as 1) and respective batch size. The final output layer would for MNIST [9] have dimensions $1 \times 1 \times 10$. At last, ConvNet architecture will reduce the full image into a single vector of class scores. ConvNet is a sequence of the layers, commonly known as Convolution Layer, Max/Avg. Pooling Layer [3] and Fully Connected layer.

We will talk about it in later sections, but a regular ConvNet could have the architecture like INPUT layer followed by CONV layer then input activations are calculated at ReLU layer. After ReLU layer we have POOL layer for max/avg pooling. Then output is taken at fully connected layer/FC layer. The layers used in ConvNet are as follows:

- INPUT [$28 \times 28 \times 1$] holds the pixel values of the original image, where height of image is 28, width is 28 and 1 represents the depth. As we are using grayscale image so the depth is 1 else it would be 3 for colored image as it represent three color channels R, G, B.
- CONV layer will compute the output of neurons connected to local regions in the input, by calculating the dot product between their weights and adding a bias value to it. This may result in volume such as [$28 \times 28 \times 16$] if we decided to use 16 filters.
- ReLU [10] layer will apply an element wise activation function, ReLU function thresholds the value at 0 and is given as max (0, input). Size of the volume remains unchanged [$28 \times 28 \times 16$].

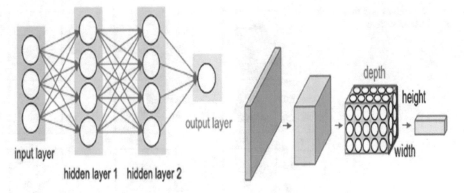

Fig. 1 Left: A four-layer neural network. Right: A ConvNet arrangement

- POOL layer will perform a down sampling operation along the spatial dimensions (width, height). For example if we use stride of 2 then it down samples the image by 2 pixels, resulting in volume such as [14 × 14 × 16].
- FC (i.e. fully connected) layer will compute the valid class scores, resulting in volume of size [1 × 1 × 10], where 10 are the number of classes in our case it is 0–9, resembling 10 categories of MNIST dataset. Each neuron in this layer will be connected to all the numbers in the previous volume.

Convolutional Neural Network regards the weights as filter sizes. Each filter is a kind of small tensor in comparison to the original image. The dimensions of original image and filter sizes are then convolved to give the output tensor which will be the input for the next convolutional layer. The inputs of the next layer are again convolved with the respective filter size taken from the receptive field of the original image and so on. So, the filter sizes play important part in feature extraction (Fig. 1).

Here In this paper we will compare the results of different filter sizes. In Sect. 2 we have discussed about MNIST dataset. Section 3 provides the architecture overview of our model. The results for different filter sizes are shown in Sect. 5. Conclusion and future work are discussed in Sects. 6 and 7 respectively.

2 Dataset

MNIST Dataset [8] is database of handwritten digits, it has total of 75,000 examples. Out of which, 60,000 examples are for training set, 10,000 examples dedicated to test set and rest 5000 for validation set.

The dataset consists of black and white (bi-level) images from NIST and were size normalized to fit in a 20 × 20 pixel box. The images were centered in

Fig. 2 Images and true labels from Mnist dataset

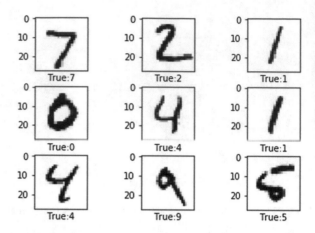

a 28 × 28 pixel image with height, width of 28 px and depth of 1. Figure 2 shows the hand written digits images in MNIST dataset with true labels.

The MNIST database [8] was constructed from NIST's Special Database. This database is divided into two types Special Database 3(contains training examples) and Special Database 1 (contains test examples) which contain binary images of handwritten digits.

3   Architecture

In this paper we are using 14 layers Convolutional Neural Network architecture. Our model is based on VGG Net [11]. The overview of our architecture is as follows:

We are using image of shape $[W, H, D]$ = [28, 28, 1] as input where 1 is regarded as number of input channels. The image has total volume of $W{\times}H{\times}D$ this means the input will be having total of 784 neurons. Figures 3, 4 and 5 shows the architectural overview of our model.

The input image is fed into the next convolutional layer. The convolutional layer contains 64 filters in first layer and like this we are using two convolutional layers consisting of 64 filter, four layers consisting of 128 filters, and four layers consisting of 512 filters. In each convolutional layer we are using ReLU [10] as our activation function. We are using Max pooling with stride of 2 for down sampling after every ascending number of filters. At the output of last convolutional layer we will be getting the shape of [3, 3, 512].

Then, we will flatten this layer to volume 3 × 3 × 512 and the flatten layer will consist of 4608 neurons.

In this model we are using three fully connected layers which will again down sample the image from 4608 to 2048 to 1024 then at last to the class score 10 (Fig. 6). Between fully connected layers we are using dropout [12, 13]. To summarize

Fig. 3 This figure shows first
two convolutional layers
which contains 64 filters

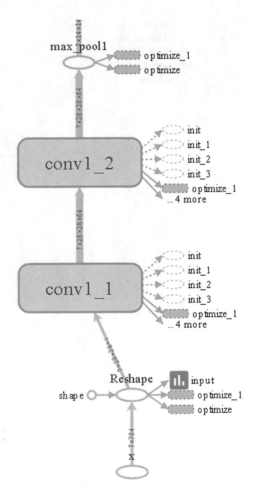

- Accepts the volume size of **W1** × **H** × **D1**
- Requires four hyper parameters

    - Number of filters **K**
    - Filter size **F**
    - Stride **S**
    - Padding **P**

- Produces a volume of size of **W2** × **H2** × **D2** here

    - **W2 = (W1 − F + 2P)/S + 1**
    - **H2 = (H1 − F + 2P)/S + 1**
    - **D2 = K**

Fig. 4 This figure shows next four convolutional layers which contains 128 filters then after max pooling the output is passed to next convNet layer

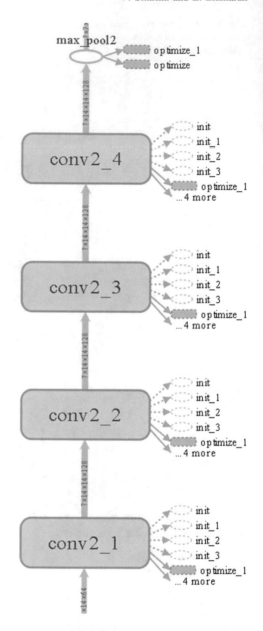

4 Factorizing Convolutions with Smaller Filter Size

Convolution with larger receptive fields or with larger filter size tends to be computationally costly (e.g., 5×5 or 7×7). Let us take one example, a 5×5 convolution compared to 3×3 convolution with same number of filter is $25/9 = 2.78$

Fig. 5 This figure shows next four convolutional layer which contains 512 filters then after max pooling the output is passed to next fully connected layer

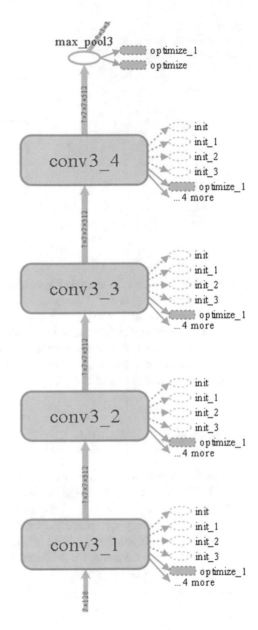

times more expensive. 5 × 5 filters takes less computational time as compared to 3 × 3 filters but loss is more in this case. Also, chances of overfitting are more [14, 15] in case of 5 × 5 filter because it covers large receptive field at a time compared to other filter sizes with lesser dimension. The other options are of choosing variable filter sizes as used in Inception v3 model [16]. This model uses

Fig. 6 Use of
multi-dimension filter layer in
Inception model [16]

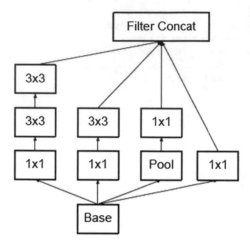

convolutional layers with variable filter sizes including 1 × 1 and 3 × 3 layers and
finally concatenating it.

1 × 1 are used before the 3 × 3 and 5 × 5 convolutions as it is the effective
way for dimensionality reduction [17]. Also, they include the use of activation
function (ReLU [10] in our case) which make them dual purpose (Fig. 7).

Here, in our work we will compare the performance of our convolutional neural
network using filter sizes of 3 × 3, 5 × 5 and 7 × 7.

5   Performance Analysis

Figures 8 and 9 shows the loss and accuracy encountered by different filter sizes
(3 × 3, 5 × 5 and 7 × 7 filters). Green, Red and Blue legends are chosen to
represent 3 × 3, 5 × 5 and 7 × 7 filters respectively. The y-axis of graph represents loss and accuracy values and the x-axis represents the number of iterations. As
training in Convolution Neural Network is time consuming and computationally
hard task. Here we are going for 50 iterations and the graph shows results for first
12 iterations (25% of 50 iterations). The steep descent in graph resembles local
minima (Table 1).

Fig. 7 This figure shows next fully connected layer. Here we are using two fully connected layers which contains 2048 and 1024 features respectively then we compute dropout and fully connected layer gives the class score as output

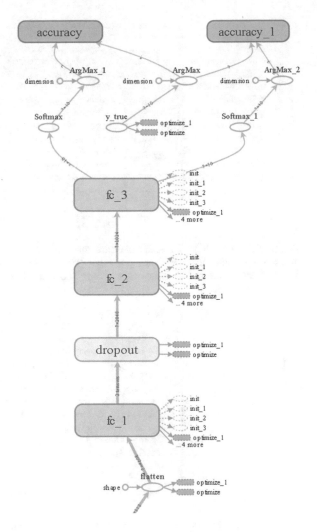

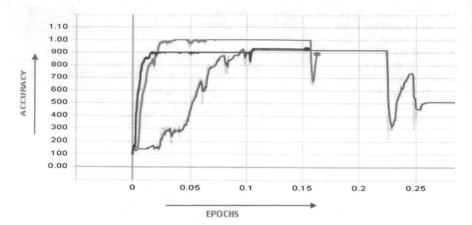

Fig. 8 This figure shows the accuracy encountered in 3 × 3, 5 × 5, and 7 × 7 filters. The Green line shows accuracy for 3 × 3 during training. The Red line shows the accuracy encountered in 5 × 5 filters. The Blue line shows the accuracy encountered in 7 × 7 filters

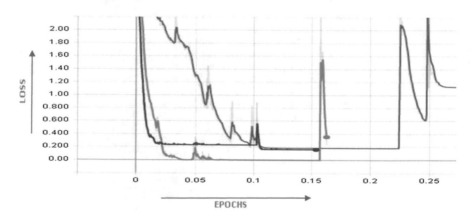

Fig. 9 This figure shows the loss encountered in 3 × 3, 5 × 5, and 7 × 7 filters. The Green line shows loss for 3 × 3 during training. The Red line shows the loss encountered in 5 × 5 filters. The Blue line shows the loss encountered in 7 × 7 filters

Table 1 Results showing loss and accuracy in training and test set with number of misclassification in 3 × 3, 5 × 5 and 7 × 7 filter respectively

Filter size	Accuracy 1500 epochs	Loss 10 epochs	Misclassification errors
3 × 3	Training: 100% Test: 98.7%	Training: 3.4% Test: 12.5%	
5 × 5	Training: 94.6% Test: 98.5%	Training: 21.4% Test: 22.8%	
7 × 7	Training: 91.2% Test: 97.8%	Training: 27.6% Test: 23.02%	

6   Conclusion

By looking at the results we have got it is clear that using filter sizes with lesser dimension provide us less loss as compared to others with higher dimensions. Figures 8 and 9 show the loss and accuracy encountered by different filter sizes. But, this comes with the cost of computation. Lesser dimension filter sizes are more computationally costly for example 3 × 3 filter size took 9:50 min for 50 iterations where as 5 × 5 filters took 7 min and 7 × 7 took 6:50 time for training. So, it is better to use filter sizes with lesser dimensions as it provides us less loss percentage but at the cost of computation time which is critical in case of larger data sets.

7   Future Work

In next phase, we will implement these results in Inception v3 model [16] and Residual learning [17] model. These models provide better image classification with loss of 6% to 7% and accuracy of up to 100% in training and test set respectively. Using the results we got, we aim at achieving much lesser loss in these models.

References

1. Zeiler, M.D., Fergus, R.: Visualizing and Understanding Convolutional Networks. arXiv:1311.2901v3 [cs.CV], 28 Nov 2013
2. Audhkhasi, K., Osoba, O., Kosko, B.: Noise enhanced convolutional neural network. **78**, 15–23 (2016) (Elsevier)
3. Lin, D., Lin, Z., Sun, L., Toh, K.A., Cao, J.: LLC encoded BoW features and softmax regression for microscopic image classification. In: 2017 IEEE International Symposium on Circuits and Systems (ISCAS)
4. Tang, J., Deng, C., Huang, G.B.: Extreme learning machine for multilayer perceptron. IEEE Trans. Neural Netw. Learn. Syst. **4**
5. Awad, M., Wang, L., Chin, Y., Khan, L., Chen, G., Chebil, F.: A framework for image classification. In: 2006 IEEE Southwest Symposium on Image Analysis and Interpretation
6. Li, J., Zhang, H., Zhang, L.: A nonlinear regression classification algorithm with small sample set for hyperspectral image
7. Sun, M., Song, Z., Jiang, X., Pan, J.: Learning pooling for convolutional neural network. Neurocomputing **224**, 96–104 (2017)
8. Lecun, Y., Cortes, C.: The MNIST database of handwritten digits
9. Cui, X., Beaver, J.M., St. Charles, J., Potok, T.E.: Dimensionality reduction particle swarm algorithm for high dimensional clustering. In: 2008 IEEE Swarm Intelligence Symposium St. Louis, MO, USA, 21–23 Sept 2008
10. Ide, H., Kurita, T.: Improvement of learning for CNN with ReLU activation by sparse regularization. In: 2017 International Joint Conference on Neural Networks (IJCNN)
11. Simonyan, K., Zisserman, A.: Very deep convolutional networks for large-scale image recognition. arXiv:1409.1556v6 [cs.CV], 10 Apr 2015

12. Srivastava, N., Hinton, G., Krizhevsky, A., Sutskever, I., Salakhutdinov, R.: Dropout: a simple way to prevent neural networks from overfitting. J. Mach. Learn. Res. **15**, 1929–1958 (2014)
13. Agarwal, S., Ranjan, P., Rajesh, R.: Dimensionality reduction methods classical and recent trends : a survey. IJCTA **9**(10), 4801–4808 (2016)
14. Zhu, Y., Mak, B.: Speeding up softmax computations in DNN-based large vocabulary speech recognition by senone weight vector selection Acoustics. In: 2017 IEEE International Conference on Speech and Signal Processing (ICASSP)
15. Li, X., Li, F., Fern, X., Raich, R.: Filter Shaping for Convolutional Neural Networks Conference Paper at ICLR 2017
16. Szegedy, C., Vanhoucke, V., Ioffe, S., Shlens, J.: Rethinking the inception architecture for computer vision. arXiv:1512.00567v3 [cs.CV], 11 Dec 2015
17. He, K., Zhang, X., Ren, S., Sun, J.: Deep residual learning for image recognition. arXiv:1512.03385v1 [cs.CV], 10 Dec 2015

Sentiment Analysis on Product Reviews Using Machine Learning Techniques

Rajkumar S. Jagdale, Vishal S. Shirsat and Sachin N. Deshmukh

Abstract Sentiment Analysis and Opinion Mining is a most popular field to analyze and find out insights from text data from various sources like Facebook, Twitter, and Amazon, etc. It plays a vital role in enabling the businesses to work actively on improving the business strategy and gain an in-depth insight of the buyer's feedback about their product. It involves computational study of behavior of an individual in terms of his buying interest and then mining his opinions about a company's business entity. This entity can be visualized as an event, individual, blog post or product experience. In this paper, Dataset has taken from Amazon which contains reviews of Camera, Laptops, Mobile phones, tablets, TVs, video surveillance. After preprocessing we applied machine learning algorithms to classify reviews that are positive or negative. This paper concludes that, Machine Learning Techniques gives best results to classify the Products Reviews. Naïve Bayes got accuracy 98.17% and Support Vector machine got accuracy 93.54% for Camera Reviews.

Keywords Sentiment analysis · Natural language processing · Product reviews
Machine learning · Support vector machine · Naïve Bayes

1 Introduction

Sentiment analysis invokes to the study of text analysis, natural language processing, computational linguistic to scientifically identify, extract and study subjective information from the textual data. Sentiment or opinion is the attitude of

R. S. Jagdale (✉) · V. S. Shirsat · S. N. Deshmukh
Department of Computer Science and IT, Dr. Babasaheb Ambedkar
Marathwada University, 431004 Aurangabad, Maharashtra, India
e-mail: rajkumarjagdale@gmail.com

V. S. Shirsat
e-mail: vss.csit@gmail.com

S. N. Deshmukh
e-mail: sndeshmukh@hotmail.com

© Springer Nature Singapore Pte Ltd. 2019
P. K. Mallick et al. (eds.), *Cognitive Informatics and Soft Computing*,
Advances in Intelligent Systems and Computing 768,
https://doi.org/10.1007/978-981-13-0617-4_61

customers comes from reviews, survey responses, online social media, healthcare media, etc. General meaning of sentiment analysis is to determine the insolence of a speaker, writer, or other subject with respect to particular topic or contextual polarity to a specific event, discussion, forum, interaction or any documents, etc. Essential task of Sentiment analysis is to determine polarity of given text at the feature, sentence, and document level. Due to increase in user of Internet every user is interested to put his opinion on the internet through different medium and this results opinioned data has generated on the internet. Sentiment analysis helps to analyze these opinioned data and extract some important insights which will help to other user to make decision. Social media data can be from different types like Product Reviews, Movie reviews, Reviews from airlines, Cricket Reviews, Hotel Reviews, employee interaction, Healthcare reviews, news and articles etc.

1.1 Data Sources

In Sentiment Analysis and Opinion Mining there are different Data Sources for generating huge amount of data on social media. Some are given below.

1.1.1 Blogs

It is website on which person can write their opinion on particular thing.

1.1.2 Datasets

Most of the researcher has uploaded different datasets online with free access. E.g. Movie reviews, Product reviews, Hotel reviews, etc.

1.1.3 Review Sites

Most of the customers are putting their opinion on E-commerce websites where they brought products. Like Amazon, CNET, epinion, zdnet, consumerreview, IMDB, etc.

1.1.4 Micro-blogging

It is popular service for sending text message shortly. Like Twitter, Tumblr, Dipity, etc. (Fig. 1).

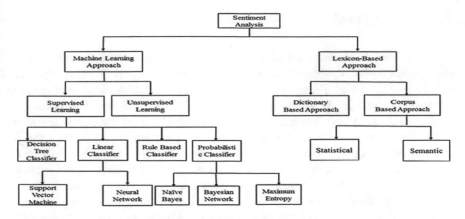

Fig. 1 Different sentiment analysis technique

2 Related Work

Sentiment Analysis is top research field under natural language processing (NLP), containing process of detecting and extracting sentiment/opinion from the text and classifying their sentiment. Sentiment analysis studies people's opinion, appraisal, emotions, and attitude towards individual, organization, products, movies, issues, events, etc.

2.1 Document Level

As per name, it analyzes the documents. In this level whole document has been analyzed and classify that whole document is [1, 2] expressing positive or negative view. In this method only one product reviews has processed and find out the opinion about same product. Opinions are expressed on only single entity. If document has multiple product reviews then this level will not work because it is not relevant to that documents which contains multiple type product reviews.

2.2 Sentence Level

In this level task goes to the sentences and analyzes then determines whether the given sentence has positive, negative, and neutral opinion. This level somehow similar to Subjectivity Classification [3] which separates the objective sentences and subjective sentences. Factual information and subjective information containing sentences called Objective sentences and Subjective sentences respectively.

2.3 Aspect Level

Earlier, Aspect level sentiment analysis is called Feature level sentiment analysis [3] contains feature-based opinion mining and summarization. To find out what exactly people like or did not like this level is very important. It is finer-grained sentiment analysis level. Aspect level directly looks at the opinion itself instead of looking to document or sentences. Output of this level will be the entity, aspect of that entity, opinion of aspect, opinion holder and time. For example, Samsung j7 has best camera quality. Now in this level, camera is aspect of entity Samsung j7 which expresses the positive opinion. In movie, particular scene, actor, acting, actions, etc., are some key points under aspect level sentiment analysis.

They compared various for Sentiment Analysis [4] by examining different methodologies and discussed Machine Learning methods like SVM, NB, and ME. They also discussed N-gram for Sentiment Analysis. Two steps method (Aspect Classification and polarity detection) has been used [5] and discussed lexical approach. Customer reviews are used as a dataset in their proposed method and in experimental result. SVM has been used and achieved 78% accuracy. Datasets consist of tweets [6] annotated for sentiment on a 2-point, 3-point, and 5-point scales. This is twitter task from SemEval 2017 and has done sentiment analysis on a two-point and on a five-point ordinal scale. They used new language, Arabic, for all subtasks. Polarity Categorization problem has been solved [7] for Sentiment Analysis. Online product reviews from Amazon are used for experimental purpose. Sentence level categorization and review-level categorization are implemented in this paper.

Hybrid approach [8] has been used, i.e., combination of Machine Learning and Lexicon based approach. In this paper, different techniques and tools have been discussed with different aspects. pSenti, combination of Lexicon-Based- and Learning-Based methods for Twitter Sentiment Analysis, SAIL, NILC_USP, combination of Lexicon-based- and Learning-based approaches for improved performance and convenience in sentiment classification, A Hybrid approach for sentiment classification of Egyptian Dialect Tweets, Sentiment Analysis: A Review and Comparative Analysis of Web Services, Alchemy API, Building Large-Scale Twitter-Specific Sentiment Lexicon: A Regression Learning Approach, Sentiment Analysis on Twitter, Sentiment Analysis using Sentiment Features, Improving Twitter Sentiment Analysis with Topic-Based Mixture Modeling and Semi-Supervised Training and MSA-COSRs are used as Hybrid tools and techniques. They surveyed on Opinion mining [9] with respect to their different levels, tools used, architecture, techniques applied, comparative study of techniques and challenges.

Sentence level sentiment analysis [10] has been done and taken live tweets from twitter using R tool. Also discussed different lexicons Like SentiWordNet, WordNet-Affect, MPQA, etc. Tweets of different events has been collected like #Budget2016, #RailBudget2016, #Freedom251, #MakeInIndia, #Oscars2016, #startup, #InternationalWomensDay, #AsiaCupT20Final, #IndvsPak and

#ProKabaddi. Each event has 10,000 tweets and classified as positive and negative tweets which gives information about people's opinion about that events. Machine learning approaches [11] used and also discussed spammed reviews and unauthenticated users. Amazon review dataset has been used in this paper. Review contains Reviewer ID, Product ID, Review Text, Rating and time of the review. They identified public perception of their product over time and discovered important areas where their products can be more improved.

3 Proposed Method

In the proposed method following preprocessing task has been completed to classify sentiment analysis from reviews.

3.1 Collection of Dataset

The dataset is collected from Amazon and it is in json format. Each json file contains number of reviews. Dataset has reviews of Camera, Laptops, Mobile phones, tablets, TVs, video surveillance.

3.2 Preprocessing

In preprocessing tokenization, stop word removal, stemming, punctuation marks removal, etc., has done. It has converted in bag of words. Preprocessing is important in sentiment analysis and opinion mining.

3.3 Score Generation

In this step, every sentence has analyzed and calculated sentiment score. To calculate sentiment score dataset has compared with opinion lexicons i.e. 2006 positive words and 4783 negative words and calculated sentiment score for every sentence.

3.4 Sentiment Classification

Using score and different features different machine learning algorithms has applied and different accuracy measurements calculated. Proposed method uses the

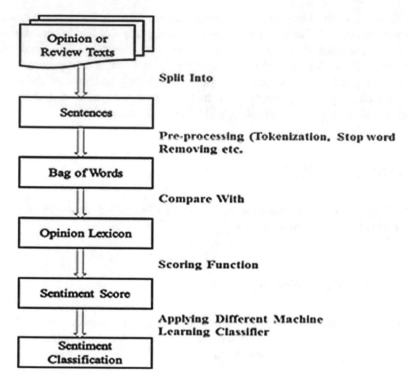

Fig. 2 Proposed methodology of sentiment analysis

following tasks to classify the sentiment analysis using machine learning techniques (Fig. 2).

4 Experimental Results

4.1 Dataset Description

Dataset contains reviews of Camera, Laptops, Mobile phones, tablets, TVs, video surveillance collected from Amazon. It is in the form of json files and each json file contains number of reviews (Fig. 3; Table 1).

4.2 Classification Results

See (Fig. 4).

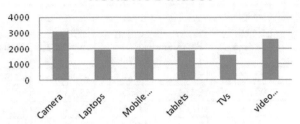

Fig. 3 Graphical representation of number of reviews

Table 1 Dataset and reviews count

Dataset name	Number of reviews
Camera	3106
Laptops	1946
Mobile phones	1918
Tablets	1894
TVs	1596
Video surveillance	2597

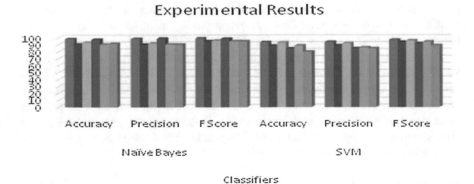

Fig. 4 Graphical representation of experimental results

Table 2 Evaluation parameters for classifiers of datasets

Dataset	Classifiers					
	Naïve Bayes			SVM		
	Accuracy	Precision	F score	Accuracy	Precision	F score
Camera	98.17	98.30	99.03	93.54	93.58	96.66
Laptops	90.22	90.01	94.74	88.16	88.52	93.71
Mobile phones	92.85	91.64	95.64	92.85	91.64	95.64
Tablets	97.17	98.73	98.31	84.12	84.31	91.37
TVs	90.16	90.17	94.72	88.49	85.56	93.89
Video surveillance	91.13	89.95	94.71	79.43	84.25	88.53

5 Conclusion and Future Work

Nowadays, Research on Sentiment Analysis and Opinion Mining is very important. Most of industries are creating different types of data and they need to analyze that data to make decisions which are benefited to industry. Social media is also generating huge amount of data and that need to analyze and find insights from that data. In this paper product reviews Dataset has taken from Amazon website containing six product reviews of Camera, Laptops, Mobilephones, Tablets, TVs, and Video Surveillance. In this proposed methodology dictionary-based approach under lexicon-based Approach has been used with machine learning techniques. Sentiment Analysis has been done on every product reviews and then classified using machine learning algorithms, i.e., NB and SVM. Table 2 shows the accuracy measurements for NB and SVM classifier for the dataset. Naïve Bayes classifier got 98.17% accuracy for Camera reviews and Support Vector Machine got 93.54% accuracy for Camera reviews. For future work, on same dataset, Aspect level sentiment analysis can improve results and from that method, we can get what exactly people liked or disliked. For example, Camera's quality, megapixel, picture size, structure, lens, picture quality, etc. Aspect level is fine-grained approach and gives good result in Sentiment Analysis.

Acknowledgements The author would like to thank to Data Analytics Research Lab, Department of Computer Science and IT, Dr. Babasaheb Ambedkar Marathwada University, Aurangabad for providing infrastructure to carry my research work. The author acknowledges the Department of Science and Technology (DST), New Delhi, India for granting financial assistance in the form of DST INSPIRE FELLOWSHIP (JRF) during this research work.

Bibliography

1. Pang, B., Lee, L., Vaithyanathan, S.: Thumbs up? Sentiment classification using machine learning techniques. In: Proceedings of the ACL-02 Conference on Empirical Methods in Natural Language Processing-Volume 10. Association for Computational Linguistics, 2002
2. Turney, P.D.: Thumbs up or thumbs down? Semantic orientation applied to unsupervised classification of reviews. In: Proceedings of the 40th Annual Meeting on Association for Computational Linguistics. Association for Computational Linguistics, 2002
3. Hu, M., Liu, B.: Mining and summarizing customer reviews. In: Proceedings of the tenth ACM SIGKDD International Conference on Knowledge Discovery and Data Mining. ACM, 2004
4. Devika, M.D., Sunitha, Cᵃ., Ganesh, A.: Sentiment analysis: a comparative study on different approaches. Proc. Comput. Sci. **87**, 44–49 (2016)
5. Bhadane, C., Dalal, H., Doshi, H.: Sentiment analysis: measuring opinions. Proc. Comput. Sci. **45**, 808–814 (2015)
6. Rosenthal, S., Farra, N., Nakov, P.: SemEval-2017 task 4: sentiment analysis in Twitter. In: Proceedings of the 11th International Workshop on Semantic Evaluation (SemEval-2017) (2017)
7. Fang, X., Zhan, J.: Sentiment analysis using product review data. J. Big Data **2**(1), 5 (2015)
8. Ahmad, M., et al.: Hybrid tools and techniques for sentiment analysis: a review. Int. J. Multidiscip. Sci. Eng. **8**(3) (2017)
9. Sharma, S., Tiwari, R., Prasad, R.: Opinion mining and sentiment analysis on customer review documents—a survey. IJARCCE **6**(2), 156–159 (2017)
10. Jagdale, R.S., Shirsat, V.S., Deshmukh, S.N.: Sentiment analysis of events from Twitter using open source tool. IJCSMC **5**(4), 475–485 (2016)
11. Kamalapurkar, D., Bagwe, N., Harikrishnan, R., Shahane, S., Gahirwal, M.: Sentiment analysis of product reviews. Int. J. Eng. Sci. Res. Technol. **6**(1), 456–460 (2017)

Vector-Controlled Induction Motor Drives Using Intelligent RST Robust Controller

Bhola Jha, M. K. Panda, V. M. Mishra and S. N. V. Ganesh

Abstract This paper addresses the vector-controlled induction motor drives using Robust Structure Theory (RST) intelligent controller. The RST controller is one of the robust cum intelligent controllers which can replace the conventional PI controller in near future. Because the performance of conventional PI controller is satisfactory in terms of accurate tracking only but its performance degrades when the machine parameters changes and sudden disturbances occur. Since the changes in machine parameters and the disturbances are inevitable so the replacement is needed. Hence the use of robust-intelligent controller now a day is encouraged. The implementation of this controllers, i.e., RST in vector-controlled induction motor drives is a novel approach meeting the all objectives such as accurate tracking, insensitiveness to the parameter variations and disturbances adaption. The performance of the proposed controller is proved better by comparing with PI controller using MATLAB/SIMULINK.

Keywords Modeling · Vector control · RST · Robust cum intelligent controller · Axis and phase transformation, etc.

B. Jha (✉) · M. K. Panda · V. M. Mishra
G. B. Pant Institute of Engineering and Teachnology, Pauri, India
e-mail: bholajhaeee@gmail.com

M. K. Panda
e-mail: pandagbpec@gmail.com

V. M. Mishra
e-mail: vmm66@rediffmail.com

S. N. V. Ganesh
Vignana Bharathi Institute of Technology, Hyderabad, India
e-mail: snvganesh@gmail.com

1 Introduction

Induction motors is known for asynchronous motors also because its speed less than synchronous speed. This motor is widely used because of robustness, less cost, almost negligible maintenance and having larger torque-to-weight ratio. Because of the absence of physical connection of stator and motor the industries have taken its advantage for achieving the vector control technique. In the olden days, separately excited dc machines were preferred for the variable speed drives because of inherent decouple nature of armature current and field flux with larger limit of speed control. Due to this nature, dc motor is having the excellent dynamic performance. But, an induction motor is not having such an inherent decouple nature. Therefore a control technique is developed called as vector control.

The vector control works on the principle of field orientation such that the characteristic of induction motor can be made same as that of dc motor. This is one of the most efficient control strategies for an asynchronous machine. Decouple control of torque, i.e., true power and flux, i.e., volt ampere reactive can easily be achieved using this method. Its performance is very high that is 0.5% in terms of the speed and 2% in terms of torque, even at standstill. This technique can achieve its objectives in all possible four quadrants modes. The modeling of induction machine [1–7] is necessary for the implementation of vector control which is found in [1–16]. The papers [8–16] describe intelligent techniques (fuzzy, sliding mode control, extended Kalman filter) based vector-controlled induction machine drives. Fuzzy [17] is one of the efficient and widely used soft computing techniques.

In this paper, a robust- intelligent RST controller is employed in speed control loop for accurate tracking, insensitiveness to parameter variations and adaptable to disturbances. The use of RST controller for the vector control of induction motor drives is a new approach. For the validation, the performances of RST-based vector control technique are measured with respect to conventional PI controller. The results are found good in line with the expectations.

The organization of paper is as follows: Sect. 1 deals the introduction, Sect. 2 describes a dynamic modeling for the vector control of motor, Sect. 3 overviews the design of RST, Sect. 4 illustrates the results and Sect. 5 ends with conclusion.

2 Dynamic Modeling of Induction Machine

A d-q model of the induction motor should be known for understanding and designing the vector-controlled drives. The instantaneous effects of varying voltage/currents, stator frequency, torque disturbance speed, etc., is taken into account for the dynamic modeling.

It is customarily to convert three-phase variables into two-phase variable using axis changes technique shown in Fig. 1. The following Eqs. (1) and (2) give the conversion formulae.

Fig. 1 Three-phase to two-phase axis transformation

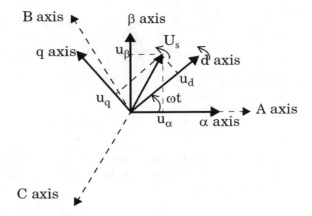

$$\begin{bmatrix} V_{qs}^s \\ V_{ds}^s \\ V_{os}^s \end{bmatrix} = \frac{2}{3} \begin{bmatrix} \mathrm{Cos}\theta & \mathrm{Cos}(\theta - 120°) & \mathrm{Cos}(\theta + 120°) \\ \mathrm{Sin}\theta & \mathrm{Sin}(\theta + 120°) & \mathrm{Sin}(\theta + 120°) \\ 0.5 & 0.5 & 0.5 \end{bmatrix} \begin{bmatrix} V_{as} \\ V_{bs} \\ V_{cs} \end{bmatrix} \tag{1}$$

$$\begin{bmatrix} V_{as} \\ V_{bs} \\ V_{cs} \end{bmatrix} = \begin{bmatrix} \mathrm{Cos}\theta & \mathrm{Sin}\theta & 1 \\ \mathrm{Cos}(\theta - 120°) & \mathrm{Sin}(\theta - 120°) & 1 \\ \mathrm{Cos}(\theta + 120°) & \mathrm{Sin}(\theta + 120°) & 1 \end{bmatrix} \begin{bmatrix} V_{qs}^s \\ V_{ds}^s \\ V_{os}^s \end{bmatrix} \tag{2}$$

In the following Eq. (3) superscripts *e* indicates the reference frames rotating with synchronous speed for which stator and rotor d-q voltages are defined.

$$\left. \begin{aligned} v_{qs}^e &= \omega_s L_s i_{ds}^{ie} + L_m p i_{qr}^{ie} + \omega_s L_m i_{dr}^{ie} + (R_s + L_s p) i_{qs}^{ie} \\ v_{ds}^e &= -\omega_s L_s i_{qs}^{ie} + L_m p i_{dr}^{ie} - \omega_s L_m i_{qr}^{ie} + (R_s + L_s p) i_{ds}^{ie} \\ v_{qr}^e &= (\omega_s - \omega_r) L_m i_{ds}^{ie} + L_m p i_{qs}^{ie} + (\omega_s - \omega_r) L_r i_{dr}^{ie} + (R_r + L_r p) i_{qr}^{ie} \\ v_{dr}^e &= -(\omega_s - \omega_r) L_m i_{qs}^{ie} + L_m p i_{ds}^{ie} - (\omega_s - \omega_r) L_r i_{qr}^{ie} + (R_r + L_r p) i_{dr}^{ie} \end{aligned} \right\} \tag{3}$$

The stator and rotor flux linkages in the synchronously rotating reference frame are given by (4) and (5).

As the rotor is shorted, so the rotor equations in terms of flux linkages are given in (6).

$$\left. \begin{aligned} \lambda_{qs}^e &= L_s i_{qs}^{ie} + L_m i_{qr}^{ie} \\ \lambda_{ds}^e &= L_s i_{ds}^{ie} + L_m i_{dr}^{ie} \\ \lambda_{qr}^e &= L_r i_{qr}^{ie} + L_m i_{qs}^{ie} \\ \lambda_{dr}^e &= L_r i_{dr}^{ie} + L_m i_{ds}^{ie} \end{aligned} \right\} \tag{4}$$

$$\left. \begin{aligned} \lambda_{qm}^e &= L_m (i_{qs}^{ie} + i_{qr}^{ie}) \\ \lambda_{dm}^e &= L_m (i_{ds}^{ie} + i_{dr}^{ie}) \end{aligned} \right\} \tag{5}$$

$$\left.\begin{array}{l} R_r i_{dr}^e + p\lambda_{qr}^e + \omega_{sl}\lambda_{dr}^e = 0 \\ R_r i_{qr}^e - \omega_{sl}\lambda_{qr}^e + p\lambda_{dr}^e = 0 \\ \text{where} \qquad \omega_{sl} = \omega_s - \omega_r \end{array}\right\} \tag{6}$$

As obvious,

$$p = \frac{d}{dt}$$

Considering resultant rotor flux linkage, λ_r lies on the direct axes for reducing the variables. Hence, aligning the d-axes with rotor flux phasor yields the following Eqs. (7), (8) and (9);

$$\lambda_r = \lambda_{dr}^e \tag{7}$$

$$\lambda_{qr}^e = 0 \tag{8}$$

$$p\lambda_{qr}^e = 0 \tag{9}$$

Now the new rotor equation as (10)

$$\left.\begin{array}{l} R_r i_{dr}^e + \omega_{sl}\lambda_r^e = 0 \\ R_r i_{qr}^e + p\lambda_r^e = 0 \end{array}\right\} \tag{10}$$

The rotor currents as a function of the stator currents are given in (11)

$$\left.\begin{array}{l} i_{qr}^e = -\frac{L}{L_r} i_{qs}^e \\ i_{dr}^e = -\frac{\lambda_r}{L_r} - \frac{L}{L_r} i_{ds}^e \end{array}\right\} \tag{11}$$

Now the flux and active current generating current are expressed as (12);

$$\left.\begin{array}{l} i_f = i_{ds}^e = \frac{[1 + T_r p]}{L_r} \lambda_r \\ i_T = i_{qs}^e = \frac{T_r \times \omega_s}{L_m} \lambda_r \end{array}\right\} \tag{12}$$

Rotor time constant is given by (13);

$$T_r = \frac{L_r}{R_r} \tag{13}$$

The induction machine's electromagnetic torque is given by (14) or (15):

$$T_e = \frac{3}{2}\frac{P}{2} L_m \left(i_{qs}^e i_{dr}^e - i_{ds}^e i_{qr}^e \right) \tag{14}$$

Or

$$T_e = \frac{3}{2}\frac{P}{2}\frac{L_m}{L_r}\left(i_{qs}^e\lambda_{dr} - i_{ds}^e\lambda_{qr}\right) = K_{te}\lambda_{dr}i_{qs}^e = K_{te}\lambda_r i_T \tag{15}$$

The electromechanical expression of an asynchronous machine is expressed by (16)

$$T_e - T_L = \frac{2}{P}J\frac{d\omega_r}{dt} \tag{16}$$

Vector Control makes the ac drives similar to dc drives, i.e., an independent control of flux and torque. Stator current phasor is resolved along the rotor flux linkages and therefore known as field producing current i_f & perpendicular to it as torque producing current component i_T. The phasor diagram is shown in Fig. 2.

The three-phase stator currents can be transformed into two-phase current (d-q axes) currents in the synchronous reference frames given by (17), where θ_f is magnetic field angle measured from stationary frame.

$$\begin{bmatrix} i_{qs}^e \\ i_{ds}^e \end{bmatrix} = \frac{2}{3}\begin{bmatrix} Sin\theta_f & Cos(\theta_f - 120°) & Cos(\theta_f + 120°) \\ Cos\theta_f & Sin(\theta_f + 120°) & Sin(\theta_f + 120°) \\ 0.5 & 0.5 & 0.5 \end{bmatrix}\begin{bmatrix} i_{as} \\ i_{bs} \\ i_{cs} \end{bmatrix} \tag{17}$$

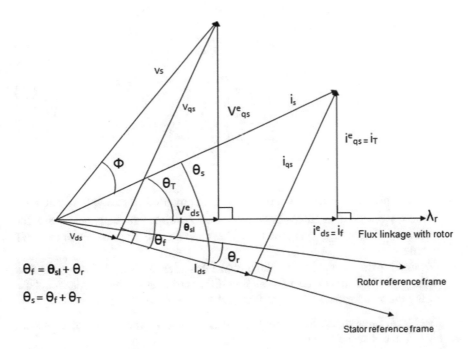

Fig. 2 Vector control phasor diagram

Now stator current phasor i_s given as (18)

$$i_s = \sqrt{\left(i_{qs}^e\right)^2 + \left(i_{ds}^e\right)^2} \tag{18}$$

Also the stator current phasor angle is given by (19)

$$\theta_s = \tan^{-1}\left(i_{qs}^e \div i_{ds}^e\right) \tag{19}$$

Now let us write rotor flux linkages and torque as a function of currents. Which are given from (20)–(26)

$$\lambda_r \alpha i_f \tag{20}$$

$$T_e \alpha \lambda_r i_T \alpha i_f i_T \tag{21}$$

It is clearly observed that i_f and i_T are dc currents because their relative speed is zero. These dc currents are ideal to use as a control vectors.

$$\theta_e = \int (\omega_r + \omega_{sl})dt = \int \omega_e dt \tag{22}$$

$$\omega_{sl} = \frac{L_m \times i_q}{T_r \times \lambda_r} \tag{23}$$

$$\lambda_r = \frac{L_m \times i_d}{(1 + T_r s)} \tag{24}$$

$$i_d^* = \frac{\lambda_r^*}{L_m} \tag{25}$$

$$i_q^* = \frac{2}{3} \times \frac{2}{P} \times \left(\frac{L_r \times T_e^*}{L_m \times \lambda_r}\right) \tag{26}$$

- The two-phase dq reference currents i_{dq}^* are converted to three-phase reference current i_{abc}^* and this three-phase reference currents are compared with the measured currents i_{abc} using the controllers for the pulses to be given to IGBT inverter.
- Obtained pulse is applied to three-phase Inverter for the desired rotor flux linkages and torque. In this way, an inverter controls both, magnitude and phase of the current for the purpose of flux and torque to be decoupled.

The block diagram and the computational flow chart of vector control are shown in Figs. 3 and 4 respectively.

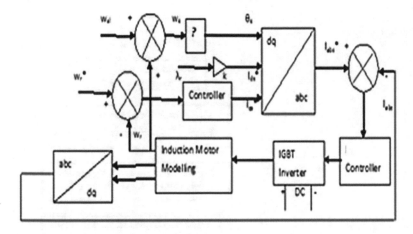

Fig. 3 Vector control block diagram

3 Design of RST Controller

The control design aims to compute the value of R, S, and T polynomials to meet the desired objectives.

$$\text{T.F} = \frac{L_m V d}{(L_r L_s - L_m^2)s + L_s R_r} = \frac{B}{A} \tag{27}$$

The induction motor having transfer function B/A [18] as given in (27) with Y_{ref} as an input and Y as an output is shown in Fig. 5.

The transfer function of this system is expressed by (28)

$$Y = \frac{BT}{AS + BR} Y_{\text{ref}} + \frac{BS}{AS + BR} \gamma \tag{28}$$

From the Bezout equation as mentioned in (29). More details of RST are found in [19–21].

$$D = AS + BR = CF, \tag{29}$$

where, C is the command polynomial and F is the filtering polynomial.

If A is a polynomial of n [(deg $(A) = n$] degree then we must have the degree of polynomials like below from (30) to (32).

$$\deg (D) = 2n + 1 \tag{30}$$

$$\deg (S) = \deg (A) + 1 \tag{31}$$

Fig. 4 Vector control for
computational flow

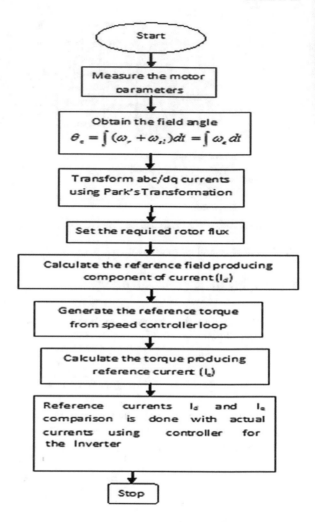

Fig. 5 Block diagram of
RST Controller

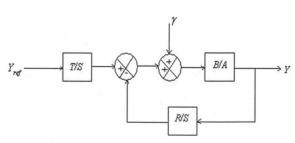

$$\deg (A) = \deg (R) \qquad (32)$$

In this case, polynomials are expressed from (33) to (37)

$$A = a_1 p + a_0 \qquad (33)$$

$$R = r_1 p + r_0 \qquad (34)$$

$$B = b_0 \qquad (35)$$

$$S = s_2 p^2 + s_1 p + s_0 \qquad (36)$$

$$D = d_3 p^3 + d_2 p^2 + d_1 p + d_0 \qquad (37)$$

In order to find the coefficients of polynomials R and S, the robust pole placement method is used. Where T_c is control variable and T_f is filtering variable are given by (38).

$$p_c = -\frac{1}{T_c} \& p_f = -\frac{1}{T_f} \qquad (38)$$

Here, P_c is the pole of C and P_f is the double pole of F. The pole P_c must accelerate the system and is generally chosen three–five times greater than the pole of A. The pole P_f is usually three times smaller than P_c. To obtain good disturbance rejections and good stability in steady state, the term $\frac{BS}{AS+BR}$ must tend to zero. The Bezout giving four equations having four variables. In which coefficients of D are related to the coefficients of polynomials R & S by the Sylvester Matrix given by (39).

$$\begin{bmatrix} d_3 \\ d_2 \\ d_1 \\ d_0 \end{bmatrix} = \begin{bmatrix} a_1 & 0 & 0 & 0 \\ 0 & a_1 & 0 & 0 \\ 0 & a_0 & b_0 & 0 \\ 0 & 0 & 0 & b_0 \end{bmatrix} \begin{bmatrix} s_2 \\ s_1 \\ r_1 \\ r_0 \end{bmatrix} \qquad (39)$$

For determining the coefficients of T, we can consider that $Y = Y_{ref}$ in steady state case. So, the term $\frac{BT}{AS+BR}$ equal to one.

4   Results and Discussion

Current, speed and torque waveforms are shown for the two different machine parameters mentioned in Table 1 as Appendix-I and Table 2 as Appendix-II. This is done in order to prove the efficacy of robustness cum intelligent. The results are shown below for the load torque of 200 Nm applied at 1 s. No load torque is

Table 1 Appendix-I
(machine parameters)

S. no.	Machine parameters	Values
1	Resistance of stator	0.1 Ω
2	Inductance of stator	51.2e−3 H
3	Inductance of rotor	51.2e−3 H
4	Resistance of rotor	0.5 Ω
5	Mutual inductance	50e−3 H
6	Moment of inertia	1.662 Kg-m^2
7	Pole pair	2
8	Frequency	60 Hz
9	DC supply	780 V
10	Friction damping coefficient	0.1

Table 2 Appendix-II
(machine parameters)

S. no.	Machine parameters	Values
1	Resistance of stator	0.087 Ω
2	Inductance of stator	35.5e−3 H
3	Inductance of rotor	35.5e−3 H
4	Resistance of rotor	0.228 Ω
5	Mutual inductance	34.7e−3 H
6	Moment of inertia	1.662 Kg-m^2
7	Pole pair	2
8	Frequency	60 Hz
9	DC Supply	780 V
10	Friction damping coefficient	0.1

applied before 1 s. From the current waveforms of Figs. 6a, b and 7a, b, it is observed that the PI controller takes more time to settle to a steady state as compared to RST controller. The dynamic performances of speed and torque are shown in Figs. 8 and 9 respectively for two different machine parameters. This is observed from Fig. 8a, b that a reference speed of 120 rad per second is achieved fastly using RST controller as compared to PI controller. At 1 s there is fall in speed due to load torque of 200 Nm. Whenever the machine parameter changes the time taken by the machine using PI controller to achieve reference speed of 120 rad per second speed is increased as compared to RST controller. The similar observations are made from the torque waveform of Fig. 9a, b.

To prove the adaptability of robust cum intelligent controller, the speed is suddenly increased at 1.5 s. The current waveforms are shown in Fig. 10a, b using PI and RST controller respectively.

This is observed from dynamic performance of speed and torque waveform of Fig. 11a, b that the RST controller is able to adapt the changes made on the motor input. This is the reason it is so called robust cum intelligent controller. So it is more sensitive to input or references changes and less sensitive to the parameter variation. The change in parameters is an internal change which usually occurs in the system

(a)

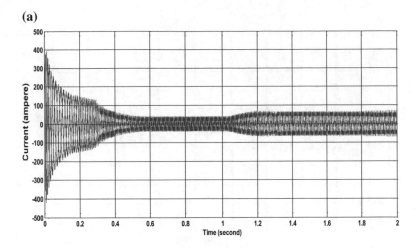

(b)

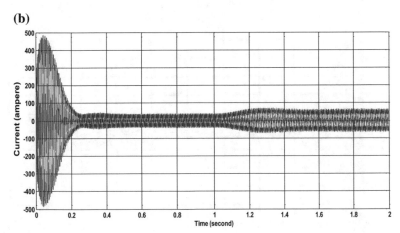

Fig. 6 **a** Current versus time using PI for Appendix-I parameter. **b** Current versus time using RST for Appendix-I parameter

due to temperature effects, aging effects, etc. This is also observed from the above figures that the starting performances of motor are good using PI controller as compared to RST controller. Therefore; the research is to be extended further for having good starting performance. The IGBT inverter output voltage fed to motor is also shown in Figs. 12 and 13. The RST and PI controller parameters are mentioned in Table 3 as Appendix-III.

(a)

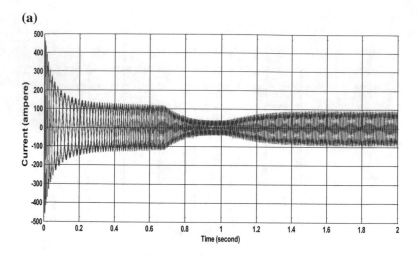

(b)

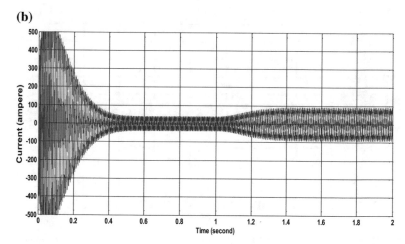

Fig. 7 **a** Current versus time using PI for Appendix-II parameter. **b** Current versus time using RST for Appendix-II parameter

(a)

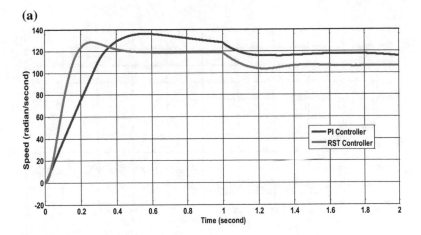

(b)

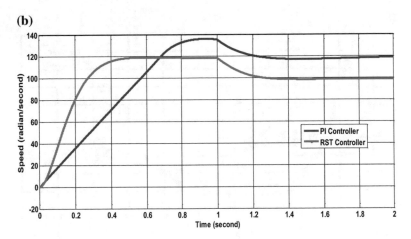

Fig. 8 a Speed versus time for Appendix-I parameter. **b** Speed versus time for Appendix-II parameter

(a)

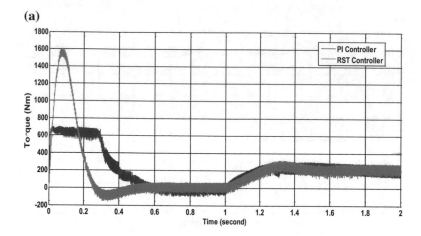

(b)

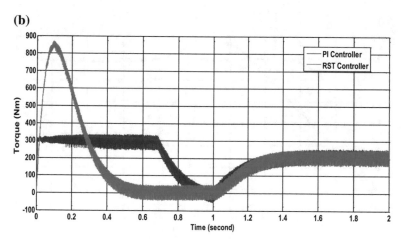

Fig. 9 a Torque versus time for Appendix-I parameter. **b** Torque versus time for Appendix-II parameter

(a)

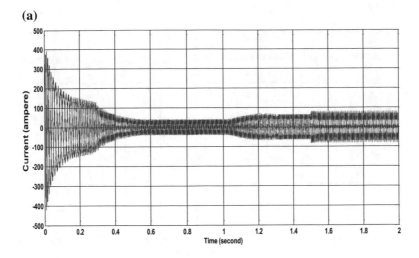

(b)

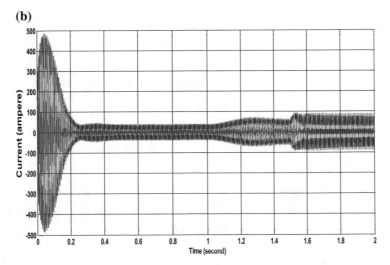

Fig. 10 **a** Current versus time using PI for sudden change in speed. **b** Current vs time using RST for speed change

(a)

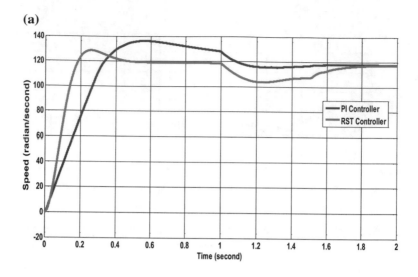

(b)

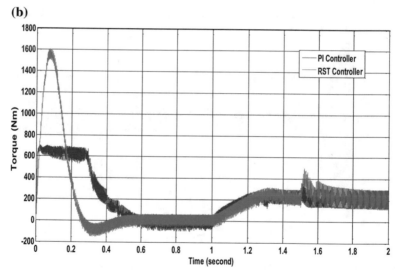

Fig. 11 **a** Speed versus time for sudden change in speed. **b** Torque versus time for sudden change in speed

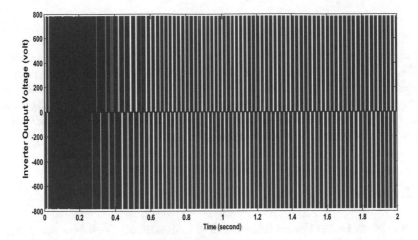

Fig. 12 Inverter output voltage using PI controller

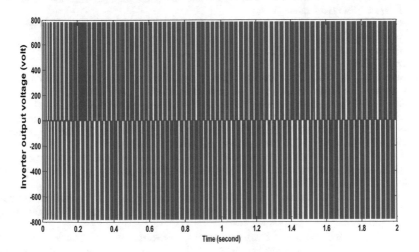

Fig. 13 Inverter output voltage using RST controller

Table 3 Appendix-III (controller parameters)

S. no.	Parameters	Values
1	$R = r_1 s + r_0$	2792.51 s + 50490.84
2	$S = s_2 s^2 + s_1 s + s_0$	38.06 s^2 + 2294.13 s + 1
3	T	50490.84
4	Kp & Ki	13 & 26

5 Conclusion

Vector Control of Induction Motor is presented using RST controller. The design of RST controller is lucidly presented. The RST controller takes less time to come to a steady state condition so it would be a better replacement of PI controller. This controller is able to adapt the changes made on motor. Robust cum intelligent RST controller performance is validated in terms of accurate tracking, insensitiveness to parameter variations, and disturbance adaptation.

References

1. Bose, B.K.: Modern Power Electronics and AC Drives. Prentice Hall, New Jersey (2002)
2. Krishnan, R.: Electric Motor Drives-Modeling, Analysis and Control. Prentice Hall, New Jersey (2001)
3. Novotney, D.W., et al.: Introduction to field orientation and high performance AC drives. In: IEEE IAS Tutorial Course (1986)
4. Kraus, P.C., Wasynczuk, O., Sudhoff, S.D.: Analysis of Electric Machinery, 2nd edn. Wiley, New Jersey (2004)
5. Slemon, G.R.: Modelling of Induction Machines for Electric Drives. IEEE Trans. Ind. Appl. **25**(6), 1126–1131 (1989)
6. Simoes, M.G., Ferret, F.A.: Alternate Energy Systems: Design and Analysis with Induction Generator, 2nd edn. CRC Press-Taylor and Francis Group, Boca Raton (2011)
7. Bhimbra, P.S.: Generalized Theory of Electrical Machinery. Khanna Publishers, Delhi (2012)
8. Menna, M., et.al.: Speed sensor less vector control of an induction motor using spiral vector model ECKF and ANN controller. In: IEEE conference on Electric Machines and Drives EMDC, May 3–5, 2007, Antalya, pp. 1165–1167 (2007)
9. Kar, B.N., et.al.: Indirect vector control of induction motor using sliding mode controller. In: IET International Conference on Sustainable Energy and Intelligent Systems, July 20–22, 2011, Chennai, pp. 507–511 (2011)
10. Liu, L., Xu, Z., Mei, M.: Sensorless vector control induction motor drive based on ADRC and flux observer. In: IEEE Control and Decision Conference, Guilin, June 17–19, 2009, pp. 245–248. (2009)
11. Zerikat, M., Mechernene, A., Chekroun, S.: High performance sensorless vector control of induction motor drives using artificial intelligence technique. In: IEEE International Conference on Automation and Robotics, Miedzyzdroje, Aug 23–26, 2010, pp. 67–75 (2010)
12. Narayan, K.B., Mohanty, K.B., Singh, M.: Indirect vector control of induction motor using fuzzy logic controller. In: IEEE, 10th International Conference on Environment and Electrical Engineering (EEEIC), 8–11, May 2011
13. Uddin, M.N., Radwan, T.S., Rahman, M.A.: Performances of fuzzy-logic-based indirect vector control for induction motor drive. In: IEEE Transaction on Industry Applications, pp. 1219–1225 (2002)
14. Tripura, P., Srinivasa Babu, Y.: Fuzzy logic speed control of three phase induction motor drive, World Academy of Science, Engineering and Technology (WASET). Int. J. Electr. Electron. Comput. Energetic Commun. Eng. **5**(12), 1769–1773 (2011)
15. Subudhi, B., et al.: d Space implementation of fuzzy logic based vector control of induction motor. In: IEEE Conference TENCON, 2008, Nov 19–21, Hyderabad, 978-1-4244-2408-5

16. Waheedabeevi, M., Suresh Kumar, A., Nair, N.S.: New online loss minimization of scalar and vector controlled induction motor drives. In: IEEE International Conference PEDES (2012)
17. Zadeh, : Fuzzy sets. Inf. Control **8**, 338–353 (1965)
18. Poitiers, F., Machmoum, M., Le Doeuff, R., Zaim, M.E.: Control of doubly-fed induction generator for wind energy conversion systems. Int. J. Renew. Energy **3**(3), 373–378 (2001)
19. Gharsallaoui, H., Ayadi, M., Benrejeb, M., Borne, P.: Flatness-based control & conventional RST polynomial control of a thermal process. Int. J. Comput. Commun. Control **IV**(1), 41–56 (2009)
20. Luo, C., Ooi, B.T.: Frequency deviation of thermal power plants due to wind farms. In: IEEE Transactions on Energy Conversion. vol. 21, no. 3, Sep. 2006, pp. 708–716 (2006)
21. Jha, B., Rao, K.R.M.: Disturbance rejection and harmonics reduction of doubly-fed induction generator using robust controller. In: 2009 Annual IEEE India Conference INDICON, Gujrat (2009)

Optimization of Electricity Bill on Study of Energy Audit by Using Grid Integration of Renewable Energy Sources

Shubham Soni, Inderpreet Kaur and Deepak Kumar

Abstract Demand of electricity increases globally day by day. Electricity consume in different each sector (industrial, commercial and institutional). Academic building energy consumption is different from residential and commercial buildings. A sincere attempt of the Energy Audit at Chandigarh University Gharuan Mohali, to estimate the Energy consumed in a day, week and month. In this audit identify the wastage of electricity in different academic buildings, campus cafe, Hostels, and hospital. The objective of Energy audit is to achieve and maintain optimum electricity utilization and minimize electricity bill without affecting load demand and quality. After auditing and analysis we suggest idea for reduce electricity and also mention the Estimate the implementation costs and payback periods for each recommended action can made. The analysis of the building data has been done with the help of power system analysis software package.

Keywords Challenge · Distribution generation · Energy audit
Integration of DG

1 Introduction

Demand of electric power increases day by day, with development of technologies we prefer electronics equipment like LED, PLC, Relays, Drives, controllers etc. all of these devices are non-liner loads in nature. Use of electronics devices in utility side at very large scale, these devices creates some unwanted signals (distortion) in waveform of voltage. For generation of electrical energy mostly conventional

S. Soni · I. Kaur (✉) · D. Kumar (✉)
Chandigarh University, Mohali, India
e-mail: hod.eee@cumail.in

D. Kumar
e-mail: dmishra.eee@gmail.com

S. Soni
e-mail: shubham.ktm94@gmail.com

sources are used like coal, petroleum and diesel. But these sources are totally gone from earth after some years if consumption rate is same as going on. Due to increase the demand of electricity, consumption of conventional sources are also increase. The Energy audit mainly focus on total energy consumption of the electrical equipment, main focus on the large available loads like air- conditioning system, electronically equipment, lighting, elevators, fans and computers etc. Field investigation of different loads available in the Chandigarh University has been carried out from December 2016 to September 2017. To collect all the data nearly three months are require for investigate the energy consumption in university from July 2017 to September 2017, including the running appliances like air-conditioning system, lighting system, Street lights, geysers and the equipment loads. Up to 2035, India will become import dependent country. The production of electricity increase up to 112% and consumption would rise up to 132%. In this paper study the management of electricity use and reduce the energy consumption. One of the most famous tool for securing our future is renewable source. Chandigarh University is located at north region of India, solar energy is available at very large scale in north region as compare to wind and other sources of energy. This is pollution free, less operating cost, easily available. Presently we cannot imagine the life without electricity. Bill of electricity depend upon the type and number of appliances used in building. If we want to reduce the electricity bill replace the high wattage appliances by low voltage appliance like replace 18 W CFL by 9 W LED. Energy audit is also helpful for industries to reduce the electricity bill.

2 Energy Audit

Energy Audit is a process through which the wastage of electrical energy can be identified and we can say saving of energy equals to generation of electrical energy.

3 Classification of Energy Audit

Energy audit has been classified into two categories

- Preliminary audit
- Detailed audit.

Preliminary audit:
Preliminary energy audit is very fast exercise it is used for estimates the possibility for saving by using the existing data, easily obtained data and helpful for identify the areas those require more detailed study.

Detailed audit:

Detail audit is further divided into three phases
Phase 1—Pre Audit Phase
Phase 2—Audit Phase
Phase 3—Post Audit Phase
This audit is much more accurate then preliminary audit because this audit is detailed audit for estimate the energy saving and cost.

4 Process of Energy Audit

The process of energy auditing has four steps: Pre-audit data collection, Detailed audit, Data analysis, Suggestion for implementation/Result.

5 Energy Audit of a Chandigarh University

Chandigarh University establish on near about 100 acrs land. This university has 10 academic blocks, 4 hostels, campus cafe, workshops, transport office building and dispensary. Each block has min 7 levels. For the manual data collection of load is being taken block wise, floor and room and note down how many fans, lights and cooler, Air conditioner and exhaust. At least one water Cooler is available at each floor of every block.

6 Data Analysis

Complete analysis of Chandigarh University load data collected was done. Energy consumption per day and per month in kWh is calculated based on each level of block-wise. Analysis of data by systemically process.

7 Analysis of Power Consumption by Different Blocks

With the help of the mathematical calculations, analysis of the power consumption by different equipments on basis of application wise as well as location wise. This is the summary of the analyze data that is represented in form of charts for better understanding.

Fig. 1 Process of energy
audit

It has been observed that from Fig. 1 that Block-1 and LC Block have maximum power consumption as compared to other buildings/blocks, electricity consumption mainly effect by electrical equipments used in building. The main equipment effects the power consumption are ACs, Geysers and Computers. Minimum electricity consumption by MBA department (block 5) is due to its small size and lesser laboratories. Library has lower consumption in spite of having lesser number of ACs and installation of CFLs and wall fans which greatly reduce the energy consumption.

8   Power Consumption Hostel Wise

There are 4 hostels in Chandigarh University. Out of these 2 are for boys and 2 for girls. All hostels have capacities ranging from 500 to 1800 seats. Most of the rooms are 4 seated. In 4 seated rooms, 4 tube lights and 2 ceiling fans are provided. In addition, each hostel has a mess, canteen, indoor games room, TV room and gym. Figure 2 shows the power consumption of hostels of Chandigarh University (Fig. 3).

This is complete data of Chandigarh University. University campus can apply renewable energy sources for generation of electrical power like PV cell at roof top of each block to supply the inside campus. But during holidays power demand is very less as compare to working days so generation by DG sources are greater than load demand then power can be supply back to utility or electricity board then board compensate it in electricity bill. This is very useful method for reduce the electricity bill.

Fig. 2 Power consumption block wise

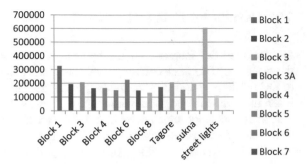

Fig. 3 Total power consumption of hostels

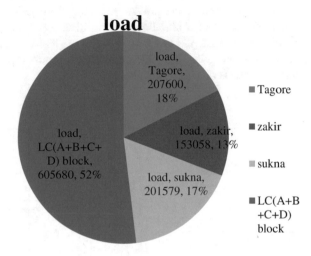

9 Renewable Energy Sources

Most of the electricity generated by conventional sources but these sources issue only for one time after that these sources are totally vanished due to that reason the cost of conventional sources increase day by day. These conventional sources are not available on earth for very long time but for generation of electrical power we move toward non-conventional (renewable) sources. Mostly concentrate on solar and wind and fuel cell for generation of electrical power.

10 Integration of Renewable Energy Sources with Grid

Recently there are many new developments are happen in electrical system one of the main development is integration of different types of power sources with grid. These sources are solar plant, wind plant, geothermal plants, fuel cell etc. By using MATLAB we check the effect of grid power if we integrate the DG sources with grid.

11 Results by Using MATLAB Software

Integration of DG sources with grid by using MATLAB is shown in Fig. 4 it shows the benefits of DG in electricity generation. During low power generating capacity of DG power is supplied to the industry from grid shown in Fig 5. In a similar way when the industries power demand is less than the DG generated power then the industries which consume power from the power grid will start supply the electrical power back to the power grid in Fig. 6. Blue line represent the power from the grid. This line in negative is feeding power to the grid. By using these waveforms trying to showing that a consumer behaves as a Prosumer (Producer + Consumer).

Green line shows the power generated from DG, blue line represent the grid power requirement if blue line is in positive region it means that amount of power require from grid, if blue line in negative region means that amount of power supplied to the grid from DG sources and red line represent the house load that is variable that is changes with time.

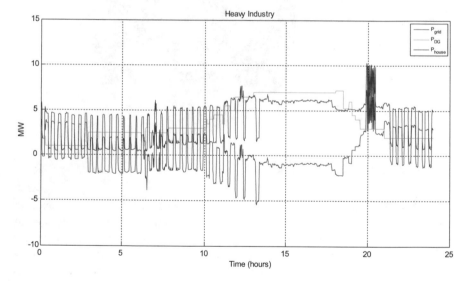

Fig. 4 Power receiving and dispatching curve

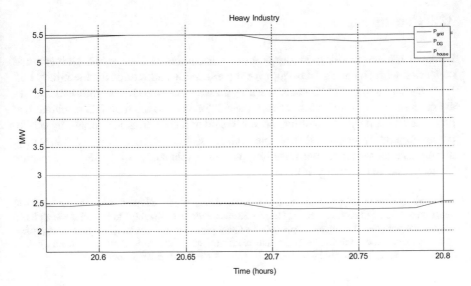

Fig. 5 Industry as a consumer

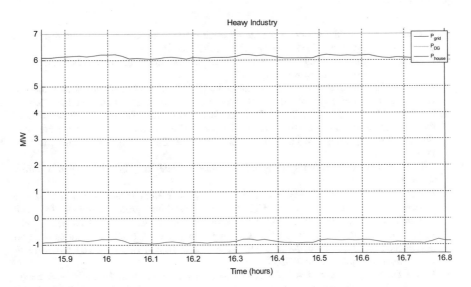

Fig. 6 Industry as a supplier

12 Conclusion

Different issues regarding energy audit are discussing in this paper. University is connected with DG and grid to improve the efficiency and reduce the electricity bill. By using integration of DG sources with grid we can save our conventional sources at very large scale. If industrial or commercial load is less the than DG production at that time DG supply the electrical power to grid and industries get benefit by selling power. Operating cost of Renewable sources is free of cost. Only investment is during installation time, after installing the renewable sources integration is require for reduction in electricity bill.

Acknowledgements My paper would not have been written without help of our electrical department teachers Inderpreet Kaur, Deepak Kumar. I am very thankful to my all teachers those helping me regarded my review paper. My teachers provided me extensive personal and professional guidance about the review paper and taught me about scientific research. I would like to thank my parents, whose guidance and love are with me in whatever I pursue.

References

1. Boutsika, T.N., Papathanassiou, S.A.: Short-circuit calculations in networks with distributed generation. Electr. Power Syst. Res. **78**, 1181–1191 (2008)
2. Bresesti, P., Cerretti, A.: SDNO: smart distribution network operation project. In IEEE Power Engineering Society General Meeting, 2007, pp. 1–4
3. Repo, S., Maki, K., Jarventausta, P., Samuelsson, O.: ADINE—EU demonstration project of active distribution network. In: IET-CIRED Seminar—SmartGrids for Distribution, 2008
4. Kanabar, P.M., Kanabar, M.G., El-Khattam, W., Sidhu, T.S., Shami, A.: Evaluation of communication technologies for IEC 61850 based distribution automation system with distributed energy resources. In: PES'09. IEEE Power & Energy Society General Meeting, 2009, pp. 1–8
5. CIGRE SC C6—distribution systems and dispersed generation. [Online]. Available: http://www.cigre-c6.org/ (2010)
6. D'Adamo, C., Jupe, S., Abbey, C.: Global survey on planning and operation of active distribution networks—update of CIGRE C6.11 working group activities In: 20th International Conference and Exhibition on Electricity Distribution, CIRED 2009, pp. 1–4
7. Kumar, D., Singh H., Reshma K.: A review on industry challenges in smart grid implementation IICPE-2016. Available at IEEE Xplore Digital Library Available at http://ieeexplore.ieee.org/document/8079395/
8. Kumar D., Singh H., Gupta A.: Voltage stability of grid integration for distributed energy resources. In: IICPE-2016 and Available at IEEE Xplore Digital Library. Available at http://ieeexplore.ieee.org/document/8079396/
9. Kumar, D., Thakur, T.: Grid integration for distributed energy resources. Int. J. Eng. Sci. Comput. **6**(5), May (2016)
10. CEA: The smart grid: a pragmatic approach, Tech. Rep., Canadian Electricity Association (2010)
11. Ministry of Power Government of India: Smart Grid Roadmap for India. India Smart Grid Forum, 2013

Computer-Aided Diagnosis of Epilepsy Based on the Time-Frequency Texture Descriptors of EEG Signals Using Wavelet Packet Decomposition and Artificial Neural Network

N. J. Sairamya, S. Thomas George, M. S. P. Subathra
and Nallapaneni Manoj Kumar

Abstract An adaptive time-frequency (t-f) representation of electroencephalo-graphic (EEG) signals with high time and frequency resolutions using wavelet packet decomposition are introduced in this paper for automated diagnosis of epilepsy. The novel texture pattern techniques namely local neighbor descriptive pattern (LNDP) and symmetric weighted LNDP (SWLNDP) are proposed to obtain distinct features from the t-f images. Proposed texture pattern techniques are insensitive to local and global variations as the consecutive neighboring pixels are compared. SWLNDP is a modified version of LNDP which improves the computational efficiency of the system by reducing the feature vector length. The histogram based features are extracted from the texture pattern of t-f images and fed into artificial neural network (ANN) for classification of signals. The obtained results show that ANN attained an accuracy of 100% using proposed techniques for classifying epileptic and normal signal. Further the performance of the proposed system was analyzed for fifteen different cases using University of Bonn EEG dataset.

Keywords Wavelet packet decomposition · Electroencephalographic (EEG)
Local neighbor descriptive pattern (LNDP) · Symmetric weighted local neighbor
descriptive pattern (SWLNDP) · Artificial neural network (ANN)

N. J. Sairamya · S. Thomas George (✉) · M. S. P. Subathra
Department of Electrical Sciences, Karunya Institute of Technology
and Sciences, Coimbatore 641114, Tamil Nadu, India
e-mail: thomasgeorge@karunya.edu

N. J. Sairamya
e-mail: sairamyanj@karunya.edu.in

M. S. P. Subathra
e-mail: subathra@karunya.edu

N. M. Kumar
Faculty of Electrical and Electronics Engineering,
Universiti Malaysia Pahang, 26600 Pekan, Pahang, Malaysia
e-mail: nallapanenichow@gmail.com

© Springer Nature Singapore Pte Ltd. 2019
P. K. Mallick et al. (eds.), *Cognitive Informatics and Soft Computing*,
Advances in Intelligent Systems and Computing 768,
https://doi.org/10.1007/978-981-13-0617-4_64

1   Introduction

Electroencephalographic (EEG) is a commonly used clinical tool for diagnosis of epilepsy, which is caused due to the sudden abnormal discharge of brain neurons [1]. EEG signals are generally recorded for long period of time and it requires manual inspection of neurologists to identify the epileptic activity in the signal, which is a tedious and inefficient process. Hence, an automatic detection of epileptic EEG signals using efficient techniques will assist the neurologist in early diagnosis of epilepsy. Currently there are various techniques like fast Fourier transform [2], discrete wavelet transforms (DWT) [3], empirical mode decomposition [4], wavelet packet decomposition (WPD) [5], local transformation techniques [6], and weighted visibility graph [7] to characterize the EEG signals. The features are extracted from each characterized signal and fed into classifiers for automatic diagnoses of epilepsy. Extraction of efficient features from the signal plays a major role in automated diagnoses of epilepsy. Hence, development of technique to extract distinct features with better computational performance is a challenging process.

Effective features representing the major changes attained due to neuronal activities of the brain are extracted from non-stationary EEG signal using techniques based on time-frequency (t-f) analysis. Recently, various t-f features based on image descriptors are proposed for the automatic identification of epileptic spikes in EEG signal [8]. In [8], the author obtained t-f image of an EEG signal using Wigner–Ville distribution (WVD) and extracted gray level co-occurrence matrix (GLCM) features from the t-f image. The extracted GLCM features were fed into support vector machine (SVM) classifier to discriminate epileptic from normal signals and attained an accuracy of 99.125%. The author combined the topographical features, texture features, morphometric features, and intensity features of a t-f image with the features extracted from t-f EEG signal and enhanced the classification accuracy to 95.33% [9].

Short time Fourier transform (STFT) [10] was used to obtain the t-f image of an EEG signal and the image processing techniques like local binary pattern (LBP), GLCM and texture feature coding method (TFCM) was used to extract the features from textured pattern t-f image. Recently a novel feature extraction based on t-f representation of EEG signal using multiscale radial basis functions (MRBF) and a modified particle swarm optimization (MPSO) was proposed to achieve a high-resolution spectral estimation result [1]. Although various techniques are proposed to extract t-f image of an EEG signal, still there is a need for effective t-f representation of EEG signal with high time and frequency resolution. In STFT, the non-stationarity of an EEG signals is extracted by employing fixed window size which limits the t-f resolution. MRBF-MPSO technique achieves a t-f image with high t-f resolution but the computational complexity is higher [1].

The proposed methodology is shown in Fig. 1. Hence, in this paper wavelet packet decomposition method is introduced to obtain t-f representation of EEG signal with high resolution. Further, effective texture pattern techniques called local

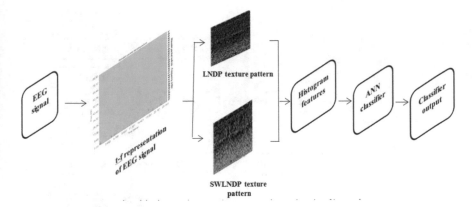

Fig. 1 The proposed system

neighbor descriptive pattern (LNDP) and symmetric weighted LNDP (SWLNDP) are proposed to analyze the t-f image of EEG signal. Both these techniques operate in two stages. In first stage, histogram based features are extracted from the transformed local pattern. In second stage, the obtained histogram features are fed into artificial neural network (ANN) classifier with 10-fold cross validation for classifying epileptic and normal signals.

2 Methodology

The t-f image representation using wavelet packet decomposition, proposed texture pattern techniques and ANN classifier are briefed in this section.

2.1 Wavelet Packet Decomposition (WPD)

WPD is the generalized form of wavelet decomposition (WD), in which the signals are transformed into a spectral data in both frequency and time domain [11]. The multiresolution analysis of a signal is achieved by using WPD. In first level of decomposition WD splits the signals into detail (high frequency components) and approximation (low-frequency components) coefficient using high and low pass filters. In second level, WD decomposes the approximation coefficients into second level of detail and approximation. This process is continued for all the other levels utilized in WD. Due to which the information from high frequency (detail) components are neglected in WD. In contrary, WPD decomposes the detail and approximate of a signal in each level. Thus, a richer range of possibilities for signal analysis are obtained using WPD. The three-level decomposition mechanism of WPD is shown in Fig. 2.

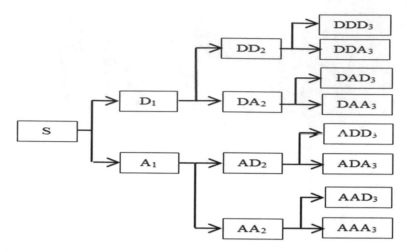

Fig. 2 Schematic diagram of WPD with 3 level decomposition. D represents the high pass detail coefficients and A represents the low pass approximation coefficients

In STFT longer windows are required to obtain a good frequency resolution, but it limits the time resolution. To identify the epileptic spikes in EEG, the signals are needed to be analyzed in high time resolution as well as frequency resolution. This is achieved in WPD as it uses discrete wavelet transformation for multiresolution analysis of a signal.

Algorithm 1. Computing LNDP descriptor
Require: Location of a given pixel (i, j) and spatial resolution, (P, R).
   Ensure: Label of LNDP descriptor, $L(i, j)$.

  1: The number of neighboring pixels (P) is obtained.
  2: Each neighboring pixel is assigned a value 0 or 1 by comparing it with the consecutive neighboring pixel, which is given as,

$$d_n = P_n - P_{n+1} \tag{1}$$

where, $n = 0$ to $P - 1$

3. LNDP code is computed as

$$L(i,j) = \sum_{n=0}^{P-1} s(d_n) 2^n \tag{2}$$

$$\text{where, } s(x) = \begin{cases} 0, & x < 0 \\ 1, & x \geq 0 \end{cases}$$

6: **return:** Label of pattern $L(i, j)$.

2.2 Local Neighboring Descriptive Pattern (LNDP)

LNDP is a powerful texture pattern technique used for detecting seizure activities in one dimensional EEG signal. LNDP computes the binary code by comparing the consecutive neighboring pixels [6]. A binary code is obtained for each pixel in an image by thresholding its value with the consecutive neighboring pixel in clockwise direction. Algorithm to compute LNDP for a two-dimensional image is given in Algorithm 1. An example for computing LNDP code is shown in Fig. 3.

2.3 Symmetric Weighted LNDP (SWLNDP)

SWLNDP is partially inspired from symmetric weighted LBP (SWLBP) [12] and binary gradient pattern (BGP) [13] method. In SWLNDP, the binary string is obtained by comparing the consecutive neighboring pixels and for encoding step symmetric weighing scheme, which is symmetric over center pixel is implemented. The direction for symmetric weighing system is incorporated from BGP technique. SWLNDP operations are shown in Fig. 4.

The algorithm for computing SWLNDP is given in Algorithm 2. Let us consider the example in Fig. 4. Thresholding process is like that of LNDP technique.

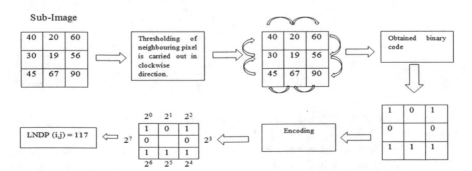

Fig. 3 LNDP code operation

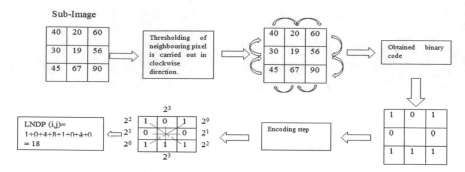

Fig. 4 SWLNDP operator

Algorithm 2. Computing SWLNDP descriptor
Require: Location of a given pixel (i, j) and spatial resolution, (P, R).
 Ensure: Label of SWLNDP descriptor, $SL (i, j)$.

1: The number of neighboring pixels (P) is obtained.
2: Each neighboring pixel is assigned a value 0 or 1 by comparing it with the consecutive neighboring pixel, which is given as,

$$d_n = P_n - P_{n+1} \qquad (3)$$

where, $n = 0$ to $P - 1$

3: Encoding step is processed using the symmetric weighing system which is symmetric over center pixel and its given as,

$$\text{SWLNDP} (i,j) = \sum_{n=0}^{\frac{P}{2}-1} s(d_n)2^n + \sum_{n=\frac{P}{2}}^{P-1} s(d_n)2^{n-\frac{P}{2}} \qquad (4)$$

where, $s(x) = \begin{cases} 0, & x < 0 \\ 1, & x \geq 0 \end{cases}$

6: **return:** Label of pattern SWLNDP (i, j).

In encoding step, the binary pattern is converted into a decimal value in following way, $1 * 2^0 + 0 * 2^1 + 1 * 2^2 + 1 * 2^3 + 1 * 2^0 + 0 * 2^1 + 1 * 2^2 + 0 * 2^3 = 18$. The major advantage of using symmetric weighing scheme is the range of SWLNDP code are limited to (0–30) for 8-bit neighborhood pixel. Hence the number of bins in histogram are reduced which in turn improves the computational performance. At the same time, SWLNDP have a good discriminating ability to distinguish the epileptic and non-epileptic texture patterns.

2.4 Artificial Neural Network (ANN)

ANN architecture is based on the structure and function of biological neural network. Like neurons in the brain ANN also consists of neurons which are arranged in various layers. Multilayer perceptron (MLP) is a popular neural network which consists of input layer to receive the external data to perform pattern recognition, output layer which gives the problem solution and hidden layer is an intermediate layer which separates the other layers. The adjacent neurons from input layer to output layer are connected through acyclic arcs. The MLP uses training algorithm to learn the datasets which modifies the neuron weights depending on the error rate between target and actual output. In general, MLP uses the backpropagation algorithm as a training algorithm to learn the datasets. In this study, the transfer function tansig was used in hidden and output layer along with the scaled conjugate gradient backpropagation algorithm. The MLP network used in this study are shown in Fig. 5.

3 Results

The performance of the proposed method is evaluated by using the publicly available University of Bonn EEG dataset [14], which consists of five sets of EEG signal namely A, B, C, D, and E. The EEG signals in sets A and B are collected

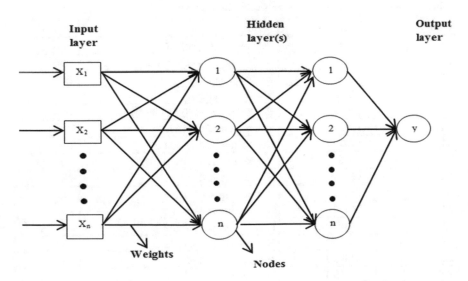

Fig. 5 Architecture of MLP neural network

from healthy persons during eyes open and closed conditions. The signals in C, D, and E are collected from seizure patients. The signals in C are recorded from hippocampal formation of opposite hemisphere of the brain, signals in D are taken from epileptogenic zone and E consists of EEG signals with seizure activity. All the signals were collected using international standard 10–20 electrode placement scheme in a 128-channel amplifier system. Each set consists of 100 single channel EEG data with a time-period of 23.6 s and a sampling frequency of 173.6 Hz. The spectral bandwidth is in the range of 0.5–85 Hz. A total of 15 different cases were considered in this study depending on different conditions shown in Table 1.

The wavelet packet spectrogram of aforesaid EEG dataset was obtained, and their respective texture pattern images are obtained using proposed texture pattern methods (LNDP and SWLNDP). The spectrogram and their respective texture patterns for sets A and E are shown in Fig. 6. The histogram features are extracted using LNDP and SWLNDP methods. The histogram features are fed into ANN classifier with 40 hidden neurons to classify epileptic and normal EEG signals. The experiments are carried out using 10-fold cross validation and the statistical parameters like accuracy, sensitivity, and specificity are computed to analyze the classifier performance. The classification performance using histogram features based on LNDP and SWLNDP techniques are shown in Table 2. For the experimental cases in Table 1, the proposed LNDP methodology attained a classification accuracy of 100, 99.95, 99.95, 100, 99.95, 100, 99.97, 100, 99.93, 100, 100, 99.96, 99.98, 99.98, and 99.93%. For the same cases SWLNDP yielded an accuracy of 100, 99.95, 100, 100, 99.90, 100, 100, 100, 99.90, 100, 100, 99.90, 99.97, 99.97, and 99.92%. As LNDP is insensitive to local and global intensity variations,

Table 1 Various experimental cases considered in this research

Case number	Cases	Description	Classes
I	A-E	Healthy (eyes open) and ictal	Two
II	B-E	Healthy (eyes closed) and ictal	Two
III	D-E	Inter-ictal and ictal	Two
IV	C-E	Inter-ictal and ictal	Two
V	A-D	Healthy (eyes open) and inter-ictal	Two
VI	AC-E	Non-seizure and seizure	Two
VII	AD-E	Non-seizure and seizure	Two
VIII	BC-E	Non-seizure and seizure	Two
IX	BD-E	Non-seizure and seizure	Two
X	AB-E	Non-seizure and seizure	Two
XI	CD-E	Non-seizure and seizure	Two
XII	ABCD-E	Non-seizure and seizure	Two
XIII	ACD-E	Non-seizure and seizure	Two
XIV	ABC-E	Non-seizure and seizure	Two
XV	BCD-E	Non-seizure and seizure	Two

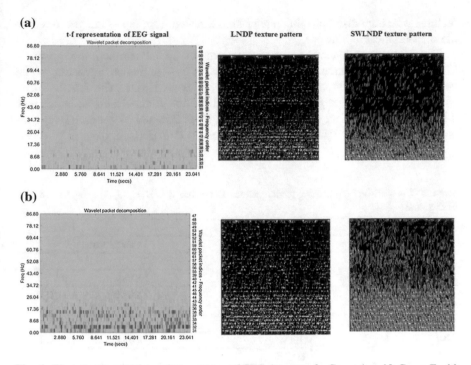

Fig. 6 Illustrates the wavelet packet spectrum of EEG data sets of **a** Group A and **b** Group E with their respective texture patterns using LNDP and SWLNDP

Table 2 Classification performance of ANN classifier using LNDP and SWLNDP techniques

S. No	Cases	LNDP			SWLNDP		
		Acc (%)	Sen (%)	Spe (%)	Acc (%)	Sen (%)	Spe (%)
I	A-E	100	100	100	100	100	100
II	B-E	99.95	99.90	100	99.95	99.90	100
III	D-E	99.95	100	99.90	100	100	100
IV	C-E	100	100	100	100	100	100
V	A-D	99.95	99.90	100	99.90	99.90	99.90
VI	AC-E	100	100	100	100	100	100
VII	AD-E	99.97	99.95	100	100	100	100
VIII	BC-E	100	100	100	100	100	100
IX	BD-E	99.93	99.90	100	99.90	99.90	99.90
X	AB-E	100	100	100	100	100	100
XI	CD-E	100	100	100	100	100	100
XII	ABCD-E	99.96	99.98	99.90	99.90	99.95	99.70
XIII	ACD-E	99.98	99.97	100	99.97	99.96	100
XIV	ABC-E	99.98	99.97	100	99.97	100	99.90
XV	BCD-E	99.93	99.93	99.90	99.92	99.93	99.90

features with high discriminating power are extracted from the t-f images to reduce the misclassification rate.

The histogram feature vector dimensions of LNDP method are given as 2P. In SWLNDP, symmetric weighing scheme is used to reduce the feature vector length with the nearly same discriminating ability of LNDP. The feature vector length for SWLNDP are given as $(2^{((P+2)/2)})-1$. Hence, for analyzing an image with 8-bit neighborhood pixel a feature vector length of 256 and 31 for each image are computed using LNDP and SWLNDP method.

Table 3 Comparison with existing techniques for the dataset {A, B, C, D, E} using t-f image of EEG signal

Author	Method			Accuracy (%)
	t-f image	Feature	Classification	
Case A-E				
Li et al. [1]	MRBF-MPSO	MRBF-MPSO-PCA	SVM	100
Boubchir et al. [16]	QTFD	LBP	SVM	99.33
Şengür et al. [10]	STFT	LBP		100
		GLCM	SVM	92.5–100
		TFCM		87
Fu et al. [8]	HHT	Histogram features	SVM	99.125
Boubchir et al. [15]	QTFD	GLCM	SVM	99.09
Boashash et al. [9]	QTFD	t-f signal and t-f image features are concatenated	SVM	95.33
Proposed method	Wavelet packet	LNDP	ANN	100
		SWLNDP		100
Case C-E				
Li et al. [1]	MRBF-MPSO	MRBF-MPSO-PCA	SVM	99.80
Proposed method	Wavelet packet	LNDP	ANN	100
		SWLNDP		100
Case D-E				
Li et al. [1]	MRBF-MPSO	MRBF-MPSO-PCA	SVM	97.60
Proposed method	Wavelet packet	LNDP	ANN	99.95
		SWLNDP		100
Case CD-E				
Li et al. [1]	MRBF-MPSO	MRBF-MPSO-PCA	SVM	98.73
Proposed method	Wavelet packet	LNDP	ANN	100
		SWLNDP		98.73

4 Discussions

A fair comparison of the proposed techniques with other methods proposed in the literature are carried out with following conditions, (i) the existing methods are evaluated with the same dataset (ii) signals were transformed into a t-f image and (iii) all the cases mentioned in the literature are evaluated in this work, shown in Table 3. In comparison to MRBF-MPSO method, the proposed method obtained a t-f image with high resolution and features with high discriminating power are extracted using proposed methods. In all the cases, the classification accuracy achieved by the proposed methods is higher and it is computationally efficient. The 15 experimental cases were evaluated based on [3], in which an accuracy of 100, 99.25, 95.12, 99.62, 99.5, 96.66, 98.25, 96.5, 99.16, 98.75, 95.85,97.31,98.68, and 95.1% was attained for the cases I-IV and VI-XV by using DWT and Naïve Bayes classifier. In all these cases the proposed methods achieved a classification accuracy of 99.9–100% which is higher than the existing method. The highest classification accuracy of the proposed method in fifteen different cases indicates that proposed WPD-LNDP-ANN and WPD-SWLNDP-ANN can be an effective and reliable technique for automated classification of epileptic EEG signals from normal EEG signals. Further WPD-SWLNDP-ANN method achieved highest classification accuracy with reduced feature vector dimensions; hence the proposed technique will utilize the hardware with less memory space. A reliable, cost effective, and efficient method for automated diagnosis of epilepsy is achieved using WPD-SWLNDP-ANN method.

5 Conclusion

A method for automatic diagnosis of epilepsy using WPD for transformation of EEG signals into t-f image and novel feature extraction based on texture pattern techniques are proposed in this research work. The t-f representations of EEG signal with high t-f resolution are obtained using WPD. The most discriminate features are extracted from t-f image using LNDP and SWLNDP techniques with a dimension of 256 for LNDP and 31 for SWLNDP. The extracted features were fed into ANN classifier for detecting epileptic and non-epileptic signals. Based on the existing literature fifteen different cases were evaluated in this work. The accuracy rate of 100% was achieved for the case A-E and for other cases the accuracy rates were in the range of 99.9%–100% for both the proposed texture pattern techniques. Further, the proposed approach outperforms recently proposed methodologies for epileptic detection. In future, the proposed methodology for EEG data with more channels will be executed and the reliability of the proposed method will be evaluated.

Acknowledgements This paper work was endorsed by the "Technology Systems Development Programme (TSDP)" under Department of Science and Technology (DST), Ministry of Science and Technology, Government of India (GoI), [Grant Number—DST/TSG/ICT/2015/54-G, 2015].

References

1. Li, Y., Wang, X., Luo, L., Li, K., Yang, X., Guo, Q.: Epileptic seizure classification of EEGs using time-frequency analysis based multiscale radial basis functions. IEEE J. Biomed. Health Inform. **99**, 1 (2017)
2. Polat, K., Güneş, S.: Classification of epileptiform EEG using a hybrid system based on decision tree classifier and fast Fourier transform. Appl. Math. Comput. **187**(2), 1017–1026 (2007)
3. Sharmila, A., Geethanjali, P.: DWT based detection of epileptic seizure from EEG signals using naive Bayes and k-NN classifiers. IEEE Access **4**, 7716–7727 (2016)
4. Riaz, F., Hassan, A., Rehman, S., Niazi, I.K., Dremstrup, K.: EMD-based temporal and spectral features for the classification of EEG signals using supervised learning. IEEE Trans. Neural Syst. Rehabil. Eng. **24**(1), 28–35 (2016)
5. Zhang, T., Chen, W., Li, M.: Fuzzy distribution entropy and its application in automated seizure detection technique. Biomed. Signal Process. Control **39**, 360–377 (2018)
6. Jaiswal, A.K., Banka, H.: Local pattern transformation based feature extraction techniques for classification of epileptic EEG signals. Biomed. Signal Process. Control **34**, 81–92 (2017)
7. Supriya, S., Siuly, S., Zhang, Y.: Automatic epilepsy detection from EEG introducing a new edge weight method in the complex network. Electron. Lett. **52**(17), 1430–1432 (2016)
8. Fu, K., Qu, J., Chai, Y., Dong, Y.: Classification of seizure based on the time-frequency image of EEG signals using HHT and SVM. Biomed. Signal Process. Control **13**, 15–22 (2014)
9. Boashash, B., Boubchir, L., Azemi, G.: A methodology for time-frequency image processing applied to the classification of non-stationary multichannel signals using instantaneous frequency descriptors with application to newborn EEG signals. EURASIP J. Adv. Signal Process. **2012**(1), 117 (2012)
10. Şengür, A., Guo, Y., Akbulut, Y.: Time–frequency texture descriptors of EEG signals for efficient detection of epileptic seizure. Brain Inform. **3**(2), 101–108 (2016)
11. Seo, Y., Kim, S.: River stage forecasting using wavelet packet decomposition and data-driven models. Procedia Eng. **154**, 1225–1230 (2016)
12. Kumar, T.S., Kanhangad, V.: Automated obstructive sleep apnoea detection using symmetrically weighted local binary patterns. Electron. Lett. **53**(4), 212–214 (2017)
13. Huang, W., Yin, H.: Robust face recognition with structural binary gradient patterns. Pattern Recogn. **68**, 126–140 (2012)
14. Andrzejak, R.G., Lehnertz, K., Rieke, C., Mormann, F., David, P., Elger, C.E.: Indications of nonlinear deterministic and finite dimensional structures in time series of brain electrical activity: dependence on recording region and brain state. Phys. Rev. **64**, 061907 (2001)
15. Boubchir, L., Al-Maadeed, S., Bouridane, A.: Haralick feature extraction from time–frequency images for epileptic seizure detection and classification of EEG data. In: The 26th IEEE International Conference on Microelectronics, pp. 32–35 (2014)
16. Boubchir, L., Al-Maadeed, S., Bouridane, A., Chérif, A.A.: Classification of EEG signals for detection of epileptic seizure activities based on LBP descriptor of time-frequency images. In: IEEE International Conference on Image Processing (ICIP), Quebec City, QC, pp. 3758–3762 (2015)

User Identification Methods in Cognitive Radio Networks

A. K. Budati, S. Kiran Babu, Ch. Suneetha, B. B. Reddy
and P. V. Rao

Abstract Spectrum sensing is the process of detection of the user whether it is present or absent. Spectrum Sensing is the key role in Cognitive Radio (CR) Networks and thus improving spectrum utilization. The major issues that may arise in spectrum sensing are Probability of false alarm (P_{fa}) and Probability of miss detection (P_{md}). An analytical comparison is proposed between two of Non-Cooperative detection methods in the spectrum sensing. In this paper, an attempt is made to identify a better detection method based on high Probability Detection (P_D).

Keywords Probability of false alarm · Probability of detection
Probability of miss detection · Cognitive radio · Matched filter
Cyclostationary feature detector

1 Introduction

A radio that alters its transmitter parameters owing to interaction with environment is named Cognitive Radio [1]. CR is understood as an improvement to spectrum use by strengthening intelligent spectrum sharing. CR Spectrum sensing networks

A. K. Budati · S. Kiran Babu (✉) · Ch.Suneetha · B. B. Reddy · P. V. Rao
Department of ECE, VBIT, Hyderabad, India
e-mail: kiranbabu009@gmail.com

A. K. Budati
e-mail: anilbudati@gmail.com

Ch.Suneetha
e-mail: suneetha25.ch@gmail.com

B. B. Reddy
e-mail: reddybb@hotmail.com

P. V. Rao
e-mail: pachararao@rediffmail.com

© Springer Nature Singapore Pte Ltd. 2019
P. K. Mallick et al. (eds.), *Cognitive Informatics and Soft Computing*,
Advances in Intelligent Systems and Computing 768,
https://doi.org/10.1007/978-981-13-0617-4_65

689

expose techniques to seek out communication opportunities in wireless spectrum for secondary users to access with low priority. There are two types of CR detection techniques namely (i). Non-Cooperative detection and (ii). Cooperative detection methods. The following three signal process strategies that are useful for spectrum sensing (a) Matched Filter detector, (b) Cyclostationary feature detector and (c) Energy detection are of Non-Cooperative detection methods [2, 3]. Cooperative detection methodology work is based on Central Fusion, where as Non-Cooperative detection methodology there's no Central Fusion and therefore the decision is taken by the individual node itself. While estimating the exact user presence, two types of issues may arise (i) Probability of false alarm (P_{fa}) and (ii). Probability of miss detection (P_{md}). As per the available literature, many of the authors have focused on (i) SNR with P_{fa} or (ii) Receiver Operating Characteristics (ROC) or P_D versus P_{fa} only. As per the available literature, not much work have been done on P_{md}. In this paper, the authors propose the work with focus on P_{md} for Matched Filter Detection with Inverse covariance matrix (MFDI) and Cyclostationary Feature Detection with Inverse covariance matrix (CFDI) and also propose the comparison between MFDI and CFDI with input power to P_D, P_{fa}, and P_{md} and also identified a better detection algorithm among two different Non-cooperative detection methods (Matched Filter and Cyclostationary Feature Detector).

2 Implementation of MFDI

2.1 Working Model

In this paper, an attempt is made to propose, algorithm of Matched Filter with Inverse covariance Matrix (MFDI-SS). The development of the matched filter is predicated on a known noise spectrum. In reality, however, the noise spectrum is typically calculated from information and hence known to a restricted exactitude. A matched filter is commonly used at the front end of the receiver to reinforce the SNR. The received input SNR is applied to MFDI-SS System block as shown in Fig. 1. The linear matched Filter improves the signal quality and suppresses the noise. In Fig. 1, $x(t)$ is the input signal power in decibels per watt, $h(t)$ is the

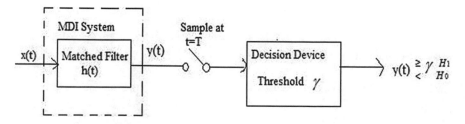

Fig. 1 Block diagram of linear MFDI

impulse response of the system and *y(t)* is the output of the MFDI-SS [16]. For the input of *x(t)* a statistical hypothesis values $(H = H_0, H_1)$ are assumed [4]. Where H_0 denotes only noise is present and H_1 denotes user signal along with noise is present. Input to the matched filter is given as sample after sample. It is assumed that in proposed system, communication environment contains non-uniform Additive White Gaussian Noise (AWGN). Thus every sample has freelance noise levels and it does not depend upon adjacent channel signals. To estimate such freelance channels inverse covariance matrix model is applied to basic matched filter and framed as MFDI model. Samples are applied to the input of MFDI, where the effective noise is suppressed and maximize the signal power. The output of this MFDI system is given as input to the decision making device, which identifies the user presence or absence based on the signal power level i.e., if the signal power is higher than the threshold value it is identified as user present (H_1), if not the user is the absence (H_0). While deciding channel occupancy two conflicts might occur (i). No user within the channel even if the detection system provides decision as user is present. Such types of decisions are known as Probability of false alarms (P_{fa}), because of the consequences of P_{fa} the assigned spectrum slot would be wasted without transmitting information. (ii). There is user within the spectrum slot but detection system provides decision as there is no user. Such types of wrong decisions are known as Probability of miss detection (P_{md}). Because of the impact of P_{md} a serious issue known as interference might arise. The interference effects of no user have benefited by utilizing the spectrum but slot is wasted and it leads to poor service. Thus, in the proposed paper an attempt has been made to identify or give exact decision using a non-cooperative-based spectrum sensing technique of Matched Filter and later comparisons was done with CFDI. In addition, input power versus P_D, P_{fa} and P_{md} are estimated.

2.2 Analytical Procedure

For estimation of channel, we assume a statistical hypothesis specified by H_1 (both Signal and noise present) and H_0 (only noise is present) [5, 6] as shown in Eq. (1)

$$H_0 = y = n(t) \tag{1}$$

$$H_1 = y = s(t) + n(t) \tag{2}$$

Here, *n(t)* is the AWGN with zero mean and Covariance of $\sigma^2 \delta(t - s)$, where σ^2 is the power density of intensity of the AWGN signal. The Inverse Covariance matrix for AWGN signal is [7]

$$P(y; H) = \frac{1}{(2\pi\sigma^2)^{N/2} \det^{1/2}(C)} \exp[-\frac{1}{2}(y - H)^T C^{-1}(y - H)] \tag{3}$$

The possibilities of two hypotheses are equal [7, 8] and can be shown as

$$\gamma = \frac{P(y; H_1)}{P(y; H_0)}, \tag{4}$$

where γ the initial threshold and y is the Output of the MFDI. However, since the exact signal is generally determined by unknown parameters that are effectively estimated (or fixed) in the filtering process. The matched filter constitutes a gen eralized maximum likelihood (test) statistic. The Binary hypotheses theory in terms of Log Probability Ratio Test statistics is established as

$$L(Y) = P(y; H_1)/P(y; H_0) \tag{5}$$

By simplifying the above Eq. (5) we get the Generalized Probability Ratio Test (GLRT) of threshold as

$$L(Y) = (\ln(\gamma) + 1/2(H_1 - H_0)^T C^{-1}(H_0 + H_1)) \tag{6}$$

The Probability of False alarm is calculated as

$$P_{\text{fa}} = Q\left(\frac{\gamma - (H_1 - H_0)^T C^{-1}(H_0)}{\sqrt{(H_1 - H_0)^T C^{-1}(H_1 + H_0)}}\right) \tag{7}$$

The Probability of Detection [4] is calculated from Eq. (7) as

$$P_D = 1 - P_{\text{fa}} \tag{8}$$

Probability of Miss Detection is calculated as [9]

$$P_{\text{md}} = 1 - P_D \tag{9}$$

3   Implementation of CFDI

3.1   Working Model

As per the dynamic spectrum access, Cyclostationary signal analysis produces several additional advantages. Coherent approaches like Matched Filter wants synchronization with the signal of interest. The Cyclostationary analysis doesn't need any time synchronization, it is going to be frequency or phase. This is often a good approach to detect unknown signals in terms of frequency and symbol timing [8]. As per the available literature it is observed that, the authors proposed their research with Cyclostationary detection methodology for arbitrarily incoming (randomly arriving) or departing signals. Lee [8] explored that there is a relation between P_{fa} versus P_{D} and SNR versus P_{D} at various SNR levels. Yang [10] planned his work on Cycloenergy detector for spectrum sensing in Cognitive radio. The plotted ROC at different SNR levels using Cyclo energy detector, and also P_{fa} versus P_{D} are plotted at different samples. However the earlier authors have conducted analysis within the same proposed area. Very little literature conducts the research to find out at which samples, P_{fa} occurs. In this letter, authors assumed the communication environment contains a non-uniform AWGN. Therefore every spectrum slot contains totally different noise levels and furthermore every channel has not relied on adjacent channels conjointly. Therefore, every slot contains totally different power levels of signal and noise. Therefore to estimate the user presence in every channel with sensible accuracy Inverse covariance matrix methodology is added for general Cyclostationary feature detector and it'll be termed as CFDI and shown in Fig. 2.

In Fig. 2, the input of $x(t)$ is statistical hypothesis sample values $(H = H_0, H_1)$ [11]. Where H_0 means only noise is presented and H_1 means user signal along with noise. The detected signal data samples $x(t)$ is applied to N-Point Fast Fourier Transform (FFT). These computed samples are fed to the correlator [6], that correlates new samples with previous samples. In case new sample and previous sample correlates, then previous sample decision is assigned to new sample. The remaining uncorrelated samples are sent for threshold comparison. If instant sample power is higher than threshold power level, we take into account instant sample value and give decision as the presence of user (H_1), otherwise there is no user (H_0). Additionally, Cyclo Autocorrelation Function (CAF) is utilized for better results [10]. In presence of incorrect decision-making effects in MFDI System, CFDI also can present a conjointly measured effect of P_{fa} and P_{md}. Therefore, within the proposed paper an attempt has been made to identify or give precise decision employing a non cooperative based spectrum sensing methodology of Cyclostationary feature detector. In previous section with MFDI the effects and P_{D} are estimated. In this Section, using CFDI the input power versus P_{D}, P_{fa} and P_{md} are estimated. Later in Result, comparison of MFDI and CFDI has been done.

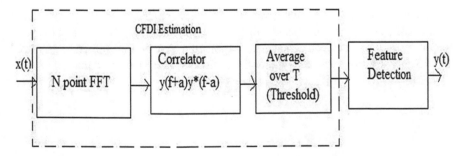

Fig. 2 CFDI block diagram

3.2 Analytical Procedure

The statistical hypothesis of the signals is similar to Matched Filter. So to estimate the signal presence or absence by CFDI scheme, the same assumptions are used. Cyclostationary method works on the principle of Autocorrelation function. Then the GLRT for the CAF of initial threshold is

$$R_{yy*}^{\alpha} = \frac{\sum_{n=0}^{N-1} \exp\left[-\frac{1}{2}(y_n \quad H_1)^T C^{-1}(y_n - H_1)\right] \exp(-j2\pi\alpha nf_s)}{\sum_{n=0}^{N-1} \exp\left[-\frac{1}{2}(y_n - H_0)^T C^{-1}(y_n - H_0)\right] \exp(-j2\pi\alpha nf_s)} \underset{<H_0}{\overset{\geq H_1}{}} \approx T, \quad (10)$$

where α is the cyclic frequency and R_{yy*} is the auto-correlation function for the input samples. The exponential function represents the Fast Fourier Transform (FFT). Simplify the above Eq. (10) as resultant equation

$$\frac{-1}{2}(y - H_1)^T C^{-1}(y - H_1) + \frac{1}{2}(y - H_0)^T C^{-1}(y - H_0) \underset{<}{\overset{\geq}{}} \ln(T) \quad (11)$$

On simplifying the above Eq. (11) and comparing with initial threshold. The threshold $\ln(T)$ is equal to predetermine threshold γ.

$$(H_1 - H_0)^T C^{-1} y - \frac{1}{2}(H_1 - H_0)^T C^{-1}(H_0 + H_1) \underset{<}{\overset{\geq}{}} \ln(T) \quad (12)$$

The negative term is moved to right hand side and the remaining term will equalize to the $T(x)$.

$$T(x) = (H_1 - H_0)^T C^{-1} y \quad (13)$$

The simplified equation is equalized to threshold of the CFDI threshold and cyclic auto correlation function, then the threshold value is

$$R_{yy*}^{\alpha} = T(x) \underset{<}{\overset{\geq}{}} \ln(T) + \frac{1}{2}(H_1 - H_0)^T C^{-1}(H_1 + H_0) \quad (14)$$

In this algorithm, it is not necessary to calculate the cyclic autocorrelation function of signal. Instead, we add a factor of exponential, which is related to cyclic frequency α, to the received signal. That exponential function is applied to threshold as FFT. The FFT is applied to the initial threshold, and then the equation is

$$T(x) = R_{yy*}^{\alpha} = FFT(\ln(T)) + \frac{1}{2}(H_1 - H_0)^T C^{-1}(H_1 + H_0) \qquad (15)$$

The sampled signal is applied to Average threshold level then the resultant equation is

$$T(x) = R_{yy*}^{\alpha} = \left(FFT(\ln(T)) + \frac{1}{2}(H_1 - H_0)^T C^{-1}(H_1 + H_0) \right)^{1/2} \qquad (16)$$

The above equation is the final threshold $T(x)$ for the Cyclostationary feature Detector. From this the Probability of False Alarm is estimated as

$$P_{fa} = \text{erfc} \left(\frac{[T(x) - [(H_1 - H_0)^T C^{-1}(H_0)]]}{\sqrt{[(H_1 - H_0)^T C^{-1}(H_1 - H_0)]}} \right) \qquad (17)$$

And the Probability of Detection is

$$P_D = 1 - P_{fa} \qquad (18)$$

Probability of miss detection is [12]

$$P_{md} = 1 - P_D \qquad (19)$$

4   Results

4.1 P_D

Monte Carlo model samples generated with non uniform noise are used for simulation. The threshold noise is calculated for every sample using of Eq. (6) for MFDI and Eq. (16) for CFDI. The input power is calculated in terms of dBW, later the power is converted again in watts. In Fig. 3, input power is ranged from −3 dBW (∼0.5012 W) to 4 dBW (∼2.501 W). In the spectrum sensing, the user presence or absence is identified based on the reverse channel in the spectrum slot. The minimum power needed is 0.2 Watts for decision-making. In this paper, P_D is calculated to the power (−3 dBW) to 4 dBW. At −3 dBW the Probability of Detection is nearly one for MFDI and 0.8600 for CFDI. For −3 to −1 dBW MFDI detection Probability is nearly one, but in CFDI it gradually decreasing to 0.7500. In MFDI system the decision of the every sample is independent. However in CFDI, present sample value is correlated with previous sample and influences the sample decision. This means, if the previous sample decision was wrong, the current sample decision also wrong. Therefore it can reduce probability of detection

Fig. 3 MFDI & CFDI P_D
versus input power

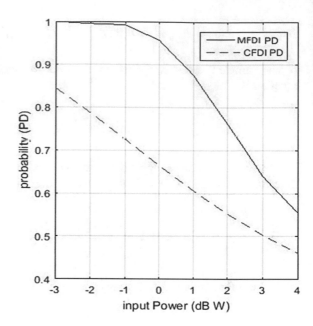

in CFDI. Thus compared to CFDI, the MFDI sensing methodology gives better Probability of Detection.

4.2 P_{fa}

In Wireless Communications, the Probability of false alarm is named as Type-1 error. Due to P_{fa}, interference might not rise, but the spectrum is going to be wasted. Thus to improve the frequency spectrum efficiency, the impact of P_{fa} must be low. The calculated P_{fa} for MFDI and CFDI is shown in Fig. 4. At power level −3 dBW, MFDI P_{fa} is nearly zero while the CFDI P_{fa} is 0.17. For power level −3 to −1 dBW, the MFDI P_{fa} is extremely low in the range from 0.01 to zero. However CFDI P_{fa} is rising step by step from 0.17 to 0.28. Thus compared to CFDI detection technique MFDI has less P_{fa}. Thus to estimate the exact detection for Type-1 error MFDI is more useful.

4.3 P_{md}

In Wireless Communications, the Probability of miss detection is termed as Type-2 error. As results of the Type-1 error spectrum slot are wasted, but it will not cause interference to users. Thus, spectrum charges may increase. Compare to Type-1 error, Type-2 error is more important.

Fig. 4 MFDI & CFDI P_{fa}
versus input power

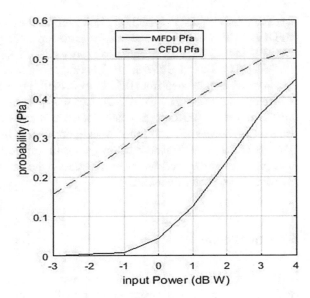

Fig. 5 MFDI & CFDI P_{md}
versus input power

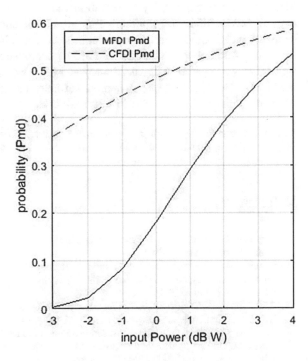

As the, user presence isn't rightly detected, device shows spectrum slot is free. Due to this, if another user is allowed to transmit, existing user and the new user get interferes each other. It causes poor service performance. Thus P_{md} becomes a necessary to consider in mobile radio environments. P_{md} impact is calculated for MFDI and CFDI detection methods for same input samples and shown in Fig. 5. at -3 dBW the MFDI, P_{md} is equal to zero while the CFDI P_{md} is 0.3583. For the power levels -3 to -1 dBW the P_{md} is ranging between 0 and 0.1. And the CFDI P_{md} is rising step by step from 0.3583 to 0.4459. For -1 to 4 dBW the MFDI, P_{md} is increasing step by step like CFDI. CFDI P_{md} is more than MFDI P_{md} because CFDI influences previous samples decision for the present period of samples decision as shown in Fig. 5. But in MFDI, correlation between present and previous samples is not done. Thus any wrong decision of previous sample does not influence present sample decision. Thus to estimate the exact detection for Type-2 error MFDI is more useful than CFDI.

5 Conclusion

Estimate of user presence and absence within the mobile radio environment is extremely key task. Any wrong detection of user presence causes a serious issue likes interfere of signals. By exploitation the non-cooperative detection strategies authors have made an attempt to identify which detection technique is more helpful to give exact decision with less probability of false alarm and Probability of miss detection. Authors have made an attempt to demonstrate that MFDI provides low P_{fa} and P_{md} compared to CFDI. Moreover, MFDI has high probability of detection than CFDI. Hence MFDI system is more suitable algorithm for identification of user presence with low P_{fa} and P_{md} in CR Networks.

Acknowledgements The work has been carried out in DST-FIST lab (SR/FST/College-209/2014) provided at Vignana Bharathi Institute of Technology, Hyderabad, Telangana.

References

1. Akyildiz, I.F., Lee, W.Y.: A survey on spectrum management in cognitive radio networks. IEEE Commun. Mag. (2008)
2. Gorcin, A., Qaraqe, K.A.: An adaptive threshold method for spectrum sensing in multi-channel cognitive radio networks. In: IEEE, 17th international conference on telecommunications (2010)
3. Anil Kumar, B., Trinatha Rao, P.: Overview of advances in communication technologies. In: INCEMIC Conference Proceedings, pp. 47–51 (2015)
4. Tertinek, S.: Optimal detection of deterministic and random signals. Adv. Sig. Process. (2002)
5. Urkowitz, H.: Energy detection of unknown deterministic signals. Proc. IEEE 55(4) (1967)
6. Alvi, S.A.: A log-probability based cooperative spectrum sensing scheme for cognitive radio networks. ELSEVIER J. Emer. Ubiquitous Syst. Pervasive Networks, Procedia Comput. Sci. **3**, 196–202 (2014)
7. Vadivelu, R.: MFDI-SS based spectrum sensing for cognitive radio at low signal to noise ratio. J. Theor. Appl. Inf. Technol. **62**, 107–113 (2014)

8. Lee, Y.: Cyclostationary based detection of randomly arriving or departing signals. ELSEVIER J. Appl. Res. Technol. **12**, 1083–1091 (2014)
9. Scott, C.: A Neyman-pearson approach to statistical learning. IEEE Trans. Inf. Theory **51**(11) (2005)
10. Yang, L.: Cyclo-energy detector for spectrum sensing in cognitive radio. ELSEVIER Article Lett. Int. J. Electron. Commun. 89–92 (2012)
11. Anil Kumar, B., Trinatha Rao, P.: CFDI-SS: Cyclostationary feature detector with inverse covariance matrix based spectrum sensing in Cognitive Radio. In: Smart tech conference proceedings (2017)
12. Skolink, M.I.: Introduction to radar principles. In: Tata McGraw hill third edition, pp. 284–285 (2008)
13. Cabric, D.: Implementation issues in spectrum sensing for cognitive radios white letter clayton scott a neyman pearson approach to statistical learning. IEEE Tran. Inf. Hypotheses **51**(11), 3806–3819 (2005)
14. http://www.dot.gov.in/sites/default/files/Annexures/Handbook.pdf

Design and Performance Analysis of Optical Signal Processing Module in Open Air and Underwater Environment

C. R. Uma Kumari and Dhanalakshmi Samiappan

Abstract Considering the characteristics of green, pervasive, license free and potentially high bandwidth, Visible Light Communication (VLC) has attracted increasing attention in the field of underwater communication. In this proposed work, it has been intended to take advantage of recent developments of Light Emitting Diode (LED) in particular with the higher light output and more precise tuning of wavelength. Photo sensor technology has been employed to perform Optical Signal Processing. Optical communication using LEDs could be conferred for scenarios where high speed, low power and less complex communication systems are desired. The proposed VLC technique aims to achieve perfect signal processing in open air and in underwater environment with high propagation speed. On-off Keying (OOK) Modulation is incorporated to achieve efficient communication. With the use of lens optics, high speed and long distance (greater than 3 m) disclosure in open air and aquatic environment can be effectuated.

Keywords Light emitting diode · Lens optics · On-off keying
Optical signal processing · Photo sensor · Visible light communication

1 Introduction

Visible Light Communication (VLC) is the burgeoning technology to provide data communication using low-cost and universal LED bulbs and PIN photodiodes. The use of the visible light spectrum which is approximately 10000 times larger than the RF spectrum has geared high data transmission by the evolution of LED bulb [1]. VLC and its associated optical signal processing is one of the promising candidates

C. R. Uma Kumari (✉) · D. Samiappan
Department of Electronics and Communication Engineering, SRM University,
Kattankulathur, Kancheepuram, Tamil Nadu, India
e-mail: umakumari.cr@gmail.com

D. Samiappan
e-mail: dhanalakshmi.s@ktr.srmuniv.ac.in

© Springer Nature Singapore Pte Ltd. 2019
P. K. Mallick et al. (eds.), *Cognitive Informatics and Soft Computing*,
Advances in Intelligent Systems and Computing 768,
https://doi.org/10.1007/978-981-13-0617-4_66

for open air and underwater communication. VLC has significant features like unlicensed channels, more bandwidth, low energy consumption and strong data encryption [2].

A German physicist, Harald Hass in the year 2004 has introduced Light Fidelity (LiFi) technology in the Technology-Entertainment-Design (TED) talk show. He conceptualizes a future where data from laptops and other mobile devices could be transmitted through a light source inside a room. Due to short-range wavelengths, optical waves can propagate only for a short distance. As distance increases, the signal gets spread out and hence Bit Error Rate (BER) increases.

Wireless VLC is a newly materialized trend that can easily surface the way for a prosperous wireless future. The scope to transmit data quickly in very secure way is the key role to many applications [2]. The fact that visible light from LED could not be spotted on the other side of a wall has considerable security advantages in deployable field. In VLC, the information is converted into bits and transmitted through blinking LEDs. The blinking of LED could not be sensed by human eye because LED's blinking rate is very high. This concept could be used in submarines and also in unmanned vehicles.

Wei [3] investigated the performance of an M-ary Under Water Optical Communication system with orbital angular momentum shift keying modulation. Vortex Laguerre–Gauss beams are employed to propagate over weak turbulent ocean. Using this scheme, Symbol Error Rate of 10^{-9} has been achieved. Orange-, yellow- and green-coloured LEDs are very rarely used in scientific studies. But these LEDs are found to predominate in chemical sensing devices [4]. These devices were integrated with LEDs in the wavelength range between 247 and 3800 nm. Such chemical sensors are adopted to detect heavy metals, toxins and toxic gases.

Photodiodes are commonly used in photo interrupters and fibre optic transmission systems. A PN photodetector provides additional bandwidth and SNR constraints [5] on VLC. A 650-nm red LED source and a CMOS-compatible reverse-biased PN photodetector receiver are employed for indoor VLC applications. Data-rate of 172 Mb/s has been achieved with a BER of 1.9×10^{-3}. In biometric module [6], detecting part consists of 2×8 LED and 2×8 photodiode array matrices. In this experiment, regardless of the wavelength, the multispectral signals provide high accuracies and false acceptance rate of 0.3% and a false rejection rate of 0.0%.

In this paper, Sects. 2 and 3 explains the design of transmitter and receiver modules respectively. This is followed by experimental setup of VLC and signal processing. Finally results are obtained for open air and underwater environment.

2 Transmitter Design

The transmitter section of the proposed project consists of an n-channel MOSFET as well as a peripheral interface controller. It has very low gate charge with fully categorized avalanche voltage and current.

In this research, a PIC 12F1840 microcontroller has been used. It has high-performance Central Processing Unit (CPU) with Harvard architecture. It has 49 instructions with a 35 MHz crystal oscillator and 125 ns instruction cycle. It has got a watchdog timer which is often used to automatically reset an embedded device that hangs because of software and hardware faults. This microcontroller has an in-built analog to digital converter with 10-bit resolution, a rail-to-rail analog comparator, five I/O ports and one input only port. Arduino Uno R3 is employed as a serial monitor. The arduino unit allows faster data transfer rate along with more memory. It has a serial clock and serial data pins in addition to RESET pin. It has got 32 k flash memory, 16 MHz clock speed, six analog inputs and 14 digital I/O pins. The transmitted text data from the laptop is transferred to the Arduino Unit. The transmitter design of the system is shown in Fig. 1. PIC in the transmitter includes the algorithm to switch ON/OFF the LED bulb based on OOK modulation. Delay is adjusted such that human eyes could not perceive the blinking of LED lamp. NRZ (non-return to zero) line code along with ADC converts the analog signal into digital values. These digital values are next transferred through the LED bulb in the form of visible rays.

3 Receiver Design

As shown in Fig. 2, a high sensitive photodiode is used in the receiver circuit to get maximum output with less noise. Operating wavelength range of photodiode is 400–1100 nm. It has a switching time of typically 5 ns and comes in 5 mm plastic LED package.

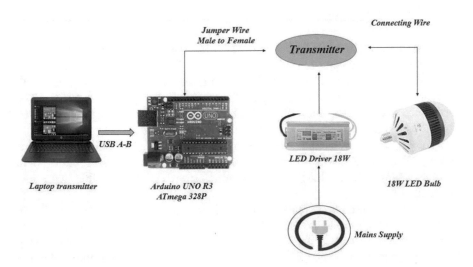

Fig. 1 Transmitter module

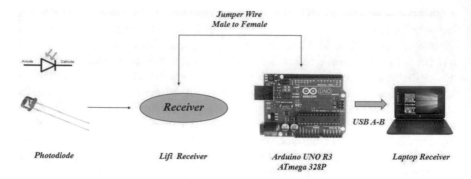

Fig. 2 Receiver setup

Its power dissipation is 100 mW with an operating temperature range from −40 to 100 °C. At a wavelength of is 850 nm, maximum sensitivity is achieved with photocurrent of 9.5 μA and dark current of 1 nA. Filter circuitry provides cancellation of noise in the received signal. Receiver section has PIC12F1840 IC along with Trans-impedance amplifier. The amplifier used here has unique features like large DC voltage gain of 100 dB, wide bandwidth of 1 MHz, internal frequency compensation for unity gain and eliminates the need for dual supplies. Operational Amplifiers can also be operated from the split power supplies.

A general-purpose transistor with minimum collector to emitter voltage of 30 V and collector current of 1 mA is used. Junction temperature of the transistor is 150 °C and storage temperature ranges from −65 to +150 °C.

4 Experimental Setup

In transmitter, a 230 V AC from the mains is given to the LED driver, which consists of in-built analog to digital converter and hence converts the 230 V AC to 30 V DC which are proffer to the transmitter module. The 6000 K LED Bulb associated with the transmitter glows by the application of this power supply. The receiver section consists of an SFH203P photodiode. When visible light falls on the photodiode, it turns ON.

Figure 3 is the overall proposed model for VLC with its optical signal processing. This model could be implemented with high-end data encryption. With On-off Keying (OOK), long range communication can be possible with low attenuation [7].

Lens optics is used to increase the transmission range and to decrease the BER by directly focusing the light rays on the lens. A convex lens system utilized for this purpose which converges the light rays at a particular point. The relationship between spectral irradiance H_r at any distance r is given by the following relation [8],

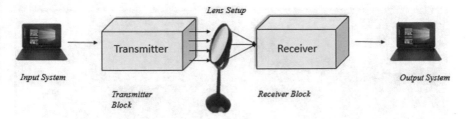

Fig. 3 Proposed model

$$H_r = \frac{J.e^{-\alpha r}}{r^2} + J.u\left(1 + v.e^{-kr}\right).\frac{K.e^{-kr}}{4\pi r} \tag{1}$$

In Eq. (1), J is the spectral radiant intensity, α is the attenuation coefficient for mono path transmission, K is the attenuation coefficient for multipath transmission and empirical constants u and v have the values as $u = 2.5$ and $v = 7$ for a particular data [8]. Main inference from the above equation is that as far as the efficiency of the system is concerned, the system should work on narrow-beam light sources to the extent possible with negligible mono path transmission.

The test was done in two different environments, open air and in underwater. From the experiment, analysis curves are drawn and comparison is done between the two types of environment. To increase the communication distance convex lens is used in between transmitter and receiver. Use of convex lens not only increases the distance but also increases the accuracy in data transmission. Actual hardware testing module of the proposed work is shown in Fig. 4.

In water environment, the experimental module looks like as in Fig. 5.

5 Results and Discussion

With cutting-edge technology, high-performance VLC transmitter and receiver provide communication in a much eco-friendly way between fighter planes and underwater submarines. In the absence of sunlight, performance of the system is much higher than that of in the presence of sunlight. The results are best visible in the form of graphs.

In the presence of sunlight (Fig. 6), distance of communication is 8 cm and in dark (Fig. 7), the distance achieved without data loss is 200 cm. With acceptable data loss, the distance achieved is 250 cm in the dark and 20 cm in the presence of sunlight. Attenuation happens due to the noise which gets added up in light rays in the presence of sunlight, causing high attenuation in the received signal.

To increase effectuation and to reduce attenuation, lens optics came into the picture. With the use of convex lens system, attenuation was reduced to some extent. Results which are published with the use of lens are much higher than

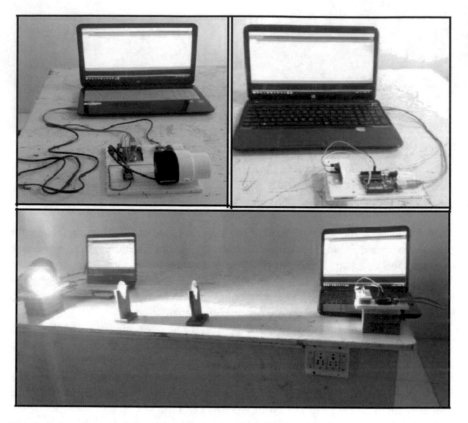

Fig. 4 Transmitter, receiver and actual working prototype in open air medium

expected. The system is working much better in underwater medium than in open air.

It is much clear that the use of convex lens increases the effective distance of communication. In the presence of sunlight (Fig. 8), the distance of communication without loss is 25 cm and with some data loss it is up to 55 cm. By incorporating lens in this module, the distance of communication in dark (Fig. 9) without data loss is found to be 340 cm and with some data loss distance achieved is around 550 cm.

Experiments are performed inside water environment and the results are as shown in Figs. 10 and 11.

The proposed system has less power consumption as compared to RF system and is much eco-friendly. Noticeable features in the proposed module are easy maintenance, simple operation and in the case of any failure bulb can be replaced directly. To gain more distance, high-power bulbs could be used.

Fig. 5 Prototype in water medium (on state)

Distance Vs % Error (In Sunlight)

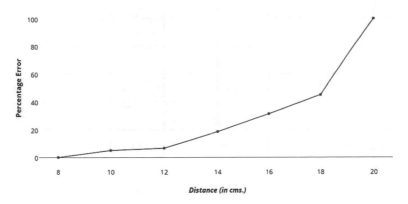

Fig. 6 Variation of error with increasing distance (in sunlight)

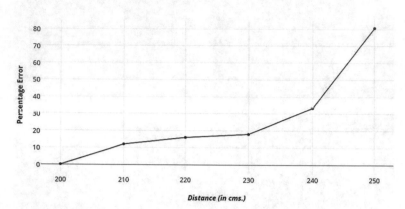

Fig. 7 Variation of error with increasing distance (in dark)

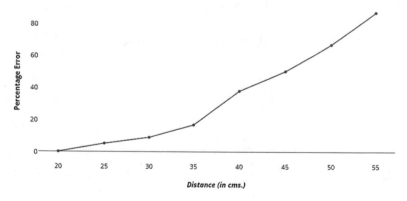

Fig. 8 Variation of error with increasing distance (in sunlight with lens)

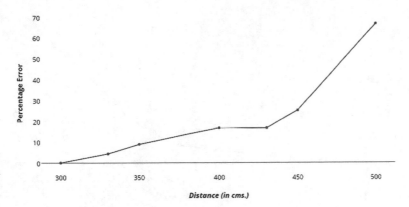

Fig. 9 Variation of error with increasing distance (in dark with lens)

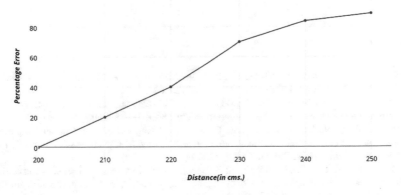

Fig. 10 Variation of error with increasing distance (underwater in the absence of sunlight)

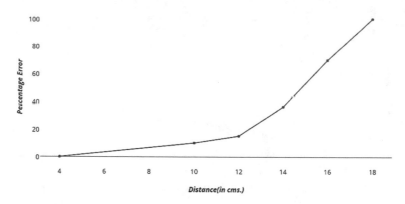

Fig. 11 Variation of error with increasing distance (underwater in the presence of sunlight)

Acknowledgements This research work is funded by SRM University under "Selective Excellence Initiative Program", SRMU/R/AR(A)/2015/126/1866.

References

1. Falcitelli, M., Pagano, P.: Visible Light Communication for Cooperative ITS. Springer International Publishing Switzerland, pp. 19–47 (2016)
2. Ullah Khan, L.: Visible light communication: applications, architecture, standardization and research challenges. Sci. Dir. Dig. Commun. Netw. **3**(2), 78–88 (2017)
3. Wei, W., Ping, W., Tian, C., Hongxin, T., Yan, Z., Lixin, G.: Performance investigation of underwater wireless optical communication system using M-ary OAMSK modulation over oceanic turbulence. IEEE Photon. J. **9**(5) (2017)
4. Yeh, P., Yeh, N., Lee, C.-H., Ding, T.-J.: Applications of LEDs in optical sensors and chemical sensing device for detection of biochemicals, heavy metals, and environmental nutrients. Renew. Sustain. Energy Rev. **75**, 461–468 (2017)
5. Hussein, A.F., Elgala, H., Fahs, B., Hella, M.M.: Experimental investigation of DCO-OFDM adaptive loading using Si PN-based receiver. In Wireless and Optical Communication Conference (2017)
6. Young, C.J., Hae, N.K., Jae H.K., Hyuck, K.H., Yeon, S.C., Suk, W.J., Sung, P.K.: Novel wearable-type biometric devices based on skin tissue optics with multispectral LED–photodiode matrix. Jpn. J. Appl. Phys. **56**(46) (2017)
7. Babar, Z., Nguyen, H.V., Botsinis, P., Alanis, D., Chandra, D., Ng, S.X., Hanzo, L.: Unity-rate codes maximize the normalized throughput of ON–OFF keying visible light communication. IEEE Photon. Technol. Lett. **29**(3) (2017)
8. Duntley, S.Q.: Measurements of the Transmission of Light From an Underwater Source Having Variable Beam-Spread. Project S F001 05 01, University of California Scripps Institution of Oceanography (1960)

Analysis of Jitter and Shimmer for Parkinson's Disease Diagnosis Using Telehealth

Harisudha Kuresan, Sam Masunda and Dhanalakshmi Samiappan

Abstract The future of telecommunications is premised on high fidelity networks with extreme precision, which in turn capacitates deployment of telediagnostic tools. Parkinson's disease (PD) clinical characterization is based on, speech problems, tremors in hands, arms, legs and face, body swelling, muscle rigidity and movement problems. Speech problems are cited as one of the earliest prodromal for PD. However, using clinical diagnosis it takes up to 5 or more years to detect PD. Therefore, with this regard speech can be used, as an early biomarker for PD. Features of interest for detecting PD will be prosodic, spectral, vocal tract and excitation source speech features. We infer from the analysis, MFFC with jitter and shimmer feature extraction provides a promising method that can help the clinicians in the diagnostic process.

Keywords Biomarker · Feature extraction · Parkinson's disease
Speech features · Speech signal · Telehealth

1 Introduction

Parkinson's disease (PD) is a neurological issue, which influences non-motor and motor actions in the human body [1]. PD is a ceaseless and dynamic development disorder, implying that side effects proceed and intensify after some time. Research has shown that in India, PD prevalence rate stands at 67/100,000 people [1], also highlighted that Parsi Community in Mumbai is among one of the worlds most

H. Kuresan (✉) · S. Masunda · D. Samiappan
Department of Electronics and Communication Engineering,
SRM Institute of Science and Technology, Kattankulathur, Tamil Nadu, India
e-mail: harisudha.k@ktr.srmuniv.ac.in

S. Masunda
e-mail: sam_masunda@srmuniv.edu.in

D. Samiappan
e-mail: dhanalakshmi.s@ktr.srmuniv.ac.in

© Springer Nature Singapore Pte Ltd. 2019
P. K. Mallick et al. (eds.), *Cognitive Informatics and Soft Computing*,
Advances in Intelligent Systems and Computing 768,
https://doi.org/10.1007/978-981-13-0617-4_67

affected with a prevalence rate of 328/100,000 population according to [1, 2]. PD causes neurodegenerative dysfunction of focal sensory system bringing about the fractional or total shutdown of motor reflexes, discourse, conduct, mental capacity, and other essential capacities [3]. Vocal impairment is cited among most punctual prodromal PD manifestations, recognizable up to five years preceding clinical analysis [4, 5]. PD patients encounter the evil impacts of talk crippling like dysphonia (blemished voice utilization), hypophonia (diminished degree of loudness), monotone (decreased pitch extend) and dysarthria (issues in sound and syllable verbalization) [6–8]. Accuracy is a very critical part of the clinical diagnosis process. Research [9], indicates that human accuracy (neurologist) in detecting PD is limited to approximately 80%, with the other 20% being misdiagnosed. Despite the fact that it is impossible to eliminate neurologist and clinicians on analysis and recommendations; [9, 10] proposed that the use of smartphones and wearable technologies for telediagnosis has high chances of improving accuracy in PD diagnosis. It has been highlighted [11] that, with the help of telediagnostic tools, neurologists and clinicians will make better-informed decisions. A primarily patient narrative about signs and symptoms is the basis on which neurologist rely on to diagnose a disease, thereafter clinical tests are performed [11]. Hence, speech signal can be used as a biomarker for the early exposure to the disease. Speech disorders can be investigated with acoustic tools thereby analyzing aperiodic vibrations in the speech signal, according to [6]. Matrices for PD progress include the Unified Parkinson's Disease Rating Scale (UPDRS) and motor sub-scale (UPDRS-III) [11–13]. UPDRS reveals the existence and severity of PD symptoms. By analyzing the above literature survey, the works done are based on the acoustic features and classification. Majority of the papers focus on acoustic features that played the main role. In this paper, we focus on the voice quality features, prosodic features and spectral features; these are of great help in detecting the healthy versus Parkinson patient.

2 Parameters

2.1 Database Parameters

The speech signals are obtained from the University of California (UC) Irvine Machine Learning repository, publicly available on their website [6, 14]. Apart from this dataset, no other participants are involved. The recorded voice tests comprise of supported vowels (a, o and u), words, numbers and short sentences [15]. In this study, MFCC was used for feature extraction because of its high accuracy and GMM classifier.

2.2 Mel-Frequency Cepstral Coefficients

According to researches [16–18], MFCCs are a very vital and prominent voice feature. Feature extraction entails collection of MFCCs and formatting their elements into vectors of similar dimensions. Feature vectors are obtained from individual frames of test speech. Figure 1 shows the MFCC process block outline.

2.3 Feature Extraction

In a similar work [6], subjects recite prearranged words and sustained vowels a, o and u, as well as numbers 1 through 9 as part of their medical examination [6]. It is from this process that speech features are extracted. Vocal tract features, prosodic features [fundamental frequency (F0), jitter, shimmer, Pitch period entropy (PPE)], formants and excitation source features are analyzed. MFCC features are extracted by a computing 25 ms Hamming window, with a 10 ms signal overlap as shown in Eq. 1. Hamming window

$$\text{Hamm}(N) = \left[0.54 - 0.460\cos\left(\frac{2(n-1)}{N-1} \right) \right] \tag{1}$$

With ($N = 160$) being the quantity of tests in a single edge and n varies from 1 to N. Then, the Mel-Cepstrum with 24 filter bank channels is prepared and ascertained using Discrete Cosine Transform (DCT). Considering the initial 12 Mel-recurrence Cepstral coefficients. Cepstral Mean Subtraction (CMS) is associated with first-and-second demand auxiliaries of these features that are processed over a setting of five (5) and nine (9) progressive edges. Linear frequency mapping to Mel-Frequency is calculated as in Eq. (2)

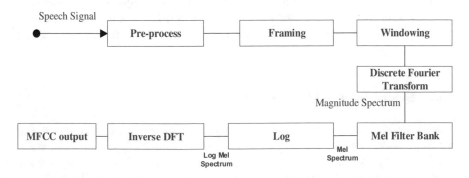

Fig. 1 MFCC block outline [18]

$$Mel(f) = 1127\text{In}(1 + f/700) \tag{2}$$

MFCC feature sets are derived from the calculated maxima, minima, mean, and median and standard deviation values of datasets used. Dimensions of individual samples vary in sizes; this makes it impossible to make a direct feature comparisons, hence the need for use of measures of central tendency in MFCCs calculation.

$$\text{mfcc} = \sum_{r=1}^{R} \log(M_r)\cos[2\pi/R(r + 1/2)m] \tag{3}$$

With m being the measure of coefficients and R as the numeral of channels. The above calculations make use of the positive range of values only. M_r, is the quantity of yield channels created after N cycles. This implies that there are R increases vital for each mfcc [17].

2.4 Classifiers

2.4.1 Support Vector Machines (SVM) Classifier

SVMs characterize a choice limit that ideally isolates two classes by boosting the separation between them. The choice limit can be portrayed as a hyperplane that is communicated regarding a straight mix of capacities parameterized by help vectors, which comprise in a subset of preparing atoms [19]. SVM algorithms use a kernel function that gives the best separating hyperplane [19].

2.4.2 K-Nearest Neighbours (KNN)

In outline affirmation, KNN is a system for orchestrating objects in perspective of closest getting ready cases in the component domain. KNN is a fundamental procedure in plan affirmation at a point where the data about the information allocation is obliged [20].

2.4.3 Hidden Markov Models (HMM)

HMMs are state machines with a double layer consisting of perception and state layers [21]. At any given time, a Markov process controls their individual states. For HMMs, by utilizing a discrete shrouded condition at the time (t) the greater part of the required data previously this time would be known thus the perception whenever depends just on its current shrouded state.

2.4.4 Artificial Neural Networks

Neurons are interconnected, their number of info is proportionate to traits and size of yield layers is proportional to masked capacities used to define choice guidelines in building classifier [22]. Peculiarly, the quantity of neurons within hidden layers is selected [22].

2.4.5 Gaussian Mixture Model

Gaussian functions give the smooth estimation to fundamental extended test conveyance of perceptions acquired from speaker articulations [23]. Gaussians reflect trademark vocal tract designs from phonetic sounds containing a man's voice.

3 Methodology

Several simulations are carried out on a variety of samples within the test set. Figure 2 depicts system flow chart. The Parkinson speech data set [6] consists of samples for voice and continuous vowels, a, o, u, together with numbers 0 to 9, words and short sentences. The recordings are stored in WMA format with the sampling frequency of 44.1 kHz and 32 bit of resolution. According to Nyquist

Fig. 2 Flow chart

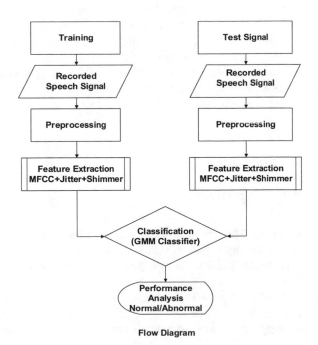

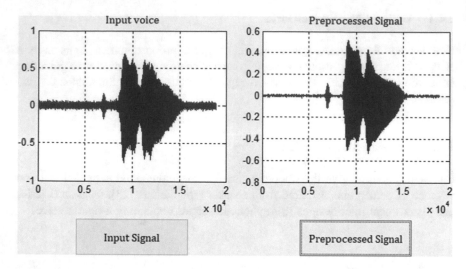

Fig. 3 Pre-processed signal by point average mean filter

sampling, 44.1 kHz allows analysis of 22 kHz signals, thus almost every aspect of voice signal is put into consideration. From every extracted syllable, 22 MFCCs are calculated by using 24 triangular filters. Analysis and interpretation of MFCC feature sets are done by calculating the MFCC mean, MFCC median, MFCC standard deviation, MFCC min and MFCC max values. The values of clean and noisy versions of the speech signals are represented using normal distribution. It has been noted that the dispersion of the component vectors impressively changes because of the effect of commotions. The progressions are seen to be distinctive for various sorts of commotions. Standard deviation, as well as mean for the typical dispersion of every variable of the MFCC highlight vectors, has been assessed for spotless and boisterous adaptations of the discourse flag. In Fig. 3, the input speech signal is shown having amplitude and time period as their axis.

4 Simulation Results

4.1 Preprocessing

In speech recording, inevitably some environmental noise is captured. Figure 3 shows the input speech signal before and after preprocessing. The preprocessing ensures that point noise is expelled from the voice motion by point normal mean filters.

Point average mean filter is used for processing the signal. When every variable is taken into consideration, the point average mean filter is utilized with the time strategy information to smooth around at this very moment changes and feature

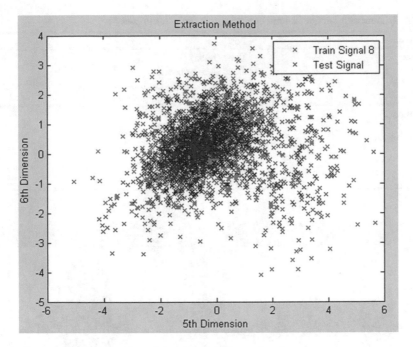

Fig. 4 Extraction of features for training signals and test signal

persevering situations. Scientifically, a roving average is somehow taken as a convolution. Thus, it can be considered as flag handling low-pass filter. Henceforth, it is proper to take it as an information smoothing process. In Fig. 4, the datasets are prepared and the highlights are separated for both trained and test signals.

4.2 Feature Extraction

For feature extraction, MFCCs, jitter and shimmer are applied to obtain speech features from preprocessed signal [24] that best in class speaker acknowledgement frameworks tend to utilize short-term spectral features only, for example, voice data. Jitter is, defined as the normal total distinction between continuous periods, partitioned by the normal time frame [25]. The feature vector is extricated for the acoustic framework in view of the ten jitter and shimmer measurements as shown in Table 1, with graphical representation on Fig. 5.

In Fig. 5, sample feature vectors obtained during the feature extraction and classification process. Here the samples are loaded and the test signal to be extracted using jitter and shimmer feature extraction process.

Table 1 Jitter and Shimer features extracted from speech samples [26]

Parameter	Features
Jitter	Jitter (local)
	Jitter (local, absolute)
	Relative Average Perturbation
	Five-point Period Perturbation
Shimmer	Shimmer (local)
	Shimmer (local, dB)
	Shimmer (apq3)
	Shimmer (apq5)
	Shimmer (apq11)
	Shimmer (ddp)

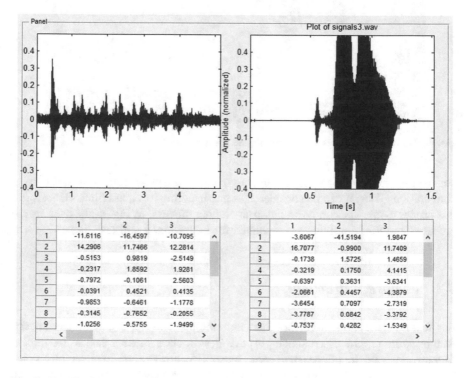

Fig. 5 Identification of normal or pathological voice using GMM classifier

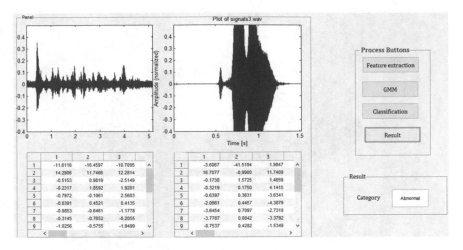

Fig. 6 3D view of the resultant speech signal

4.3 Classification

A Gaussian Mixture Model (GMM) accepts every one of the information focus is created from a blend of a limited number of Gaussian circulations with obscure parameters and is characterized as a probabilistic model. One can consider blend models as summing up k-means grouping to consolidate data about the covariance structure of the information and also the focuses of the inert Gaussians [27].

- GMMs are parametric likelihood thickness work addressed as a weighted total of Gaussian section densities.
- GMMs are frequently utilized in biometric frameworks, most strikingly in speaker recognition frameworks, due to their ability to speak to an extensive class of test conveyances [27].
- GMMs have a desirable feature lying in their ability to frame smooth approximations to discretionary moulded densities.

The utilization of a GMM for speaking to include conveyances in a biometric framework may likewise be roused by the instinctive thought of individual densities showing a fundamental arrangement of shrouded classes [27].

In Fig. 6, the test signal features are extracted, trained, and compared with the dataset and the result is displayed, it indicates whether the speech has been classified as either as normal or abnormal, using GMM classifier. Then the 3D representation of the signal is as shown in Fig. 7. Here the person's name, to whom the speech signal belongs, is displayed.

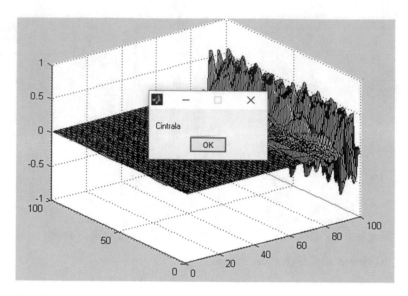

Fig. 7 3D view of the resultant speech signal

5 Conclusions

GMM is widely used in acoustics modelling involving phonemes and words. It is among the best classifiers for the following speech phonetic parameters: pitch, pitch range, average pitch. Because of vibration of human vocal tract during pronunciation of vowels, single words and numbers, frequencies of the vocal tract, formants, also changes. In this study a combination of GMM classifier, jitter and shimmer, voice features are extracted and used for machine learning purposes in diagnosis of Parkinson's disease. The result shows around 97.6% efficiency in combining MFCC, Jitter and Shimmer feature extraction and GMM classifier.

References

1. Verma, A.K., Raj, J., Sharma, V., Singh, T.B., Srivastava, S., Srivastava, R.: Epidemiology and associated risk factors of Parkinson's disease among the north Indian population. Clin. Epidemiol. Glob. Heal. **5**(1), 8–13 (2017)
2. Bala, A., Gupta, B.: Parkinson's disease in India: an analysis of publications output during 2002–2011. Int. J. Nutr. Pharmacol. Neurol. Dis. **3**(3), 254 (2013)
3. Jankovic, J.: Parkinson's disease: clinical features and diagnosis. J. Neurol. Neurosurg. Psychiatry **79**(4), 368–376 (2008)
4. Tsanas, A., Little, M.A., McSharry, P.E., Spielman, J., Ramig, L.O.: Novel speech signal processing algorithms for high-accuracy classification of Parkinsons disease. IEEE Trans. Biomed. Eng. **59**(5), 1264–1271 (2012)

5. Harel, B., Cannizzaro, M., Snyder, P.J.: Variability in fundamental frequency during speech in prodromal and incipient Parkinson's disease: A longitudinal case study. Brain Cogn. **56**(1), 24–29 (2004)
6. Sakar, B.E., et al.: Collection and analysis of a Parkinson speech dataset with multiple types of sound recordings. IEEE J. Biomed. Heal. Informatics **17**(4), 828–834 (2013)
7. Cernak, M., Orozco-Arroyave, J.R., Rudzicz, F., Christensen, H., Vásquez-Correa, J.C., Nöth, E.: Characterisation of voice quality of Parkinson's disease using differential phonological posterior features. Comput. Speech Lang. **46**, 196–208 (2017)
8. Benba, A., Jilbab, A., Hammouch, A.: Hybridization of best acoustic cues for detecting persons with Parkinson's disease. 2nd World Conf. Complex Syst. WCCS **2014**, 622–625 (2015)
9. Arora, S., et al.: Detecting and monitoring the symptoms of Parkinson's disease using smartphones: a pilot study. Park. Relat. Disord. **21**(6), 650–653 (2015)
10. Sharma, V. et al.: Spark: Personalized parkinson disease interventions through synergy between a smartphone and a smartwatch. Lect. Notes Comput. Sci. (including Subser. Lect. Notes Artif. Intell. Lect. Notes Bioinformatics), pp. 103–114 (2014)
11. Wicks, P., Stamford, J., Grootenhuis, M.A., Haverman, L., Ahmed, S.: Innovations in e-health. Qual. Life Res. **23**(1), 195–203 (2014)
12. Stebbins, G.T., Goetz, C.G.: Factor structure of the Unified Parkinson's disease rating scale: motor examination section. Mov. Disord. **13**(4), 633–636 (1998)
13. Qian, L., et al.: Frequency specific brain networks in Parkinson's disease and comorbid depression. Brain Imaging Behav. **11**(1), 224–239 (2017)
14. Index of /ml/machine-learning-databases/00301. [Online]. Available: https://archive.ics.uci.edu/ml/machine-learning-databases/00301/. Accessed 24 Oct 2017
15. Meghraoui, D., Boudraa, B., Meksen, T.M., Boudraa, M.: Features dimensionality reduction and multi dimensional voice processing program to parkinson, pp. 16–18 (2016)
16. Han, W., Chan, C.F., Choy, C.S., Pun, K.P.: An efficient MFCC extraction method in speech recognition. In: *IEEE Int. Symp. Circuits Syst.*, pp. 145–148, 2006
17. Patel, I., Rao, Y.S.: Speech recognition using hidden markov model with MFCC-subband technique. In: 2010 International conference recent trends information, telecommunication computer, pp. 168–172 (2010)
18. Tirumala, S.S., Shahamiri, S.R., Garhwal, A.S., Wang, R.: Speaker identification features extraction methods: a systematic review. Expert Syst. Appl. **90**, 250–271 (2017)
19. Bourouhou, A., Jilbab, A., Nacir, C., Hammouch, A.: Comparison of classification methods to detect the parkinson disease. In: 2016 international conference on electrical and information technologies (ICEIT), pp. 6–8 (2016)
20. Shirvan, R., Tahami, E.: Voice analysis for detecting Parkinson's disease using genetic algorithm and KNN classification method. Biomed. Eng. (ICBME), 14–16 (2011)
21. Khorasani, A., Daliri, M.R.: HMM for classification of Parkinson's disease based on the raw Gait data. J. Med. Syst. **38**, 147 (2014)
22. Gopi, E.S.: Digital speech processing using matlab. Springer India, New Delhi (2014)
23. Reynolds, D.A., Rose, R.C.: Robust text-Independent speaker identification using gaussian mixture speaker models. IEEE Trans. Speech Audio Process **3**, 72–83 (1995)
24. Aggoun, A., Almaadeed, N., Amira, A.: Speaker identification using multimodal neural networks and wavelet analysis. IET Biometrics **4**(1), 18–28 (2015)
25. Boersma, P., Weenink, D.: Praat: doing Phonetics by Computer. [Online]. Available: http://www.fon.hum.uva.nl/praat/ (2017). Accessed: 24 Oct 2017
26. Boersma, P., Weenink, D.: Praat: doing phonetics by computer (Version 4.5.)[Computer program], Retrieved from www.praat.org, (2007)
27. Kilaru, V., Amin, M.G., Ahmad, F., Sévigny, P., Difilippo, D.: Gaussian mixture model based features for stationary human identification in urban radar imagery. Rad. Conf. IEEE **2014**, 426–430 (2014)

Real-Time Biomedical Recursive Images Detection Algorithm for Indian Telemedicine Environment

Syed Thouheed Ahmed and M. Sandhya

Abstract India is a developing nation with its literacy rate slowly progressing with respect to technological enhancements and services. In India, Telemedicine is blooming research concern, as client-server direct communication is involved. In telemedicine environment, image samples are uploaded at various levels creating a heap of similar images. In this paper, a novel technique is proposed and discussed on reduction of recursive images in Cloud/Server. The technique is based on pixel value density matching with edge extraction under proposed (Real-Time Biomedical Images Recursion Detection (RT-BIRD). The proposed technique is simulated on 12,800 unrepeated UCL machine learning repositories samples with 97.8% of accuracy rate. The results also focus on time for evaluation and processing. This technique is implemented and calibrated using HADOOP platform for optimizing big data infrastructure.

Keywords Telemedicine · Big data · Image recursion detection
Time optimization · Space optimization

1 Introduction

India is marked among fastest growing economy in the world with Indian population reaching second after China. Medical attention and services are major revolutionary policies for any developing nation as India. Today, medical services have reached doorsteps of villages, towns, and cities but a lack of information, awareness and trust-building is a challenging task to make telemedicine successful. India's 78% of population belong to rural development authorities which includes more than 45% of under privileged community as per latest research from

S. Thouheed Ahmed (✉) · M. Sandhya
School of Computer, Information and Mathematical Sciences,
BSA Crescent Institute of Science and Technology,
(A, Deemed to Be University), Chennai, India
e-mail: thouheed_cse_phd_17@bsauniv.ac.in

© Springer Nature Singapore Pte Ltd. 2019
P. K. Mallick et al. (eds.), *Cognitive Informatics and Soft Computing*,
Advances in Intelligent Systems and Computing 768,
https://doi.org/10.1007/978-981-13-0617-4_68

government of India. Telemedicine can be a bloom if and only if the under-privileged community is benefitted. Indian medical policies and government has planned rural medical camps connecting small villages to a town and there by towns to cities. Unfortunately, population with respect to resources availability is a major set-back today causing a heap of data and contributing to big data era or large data management.

Big data era or large data management is a major challenge for any infrastructure, retrieving information, images, and business operations. Telemedicine holds well with attributes for contributing Big data era. In India, Telemedicine environment can be well understood as a network for connecting rural India to its facilities in urban cities and on a hold, assume if a 100 records are uploaded in an interval of 60 min, on an average 500 MB of cloud space is occupied, the factor is expanded with respect to incoming records and the complexity of cloud storage is thus influenced with expansion. Major storage companies provide a dedicated sector of bandwidth and memory space for commercial operations. Thus it is necessary for an enterprise or infrastructure to design a space optimizing technique.

Most traditional techniques are focused on space optimization with respect to quality compromise, subsequently resulting in false positive datasets for biomedical processing. Thus the novel technique for image recursion detection is proposed in this paper. Owing high quality of space optimization and improving processing time consumption on a LINUX server. In RT-BIRD proposed technique, the image samples are processed with pixel density evaluation and edge extraction. The most frequently repeated pattern of pixels are recorded in a time series log and thus a cross over comparison is computed. Moreover, an additional task is required to redefine a final conclusion. The test is sampled over 500 UCL datasets generating 97.8% of accuracy under mixed and individual operation for recursion detection.

RT-BIRD technique targets datasets (UCL) stored on LINUX cloud servers by a balanced processing mechanism of extracting patters from least level of pixel density (X-Ray/CT Images) up to color image slices. The designed system is calibrated and implemented using Hadoop package. Experiments on UCL repository for Machine Learning [1] and cancerdata.org [2] shows it is scaling larger datasets streams with a precision of 97.8% of accuracy with respect to time of processing, space optimization and higher degree of indexing.

2 Protocol Design

Each phase, data is collected, calibrated, and stored under telemedicine environment. The mass accumulation of data is resulting towards declined resource utilization and poor storage cum maintained approaches. The primary resource protocol is shown in Fig. 1, demonstrating an overview for data collection and origin. This input string is recognized as bottleneck for telemedicine researchers to design and develop effective techniques towards optimization, repeated pattern detection, or recursion detection.

Fig. 1 Primary source for data collection

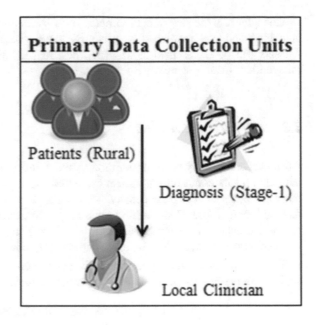

Case Study: Considering, local clinician as most common point for data sample uploading to telemedicine servers, an overload of huge updating logs and address recollecting logs are queued causing poor storage reallocation. The servers are then stacked with similar data samples stored in various unique locations. On an average, 33% of Indian telemedicine data is either recursive or duplicated under common copy paste operations. Thus in this approach, a dedicated real-time biomedical image recursion detection is designed and implemented using HADOOP infrastructure.

3 Real-Time Image Recursion Detection Algorithm

Typically, the proposed algorithm is segmented into Data acquisition mode, Attribute extraction, Image Indexing and Stacking and mapping and optimization. The proposed technique is independent of image type and resolution.

3.1 Data Acquisition Mode

Collection of data samples and datasets from various repositories, independent contributors, regional centers and many more for building telemedicine server for the detection of recursion reduction.

Consider n is the size of data samples of M bytes under which M_K bytes of data is contributed and renowned as recursive data. (i.e.,) on expansion, the given datasets can be represented as $M = \{M_1, M_2, M_3, \ldots, M_n\}$, where $(1, \ldots, n)$ is an independent dataset sample with M byte Size.

3.2   Attribute Extraction

Attributes are primary units for optimizing and identifying an independent source of datasets and its count on multiplication. Thus for each sample M_i there exist primary attributes such as pixel density, color ratio, dataset type, memory occupied on disk and many more. Considering in general, A (attribute set)=$\{A_1, A_2, A_3, \ldots, A_m\}$, where, for each considered dataset (A) there has m number of attributes at least.

Each attribute $\underline{AdefM_i}$ with each pattern (M) are linked with interdependent attributes (A). Since the optimization is performed, the major consideration of attribute is pixel density. And this is represented under (1)

$$PDm = \text{Pattern Occurried} * \text{Region Elimated}. \tag{1}$$

3.2.1   Pattern Occurred Extraction

Thus, in attribute extraction, patterns play an important and significant role. The patterns are considered with respect to the pixel changing ratio of an image and thus it can be expressed with one or more patters formation in a single dataset. $P_o = \{Po_1, Po_2, Po_3, \ldots, Po_K\}$ with each dataset (M) can contribute K patterns in the process.

$$\text{Pattern } (P_o) = \int_0^{\text{pixel}_{\text{size}}} \frac{\partial(\text{Pixel density})}{\partial(\text{Image Size})} * K_{\text{mean}} \tag{2}$$

The patterns extracted in (2) are represented by coefficient attributes of pixel density and thus reframing the overall (3) can be demonstrated.

$$P_o = \sum_{i=0}^{K} \int_0^{\text{pixel}_{\text{size}}} \frac{\partial(\text{Pixel density})}{\partial(\text{Image Size})} * K_{\text{mean}} \tag{3}$$

3.2.2 Region elimination extraction

The pixel density for the extracted region is in combination as shown in (1). Thus the region of elimination for a given data pattern ratio (R_d) is represented in (4)

$$(\text{Image Pixel density} - P_o) \overset{\text{yields}}{\rightarrow} R_d \tag{4}$$

Thus pixel density of image (I_D) is $(m * n)$ matrix value for the image. With bottom-up tracking, the medical dataset image for an instance up to K's patterns can be extracted and archived.

3.3 *Image Indexing and Stacking*

Biomedical datasets consist of most influential attributes, storing, and retraining the attributes during processing is a major challenge. In this real-time infrastructure, a dedicated design for image indexing and sharing is proposed. This technique is improvised with additional parameters on dual head indexing and thus accessing stack logs with respect to dual parameters makes the system dynamic. In design, the dual parameters are attribute level extraction indexing and data mining pattern extraction indexing, a detailed description is shown in Fig. 2.

From Fig. 2, a detailed descriptive agenda on logical representation is presented. The datasets are evenly considered and distributed to retrieve the attributes and thus perform pattern extraction operations as discussed in Eqs. (1)–(4). Thus the extraction of data stack is represented in Eq. (5)

$$D_S = \sum_{i=0}^{n} \frac{\partial(Rd_i)}{\partial(M_i)} \tag{5}$$

The stack value of each pattern is segregated and stored with respect to radio density. Thus the overall count stack is increment with independent notifications of dataset under C_s in Eq. (6)

$$C_S = \sum_{i=0}^{n} Ds_i - \sum_{j=i+1}^{n} Ds_j \tag{6}$$

(i.e.) for the data stack value of ith dataset of M_i with other datasets values of $(i + 1)$ is represented with j up to nth value count of C_{mi} is incremented in proper order is shown in (6).

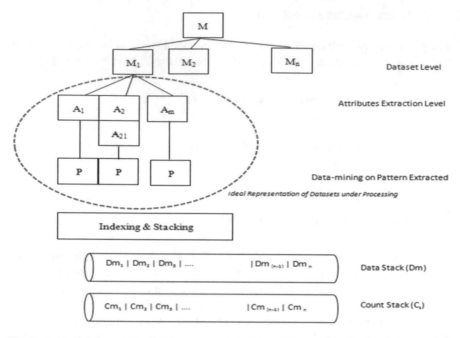

Fig. 2 Logical representation of dataset segmentation, attribute extraction, pattern extraction, indexing and stacking

3.4 Mapping and Optimizations

On stacking and indexing the system is introduced with mapping operations. The mapping approach synchronized the overall indexing log file with count and optimization time. The major achievement of proposed technique is time optimization. The overall optimization is shown in Table 1.

Table 1 Time optimization of proposed technique to detect recursive images in HADOOP and JAVA environment

Sample no	JAVA execution time (ms)	Hadoop execution time (ms)	Data size (MB)
1	2100	2579	1
100	2850	2078	3
200	2400	1147	5
400	7250	2085	11
800	5500	2039	22
1600	8500	2010	46
3200	13,600	2600	93
6400	36,800	4338	186

Table 2 Results of data samples under processing for Telemedicine environment

Sample	Input image	Gray scale image	Pixel density cum edge extraction	Processed pattern image
1				
2				
3				
4				

4 Discussion and Results

The proposed technique has successfully detected the images recursion value under the optimized time constrains as shown in Table 1. The real-time approach of framework makes it simplified to process and analyses time and space complexities of the server system. The overall approach is implemented under HADOOP environment. Table 2 demonstrates experimental results for various scenarios of telemedical data samples in Indian environment.

5 Conclusion

The proposed technique has achieved 60% time optimization with reduction of initial processing by real-time infrastructure. The datasets were considered as shown in Table 2 and processed with a sampling rate of 6400 samples per iterations under the worst case efficiency scenarios. The proposed system has achieved 97.8% of efficiency under time and space optimization. In general, the proposed scheme can reduce 34.5% of unused and recursive files under telemedicine cloud. This technique is initially the first of its kind in Indian Telemedicine system dedicated towards the improvisation of current challenges. The proposed approach has successfully targeted low bandwidth image processing and has retained its overall quality of data samples under processing. The detailed processing is archived in Table 2.

In near future, the technique is likely to be featured under machine learning objectives such as auto recognition of patterns on dynamic web portals and thus making the overall synchronization scheduling easier and efficient. In the proposed technique, the manual interventions can be replaced in a large wide towards inclusion of API design. In reality, the proposed system can also contribute towards the reformation of current Indian Telemedicine policies and Government norms.

References

1. Patašius, M., Marozas, V., Jegelevièius, D., Lukoševièius, A.: Recursive algorithm for blood vessel detection in eye fundus images: preliminary results. In: Dössel, O., Schlegel, W.C. (eds.) World congress on medical physics and biomedical engineering, Munich, Germany. IFMBE proceedings, vol. 25/11, Springer, Berlin, Heidelberg (2009)
2. http://www.cancerdata.org. Accessed on 15/10/2017
3. Telemedicine Opportunities and developments in Member States, Report on the second global survey on eHealth WHO (ISSN 2220-5462 2010)

4. Patil, K.K., Ahmed, S.T.: Digital tele mammography services for rural India, software components and design protocol. In: International conference on advances in electronics, computers and communications (ICAECC), pp. 1–5 (2014)
5. Ferguson, E.W., et.al.: Survey of global telemedicine. J. Med. Syst. **19**(1) (1995)
6. http://www.telemedindia.org/dit.html. Accessed on 23/10/2017
7. http://narayanahealth.org/telemedicine. Accessed on 23/10/2017

From Nonlinear Digital Filters to Shearlet Transform: A Comparative Evaluation of Denoising Filters Applied on Ultrasound Images

S. Latha and Dhanalakshmi Samiappan

Abstract Ultrasound images suffer from poor bone and air penetration, lighting conditions, light scattering, bending, absorption, and reflection. The blur and noise present in the image may be removed by suitable denoising algorithms, so that the preprocessed image will provide better results in further processing. Various denoising algorithms are analyzed, and the results are compared with denoising performance evaluation metrics like PSNR, Mean Square Error, Structural Similarity, and Correlation.

Keywords (5–6) curvelet · Denoise · Homomorphic · Speckle
Ultrasound

1 Introduction

Acquired based on the operator efficiency, ultrasound images are prone to all kinds of image noises like Poisson, Gaussian, salt & pepper, speckle, etc. Absorption removes the intensity of the light and the direction of the focused light path is altered by scattering property. The density factor and other factors, lead to a degraded image, which needs cure algorithms to be imposed. Ultrasound being a low-cost diagnosis tool is the most frequently used tool in medical imaging. Less amount of radiation emission and fast results makes it suitable for early identification of diseases like stroke and heart attack.

Denoising is performed by grouping the noise areas into disjoint regions with nearly same spatial characteristics and further filtering is performed [1]. The method is patch-based and prominent principal components of the patches are identified

S. Latha · D. Samiappan (✉)
Department of Electronics and Communication Engineering, SRM Institute of Science
and Technology, Kattankulathur, Kancheepuram, Tamil Nadu, India
e-mail: dhanalakshmi.s@ktr.srmuniv.ac.in

S. Latha
e-mail: latha.su@ktr.srmuniv.ac.in

© Springer Nature Singapore Pte Ltd. 2019
P. K. Mallick et al. (eds.), *Cognitive Informatics and Soft Computing*,
Advances in Intelligent Systems and Computing 768,
https://doi.org/10.1007/978-981-13-0617-4_69

with least length factor, and is given as input to the k means clustering algorithm. The denoised patches are recombined to reconstruct the original image. The advantage of hybridizing transform domain and spatial domain is better directionality, consideration of geometric and photometric distance of the pixels, thus noise reduction and edge preservation are achieved. Bilateral filtering based on an optimal expansion of filter kennel into a sum of factorized terms. By reducing the expansion error in Mean Square Error, the terms are computed [2].

Dual tree complex wavelet transform based Levy-Shrink algorithm with Bayesian map is used to reduce the speckle content in ultrasound images [3]. Nonlocal self-similarity and low-rank approximation property of images can be used for better speckle reduction. Similar patch groups are accumulated by a patch matching algorithm and the similar patch groups are subject to Singular Value Decomposition, with energy compaction in least square idea [4]. Denoising the ultrasound images is still a gap in research which is not much dealt in the literature. Removing the artifacts enhances the performance of the proceeding segmentation algorithms.

2 Denoising Filters

2.1 Speckle Modeling

Speckle noise is a granular multiplicative type of noise, which degrades the image quality, and is deterministic. On acquiring images in similar conditions, similar quantity of speckle may be present in the image. The speckle will comprise of texture information, which needs to be preserved [5].

$$I(x, y) = R(x, y)S(x, y) + A(x, y), \tag{1}$$

where I is the observed image, R is the actual signal or the reflectivity component, S is the speckle noise content, A is the additive noise and x, y are the spatial coordinates. Considering the additive noise negligible, and on applying logarithm,

$$i = r + s \tag{2}$$

The logarithm is removed by an exponential operation after all the frequency level transformations [6]. This circularly symmetric curve shaped, is centered at (0,0) in frequency domain.

$$H(u, v) = ['\Upsilon_H - '\Upsilon_L] \tag{3}$$

2.2 Speckle Reducing Filters

Median filter is an order statistics filter, more suitable for Gaussian and salt & pepper noise. It replaces the intensity of a pixel by median of its neighboring pixels. Hybrid median filter preserves edges better by estimating three median values from and $N \times N$ pixel backing an advantage that different spatial directions are ranked separately. The local density of the edges in the images is estimated and given to a morphological filter to remove noise and preserve edges.

Lee and Kuan filters are derived from the Minimum Mean Square Error (MMSE) criteria, with a local statistics method.

$$U(x, y) = I(x, y) + J(x, y)(1 - F(x, y)) \tag{4}$$

J is the average of the intensities in the filter window and F is the adaptive filter coefficient.

$$F(x, y) = 1 - \frac{C_B^2}{C_B^2 + C_I^2} \quad \text{for Lee filter} \tag{5}$$

$$F(x, y) = \frac{1 - C_B^2/C_I^2}{1 + C_B^2} \quad \text{for Kuan filter,} \tag{6}$$

where C_I is the coefficient of variation of the noised image and C_B is the noise coefficient of variation. The filter coefficient value nears to zero in uniform areas and nears to 1 at edges. This modifies the edge pixels, thus reduces the noise content. Frost filter is an exponentially varying weighted mean filter based on the variation coefficient, in which the degraded image local standard deviation is divided by the local mean value.

$$F(x, y) = e^{-kC_I^2(a,b)|(x,y)|}, \tag{7}$$

where the constant k controls the damping rate of the impulse response function, (a,b) are the pixels yet to be filtered. On small CI values, the filter works as a low pass filter, reducing speckle content and at high values, preserves the actual observed image. Diffusion which means iteratively updating the data is applied on Lee and Frost filter. But the quality of the image reduces slightly by iteration instead of enhancement.

$$\frac{dI}{dt} = f(I, \nabla I, \nabla^2 I, r), \tag{8}$$

where f is the scalar function of the image. Scale space is generation by the diffusion PDE, thus modifies the performance characteristics.

Adaptive wiener filter is based on local mean and variance of each pixel in a noisy image. The denoised image is given by

$$f(n_a, n_b) = \mu + \frac{\sigma^2 - v^2}{\sigma^2} (f(n_a, n_b) - \mu) \tag{9}$$

v is the variance of the noise which is estimated by the mean of all the locally generated variances [6]. A denoising weight ∂ is applied, which denoise the phase optimally by using a median patch.

$$\partial_{xy} = \frac{1}{\sigma^2} \exp\left\{ -\frac{||y_i - y_j||^2 + (1 - |\gamma_j|)}{h^2} \right\}, \tag{10}$$

where σ^2 is the noise variance, $h^2 = \sigma^2 N$, is the flatness constraint, N pixel count in each patch and γ_j is the coefficient of coherence of the pixel in the patch j [7].

Speckle Reducing Anisotropic Diffusion filter (SRAD) and Detail Preserving Anisotropic Diffusion filter (DPAD) generate piece-wise constant images. The SRAD filter is given by $\frac{\partial v(x,t)}{\partial t} = \emptyset.(c(q)\emptyset v(x, t))$

$$v(x)_{at\, t=0} = v_0(x), \frac{\partial v(x, t)}{\partial n} |\partial\Omega = 0 \tag{11}$$

The diffusion factor is a declining function of the instant figure of variation. In scale space, parameterized group of consecutive blur images based on diffusion are generated. The consequential image is an amalgamation of the actual image and a filter dependent on the local image content [5].

Discarding noise intensities, the image is reconstructed with the remaining coefficients with wavelet thresholding methods. Over or under thresholding is ruled out by choosing an optimal threshold value, which reduces the Mean Square Error. Hard thresholding is a keep or kill procedure and soft thresholding gives a smoother resultant image [8]. A pixel is replaced by the weighted average of its neighborhood pixels. The weights are identified as a likeness function among the values and the pixel similarities. The probabilistic model is given by

$$g(x) = m(x) + \sqrt{m(x)}n(x) \tag{12}$$

n is the Gaussian noise with mean value zero and a constant standard deviation [9]. The similarity between the image patches x and y are given by

$$w_{xy} = \exp\left\{ -\sum_{\partial \in P} \frac{G_\alpha(\partial)(y_{x+\partial} - y_{y+\partial})^2}{2h^2} \right\} \tag{13}$$

h is a smoothing parameter, $G_\propto$ is the Gaussian kennel with a radius $\propto$, and P is the image patch [10].

Similar to wavelet transform, curvelet transform offers multiscale analysis on images. Sub-band decomposition, even subdividing, renormalization and Ridgelet analysis are undergone along with thresholding. Curvelet transform based on fast fourier transform is useful to progress the denoising efficiency with better edge preservation [11]. Ridgelet transform, which is a better signal representer is a sparse expansion for functions on continuous spaces which smoothes the discontinuities along lines and corners. They define the direction and the scale position [12].

$$\varphi_\tau(x) = a^{-1/2}\varphi\left(\frac{u.x-b}{a}\right) \quad K_\varphi = \int \frac{|\varphi(\varepsilon)|^2}{|\varepsilon|^d} d\varepsilon\varepsilon < \text{infinity} \qquad (14)$$

where φ is the neural function, a, u, b the scale index, alignment and position of Ridgelet respectively and S is sobolev space. Purelet-Poisson Unbiased Risk Estimate—Linear Expansion of Thresholds denoises shot noise efficiently. Shearlets, which are extensions of wavelets are multiscale framework which encodes the anisotropic features in multivariate problem classes. Parabolic scaling, shearing, and translation applied in generating functions is executed to generate ridges. They provide optimally sparse optimizations which improves the denoising efficiency. Based on Laplacian pyramid and directional filter banks, contourlet transform was developed with up-sampling and down-sampling done in directional filter banks and Laplacian pyramid.

2.3 Performance Measures

The denoising filters are applied on a speckled underwater image of variance 0.02. The original image is compared with the denoised image with the performance evaluation metrics of Table 1. Lesser the error values, and more the other components like Signal-to-Noise Ratio, similarity, quality index and correlation, better is the filter in terms of denoising performance.

3 Results and Discussions

Gray-scale ultrasound images are subjected to denoising filters and the performance is evaluated. From Figs. 1, 2 and 3, it is seen that contourlet and shearlet transform provides better results in terms of the performance criteria (Fig. 4).

Table 1 Performance evaluation metrics

Sl. no.	Quality metrics	Formula and remarks
1	Signal to Noise Ratio (SNR)	μ/σ; Level of the background noise of image can be compared; μ—signal mean; σ—standard deviation of the noise
2	Peak Signal to Noise Ratio (PSNR)	$10\log10\ (255^2/MSE)$; Approximation to human perception of reconstruction quality. M, N—dimensions of the image
3	Mean Square Error (MSE)	$\frac{1}{MN}\sum_{i=1}^{M}\sum_{j=1}^{N}(X(i,j)-Y(i,j))^2$ Collective squared error between the denoised and the original image
4	Mean Absolute Error (MAE)	$\frac{1}{MN}\sum_{i=1}^{M}\sum_{j=1}^{N}|X(i,j)-Y(i,j)|$ Average of the absolute errors
5	Root Mean Square Error (RMSE)	$\sqrt{\frac{1}{MN}\sum_{i=1}^{M}\sum_{j=1}^{N}[X(i,j)-Y(i,j)]^2}$ Square root of the squared error averaged in the window
6	Mean Structural Similarity Index Metrics (MSSIM)	$\frac{(2\mu_x\mu_y+C_1)(2\sigma_{xy}+C_2)}{(\mu_x^2+\mu_y^2+C_1)(\sigma_x^2+\sigma_y^2+C_2)}$ $C_1=(K_1L)^2$; $C_2=(K_2L)^2$ Dynamic range of pixel $L=2^{\text{bitsperpixel}}-1$ $K_1=0.01$; $K_2=0.03$ by default Measures similarity between two images consistent with visual perception
7	Image Quality Index (IMGQ)	$\frac{\sigma_{xy}}{\sigma_x\sigma_y}\ \frac{2\mu_y\mu_x}{\mu_y^2+\mu_x^2}\ \frac{2\sigma_x\sigma_y}{\sigma_x^2+\sigma_y^2}$ Gives performance based on visual interpretation of the denoised image
8	Correlation (r)	$\sum_{i=-k}^{k}\sum_{j=-k}^{k}X(i,j)Y(x-i,y-j)$ Correlation between the images

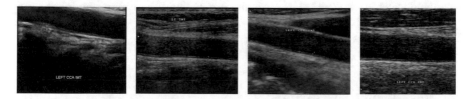

Fig. 1 Sample ultrasound images

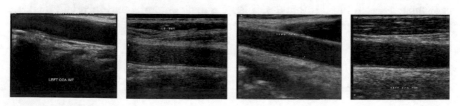

Fig. 2 Speckled images

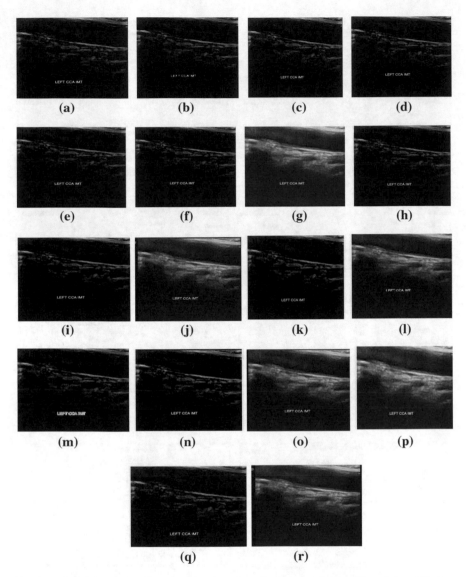

Fig. 3 **a–d** Hybrid median filter; Homlog filter; Local enhancement; Lee filter **e–h** Lee diffusion; Frost; Frost diffusion; Kuan filters **i–l** Additive; Wiener; SRAD; Wavelet decomposition filters **m–p** NLM; Curvelet decomposition; Purelet; Ridgelet filter **q–r** Contourlet; Shearlet filter

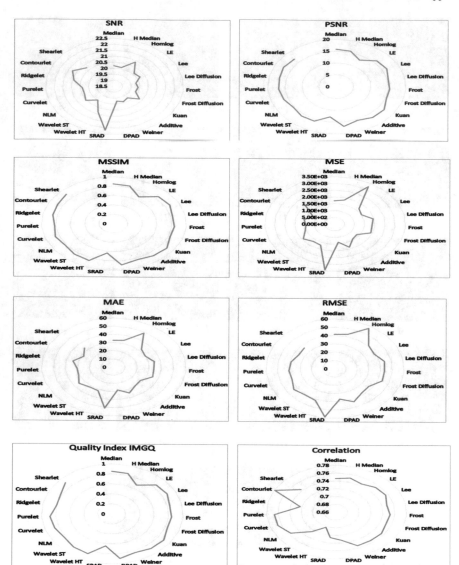

Fig. 4 Performance metrics for the filters

4 Conclusion

Ultrasound images suffer with extensive quantity of speckle noise, which may also have useful information. Denoising ultrasound images is an important area which needs to be addressed, so that further analysis with the images will be effective. Multiplicative speckle noise blurs the image which is converted to additive by homomorphic filtering. Denoising methods are applied on the speckled images and their performance is compared with performance evaluation metrics. The simulation results prove that there are considerable scope of research in the area of ultrasound image denoising.

Acknowledgements The authors thankfully acknowledges the financial support provided by The Institution of Engineers (India) for carrying out Research & Development work in this subject.

References

1. Xu, L., Li, J., Shu, Y., Peng, J.: SAR image denoising via clustering based principal component analysis. IEEE Trans. Geosci. Remote Sens. **52**(11), 6858–6869 (2014)
2. Papari, G., Idowu, N., Varslot, T.: Fast bilateral filtering for denoising large 3D images. IEEE Trans. Image Process. **26**(1), 251–261 (2017). http://ieeexplore.ieee.org/search/searchresult.jsp
3. Ranjani, J.J., Chithra, M.S.: Bayesian denoising of ultrasound images using heavy-tailed Levy distribution. IET Image Proc. **9**(4), 338–345 (2014)
4. Guo, Q., Zhang, C., Zhang, Y., Liu, H.: An efficient SVD-based method for image denoising. IEEE Trans. Circuits Syst. Video Technol. **26**(5), 868–880 (2016)
5. Ramos-Llorden, G., Vegas-Sanchez-Ferrero, G., Martin-Fernandez, M., Alberola-Lopez, C., Aja Fernandez, S.: Anisotropic diffusion filter with memory based on speckle statistics for ultrasound images. IEEE Trans. Image Proc. **24**(1), 345–358 (2015)
6. Lahmiri, S.: Denoising techniques in adaptive multi-resolution domains with applications to biomedical images. Healthc. Technol. Lett. **4**(1), 25–29 (2016)
7. Cao, M., Li, S., Wang, R., Li, N.: Interferometric phase denoising by median patch-based locally optimal wiener filter. IEEE Geosci. Remote Sens. Lett. **12**(8), 1730–1734 (2015)
8. Devi Priyaa, K., Sasibhushana Raob, G., Subba Rao, P.S.V.: Comparative analysis of wavelet thresholding techniques with wavelet-wiener filter on ECG signal. Procedia Comput. Sci. **87**, 178–183 (2016)
9. Lu, L., Jin, W., Wang, X.: Non-local means image denoising with a soft threshold. IEEE Signal Process. Lett. **22**(7), 833–837 (2015)
10. Fedorov, V, Ballester, C.: Affine non-local means image denoising. IEEE Trans. Image Process. **26**(5), 2137–2148 (2017)
11. Qiao, T., Ren, J., Wang, Z., Zabalza, J., Sun, M., Zhao, H., Li, S., Benediktsson, J.A., Dai, Q., Marshall, S.: Effective denoising and classification of hyperspectral images using curvelet transform and singular spectrum analysis. IEEE Trans. Geosci. Remote Sens. **55**(1), 119–133 (2017)
12. Yang, S., Min, W., Zhao, L., Wang, Z.: Image noise reduction via geometric multiscale ridgelet support vector transform and dictionary learning. IEEE Trans. Image Process. **22**(11), 4161–4169 (2013)

Application of Multi-domain Fusion Methods for Detecting Epilepsy from Electroencephalogram Using Classification Methods

**L. Susmitha, S. Thomas George, M. S. P. Subathra
and Nallapaneni Manoj Kumar**

Abstract Electroencephalogram (EEG) signal is a time series delineative signal which contains the useful knowledge about the state of the brain. It has high temporal resolution for detection of chronic brain disorders such as epilepsy/ seizure, dementia, etc. Technically, a feature mainly targets to capture the significant and typical characteristics hidden in EEG signals. In view of the low accuracy of commonly used methods for discrimination of EEG signals, this paper presents an efficient multi-domain fusion method to enhance classification performance of EEG signals. Features are extracted using autoregressive method (AR) employing Yule-Walker and Burg's algorithms respectively to generate feature from EEG. This paper implements two schemes of multi-domain fusion methods, the first one is AR method and wavelet packet decomposition (WPD) and the second one is AR method and Sample entropy (SampEn). Next, classification of extracted features is performed by different classifiers like Support vector machine (SVM) classifier, Linear Discriminant Analysis (LDA) classifier, Artificial neural network (ANN) classifier, K-nearest neighbor (KNN) and Ensemble classifier. Compared to AR-based method, fusion methods are yielding high accuracies. The ANN classifier has obtained the highest classification accuracy of 98.12% with the feature AR Burg-WPD combination compared to other classifiers in multi-domain fusion methods.

L. Susmitha · S. Thomas George (✉) · M. S. P. Subathra
Department of Electrical Sciences, Karunya Institute of Technology
and Sciences, Coimbatore 641114, Tamil Nadu, India
e-mail: thomasgeorge@karunya.edu

L. Susmitha
e-mail: lsusmitha94@gmail.com

M. S. P. Subathra
e-mail: subathra@karunya.edu

N. M. Kumar
Faculty of Electrical and Electronics Engineering,
Universiti Malaysia Pahang, 26600 Pekan, Pahang, Malaysia
e-mail: nallapanenichow@gmail.com

© Springer Nature Singapore Pte Ltd. 2019
P. K. Mallick et al. (eds.), *Cognitive Informatics and Soft Computing*,
Advances in Intelligent Systems and Computing 768,
https://doi.org/10.1007/978-981-13-0617-4_70

743

Keywords Electroencephalograph · AR method · WPD · SampEn ANN

1 Introduction

EEG displays a graphical record of electrical activity of the brain, which represents the status of the body and thoughts of the mind. Production of electrical signals due to neural activity of the brain starts from the 17th week of prenatal development. While, EEG signal is so specific, feeble, non-stationary and encloses a lot of noise. Consequently, to process EEG signals is difficult. The electrical signals of brain will strike the scalp, it can be easily detected by localizing electrodes over the scalp using 10–20 International system which is well recognized in whole world.

For better interpretation of EEG signals, the extraction of features is essential to separate one EEG category from another. To assist better analysis of FFT, wavelet, eigenvectors, autoregressive and time-frequency distribution methods in both frequency domain and time-frequency domain is evaluated. So, it is essential to have a clear idea regarding the signal to be examined for application of several methods and functioning of analyzing method is reviewed [1, 2]. The Parametric methods can effectively apprehend the input process dynamics. The spectral features are extracted by ARMA modeling, where the EEG signals are recorded from every individual channel provides better perception for each control task between two subject groups [3]. The classification model based on nonlinear analysis and multi-domain feature is discussed and it can differentiate the seizure signals ten-fold cross-validation with an overall average classification accuracy reaching 99.25% [4, 5]. The suggested model uses approximate and sample entropies to be applied on WPD to determine the feature values, which acts as inputs for classifiers. The collaboration of Sample Entropy and ELM has depicted extreme execution of classification accuracy [6]. Approximate entropy, autoregressive coefficients are estimated separately, and by assembling them the features are obtained. Extensive empirical comparison has been executed between proposed combination method and AR method [7], sample entropy is used instead of approximate entropy in [8]. Experimental results show that the combination of methods is capable to improve the classification performance [9]. In view of low accuracies obtained by independent domain methods, this paper adopts multi-domain fusion methods for higher classification accuracies.

2 Methodology

In this section, a brief sketch of aimed model for EEG signal feature extraction and classification is described and introduces different feature extraction methods like AR model based on Yule-Walker and Burg method, SampEn and WPD.

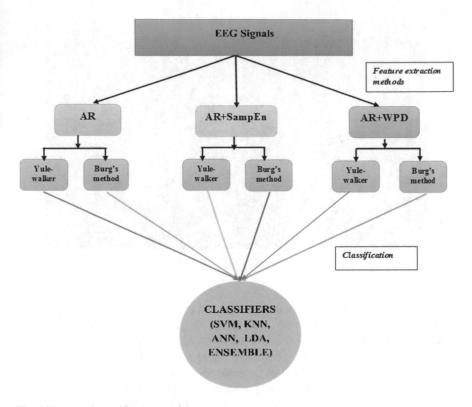

Fig. 1 Proposed classification model

Multi-domain fusion of feature extraction is introduced, and it is applied on normal and focal EEG data for effective classification.

The pictorial representation of prospective EEG signal classification model is shown in Fig. 1. The proposed model is based on combination of two features (AR + SampEn and AR + WPD) methods, referred as multi-domain based in approach. The classification will be implemented when these feature vectors are introduced into different classifiers like SVM, LDA, KNN, Ensemble and ANN.

2.1 Data Sets

The dataset used for this system is the EEG signal recorded from different persons which contains normal and epilepsy detected patients, by using the 16 channels. The 16-channel EEG signals from the patients was measured based on 10–20 electrode placement (standard international system) from neonate, infant, and adults using bipolar setting. For EEG recording the scalp of the patients were carefully prepared to have a contact impedance less than 5 kΩ and the signals are acquired

Fig. 2 EEG signal acquisition

with a sampling rate of 256 Hz using analog pass band ranges from 0.01 to 100 Hz. Each EEG segment consists of 2560 sampling points; hence in the order of milliseconds epileptic seizure activities were captured with high temporal resolution. EEG dataset of focal and generalized epileptic patients has been accumulated from a diagnostic center, located at Coimbatore, Southern India were used in this research [10]. Apart from that, experiments were performed to record 16-channel EEG data with a healthy subject by using the g. EEGsys research system which is shown in Fig. 2.

The artifacts are removed via visual examination of all the EEG signals that last for 10 s duration. These signals are sampled at a frequency of 256 Hz with 24-bit resolution. Hence, each signal consists of $10 * 256 = 2560$ sample points. Interested readers can have access to this public dataset at the specified URL in the brackets (http://www.karunya.edu/research/EEGdatabase/public/index.php), and which was in Selvaraj et al. [11].

2.2 Feature Extraction

Feature extraction has the direct effect on the classification accuracy. Taking complexity of EEG signals into consideration, this paper adopts different methods to extract features and notions of them are stated in following.

Autoregressive Method. Autoregressive (AR) model measures power spectrum density (PSD) of the EEG by applying parametric approach. AR method has less spectral leakage problem, unlike nonparametric approach it results enhanced frequency resolution. Estimation of PSD is accomplished by estimating their coefficients, which are considered as the parameters of the linear system. The explicit equation is:

$$g(t) = \sum_{k=1}^{s} \emptyset_i g(t-k) + \epsilon_t \qquad (1)$$

whereas in Eq. (1), $g(t)$ is the modelled time series, k is model coefficient, ϵ_t is independent white noise attained of former points and parameter s is the order of AR model. Different categories of techniques exist to deal with existed model coefficients, such as Burg method, Yule-Walker method and least-squares method. Firstly, choosing an appropriate model order is very important. After analyzing previous studies [8, 9], model order has been set by varying 1–6. Second, Burg algorithm estimates the better spectrum quality and low complexity compared to Yule-Walker method.

Wavelet Pocket Decomposition (WPD). WPD bestows level-by-level transformation of a signal from the time domain into the frequency domain. It is computed using a recursion of filter-decimation operations, which results to diminish in time resolution and rise in frequency resolution. In contrary, WPD decomposes the detail and approximate of a signal in each level. Thus, a richer range of possibilities for signal analysis are obtained using WPD with better frequency resolutions, where as in discrete wavelet transform, the approximation coefficient will further itself split into a second-level approximation coefficients and detail coefficients, and repetition is done.

WPD splits into both detail and approximation coefficients, which contributes a structure of a complete wavelet packet tree as shown in Fig. 3. For m levels of decomposition, the discrete wavelet transform develops only $(m + 1)$ sets whereas the WPD produces 2^m different sets of coefficients.

Sample Entropy (SampEn). SampEn diverges with approximate entropy. SampEn is intended as a measure of the order in a time series. SampEn expels self-counting matches and adopts logarithm of the sum of conditional probabilities.

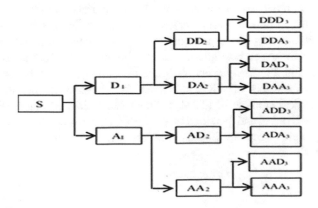

Fig. 3 Wavelet packet decomposition for 3 levels

Assume data points of length N, where m determines the length of vectors $(N - m + 1)$, namely $y(i) = \{v(i), v(i+1), \ldots, v(i+m-1)\}$, where i is from 1 to $(N - m + 1)$. Then, for each $i(i = 1, \ldots, N - m + 1)$, the value of $C_i^m(r)$ will be computed by the following equation.

$$C_i^m(r) = \frac{\sum_j \{1[y(i), y(j)] < r\}}{N - m + 1} \qquad (?)$$

Whereas in Eq. (2), $C_i^m(r)$ is ratio of vector pairs holding $d[y(i), y(j)] \leq r$ to the value $(N - m + 1)$. Typically, value of r is set to be some percentage of the standard deviation. This effectively rescales the series to have similar dynamic scales, and will generally give less weight to spikes. The meaning of the numerator in Eq. (2) is number of j such that the distance is not greater than r. The sample entropy, SampEn (m, r, N) is expressed in Eq. (3).

$$\text{SampEn}\,(m, r, N) = \ln\left[\frac{C^{m+1}(r)}{C^m(r)}\right] \qquad (3)$$

2.3 Multi-domain Fusion Methods

According to above-mentioned feature extraction methods, this paper constitutes two fusion scenarios of features extraction from EEG signals as discussed in [8]. The first multi-domain approach combines AR model with SampEn termed as AR + SampEn method. In fusion method, examines the EEG signals as time series and estimates AR coefficients from every sector. There exist 16 clusters of AR coefficients for 16 channels in similar part. Hence, $(16 * s)$ features are yielded, where s is order of AR model. Next, SampEn computes entropies of every channel to yield 16 features. Finally, both AR coefficients and entropies are fused to form $(16 * s + 16)$ structural feature vectors and fed into classifiers. The second multi-domain approach is fusion of WPD with AR model termed as AR + WPD method. Firstly, fragment every section consists of EEG signals into distinct sub-bands by utilizing WPD and the AR coefficients estimated from various sub-bands. For the decomposition of n level sub-bands, WPD develops 2^m different series coefficients. For every array, estimate its s-order AR coefficients. Likewise, there exists 16 clusters of AR coefficients for 16 channels belongs to similar part. Every radical consists of $(2^m * s)$ dimensional feature vectors. By choosing appropriate AR order and parameters of WPD, this fusion method could attain significantly higher accuracy in classification when compared with other feature extraction method.

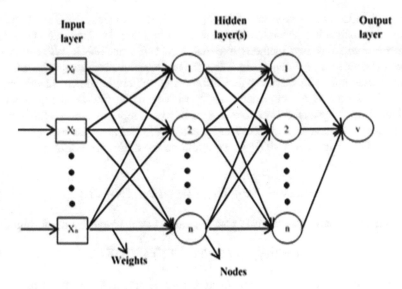

Fig. 4 Artificial neural network (ANN) architecture

2.4 Classification Techniques

A classifier is a technique which exploits various independent variable values (features) as input and terminates to which the independent variable is inherited in corresponding class. In analysis of EEG signal, features can be any sort of extracted data from the signal and class can be any type of task or the stimulus involved during the recording. Artificial neural network is a mathematical tool which mimics some functional aspects of a biological neural network. The computing systems constitutes huge statistics of simple, highly interconnected processing elements (called nodes) which outline, imitate system and procedure of biological nervous system. In this paper, neural network relevant for classification of EEG data is Multilayer Perceptron (MLPNN) shown in Fig. 4. The main goal of this algorithm is reduction of errors and to train the network without any interpretation until it learns the data. The ANN classifier gave higher accuracy compared to other mentioned classifiers.

3   Results and Discussions

The experiments were run on Intel core i5-3210M (2.50 GHz) processor with 4 GB RAM using the MATLAB (Math Work R2017a) software. The classifiers like SVM, ANN, KNN, ensemble and LDA has been used for classification of extracted features and the obtained accuracies are shown below. The classification is done in groups of EEG data as Normal EEG data and Focal Epilepsy EEG data. As the obtained classification accuracies for AR burg and Yule-walker method is less, hence by combining two methods like AR + SampEn and AR + WPD to obtain higher classification accuracies. In Table 1, the classification accuracies of three feature extraction methods using five different classifiers is shown. In this paper, focal EEG segments are considered as "positive" while non-focal segments are considered as "negative". Therefore, for each test sample, a binary classifier has four possible outcomes such as True positive (TP), False positive (FP), True negative (TN), False negative (FN). To evaluate the algorithm performance, these parameters are computed.

$$\text{Accuracy} = (\text{TP} + \text{TN})/(\text{TP} + \text{FP} + \text{TN} + \text{FN}) \tag{4}$$

Sensitivity is termed as the true positive ratio, is calculated by the formula

$$\text{Sensitivity} = \text{TP}/(\text{TP} + \text{FN}) \times 100\% \tag{5}$$

Specificity is termed as the true negative ratio is calculated by the formula:

$$\text{Specificity} = \text{TN}/(\text{TN} + \text{FP}) \times 100\% \tag{6}$$

The three parameters are calculated using Eqs. (4), (5), and (6). For AR + SampEn method, suppose s as order of AR model, pattern length m and threshold r of SampEn.

In significance to accuracy obtained for classification of EEG signals is acquiring better results since order s increases, where computational cost also raises. Appropriate values of m and r also enhances classification accuracy of EEG signals,

Table 1 Classification accuracies in % obtained using AR, AR + SampEn, AR + WPD

Feature extraction method	Accuracy				
	SVM	KNN	LDA	ENSEMBLE	ANN
AR method (a) *Yule walker* Normal and focal epilepsy	55.7	53.5	47	63.5	90
(b) *Burg's method* Normal and focal epilepsy	54.2	53.8	46.7	55	78.6
AR + SampEn (a) *Yule walker* Normal and focal epilepsy	57.6	60.1	52.4	61.8	96.2
(b) *Burg's method* Normal and focal epilepsy	61.1	53.8	46.7	55	97.54
AR + WPD (a) *Yule walker* Normal and focal epilepsy	73.8	68.4	69.5	70.7	92.8
(b) *Burg's Method* Normal and focal epilepsy	78.6	72.4	74	70.8	98.12

the parameter m is fixed to be 2 as reported by many researchers and the assessment of classification performance is done by different parameter combinations of (s, r). The s value is varied from 1 to 6 and r values are taken as 0.12, 0.14, 0.16, 0.18, and 0.20. The three AR-based feature extraction methods are computed at different parameter combinations of d and r and classification is done using ANN classifier. The average classification accuracies, sensitivities, and specificities of three feature extraction methods are stored in the Tables 2, 3, and 4 for all the subjects is shown.

Table 2 Average classification accuracies obtained using AR, AR + SampEn, and AR + WPD by varying order values

Feature extraction	$s = 1$	$s = 2$	$s = 3$	$s = 4$	$s = 5$	$s = 6$
AR Yule-walker method	76.3	77.84	78.59	82.67	84.94	90.62
AR Burg method	78.94	79.67	80.98	84.73	84.78	91.25
AR Yule + WPD	81.57	83.95	85.89	87.72	88.75	91.37
AR Burg + WPD	85.39	87.48	89.26	91.51	93.64	98.12
AR Yule + SampEn ($r = 0.12$)	81.63	82.34	84.75	83.7	87.56	89.45
AR Yule + SampEn ($r = 0.14$)	82.56	84.72	85.26	85.98	88.93	91.42
AR Yule + SampEn ($r = 0.16$)	85.92	85.29	83.3	83.7	88.34	93.77
AR Yule + SampEn ($r = 0.18$)	86.48	86.94	85.27	84.63	89.56	95.82
AR Yule + SampEn ($r = 0.20$)	88.43	89.39	91.59	93.45	95.47	96.2
AR Burg + SampEn ($r = 0.12$)	82.35	83.19	84.71	86.32	88.68	91.25
AR Burg + SampEn ($r = 0.14$)	83.37	84.29	85.43	86.79	89.57	92.39
AR Burg + SampEn ($r = 0.16$)	86.74	85.78	86.92	91.07	92.64	94.52
AR Burg + SampEn ($r = 0.18$)	87.39	87.36	88.26	89.92	91.75	96.25
AR Burg + SampEn ($r = 0.20$)	88.25	89.77	92.14	94.58	96.79	97.40

Table 3 Average classification sensitives obtained using AR, AR + SampEn, AR + WPD by varying order values

Feature extraction	$s = 1$	$s = 2$	$s = 3$	$s = 4$	$s = 5$	$s = 6$
AR Yule-walker method	63.46	60.7	67.5	78.08	79.31	92.20
AR Burg method	70.23	68.36	66.67	81.25	78.16	93.42
AR yule + WPD	65.38	76.74	66.67	80.55	82.92	92.8
AR Burg + WPD	72.72	70.37	73.23	76.27	81.11	96.67
AR Yule + SampEn ($r = 0.12$)	75.5	76.75	78.34	84.23	86.15	89.90
AR Yule + SampEn ($r = 0.14$)	72.54	74.5	77.43	85.39	87.32	90.35
AR Yule + SampEn ($r = 0.16$)	75	75.32	76.5	85.39	87.54	90.78
AR Yule + SampEn ($r = 0.18$)	75.55	75.5	77.2	85.30	88.98	92.87
AR Yule + SampEn ($r = 0.20$)	74.73	73.2	78.2	86.13	89.01	94.23
AR Burg + SampEn ($r = 0.12$)	77.50	78.45	79.18	81.63	88.43	93.42
AR Burg + SampEn ($r = 0.14$)	75.21	74.83	78.12	84.59	88.62	93.89
AR Burg + SampEn ($r = 0.16$)	75.43	77.18	78.39	85.78	87.67	94.12
AR Burg + SampEn ($r = 0.18$)	76.69	76.31	78.79	86.34	88.65	94.45
AR Burg + SampEn ($r = 0.20$)	77.17	78.80	79.15	86.32	89.95	94.73

Table 4 Average classification specificities obtained using AR, AR + SampEn, AR + WPD by varying order values

Feature extraction	$s = 1$	$s = 2$	$s = 3$	$s = 4$	$s = 5$	$s = 6$
AR Yule-walker method	75	68.96	67.5	73.56	84.93	89.15
AR Burg method	72.36	79.03	69.86	70.83	83.56	89.28
AR yule + WPD	78.57	81.08	65.85	75	84.61	88.88
AR burg + WPD	61.90	70.88	68.53	65.34	90	92.59
AR Yule + SampEn ($r = 0.12$)	82.85	83.78	79.64	84.83	87.01	88.76
AR Yule + SampEn ($r = 0.14$)	89.65	85.94	86.3	87.51	87.29	89.47
AR Yule + SampEn ($r = 0.16$)	83.82	84.29	85.91	87.37	87.71	91.07
AR Yule + SampEn ($r = 0.18$)	82.85	83.7	85.53	88.69	88.49	89.31
AR Yule + SampEn ($r = 0.20$)	86.15	86.43	87.26	89.45	89.97	91.42
AR Burg + SampEn ($r = 0.12$)	83.91	84.05	85.59	86.62	88.41	89.28
AR Burg + SampEn ($r = 0.14$)	84.23	84.89	86.91	87.62	88.79	89.95
AR Burg + SampEn ($r = 0.16$)	85.36	85.59	86.45	87.95	89.06	90.73
AR Burg + SampEn ($r = 0.18$)	86.05	86.73	87.80	88.58	89.45	91.69
AR Burg + SampEn ($r = 0.20$)	86.76	89.03	89.92	89.34	89.67	92.5

Based on the attained results, Burg's method estimates the better spectrum quality, with low complexity and provides higher classification accuracy in comparison to Yule-Walker method.

Figures 5, 6, and 7 provide a perceptive comparability on average accuracies attained for each feature extraction method applied to all the subjects. The s value is varied from 1 to 6, where at order is 6 high classification accuracies are obtained in three AR-based methods. As the order increases computational complexity will be elevated. The fusion of AR model and WPD contributes much useful potentiality for detection of epilepsy than that of another combination method.

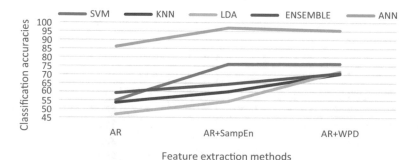

Fig. 5 Comparison of accuracies obtained from 5 different classifiers

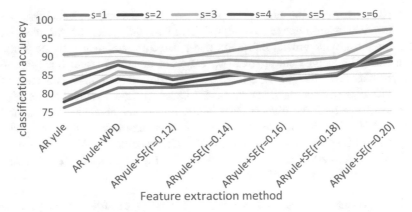

Fig. 6 Comparison of accuracies obtained by varying order of AR Yule based methods

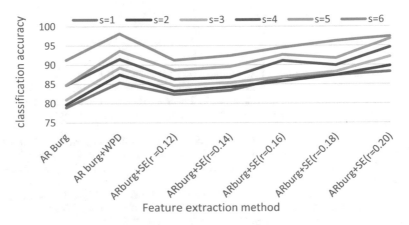

Fig. 7 Comparison of accuracies obtained by varying order of AR Burg based methods

4 Conclusion

For the classification of EEG signals, features are taken from the original data are commonly used. This paper constitutes of three different methods of feature extraction. The features of Normal and Focal EEG signals using three independent feature domain methods like AR method, AR + SampEn, and AR + WPD are extracted. The AR model has obtained low accuracy compared to other two methods, so multi-domain fusion method is applied for AR model using different combinations. The first method is the fusion of AR model with sample entropy used to perform as feature vectors, the other method, fusion of AR model with WPD is done particularly splitting EEG signals into sub-bands by WPD and by computing AR coefficients represents feature vectors. The feature vectors are employed into the classifier for classification. This paper executes data based validation, which proclaims that feature extraction method based on the fusion of AR model with WPD, could provide crucial information for detecting epilepsy than the fusion of AR model with sample entropy. Classification of extracted features is done using different classifiers like SVM, KNN, LDA, Ensemble Classifier and ANN. The accuracies obtained for classification of AR based methods gradually rises with the increase of s value. The highest classification accuracy obtained is compared by using ANN classifier in all the three feature extraction methods and AR Burg + SampEn method obtains the classification accuracy of 97.4% for $s = 6$ and $r = 0.20$, whereas AR Burg + WPD achieves the highest classification accuracy of 98.12% by increasing the order value.

Acknowledgements This paper work was endorsed by the "Technology Systems Development Programme (TSDP)" under Department of Science and Technology (DST), Ministry of Science and Technology, Government of India (GoI), [Grant Number—DST/TSG/ICT/2015/54-G, 2015].

References

1. Al-Fahoum, A.S.,Al-Fraihat, A.A.: Methods of EEG signal features extraction using linear analysis in frequency and time-frequency domains. ISRN Neurosci. **2014**, 1–7 (2014)
2. Cheong, L.C., Sudirman, R., Hussin, S.S.: Feature extraction of EEG signal using 256 wavelet transform for autism classification. ARPN J. Eng. Appl. Sci. **10**(19), 8533–8540 (2015)
3. Pinzon-Morales, R.D., Orozco-Gutierrez, A., Castellanos-Dominguez, G.: EEG seizure identification by using optimized wavelet decomposition. In: Engineering in Medicine and Biology Society, EMBC, 2011 Annual International Conference of the IEEE, pp. 2675–2678 (2011)
4. Sakkalis, V., Cassar, T., Zervakis, M., Camilleri, K.P., Fabri, S.G., Bigan, C., Micheloyannis, S.: Parametric and nonparametric EEG analysis for the evaluation of EEG activity in young children with controlled epilepsy. Comput. Intell. Neurosci. **2008**, 1 (2008)
5. Vidyaratne, L.S., Iftekharuddin, K.M.: Real-time epileptic seizure detection using EEG. IEEE Trans. Neural Syst. Rehabil. Eng. **25**(11), 2146–2156 (2017)
6. Wang, L., Xue, W., Li, Y., Luo, M., Huang, J., Cui, W., Huang, C.: Automatic epileptic seizure detection in EEG signals using multi-domain feature extraction and nonlinear analysis. Entropy **19**(6), 222 (2017)
7. Zhang, Y., Zhang, Y., Wang, J., Zheng, X.: Comparison of classification methods on EEG signals based on wavelet packet decomposition. Neural Comput. Appl. **26**(5), 1217–1225 (2015)
8. Zhang, Y., Ji, X., Liu, B., Huang, D., Xie, F., Zhang, Y.: Combined feature extraction method for classification of EEG signals. Neural Comput. Appl. **28**(11), 3153–3161 (2017)
9. Zhang, Y., Liu, B., Ji, X., Huang, D.: Classification of EEG signals based on autoregressive model and wavelet packet decomposition. Neural Process. Lett. **45**(2), 365–378 (2017)
10. Selvan, S.E., George, S.T., Balakrishnan, R.: Range-based ICA using a nonsmooth quasi-newton optimizer for electroencephalographic source localization in focal epilepsy. Neural comput. **27**(3), 628–671 (2015)
11. Selvaraj, T.G., Ramasamy, B., Jeyaraj, S.J., Suviseshamuthu, E.S.: EEG database of seizure disorders for experts and application developers. Clin. EEG Neurosci. **45**(4), 304–309 (2014)

In Silico Studies of Charge Transfer Complexes of 6-(Trifluoromethyl)furo [2,3-b]pyridine-2-carbohydrazides with Iodine as σ-Acceptor

Shirish Kumar Kodadi and Parthasarathy Tigulla

Abstract Comprehending the charge transfer process between bioactive molecules is decisive as this interface can be used to construe the bioactive molecules–receptor interactions. In view of the increasing importance of computational software and considerable progress in ab initio method development, computational software can now be used to study the charge transfer process between bioactive molecules and the acceptor molecules. Accordingly, using General Atomic and Molecular Electronic Structure System (GAMESS) computations charge transfer (CT) complexation between 6-(trifluoromethyl)furo[2,3-b]pyridine-2-carbohydrazides (FP2C) and sigma acceptor iodine (I_2) has been studied.

Keywords FP2C · I_2 · Charge transfer complexation · GAMESS and in silico study

1 Introduction

GAMESS (general atomic molecular electronic structure system) is a program for ab initio molecular quantum chemistry that can compute Self-consistent field (SCF) wave functions ranging from Restricted Hartree–Fock (RHF), Restricted open-shell Hartree–Fock (ROHF), Unrestricted Hartree–Fock (UHF) and

S. K. Kodadi (✉)
Department of Chemistry, Vignana Bharathi Institute of Engineering
and Technology, Hyderabad 501301, Telangana, India
e-mail: sirikumar834@gmail.com

P. Tigulla
Department of Chemistry, Osmania University, Hyderabad 500007,
Telangana, India
e-mail: sarathychem@gmail.com

© Springer Nature Singapore Pte Ltd. 2019
P. K. Mallick et al. (eds.), *Cognitive Informatics and Soft Computing*,
Advances in Intelligent Systems and Computing 768,
https://doi.org/10.1007/978-981-13-0617-4_71

Multi-configurational self-consistent field (MCSCF). The correlation corrections to these SCF wave functions comprise configuration interaction, second-order perturbation theory and coupled cluster approach as well as the density function approximation theory. Basic molecular modeling computations such as alignment, stochastic conformational sampling, dihedral driver and Molecular Mechanics (MM2) experiments can be successfully enabled by using this software.

GAMESS computations are used for geometry optimization, computation of ligand binding energy and study of the bioactive molecule–receptor interactions with the aim of scheming new bioactive molecules with greater efficacy.

The 6-(trifluoromethyl)furo[2,3-b]pyridine-2-carbohydrazides (FP2C) are bio-active molecules that wield their principle pharmacological and therapeutic effects by acting at peripheral sites to decrease the activity of components of the sympathetic division on autonomous nervous system [1]. Considering the significance of these compounds a series of novel trifluoromethyl substituted pyrido pyrimidine derivatives were synthesized by our collaborative research group. All the compounds were screened for cytotoxic activity against breast carcinoma MDA-MB 231(aggressive) cell lines at 10 μm concentration and promising compounds have been identified [2].

FP2C analogues are known to form CT complexes with electron acceptors. σ-electron acceptors have electrons localized in the σ part of a bond and single bond electrons. I_2 is one of the good σ-electron acceptor with electron affinity of 3.059038 eV has been considered for the study.

Charge transfer (CT) complexes are formed when electron donor and electron acceptor interact, which is evident from Mulliken theory [3].

The main objective, therefore, of the present article is to study the charge transfer process between FP2C (donor) and I_2 (σ-acceptor).

FP2C analogues used in the study

3-Amino-N'-heptanoyl-6-(trifluoromethyl) furo [2,3-b] pyridine-2-carbohydrazide (S1)

3-Amino-N'-hexanoyl-6-(trifluoromethyl)furo[2,3-b]pyridine-2-carbohydrazide (S2)

N-(3-isobutyl-1-oxo-7-(trifluoromethyl)pyrido[3',2':4,5]furo[3,2-d]pyrimidin-2(1H)
-yl)-3-methylbutanamide (S3)

2 Computational Study

2.1 Molecular Structure and Optimized Geometry

In silico study has been carried out using GAMESS [4, 5] computations as a package of ChemBio3D Ultra 12.0 [6]. MM2 method has been applied for optimization of the energy in all the instances.

i. S1-I₂ CT complex

15.2580 and 0.0 kcal/mol are the respective recorded energy values of the donor S1 and the acceptor I_2. Whereas, energy of the formed complex recorded 9.4236 kcal/mol, ascertaining the noteworthy stability of the S1-I_2 complex.

Figures 1 and 2 represent the optimized structures of all the three substances, i.e., I_2 (acceptor), S1 (donor) and S1-I_2 CT complex. Table 1 comprises the bond lengths of the S1 and S1-I_2 CT complex respectively.

Fig. 1 Energy-optimized molecular structure of iodine (acceptor)

Fig. 2 Energy-optimized molecular structure of S1 (donor) and S1-I_2 CT complex

Table 1 Bond lengths of S1 and S1-I_2 CT complex

S1		S1-I_2 CT complex		S1		S1-I_2 CT complex	
Atoms	Bond length $\times 10^{-3}$ (A^0)	Atoms	Bond length $\times 10^{-3}$ (A^0)	Atoms	Bond length $\times 10^{-3}$ (A^0)	Atoms	Bond length $\times 10^{-3}$ (A^0)
O(11)-Lp(49)	596	O(11)-Lp(51)	595	C(9)-C(8)	1347	C(9)-C(8)	1347
O(7)-Lp(47)	598	O(7)-Lp(49)	598	C(1)-O(7)	1231	C(1)-O(7)	1230
N(26)-H(45)	1049	N(26)-H(47)	1047	C(5)-N(6)	1269	C(5)-N(6)	1269
N(26)-H(44)	1049	N(26)-H(46)	1048	C(4)-C(5)	1347	C(4)-C(5)	1347
C(24)-H(43)	1114	C(24)-H(45)	1114	C(8)-N(26)	1264	I(27)-I(28)	2662
C(19)-H(31)	1116	C(19)-H(33)	1115	C(14)-F(15)	1324	C(14)-F(16)	1325
N(12)-H(29)	1011	N(12)-H(31)	101	C(10)-N(12)	1366	C(9)-C(10)	1364
C(3)-H(27)	1101	C(3)-H(29)	1100	N(12)-N(13)	1352	C(10)-O(11)	1208

i. (a) **Optimized geometrical structural parameters**

Bond length of I_2 (acceptor) $2660 \times 10^{-3}\,A^0$

Note: Important bond length, electron density, Mulliken and Huckel charge values are listed.

As far as bond lengths are concerned (Table 1), the bond length of I_2 which was **$2660 \times 10^{-3}\,A^0$** before the formation of complex with S1 increased to **$2662 \times 10^{-3}\,A^0$** upon complexation with S1. The results of this observation can be due to the n-electron movement from the HOMO of S1 to the LUMO of I_2. The said movement resulted in the increase of bond lengths of S1. This phenomenon is because of the electron density increase of I_2 in the S1-I_2 CT complex vis-a-vis I_2 alone. On the other hand electron density decrease on S1 (donor) of the complex resulted in the bond lengths decrease vis-à-vis donor alone (Table 1), this decrease in bond lengths is more pronounced on the nitrogen linked bonds, which disclose the significant involvement of these bonds in CT process, Moreover, slight decrease of bond lengths related to other atoms can be observed in Table 1, which unveils even their involvement in CT process but their contribution is less compared to nitrogen linked bonds.

The S1-I_2 CT complex is further corroborated from the decrease in its electron density values which were compiled in Table 2.

Table 2 Electron density values of S1 and S1-I_2 CT complex

S1		S1-I_2 CT complex	
Atom	Electron density $\times\,10^{-3}$	Atom	Electron density $\times\,10^{-3}$
1 C	3291	1 C	3267
3 C	3678	3 C	3677
4 C	3657	4 C	3653
5 C	3569	5 C	3564
8 C	3545	8 C	3490
13 N	2687	13 N	2661
18 C	3467	18 C	3419
21 C	3601	21 C	3593
22 C	3605	22 C	3596
23 C	3622	23 C	3612
24 C	3675	24 C	3663
26 N	2581	26 N	2573
31 H	899	31 H	829
32 H	903	32 H	833
33 H	918	33 H	896
34 H	911	34 H	904
35 H	918	35 H	906
43 H	926	43 H	916

i. (b) **Atomic charges**

Adding more corroboration to the CT complex formed between S1 and I_2, Mulliken atomic charges and Huckel atomic charges were computed and the consequent observations are shown in Table 3. The surge in charge on the iodine atoms (acceptor) from negligible values of **1.55 × 10⁻¹⁵** and **−4.44 × 10⁻¹⁶** to the considerable **61.7 × 10⁻⁵** and **−74.3 × 10⁻⁴** respectively, can be noticed from Table 3, affirming the CT process. As far as the donor S1 is concerned, the decrease in atomic charges of its atoms than the free donor molecule (S1) can be observed, confirming the CT from S1 to I_2. This change in Mulliken atomic charge is clearer on N(6) of pyridine ring whose value decreased from **−82.365 × 10⁻²** (S1) to **−83.483 × 10⁻²** (S1-I_2 CT complex). Furthermore, apparent decrease in atomic charge of **C(2)**, **C(3)** and **C(4)** unveil the involvement of π-electrons pertaining to benzene ring of S1 in CT process. Table 3 which also comprises Huckel charges are in concordance with Mulliken charges.

ii. **S2-I_2 CT complex**

The energy of donor S2 and the acceptor I_2 energies were recorded as 14.6149 and 0.0 kcal/mol, respectively. Whereas, energy of the formed complex recorded 9.5046 kcal/mol, making it clear that the obtained complex is reasonably stable.

The optimized structures of S2 (donor) and S2-I_2 CT complex with atom number are shown in Fig. 3. The bond length values are compiled in Table 4 for the S2 and S2-I_2 CT complex respectively.

ii. (a) **Optimized geometrical structural parameters**

As far as bond lengths are concerned (Table 4), the bond length of I_2 which was **2660 × 10⁻³ A⁰** before the formation of complex with S2 increased to **2661 × 10⁻³ A⁰** upon complexation with S2. The results of this observation can be due to the n-electron movement from the HOMO of S2 to the LUMO of I_2. The said movement resulted in the increase of bond lengths. This phenomenon is because of the electron density increase of I_2 in the S2-I_2 CT complex vis-à-vis I_2 alone. On the other hand electron density decrease on S2 (donor) of the complex resulted in the bond lengths decrease vis-à-vis donor alone (Table 4). However there are slight changes (Table 4) hence, these slight changes could not disclose full data about the CT process. Sequentially to ascertain the CT process further computations were made.

The S2-I_2 CT complex is further confirmed from the decrease in its electron density values which were compiled in Table 5.

ii. (b) **Atomic charges**

To ascertain the CT process to the formed S2-I_2 CT complex, Mulliken atomic charges and Huckel atomic charges were computed and the consequent observations are shown in Table 6. The surge in charge on the iodine atoms (acceptor) from

Table 3 Mulliken and Huckel charges of S1 and S1-I$_2$ CT complex

S1				S1-I$_2$ CT complex			
Atom	Atom type	Huckel charge	Mulliken charge	Atom	Atom type	Huckel charge	Mulliken charge
C(2)	C Alkene	-10.992×10^{-2}	-17.181×10^{-2}	C(2)	C Alkene	-11.033×10^{-2}	-24.966×10^{-2}
C(3)	C Alkene	-43×10^{-3}	-65.88×10^{-3}	C(3)	C Alkene	-42.83×10^{-3}	-67.03×10^{-3}
C(4)	C Alkene	-15.495×10^{-2}	-26.752×10^{-2}	C(4)	C Alkene	-15.51×10^{-2}	-27.165×10^{-2}
N(6)	N Pyridine	-28.512×10^{-2}	-82.365×10^{-2}	N(6)	N Pyridine	-28.415×10^{-2}	-83.483×10^{-2}
O(7)	O Furan	-27.53×10^{-3}	-81.02×10^{-2}	O(7)	O Furan	-28.07×10^{-3}	-82.281×10^{-2}
C(10)	C Carbonyl	25.6839×10^{-2}	97.4698×10^{-2}	C(10)	C Carbonyl	26.3379×10^{-2}	95.6106×10^{-2}
C(14)	C Alkane	83.7976×10^{-2}	12.5243×10^{-1}	C(14)	C Alkane	83.7987×10^{-2}	12.4916×10^{-1}
C(19)	C Alkane	-10.421×10^{-2}	-52.91×10^{-2}	C(19)	C Alkane	-10.655×10^{-2}	-54.096×10^{-2}
C(20)	C Alkane	-55.16×10^{-3}	-35.596×10^{-2}	C(20)	C Alkane	-53.07×10^{-3}	-37.035×10^{-2}
C(21)	C Alkane	-58.9×10^{-3}	-38.086×10^{-2}	C(21)	C Alkane	-59.68×10^{-3}	-40.218×10^{-2}
C(24)	C Alkane	-12.94×10^{-2}	-56.443×10^{-2}	C(24)	C Alkane	-12.941×10^{-2}	-56.938×10^{-2}
H(27)	H	15.488×10^{-3}	26.2231×10^{-2}	I(27)	I	42.26×10^{-4}	-74.3×10^{-4}
H(28)	H	18.628×10^{-3}	28.909×10^{-2}	I(28)	I	25.22×10^{-4}	61.7×10^{-5}
H(29)	H Amide	94.183×10^{-3}	41.425×10^{-2}	H(29)	H	15.503×10^{-3}	26.6339×10^{-2}
H(30)	H Amide	10.0307×10^{-2}	41.7994×10^{-2}	H(30)	H	18.633×10^{-3}	28.7201×10^{-2}
H(31)	H	40.91×10^{-3}	23.1812×10^{-2}	H(31)	H Amide	93.85×10^{-3}	40.2571×10^{-2}
H(32)	H	41.987×10^{-3}	23.7833×10^{-2}	H(32)	H Amide	98.037×10^{-5}	39.7688×10^{-2}
H(44)	H Amine	93.447×10^{-3}	33.9031×10^{-2}	H(44)	H	39.152×10^{-3}	19.8119×10^{-2}
H(45)	H Amine	95.233×10^{-3}	39.3904×10^{-2}	H(45)	H	39.147×10^{-3}	18.2713×10^{-2}
				H(46)	H Amine	93.461×10^{-3}	33.8688×10^{-2}

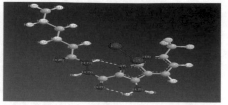

Fig. 3 Energy optimized molecular structure of S2 (donor) and S2-I$_2$ CT complex

Table 4 Bond lengths of S2 and S2-I$_2$ CT complex

S2		S2-I$_2$ CT complex		S2		S2-I$_2$ CT complex	
Atoms	Bond length × 10^{-3} (A^0)	Atoms	Bond length × 10^{-3} (A^0)	Atoms	Bond length × 10^{-3} (A^0)	Atoms	Bond length × 10^{-3} (A^0)
O(24)-Lp(48)	600	O(24)-Lp(50)	600	C(1)-C(2)	1339	C(1)-C(2)	1338
O(11)-Lp(46)	596	O(11)-Lp(48)	595	C(9)-C(8)	1346	C(9)-C(8)	1347
O(11)-Lp(45)	601	O(11)-Lp(47)	601	O(7)-C(9)	1235	O(7)-C(9)	1235
O(7)-Lp(44)	599	O(7)-Lp(46)	598	C(1)-O(7)	1230	C(1)-O(7)	1230
N(6)-Lp(43)	600	N(6)-Lp(45)	600	N(6)-C(1)	1265	N(6)-C(1)	1264
N(25)-H(42)	1049	N(25)-H(44)	1047	C(5)-N(6)	1269	C(5)-N(6)	1269
C(23)-H(40)	1114	C(23)-H(42)	1114	C(8)-N(25)	1264	I(26)-I(27)	2661
C(21)-H(34)	1116	C(21)-H(36)	1116	C(19)-C(20)	1537	C(20)-C(21)	1538
C(20)-H(33)	1116	C(20)-H(35)	1116	C(18)-C(19)	1519	C(19)-C(20)	1536
C(19)-H(30)	1116	C(19)-H(32)	1115	C(14)-F(16)	1325	C(14)-F(17)	1325
C(3)-H(26)	1101	C(3)-H(28)	1100	C(10)-O(11)	1208	C(10)-N(12)	1366

negligible values of 1.55×10^{-15} and -4.44×10^{-16} to the considerable 66.74×10^{-4} and -11.696×10^{-3} respectively, can be noticed from Table 6, affirming the CT process. As far as the donor S2 is concerned, the decrease in atomic charges of its atoms than the free donor molecule (S2) can be observed, confirming the CT from S2 to I$_2$. This change in Mulliken atomic charge is clearer on N(12) whose value decreased from -67.9865×10^{-2} (S2) to -70.3541×10^{-2} (S2-I$_2$ CT complex). Furthermore, decrease in effective atomic charge of C(2), C(3) and C(4) from -20.7286×10^{-2}, -56×10^{-3} and -26.9105×10^{-2} to -22.437×10^{-2}, -58.787×10^{-3} and -27.5036×10^{-2} respectively, disclose the remarkable participation of π-electrons related to benzene ring of S2 in CT process. Table 6 which also comprises Huckel charges are in concordance with Mulliken charges.

Table 5 Electron density values of S2 and S2-I_2 CT complex atoms

S2		S2-I_2 CT complex	
Atom	Electron density $\times 10^{-3}$	Atom	Electron density $\times 10^{-3}$
1 C	3310	1 C	3284
3 C	3683	3 C	3679
4 C	3658	4 C	3656
7 O	1692	7 O	1680
10 C	3639	10 C	3629
11 O	1904	11 O	1879
12 N	2680	12 N	2677
14 C	3496	14 C	3492
16 F	947	16 F	941
17 F	944	17 F	938
18 C	3442	18 C	3425
21 C	3599	21 C	3585
22 C	3619	22 C	3595
26 H	908	26 I	923
27 H	909	27 I	912
28 H	834	28 H	911
29 H	832	29 H	911
30 H	909	30 H	825
31 H	913	31 H	830
32 H	924	32 H	902
34 H	0.927	34 H	904
35 H	922	35 H	913
36 H	0.905	36 H	0.924
37 H	0.917	37 H	0.918
41 H	0.875	41 H	0.916

iii. S3-I_2 CT complex

The energy of the donor S3 and the acceptor I_2 recorded 14.5674 and 0.0 kcal/mol, respectively. Whereas, the energy of the formed complex recorded 9.1275 kcal/mol, making it clear that the obtained complex is substantially stable.

The optimized structures of S3 (donor) and S3-I_2 CT complex with atom number are shown in Fig. 4. The bond lengths are compiled in Table 7 for the S3 and S3-I_2 CT complex respectively.

Table 6 Atomic charges of S2 and S2-I$_2$ CT complex

S2				S2-I$_2$ CT complex			
Atom	Atom type	Huckel charge	Mulliken charge	Atom	Atom type	Huckel charge	Mulliken charge
C(2)	C Alkene	-10.9914×10^{-2}	-20.7286×10^{-2}	C(2)	C Alkene	-11.0689×10^{-2}	-22.437×10^{-2}
C(3)	C Alkene	-42.9842×10^{-3}	-56.322×10^{-3}	C(3)	C Alkene	-42.949×10^{-3}	-58.787×10^{-3}
C(4)	C Alkene	-15.4992×10^{-2}	-26.9105×10^{-2}	C(4)	C Alkene	-15.6065×10^{-2}	-27.5036×10^{-2}
N(6)	N Pyridine	-28.5142×10^{-2}	-82.239×10^{-2}	N(6)	N Pyridine	-28.6026×10^{-2}	-83.5348×10^{-2}
O(7)	O Furan	-27.6731×10^{-3}	-82.1037×10^{-2}	O(7)	O Furan	-028.6435×10^{-3}	-82.6862×10^{-2}
C(8)	C Alkene	79.7362×10^{-3}	48.1687×10^{-2}	C(8)	C Alkene	80.3452×10^{-3}	44.269×10^{-2}
N(12)	N Amide	36.1341×10^{-2}	-67.9865×10^{-2}	N(12)	N Amide	35.7358×10^{-2}	-70.3541×10^{-2}
N(13)	N Amide	41.7593×10^{-3}	-70.6009×10^{-2}	N(13)	N Amide	41.5264×10^{-2}	-69.4409×10^{-2}
F(15)	F	-26.6309×10^{-2}	-39.8091×10^{-2}	F(15)	F	-26.6247×10^{-2}	-39.8174×10^{-2}
F(16)	F	-26.2273×10^{-2}	-38.189×10^{-2}	F(16)	F	-26.2441×10^{-2}	-38.6471×10^{-2}
F(17)	F	-26.2433×10^{-2}	-38.3612×10^{-2}	F(17)	F	-26.2354×10^{-2}	-38.7346×10^{-2}
C(19)	C Alkane	-10.4177×10^{-2}	-51.7891×10^{-2}	C(19)	C Alkane	-10.581×10^{-2}	-53.4578×10^{-2}
C(20)	C Alkane	-55.1948×10^{-3}	-37.597×10^{-2}	C(20)	C Alkane	-54.7357×10^{-3}	-38.3722×10^{-2}
C(21)	C Alkane	-57.5006×10^{-3}	-37.3503×10^{-2}	C(21)	C Alkane	-57.4672×10^{-3}	-38.8597×10^{-2}
H(26)	H	15.4823×10^{-3}	27.0564×10^{-2}	I(26)	I	19.9943×10^{-4}	66.74×10^{-4}
H(27)	H	18.6256×10^{-3}	28.3237×10^{-2}	I(27)	I	15.8933×10^{-4}	-11.696×10^{-3}
H(28)	H Amide	94.1753×10^{-3}	40.2725×10^{-2}	H(28)	H	15.4894×10^{-3}	26.4519×10^{-2}
H(29)	H Amide	10.0248×10^{-2}	40.1799×10^{-2}	H(29)	H	18.6327×10^{-3}	27.8797×10^{-2}
H(30)	H	40.9936×10^{-3}	23.1542×10^{-2}	H(30)	H Amide	94.0951×10^{-3}	41.559×10^{-2}
H(31)	H	41.8772×10^{-3}	22.1709×10^{-2}	H(31)	H Amide	99.6127×10^{-3}	40.2576×10^{-2}

Fig. 4 Energy-optimized molecular structure of S3 (donor) and S3-I$_2$ CT complex

iii. (a) **Optimized geometrical structural parameters**

As far as bond lengths are concerned (Table 7), the bond length of I$_2$ which was **2660 × 10^{-3} A^0** before the formation of complex with S3 increased to **2661 × 10^{-3} A^0** upon complexation with S3. The results of this observation can be due to the n-electron movement from the HOMO of S3 to the LUMO of I$_2$. The said movement resulted in the increase of bond lengths. This phenomenon is because of the electron density increase of I$_2$ in the S3-I$_2$ CT complex vis-à-vis I$_2$ alone. On the other hand electron density decrease on S3 (donor) of the complex resulted in the bond lengths decrease vis-à-vis donor alone (Table 7). However there are slight changes (Table 7) hence, these slight changes could not reveal full data about the CT process. Sequentially to ascertain the CT process further computations were made.

iii. (b) **Atomic charges**

The S3-I$_2$ CT complex is further corroborated from the decrease in its electron density values which were compiled in Table 8.

To ascertain the CT process to the formed S3-I$_2$ CT complex, Mulliken atomic charges and Huckel atomic charges were computed and the consequent

Table 7 Bond lengths of S3 and S3-I$_2$ CT complex

S3		S3-I$_2$ CT complex		S3		S3-I$_2$ CT complex	
Atoms	Bond length × 10^{-3} (A^0)	Atoms	Bond length × 10^{-3} (A^0)	Atoms	Bond length × 10^{-3} (A^0)	Atoms	Bond length × 10^{-3} (A^0)
C(28)-H(48)	1114	C(28)-H(50)	1113	C(13)-C(15)	1517	N(19)-C(20)	1372
C(28)-H(47)	1114	C(28)-H(49)	1114	C(15)-F(18)	1325	C(13)-C(15)	1518
C(27)-H(46)	1120	C(27)-H(48)	1119	C(15)-F(17)	1325	C(15)-F(18)	1325
C(26)-H(44)	1114	C(26)-H(46)	1114	O(8)-C(5)	1431	C(15)-F(16)	1324
C(21)-H(38)	1115	C(21)-H(40)	1115	C(11)-C(12)	1342	C(12)-C(13)	1346
C(21)-H(37)	1115	C(21)-H(39)	1115	O(8)-C(10)	1232	C(11)-C(12)	1341
C(11)-H(34)	1101	C(11)-H(36)	1101	C(6)-N(7)	1369	C(2)-N(7)	1350
C(5)-H(33)	1114	C(5)-H(35)	1117	C(5)-C(6)	1518	C(6)-N(7)	1364
C(4)-H(32)	1115	C(4)-H(34)	1113	C(4)-C(5)	1537	C(5)-C(6)	1510
C(1)-H(31)	1116	C(1)-H(33)	1115	N(3)-C(4)	1469	C(4)-C(5)	1533
C(1)-H(30)	1114	C(1)-H(32)	1114	C(2)-N(3)	1262	N(3)-C(4)	1472
C(27)-C(29)	1538	I(30)-I(31)	2661			C(2)-N(3)	1264

Table 8 Electron density values of S3 and S3-I_2 CT complex

S3		S3-I_2 CT complex	
Atom	Electron density $\times 10^{-3}$	Atom	Electron density $\times 10^{-3}$
3 N	2742	3 N	2738
4 C	3530	4 C	3511
10 C	3225	10 C	3192
11 C	3705	11 C	3699
12 C	3667	12 C	3636
15 C	3493	15 C	3483
17 F	945	17 F	937
18 F	939	18 F	938
19 N	2631	19 N	2601
23 O	1987	23 O	1973
34 H	906	34 H	891
35 H	916	35 H	889
38 H	912	38 H	837
40 H	937	40 H	921
41 H	919	41 H	914
44 H	930	44 H	917
45 H	924	45 H	923

observations are shown in Table 9. The surge in charge on the iodine atoms (acceptor) from negligible values of 1.55×10^{-15} and -4.44×10^{-16} to the considerable 32.832×10^{-3} and -40.516×10^{-3} respectively, can be noticed from Table 9, affirming the CT process. As far as the donor S3 is concerned, the decrease in atomic charges of its atoms than the free donor molecule (S3) can be observed, confirming the CT from S3 to I_2. This change in Mulliken atomic charge is clearer on N(7) whose value decreased from -83.7064×10^{-2} (S3) to -86.4762×10^{-2} (S3-I_2 CT complex). Moreover, atomic charge decrease of C(11), C(12) and C(13) from -11.7512×10^{-2}, -27.5694×10^{-2} and 18.0012×10^{-2} to -11.9057×10^{-2}, -28.9988×10^{-2} and 15.7474×10^{-2}, clearly indicate the involvement of π-electrons pertaining to benzene ring of S3 in CT process. Table 9 which also comprises Huckel charges are in concordance with Mulliken charges.

2.2 Electronic Properties

Figures 5, 6, and 7 represent the HOMO and LUMO of S1, S2 and S3 CT complexes respectively. The above-said figures show that the HOMO in all the three instances is spread evenly on the iodine suggesting the CT from S1, S2 and S3 (donors) to I_2 (acceptor).

Table 9 Atomic charges of S3 and S3-I$_2$ CT complex

S3				S3-I$_2$ CT complex			
Atom	Atom type	Huckel charge	Mulliken charge	Atom	Atom type	Huckel charge	Mulliken charge
C(1)	C Alkane	-94.3127×10^{-3}	-37.8805×10^{-2}	C(1)	C Alkane	-98.581×10^{-3}	-39.5085×10^{-2}
C(2)	C Alkene	29.1294×10^{-2}	82.4982×10^{-2}	C(2)	C Alkene	26.5689×10^{-2}	82.7587×10^{-2}
C(4)	C Alkane	50.3223×10^{-3}	-12.3439×10^{-2}	C(4)	C Alkane	50.4252×10^{-3}	-14.7058×10^{-2}
N(7)	N Amide	44.8259×10^{-2}	-83.7064×10^{-2}	N(7)	N Amide	50.1173×10^{-2}	-86.4762×10^{-2}
O(8)	O Furan	-16.7631×10^{-2}	-72.9312×10^{-2}	O(8)	O Furan	-16.9177×10^{-2}	-72.1817×10^{-2}
C(11)	C Alkene	-48.3235×10^{-3}	-11.7512×10^{-2}	C(11)	C Alkene	-49.4995×10^{-3}	-11.9057×10^{-2}
C(12)	C Alkene	-16.9944×10^{-2}	-27.5694×10^{-2}	C(12)	C Alkene	-16.9401×10^{-2}	-28.9988×10^{-2}
C(13)	C Alkene	10.6687×10^{-2}	18.0012×10^{-2}	C(13)	C Alkene	10.4899×10^{-2}	15.7474×10^{-2}
N(14)	N Pyridine	-30.5473×10^{-3}	-83.7859×10^{-2}	N(14)	N Pyridine	-30.8419×10^{-2}	-82.0378×10^{-2}
C(21)	C Alkane	-10.7823×10^{-2}	-47.038×10^{-2}	C(21)	C Alkane	-10.9283×10^{-2}	-47.4067×10^{-2}
C(25)	C Alkane	-13.7504×10^{-2}	-52.9512×10^{-2}	C(25)	C Alkane	-13.5602×10^{-2}	-54.1336×10^{-2}
C(26)	C Alkane	-13.4775×10^{-2}	-51.7803×10^{-2}	C(26)	C Alkane	-13.4619×10^{-2}	-52.3206×10^{-2}
C(28)	C Alkane	-13.583×10^{-2}	-51.9484×10^{-2}	C(28)	C Alkane	-13.6421×10^{-2}	-53.4968×10^{-2}
C(29)	C Alkane	-13.4587×10^{-2}	-53.5743×10^{-2}	C(29)	C Alkane	-13.4257×10^{-2}	-53.8546×10^{-2}
H(30)	H	28.501×10^{-3}	20.7402×10^{-2}	I(30)	I	15.2832×10^{-4}	32.832×10^{-3}
H(31)	H	36.8178×10^{-3}	26.1924×10^{-2}	I(31)	I	59.0237×10^{-4}	-40.516×10^{-3}
H(36)	H Amide	91.4038×10^{-3}	39.2638×10^{-2}	H(36)	H	16.9658×10^{-3}	28.1029×10^{-2}
H(37)	H	32.5506×10^{-3}	24.9133×10^{-2}	H(37)	H	18.9951×10^{-3}	27.3974×10^{-2}
H(38)	H	40.5949×10^{-2}	22.3926×10^{-2}	H(38)	H Amide	94.8165×10^{-3}	38.9312×10^{-2}
H(51)	H	37.6027×10^{-3}	18.3774×10^{-2}	H(51)	H	38.4492×10^{-3}	17.31×10^{-2}
H(52)	H	38.1379×10^{-3}	18.2751×10^{-2}	H(52)	H	37.888×10^{-3}	19.3311×10^{-2}

Fig. 5 LUMO, HOMO; LUMO and HOMO molecular orbital surfaces of I$_2$, S1; S1-I$_2$ CT complex respectively

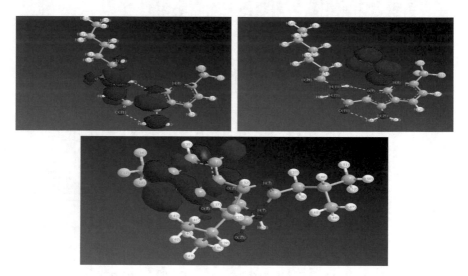

Fig. 6 HOMO; LUMO and HOMO molecular orbital surfaces of S2; S2-I$_2$ CT complex respectively

The HOMO and LUMO figures in case of CT complexes which were studied, clearly suggests that the potential oscillate notably on nitrogen, considerably on benzene ring, by noticing these, one can easily conclude that how significantly the nitrogen atoms of S1, S2 and S3 (donors) transfer the charge associated with them to I$_2$ during CT process.

Fig. 7 HOMO; LUMO and HOMO molecular orbital surfaces of S3; S3-I$_2$ CT complex respectively

Table 10 Calculated optimized energies of the formed FP2C-I2 CT complexes

CT complex	Calculated optimized energy (eV)	Electronic transition
S1-I$_2$ CT complex	7.03	HOMO-2 to LUMO
S2-I$_2$ CT complex	6.45	HOMO-2 to LUMO
S3-I$_2$ CT complex	7.574	HOMO-2 to LUMO

The calculated optimized energies for all the CT complexes are tabulated in Table 10.

3 Conclusion

Ab initio molecular quantum chemistry is now a key tool in designing bioactive molecules with greater efficacy. This article shows the reliability of computational tool like GAMESS which is based on QM/MM methods. The GAMESS program has been successfully used to study the CT from FP2C to I$_2$. Based upon the optimized parameters values of S1-I$_2$, S2-I$_2$ and S3-I$_2$ complexes it can be inferred that the electron donating ability of FP2C is in the order of S1 > S3 > S2 with I$_2$, which disclose that among S1, S2 and S3, S1 interacts efficiently with acidic receptors followed by S3. These types of studies will help us to understand more about the mechanism of the FP2C action in real pharmacokinetic study.

References

1. Delagado, J.N., Remers, W.A. (eds.): Wilson and Gisfold's Textbook of "Organic Medicinal and Pharmaceutical Chemistry", vol. 413, 9th edn. Lippincott-Raven Publishers, Philadelphia
2. Reddy, A.C.S., Narsaiah, B., Venkataratnam, R.V.: Fluoro organics: synthesis of novel fluorinated pyrido[3', 2':4, 5] furo[3,2-d] (1, 3) oxazines and reactions. J. Fluorine Chem. **74**, 1–7 (1995). https://doi.org/10.1016/0022-1139(95)03239-a
3. Murata, T., Morata, Y., Fukui, K., Sato, K., Shiomi, D., Takui, T., Maesato, M., Yamochi, H, Saito, G., Nakasujl, K.: A purely organic molecular metal based on a hydrogen-bonded charge-transfer complex: crystal structure and electronic properties of TTF-imidazole–p-chloranil. Angew. Chem. Int. Ed. Engl. **43**, 6343–6346 (2004). https://doi.org/10.1002/anie.200460801
4. Shirish Kumar, K., Parthasarathy, T.: Synthesis, spectroscopic and computational studies of CT complexes of amino acids with iodine as σ-acceptor. J. Sol. Chem. **46**, 1364–1403 (2017). https://doi.org/10.1007/s10953-017-0643-6
5. Schmidt, M.W., Baldridge, K.K., Boatz, J.A., Elbert, S.T., Gordon, M.S., Jensen, J.H., Koseki, S., Matsunaga, N., Nguyen, K.A., Su, S., Windus, T.L., Dupuis, M., Montgomery, J.A.: General atomic and molecular electronic structure system. J. Comput. Chem. **14**, 1347–1363 (1993). https://doi.org/10.1002/jcc.540141112
6. Khan, Z.F.: ChemBio3D Ultra12.0 with GAMESS interface. The Islamic 830 University, issue: 20.2

A Cognitive Approach for Design of Smart Toilet in Healthcare Units

Mohan Debarchan Mohanty and Mihir Narayan Mohanty

Abstract In this ultramodern era, there is an increase in demand of smart systems. It is quite observable that the bedridden patients face problems in defecating and urinating. Taking the problems into account, we have taken an approach to design the user-friendly toilet to support the hospitals and the patients. Based on the cognitive science, fuzzy-based smart toilet is designed so that maximum ICU patients can be benefited. The toilet is fixed with the bed and can be used by the patients with a simple switch. The FL-based PID controller is designed to slide the pan cover as well as water supply to the toilet. It can clean the body part of the patient along with the toilet. Rule based fuzzy is applied to design the system and defuzzification is done using COG method. According to a performed survey, the proposed idea is a type of big data analysis and can be widely used for the betterment of hospitals and old-age homes.

Keywords Cognitive technology · Smart toilet · PI · PD · PID
Fuzzy system · Intelligent control

1 Introduction

Cognitive Informatics (CI) investigates human information processing mechanisms and processes. The engineering aspects can be applied in terms of computation for development of the smart system. It is a paradigm of intelligent computing methodologies and systems based on cognitive informatics that implements computational intelligence by autonomous inferences. Enough developments in

M. D. Mohanty
Department of I&EE, CET, BPUT, Bhubaneswar, Odisha, India
e-mail: mohan.debarchan97@gmail.com

M. N. Mohanty (✉)
Department of Electronics and Communication Engineering,
Siksha 'O' Anusandhan University, Bhubaneswar, India
e-mail: mihir.n.mohanty@gmail.com

© Springer Nature Singapore Pte Ltd. 2019 771
P. K. Mallick et al. (eds.), *Cognitive Informatics and Soft Computing*,
Advances in Intelligent Systems and Computing 768,
https://doi.org/10.1007/978-981-13-0617-4_72

healthcare information technologies in computing over last two decades has used in clinical practice. Some of the main topics of artificial intelligence are fuzzy logic and possibility theory based on representation of knowledge and approximation of reasoning. Many more attempts are being made to develop machines that could act not only autonomously, but also in an increasingly intelligent and cognitive manner [1–7].

In the modern society, there is always a need of a proper care of the patients which can be done by improvising the healthcare units. Taking the need into account, many e-healthcare units have already been developed. An effective healthcare system consists of proper patient entry, patient consultation, test and diagnosis. For the system to be smart enough, there is a requirement of specialized physicians in specific fields along with their proper testing laboratories. Some of the works have been proposed earlier for the above mentioned components [7–10].

Similarly, the patients admitted in the hospitals need a healthy environment condition and that also plays an important role in the fast treatment. One such aspect in this work is the distribution of the water in the storage tanks of the healthcare units into the toilets after reaching a certain specified level in the tank. The fuzzy-based tuning of the PID Controller to control the inflow and outflow rate makes the system intelligent. In this project we considered a toilet fitted in the bed itself with automatic water supply for the flush and cleaning purpose. Considering the combination of both the concepts, it can help in improvising the modern healthcare systems.

To control a steam engine and boiler combination, the mamdani fuzzy inference system has been explored using linguistic rules [9, 10]. In this they developed a system with if-then rules corresponding to the human language. The authors in [5] have developed a system using PD controller for the distribution of water.

Several works have been done in the design of PID controller with the help of a fuzzy controller [11–14]. The authors in [15] have designed a sliding toilet for the patients that is helpful for the bathing purposes of the patients. The authors in [9] have focussed on the auto-tuning of the PID controller where they compared the tuning methods- trial and error method, ziegler nichols and the fuzzy-based tuning. They found the fuzzy based tuning to be more effective than the other methods.

Works have been explored in the level control of coupled tanks using a PID Controller [16, 17]. The authors in [14] have done the level control using P, PI, PD, and PID controller. In [16], authors have extracted the requirements of a controller for the level measurement.

Since no mathematical model is required to design a FL controller, hence it maintains its robustness. Ease of application of FL controller makes it popular in the industry. Nevertheless, knowledge database or expert experience is desired in deciding the rules or the membership functions (MF).

2 Proposed Method

Mostly the design of smart toilet in huge scale can be used in the healthcare units due to the intelligent control of opening and closing of the toilet as well as the control of water flow. The patients in the intensive care units can avail such facilities. The approach is partially cognitive and Fuzzy logic based. Conventional control algorithms relying on linear system need to be linearized before applied to systems, although the performance is not guaranteed, nonlinear controllers can accommodate nonlinearities; however the lack of a general structure recreates in their designing [14]. Thus linear or nonlinear control algorithm solutions are mostly developed using precise mathematical system models. Difficulties in describing these systems using traditional mathematical relations happened to provide unsatisfactory design models or solutions [10]. This is a motivating factor to be used for system design which can employ approximate reasoning to resemble the human decision process [18].

Fuzzy has been proposed by Zadeh in 1965 and provides a specific Artificial Intelligence (AI) level to the traditional proportional, integral and derivative (PD) controllers. Its use in the control systems provides better dynamic response, rejects disturbance, allows low parameter sensitive variation and removes external influences. The algorithm is robust, simple, and efficient. Experiments of the proposed control system have shown remarkable tracking performance and demonstrate the usefulness and validity of the hybrid fuzzy controller convincingly, with high performance under parameter variations and load uncertainties. The performance has been satisfactory for most of the reference tracks [15, 18, 19].

The block diagram of the proposed system is shown in Fig. 1. In this figure the water level variation is also shown. The linguistic rules are used to define the relation between the input and the output. Tuning is done in such a way that PI control of the controller will control the level of water in the tanks and the PD control of the controller will control the distribution of the water supply.

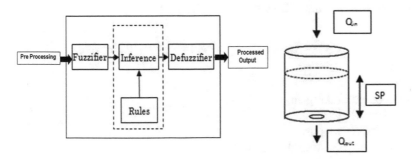

Fig. 1 Block diagram of the proposed system and water level marker

The distribution of flush and cleaning water is implemented on microcontroller based smart toilet; a toilet designed for the bedridden people which basically works on a switching principle with some delay.

2.1 PID Controller Tuning with Fuzzy Logic

The proposed method has the four stages of processing. First stage is the Fuzzification. Fuzzification implies the process where the preprocessing inputs are transformed into the fuzzy domain. The second stage is the knowledge base. The knowledge base of Fuzzy Logic controller is based upon data base and rule base. Data Base provides major data for the functioning of the Fuzzification and defuzzification modules and their respective rule base. Rule Base has the function to represent the control policy in a structured manner. The third stage is the fuzzy inference system. Fuzzy inference system has a simple I/O relationship. The input is processed from the external world by the system through a series of events called as fuzzy inference algorithm. Mamdani fuzzy is one of the widely used fuzzy inference system. And the final stage is the defuzzification. Defuzzification is the process in which the fuzzy sets assigned to a control output variable are transformed into a processed value. There are several methods for defuzzification but our work is done by the (COG) method. This method provides a processed value based on the center of gravity of the fuzzy set. The total area of the membership function distribution used to represent the combined control action which is divided into a number of sub-areas. The area and the center of gravity of each sub-area are calculated and then the summation of all these sub-areas is taken to find the defuzzified value for a discrete fuzzy set.

Let A be a fuzzy set with membership function $\mu(x)$ defined over $x \in X$, where X is a universe of discourse. The defuzzified value is let say $d(A)$ of a fuzzy set and this can be defined in two forms—continuous and discrete. The nine rules used for the design of controller is given in (Table 1).

Mathematically, the PID controller is given by:

$$o(t) = i(t) + \mathrm{Kp} * e(t) + \mathrm{Kd} * \frac{de(t)}{dt} + \mathrm{Ki} \int e(x) \, . \, dx \qquad (1)$$

where,

Kp—proportional gain
Ki—integral gain
Kd—derivative gain
e(t)—error present in the controller
i(t)—initial response
t—instantaneous time
x—variable of integration

Table 1 9 basic rules of FL

Δ|E||E|	POS	ZER	NEG
POS	POS	ZER	NEG
ZER	ZER	ZER	NEG
NEG	NEG	ZER	POS

POS Positive, *NEG* Negative, *ZER* Zero

For discrete membership function,

$$d(A) = \frac{\sum_{i=1}^{n} (xi) \cdot (\mu(xi))}{\sum_{i=1}^{n} (\mu(xi))} \tag{2}$$

For continuous membership function,

$$d(A) = \frac{\int x \cdot \mu(x) dx}{\int \mu(x) dx} \tag{3}$$

The important rules following the fuzzy techniques are

1. If the controller is ON and the motor is ON, then there will be both the opening of the bed gap as well as flow of flush and cleaning water in the smart toilet.
2. If the controller is OFF and the motor is ON, then the cover of the bed gap will only open.

If the controller is OFF and also the motor is OFF, then the bed gap closes and the flow of water for the flush and cleaning purpose stops.

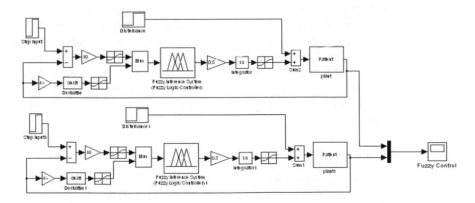

Fig. 2 Schematic diagram of fuzzy control circuit

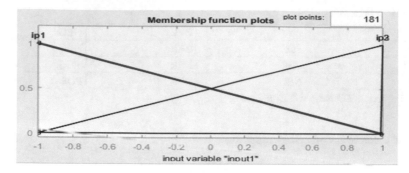

Fig. 3 Membership function for the controller input

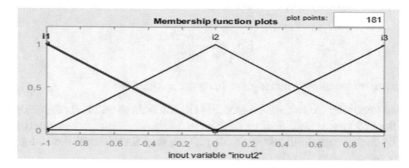

Fig. 4 Membership function for the motor input

2.2 PI Control for Tank Level

Figure 2 shows the circuit model for the controller. The membership functions used in this work are shown in Fig. 3 through Fig. 4. The PI control output consists of only the sum of the product of errors of the proportional gain and the integral gain only. The use of integral control is used for removing the steady-state error which results in an improved transient response, but also increases the system settling time. The system consists of a set point (SP) and a process variable (PV). Set Point is defined as the desired level of the tank and the Process Variable is defined as the actual level of the tank. The error is given by

$$e(t) = \text{SP} - \text{SV} \tag{4}$$

Now for PI control, putting Kd = 0 in Eq. (1), we get,

$$o(t) = \text{PV} + \text{Kp} * e(t) + \text{Ki} \int e(t) \mathrm{d}t \tag{5}$$

Consider a tank with cross-sectional area A. Let the input flow rate be Q_{in} cm³/s. The output flow Q_{out} cm³/s in a water tank occurs through a hole in the bottom of the tank of area α [17]. The flow rate of water through the hole follows the Bernoulli equation given by

$$Q_{in} - \alpha\sqrt{2g(SP)} \tag{6}$$

From conservation of mass property we get,

$$A\frac{d(SP)}{dt} Q_{in} - Q_{out} = Q_{in} - \alpha\sqrt{2g(SP)} \tag{7}$$

3 Results

The results of this work are obtained for this intelligent controller. The controller is implemented for the main water tank and also in the toilet attached to the bed. Figure 5 shows the output membership function of the fuzzy logic used. The controller used in the toilet is through a motor that works for shiding of the cover of the toilet to open/close. Once the lid of the toilet opened the water supply is also controlled. Figure 6 shows the effect of the motor for opening.

The error rate in 3-D view is shown in Fig. 7. The implemented controller circuit is shown on Fig. 8 as the snapshot. Also the snapshot of controller response for this toilet is shown in Fig. 9.

The proposed method is experimentally tested and verified. The implemented prototype may be used in the hospitals, though further development is required.

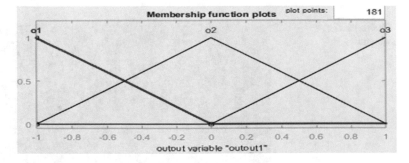

Fig. 5 Output membership values of the FL

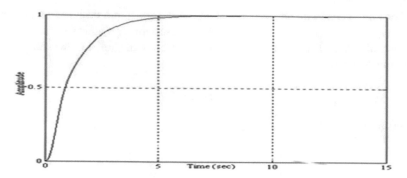

Fig. 6 Effect of motor in opening the bed gap

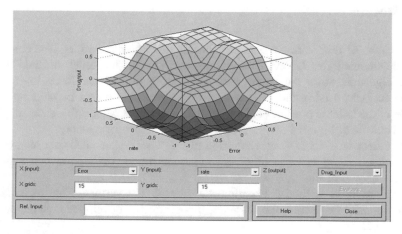

Fig. 7 3-D view of error and rate

Fig. 8 Controller circuit
implementation

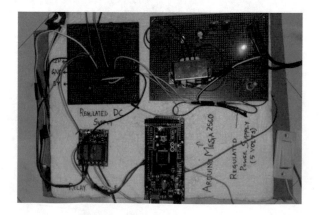

Fig. 9 Tuned PID controller
response

4   Conclusion

In this article, author has attached to develop this toilet system for ICU patients in the hospitals. The same can be useful for bedridden older people in the home. The smart technology based on cognitive informatics has been applied. The fuzzy-based controller is designed and implemented using the sensor technology and kept for the future work.

References

1. Witold, K.: Towards cognitive machines: multiscale measures and analysis. Intern. J. Cogn. Inform. Nat. Intell. **1**(1), 28–38 (2007)
2. Wang, Y. (ed.): Novel Approaches in Cognitive Informatics and Natural Intelligence. Ch. 13, pp. 188–199. IGI Global, Hershey (2009). ISBN 978-1-60566-170-4
3. Haykin, S., Kosko, B. (eds.): Intelligent Signal Processing. IEEE Press, Piscataway, NJ (2001)
4. Kinsner, W.: Challenges in the design of adaptive, intelligent and cognitive systems. Intern. J. Softw. Sci. Comput. Intell. **1**(3), 16–35 (2009)
5. Kosko, B.: Neural Networks and Fuzzy Systems. Prentice-Hall, New Jersey (1992)
6. Mohanty, M.D., Mohanty, M.N.: Intelligent PD Controller for Water Supply in Healthcare Units. ICITKM (2017)
7. Sarangi, L., Mohanty, M.N., Patnaik. S.: Detection of abnormal cardiac condition using fuzzy inference system. Int. J. Autom. Control **11**, 372–383 (2017)
8. Sarnagi, L., Mohanty, M.N., Patnaik, S.: Design of ANFIS based e-health care system for cardio vascular disease detection. In: Recent Developments in intelligent Systems and Interactive Applications, pp. 445–453. Springer International Publishing, New York (2016)
9. Sarnagi, L., Mohanty, M.N., Patnaik, S.: An intelligent decision support system for cardiac disease detection. IJCTA **8**(5) (2015)

10. Yen, J., Langari, R.: Fuzzy Logic. Intelligence, Control, Control, and Information. Prentice-Hall, USA (1999)
11. Yukawa, T., Nakata, N., Obinata, G., Makino, T.: Assistance system for bedridden patients to reduce the burden of nursing care (first report—development of a multifunctional electric wheelchair, portable bath, lift, and mobile robot with portable toilet). In: IEEE/SICE International Symposium on System Integration (SII), pp. 132–139 (2010)
12. Venugopal, P., Ganguly, A., Singh, P.: Design of tuning methods of PID controller using fuzzy logic. Int. J. Emerg. Trends Eng. Dev. 5(3), 239–248 (2013)
13. Verbruggen, H.B., Bruiji, P.M.: Fuzzy control and conventional control. What is (and can be) the real contribution of fuzzy systems. Fuzzy Sets Syst. 90, 151–160 (1997)
14. Kowalska, T.O., Szabat, K., Jaszeznk, K.: The influence of parameters and structure of PI-type fuzzy-logic controller on DC drive system dyanamic. Fuzzy Sets Syst. 131, 251 264 (2002)
15. Liu, B.D.: Design and implementation of the tree-based fuzzy logic controller. IEEE Trans. Syst. Man Cybern. B Cybern. 27(3), 475–487 (1997)
16. Ahmed, M.S., Bhatti, U.L., Al-Sunni, F.M., El-Shafci, M.: Design of a fuzzy servo-controller. Fuzzy Sets Syst. 124, 231–247 (2001)
17. Zilonchian, A., Juliano, M., Healy, T.: Design of fuzzy logic controller for a jet engine fuel system. Control Eng. Pract. 8(8), 873–883 (2000)
18. Zhiquiang, G.: A stable self-tuning fuzzy logic control system for industrial temperature regulation. IEEE 1886 Trans. Ind. Appl. 38(2), 14–424 (2002)
19. Zadeh, L.A.: Fuzzy sets. Inf. Control 8, 339–353 (1965)

Autonomic Nervous System for Sympathetic and Parasympathetic for Cardiac Event Coherence

Noel G. Tavares, R. S. Gad, A. M. Hernandez, Uday Kakodkar and G. M. Naik

Abstract Human body physiology is regulated through the central neural control (CNS) which takes signal from the respiratory system and ambiance which signifies atmospheric pressure, temperature and various gases in the environment. The central nervous system then controls the metabolic control of various organs through the afferent nerves and the efferent nerves reflecting the various reflex of the organs back to the CNS, which regulates the cardiovascular system (CVS) for the stroke volume (SV) of the blood and heart rate (HR). The SV and HR collectively synthesize the cardiac output of the heart balancing the body for the coherence or non-coherence states. We have defined and simulated here in this paper the Neural Mass Model (NMM), which is one of the component which feeds the CNS and controls the cardiovascular system for the human blood pressure (ABP) and heart rate. We have defined and simulated arterial blood pressure model, i.e., Windkessel model; describing the arterial blood pressure for the particular input volume of the blood and ECG model for the computing heart rate and heart rate variability (HRV). The integration of CNS, Windkessel and EEG model has thrown light on some aspects of sympathetic and parasympathetics of ANS for further improvisation and experimentations.

N. G. Tavares · R. S. Gad (✉) · G. M. Naik
Department of Electronics, Goa University, Taleigao Plateau, Goa, India
e-mail: rsgad@unigoa.ac.in

N. G. Tavares
e-mail: elect.noel@unigoa.ac.in

G. M. Naik
e-mail: gmnaik@unigoa.ac.in

A. M. Hernandez
Bioinstrumentation and Clinical Engineering Research Group (GIBIC),
Universidad de Antioquia in Medellín, Medellín, Colombia
e-mail: mauricio.hernandez@udea.edu.co

U. Kakodkar
Pulmonary Medicine, Goa Medical College and Hospital,
Bambolim, Goa, India
e-mail: udaykakodkar@hotmail.com

© Springer Nature Singapore Pte Ltd. 2019   781
P. K. Mallick et al. (eds.), *Cognitive Informatics and Soft Computing*,
Advances in Intelligent Systems and Computing 768,
https://doi.org/10.1007/978-981-13-0617-4_73

Keywords ANS · Neural mass · Human blood pressure · EEG · HRV

1 Introduction

Computer modeling of a human physiological system centered around the autonomic Nervous System (ANS) is discussed. The ANS usually consists of two major
components Central Nervous System (CNS) and Cardiovascular system which
collectively generate body blood pressure indirectly dictated by the HRV of the EEG
signals generated at the SA node of the heart. Here, we present CNS, Cardiovascular
and EEG signal mathematical models and further present the integration of them to
demonstrate the sympathetic and parasympathetic signal of the ganglion which
regulated the body physiology through the control of various organs.

2 Neural Mass Model

Neuronal models have gathered a lot of interest over years to study cognitive
processes of the brain. The bases of these models are usually the differential
equations associated with state of the variable of the system describing it, which
explain the dynamics of the brain. Physically it corresponds to membrane potentials
of each neural of interest. In a macroscopic sense the state variables are the variables which describe the model state while at the abstract level it is the membrane
potential. The differential equations describe the flow of state in time domain
between the interplay of the system equations. Such time flows form orbits, thus
producing time series for all of the states. These orbits capture the features of the
system like steady state, periodic, quasi-periodic, and chaotic. The main variable of
interest of such ensembles describing the NMM is the mean and variance state of
membrane potential for the set of ensemble of neuron's synapses in space. The
mean membrane potential increases and decreases in response towards the collective synaptic inputs and this increases or decreases depends on the variation of
afferent inputs. Variance can be ignored or minimized if the ensemble activity is
coherent; which therefore reduces the number of dimensions so that the interesting
local population can be modeled for excitatory and inhibitory neurons using small
numbers of equations (as seen in Fig. 1) representing the regions over said neural
mass. This approach is usually used to described the neural mass models (NMM).

2.1 NMM Using Differential Equations

Modeling brain neural activities by mathematical models usually give simplification
of the real brain activity with a useful insight into the processes that generate neural

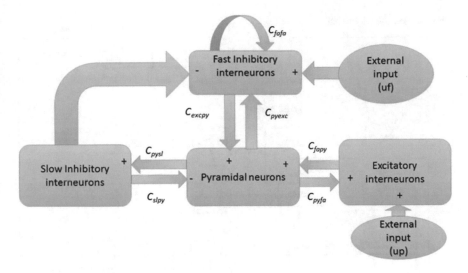

Fig. 1 Schematic of neural mass model

activity. We can also change the system's parameters to obtain different state variables indicating specific behaviors which might not be possible in real-world experiments. Microscopic neural mass models are based on the coupling of several single neurons and hence computational expensive since there involves large no of parameters. Macroscopic neural mass models are less detailed and involve an average pattern of a large group of neurons similar to EEG (Electroencephalogram) signals.

Here in Fig. 2 illustrates the multiple interacting local populations, such as excitatory and inhibitory neurons in different layers of the cortex, to be modeled by a small number of equations. Here we present a macroscopic Wendling NMM [1, 2] which is developed from the original works of Lopes DaSilva [3] and Freeman [4]. The extended Wendling model [5] can be described by a set of ten differential equations given below.

Pyramidal neuronal population

$$\frac{dy_{py}(t)}{dt} = x_{py}(t) \tag{1}$$

$$\frac{dx_{py}(t)}{dt} = G_e w_e \text{Sig}(z_{py}(t)) - 2w_{exc}y_{py}(t) - w_{exc}^2 x_{py}(t) \tag{2}$$

$$z_{py}(t) = \frac{2e_0}{1 + e^{-rv_{py}}} - e_0 \tag{3}$$

$$v_{py}(t) = C_{pyexc}y_{exc}(t) - C_{pysl}y_{sl}(t) - C_{pyfa}y_{fa}(t) \tag{4}$$

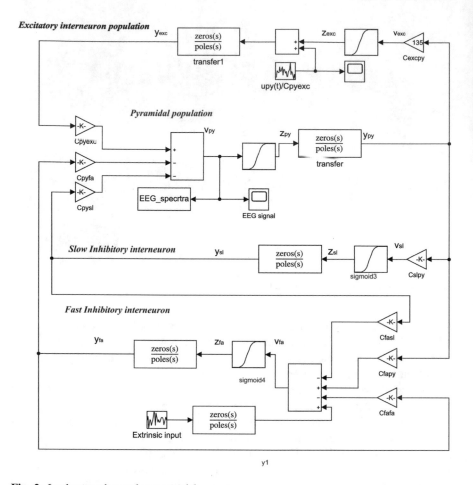

Fig. 2 Implemented neural mass model

Excitatory neuronal population

$$\frac{dy_{exc}(t)}{dt} = x_{exc}(t) \tag{5}$$

$$\frac{dx_{exc}(t)}{dt} = G_{exc}w_{exc}\left(z_{exc}(t) + \frac{u_p(t)}{C_{pyexc}}\right) - 2w_{exc}x_{exc}(t) - w_{exc}^2 y_{exc}(t) \tag{6}$$

$$z_{exc}(t) = \frac{2e_0}{1 + e^{-rvexc}} - e_0 \tag{7}$$

$$v_{exc}(t) = C_{excpy}y_{py}(t) \tag{8}$$

Slow inhibitory neuronal population

$$\frac{dy_{sl}}{dt}(t) = x_{sl}(t) \tag{9}$$

$$\frac{dx_{sl}(t)}{dt} = G_{sl}w_{sl}z_{sl}(t) - 2w_{sl}x_{sl}(t) - w_{sl}^2 y_{sl}(t) \tag{10}$$

$$z_{sl}(t) = \frac{2e_0}{1 + e^{-rvsl}} - e_0 \tag{11}$$

$$v_{sl}(t) = C_{slpy}y_{py}(t) \tag{12}$$

Fast inhibitory neuronal population

$$\frac{dy_{fa}(t)}{dt} = x_{fa} \tag{13}$$

$$\frac{dx_{fa}(t)}{dt} = G_{fa}w_{fa}z_{fa}(t) - 2w_{fa}x_{fa}(t) - w_{fa}^2 y_{fa}(t) \tag{14}$$

$$\frac{dy_{lo}(t)}{dt} = x_{lo} \tag{15}$$

$$\frac{dx_{lo}(t)}{dt} = G_{exc}w_e u_{fa}(t) - 2w_{exc}x_{exc}(t) - w_{lo}^2 y_{lo}(t) \tag{16}$$

$$z_{fa}(t) = \frac{2e_0}{1 + e^{-rvfa}} - e_0 \tag{17}$$

$$v_{fa}(t) = C_{fapy}y_{py}(t) - C_{fasl}y_{sl}(t) - C_{fafa}y_{fa}(t) + y_{lo}(t) \tag{18}$$

The neural mass model as shown in Fig. 2 has been implemented in (MATLAB 2016b) to generate the alpha, beta, and gamma brain waves as shown in Fig. 3a–c respectively using Runge-Kutta (RK4) method. The input uf and up are given as Gaussian white noise with standard deviation sd = 3 and mean as $m = 90$.

The model parameter values are shown in Table 1.

The impact of dynamical models of brain activity has sluggish; and is still a mystery to neuroscientists around the world. Prediction testing of these models may be one of the technical reasons for this.

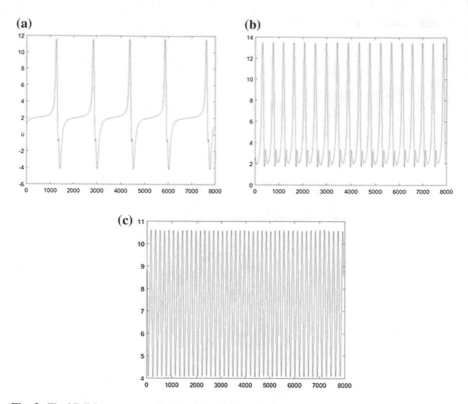

Fig. 3 The NMM to generate the (**a**) alpha (8 Hz), (**b**) beta (14 Hz) and (**c**) gamma (30 Hz)

3   Arterial Blood Pressure (ABP) Model

Located in the medulla of the brainstem is a cardiovascular center, which integrates information which is sensory in nature received from mechanoreceptors (also called baroreceptors), proprioceptors (limb position), chemoreceptors (blood chemistry) and from the heart and information from the cerebral cortex and limbic system.

3.1   Windkessel ABP Model Using Differential Equations

German physiologist Otto Frank published an article describing a human blood pressure model in 1899 [6]. The capability of an arterial wall to compress or expand to accommodate changes in pressure is necessary for proper functioning of the heart. This ability of the arterial walls to distend is called arterial compliance. This is similar to the compressibility of the air pocket and it simulates the elasticity and extensibility of the major artery. The capacitor is analogous to the compliance of the

Table 1 Model system parameters

System parameters	Description	Value
A	Average excitatory synaptic gain	3.35 mV
B	Average slow inhibitory synaptic gain	22 mV
G	Average fast inhibitory synaptic gain	10 mV
a	Reciprocal of excitatory time constant	100 Hz
b	Reciprocal of slow inhibitory time constant	50 Hz
g	Reciprocal of fast inhibitory time constant	500 Hz
C	General connectivity constant	135
C_{pyexc}	Connectivity: pyramidal to excitatory cells	1
C_{exppy}	Connectivity: excitatory to pyramidal cells	0.8
C_{pysl}	Connectivity: pyramidal to slow inhibitory cells	0.25
C_{slpy}	Connectivity: slow inhibitory to pyramidal cells	0.25
C_{pyfa}	Connectivity: pyramidal to fast inhibitory cells	0.3
C_{slfa}	Connectivity: slow to fast inhibitory cells	0.1
C_{fapy}	Connectivity: fast inhibitory to pyramidal cells	0.8
v_0	Sigmoid function: potential at half of max firing rate	6 mV
e_0	Sigmoid function: half of maximum firing rate	2.5 Hz
r	Sigmoid function: steepness parameter	0.56 mV^{-1}

blood vessel. The resistance emulates the systemic vascular resistance to arterial blood flow. The inductance L is the inertia of blood flow in vascular system.

Blood flow occurs due to a pressure gradient ($p1 - p2$) across the length of the systemic vasculature. $p1$ is the pressure at the start of the artery and $p2$ is at the end.

The Windkessel arterial blood pressure is described by the following set of equations:

$$\frac{dq}{dt} = \frac{1}{L}(p_1 - p_2) \tag{19}$$

$$\frac{dp_1}{dt} = \frac{1}{C1}(q_{in} - q) \tag{20}$$

$$\frac{dp_2}{dt} = \frac{1}{C_2}\left(q - \frac{p_2}{R}\right) \tag{21}$$

The four element Windkessel model [7] is shown in Fig. 4a. The output arterial blood pressure waveform is shown in Fig. 4b and the input volume to the model is shown in Fig. 4c which is the left ventricular ejection in one cardiac cycle.

The input q_{in} is defined by the following equations:

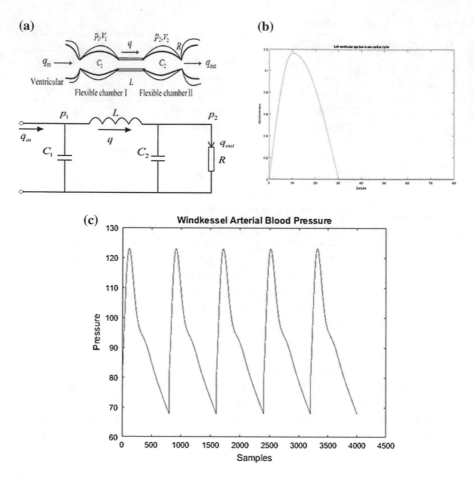

Fig. 4 The blood pressure model. **a** Windkessel model, **b** input volume, **c** arterial blood pressure

$$q_{in}(t) = \begin{cases} q_0 \sin\left(\frac{\pi t}{2\alpha T_s}\right) & 0 \le t \le \alpha T_s \\ q_0 \cos\frac{\pi}{4\alpha T_s}(t - \alpha T_s) & \alpha T_s \le t \le T \\ 0 & T_s \le t \le T \end{cases} \tag{22}$$

4 ECG and HRV Model

The sinoatrial (SA) node and atrioventricular (AV) node are responsible for the repeated contraction of the heart. They are known as the pacemakers of the heart. The SA node is located in the upper wall of the right atrium. The firing rate of the SA node is usually around 60–100 action potentials per minute. Sympathetic nervous system (SNS) and parasympathetic nervous system (PNS) compete with each other to maintain overall sympathovagal balance. The PNS is dominant during state off rest or relaxation which generates an average heart rate of 75 beats per minute (bpm). This is comparatively lower than the SA node actual heart rate of about 107 bpm at 20 years to 90 bpm at 50 years [8]. The PNS can decrease the heart rate to as low as 20 bpm.

4.1   ECG Model by Sharrys

The dynamical equations of motion are given by a set of three differential equations adapted from Mc Sharry's model [9].

$$\dot{x} = ax - wy \tag{23}$$

$$\dot{y} = ay + wx \tag{24}$$

$$\dot{z} = -\sum_{I \in P,Q,R,S,T} a_i \Delta_i \exp\left(-\Delta \frac{\theta_i^2}{2b_i^2}\right) - (z - z_0), \tag{25}$$

where; $a = 1 - \sqrt{x^2 + y^2}$, $\Delta\theta_i = \theta - \theta_i \bmod 2\Pi$, $\theta = a \tan 2(y, x)$.

The ECG wave varies as a function of heart rate as shown in Fig. 5a–c. The model parameters of the system implemented are shown in Table 2.

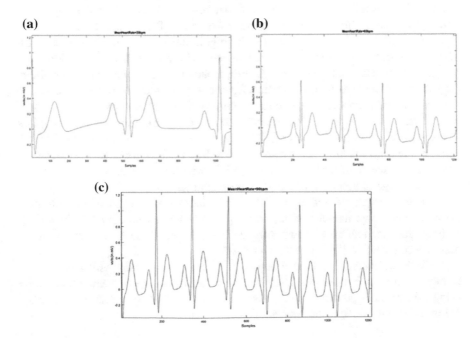

Fig. 5 The ECG model to generate the **a** mean HR 30 bpm, **b** mean HR 60 bpm and **c** mean HR 90 bpm

Table 2 Model system parameters

Index (i)	P	Q	R	S	T
Time (s)	−0.2	−0.05	0	0.05	0.3
θ_i	−1/3 π	−1/12 π	0	1/12 π	1/2 π
a_i	1.2	−5	30	−7.5	0.75
b_i	0.25	0.1	0.1	0.1	0.4

5   Autonomic Nervous System (ANS)

The autonomic nerves regulate the cardiac and vascular function. Parasympathetic and sympathetic efferent nerves control heart and vasculature which are found in the cell bodies of medulla oblongata. The central nervous system receives input from baroreceptor and chemoreceptors which are responsible to maintain blood pressure and chemistry respectively. There are billions of neurons in the brain which receive and transport neurotransmitters in the brain. Each of these neurons are classified as sensory, motor, or interneurons. Sensory neurons are neurons which detect changes in the external environment and generate impulses which are sent to the brain. Motor neurons activate muscle cells and control motor activity like walking, speaking, etc. Various cardiovascular models have been described in the literature to have a better understanding of the underlying mechanisms of the heart and vasculature.

Cardiovascular systems are proposed in the literature in the form of empirical and functional models. Extended models are also available which include the respiratory system interaction through an autonomic neural controller [10]. The various components of cardiovascular system help in regulating cardiac output (see Fig. 6). The cardiac output is a function of Heart Rate and stroke volume. Heart Rate is a complex parameter obtained through β-sympathetic (ftbs) and parasympathetic (ftp) peripheral nerve response. The ftbs, ftp and α sympathetic(ftas) nerve responses are generated through central autonomic control from the complex integration of central respiratory control (Nt), chemoreflexes (Fchem), lung stretch receptors reflexes (fls), baroreflexes (fcs) and CNS responses [11, 12]. The ftas signal which signifies the α-sympathetic nerve response regulates the total peripheral resistance (α TPR) to produce arterial blood pressure (ABP) which further generates stroke volume via change in venous return and cardiac contractibility.

Short-term rhythmic activity in HRV calculations is a result of the interplay between autonomic neural system, blood pressure and respiratory activities [13]. A heart rate tachogram is commonly used to notice these short term changes between consecutive RR intervals over time.

HRV is considered thus as a function of heart and brain which reflect the changes in the ANS dynamics. HRV can be measured in the frequency domain by using FFT of other power spectral density algorithms. The RR interval of the ECG are transformed into specific frequency components.

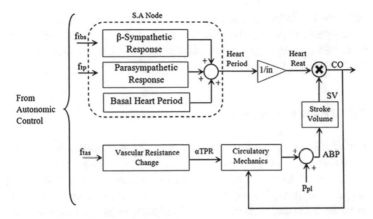

Fig. 6 Illustrates the nervous system link between the heart and brain. The sympathetic branch increases the heart flow while the parasympathetic slows the rate

Power spectral analysis plots of HR oscillations revels specific frequency components wherein heart rate cycles are classified as HF (0.15–0.40 Hz); LF (0.04–0.15 Hz); very LF (0.0033–0.04 Hz); and Ultra Low-Frequency (below 0.0033–0.2 Hz).

5.1 ANS for Coherence

Well-being of an organism is determined by the healthy functioning of the regulatory systems. For this to happen there should be optimal variability in the functioning of the regulatory systems. Too much variability is harmful for effective physiological processes, too little variation means deterioration or diseases.

Evidence shows age of a patient reflects the relationship between regulatory capacity and reduced IIRV.

As a person gets older the HRV decreases due to loss of neurons in the spinal cord which reduces signal transmission [11] and reduce regulatory capacity. Functional gastrointestinal disorders, inflammation, and hypertension are caused due to lower HRV. While patients with functional gastrointestinal disorders often have reduced HRV [14], HRVB has increased vagal tone and improved symptom ratings in these patients [15].

5.1.1 High-Frequency Band

The power in HF varies over 0.15–0.4 Hz band of frequencies. This band is known as the respiratory band due to HR perturbations related to the respiratory cycle. It reflects the vagal or parasympathetic activity of the ANS. Parasympathetic activity is often lowered in patients under stress or if diagnosed with stress. Also as a person ages the parasympathetic activity reduces thereby reducing HRV [16]. Normally the parasympathetic activity increases at night and decreases during daytime for a healthy individual [17].

5.1.2 Low-Frequency Band

The power in the LF varies over 0.04–0.15 Hz band of frequencies. This band is called the baroreceptor range since it reflects the baroreceptor activity at rest [18]. The afferent signals from the heart and other visceral organs along with barore-ceptor signals are passed onto the brain via the vagus nerves. Located in the chambers of the heart and vena cava, carotid sinuses (which contain the most fragile mechanoreceptors), and the aortic arch are baroreceptors which are stretch fragile mechanoreceptors. The baroreflex reduces blood pressure when the blood pressure is observed too high by inhibiting sympathetic activation along with parasympa-thetic activation. Peripheral resistance is lowered by sympathetic inhibition, whereas activation of parasympathetic branch reduces heart rate and contractility. Cycle to cycle differences in heart rate per unit change in blood pressure is cal-culated as the baroreflex gain. Aging and weakened regulatory ability causes a decrease in baroreflex gain.

5.1.3 Autonomic Balance and the LF/HF Ratio

One can accept that the SNS and PNS are the two branches of ANS. These branches can be simultaneously active to regulate the SA node firing, i.e., where increase in SNS activity is coupled with decrease in PNS activity. Certain orthostatic chal-lenges sometime disturbed the SNS activities and vagal withdrawal. Also psy-chological stresses induced changes in SNS and PNS activity [19]. Hence, relation of SNS and PNS in LF power spectral density is complex nonlinear and solely depends on the experimental paradigm employed to the subject [20, 21]. The LF/HF is a ratio of LF and HF power is used to reflect sympathetics and parasym-pathetics activities. The low ratio gives idea of energy conservation and engaging in tend-and-befriend behaviors [22]. The high value of ratio indicates high sympa-thetics activities than parasympathetics which is seen when subjects is meeting challenges which demands high SNS activities. But due to above cited reasons in LF power the said ratio has to be used with the caution.

6 Conclusion

Researchers are keen in understanding interaction between CVS and ANS which are predictors of adverse cardiovascular events. The source of such events has to be diagnosed to pinpoint abnormality of ANS or a pathological organ response. Also other factors like age, lipid profile, smoking status, and family history will also play role in the prediction of CV events. The coherence model of psycho-physiological system proposed by Institute of Hearth-math focuses on increasing self-regulatory capacity of an individual, which reflects in maintaining the heart rhythms. These rhythmic activities reflect the perception over cognitive, biological, social and environment networks in living systems. Here afferent pathways which collect basal regulatory information from the cardiovascular system are given more relevance in this model. They claim that to have improved performance better self regulation and well being of living system, the system should go through positive emotion which induces coherence and harmonious physiological mode. This is very often known as physiological coherence which describes the orderly and stable system rhythm generated by living beings which is quantified by maximum peak 0.04–0.26 Hz of the HRV power spectrum.

References

1. Wendling, F., Bartolomei, F., Mina, F., Huneau, C., Benquet, P.: Interictal spikes, fast ripples and seizures in partial epilepsies combining multi-level computational models with experimental data. Eur. J. Neurosci. **36**(2), 21642177 (2012)
2. Wendling, F., Bartolomei, F., Bellanger, J.J., Chauvel, P.: Epileptic fast activity can be explained by a model of impaired gabaergic dendritic inhibition. Eur. J. Neurosci. **15**(9), 14991508 (2002)
3. Lopes da Silva, F.H., Hoeks, A., Smits, H., Zetterberg, L.H.: Model of brain rhythmic activity, the alpha rhythm of the thalamus. Kybernetik **15**(1), 2737 (1974)
4. Freeman, W.J.: Models of the dynamics of neural populations. Electroencephalogr. Clin. Neurophysiol. Suppl. (34), 9–18 (1977)
5. Zavalgia, M., Cona, F., Ursino, M.: A neural mass model to simulate different rhythms in a cortical region. Comput. Intell. Neurosci. Hindawi **10** (2010)
6. Otto, F.: Die Grundform des arteriellen Pulses. Zeitung für Biologie **37**, 483–586 (1899)
7. Wang, L., Xu, L., Zhou, S., Wang, H., Yao, Y., Hao, L., Li, B.N., Qi, L.: Design and implementation of a pulse wave generator based on Windkessel model using field programmable gate array technology. Biomed. Sig. Process. Control **36**, 93–101 (2017)
8. Opthof, T.: The normal range and determinants of the intrinsic heart rate in man. Cardiovasc. Res. **45**(1), 177–184 (2000)
9. McSharry, P.E., Clifford, G.D., Tarassenko, L., Smith, L.A.: A dynamical model for generating synthetic electrocardiogram signals. IEEE Trans. Biomed. Eng. **50**(3) (2003)
10. Lewis, M. J., Short, A. L.: Autonomic nervous system control of the cardiovascular and respiratory systems in asthma. Respir. Med. **100**(10), 1688–1705 (2006)
11. Shaffer, F., McCraty, R., Zarr, L.: A healthy heart is not a metronome: an integrative review of the heart's anatomy and heart rate variability. Front. Psychol. **5**, 1040 (2014)

12. Ursino, M.: Interaction between carotid baroregulation and the pulsating heart: a mathematical model. AMJ Physiol. **275**(5), H1733–H1747 (1998)
13. McCraty, R., Shaffer, F.: Heart rate variability: new perspectives on physiological mechanisms, assessment of self-regulatory capacity, and health risk. Glob. Adv. Health Med. **4**(1), 46–61 (2015)
14. Jang, A., Hwang, S.-K., Padhye, N.S., Meininger, J.C.: Effects of cognitive behavior therapy on heart rate variability in young females with constipation-predominant irritable bowel syndrome: a parallel-group trial. J. Neurogastroenterol. Motil. **23**(3), 435 (2017)
15 Sowder, E.: Restoration of vagal tone, "a possible mechanism for functional abdominal pain". Appl. Psychophysiol. Biofeedback **35**(3), 199–206 (2010)
16. Reardom, M., Malik, M.: Changes in heart rate variability with age. Pacing Clin. Electrophysiol. **19**(11), 1863–1866 (1996)
17. Elsenbruch, S., Harnish, M.J.: Heart rate variability during waking and sleep in healthy males and females. Sleep **22**(8), 1067–1071 (1999)
18. Cerutti, C., Barres, C., Paultre, C.: Baroreflex modulation of blood pressure and heart rate variabilities in rats: assessment by spectral analysis. Am. J. Physiol. **266**(5), H1993–H2000 (1994)
19. Cacioppo, J.T., Berntson, G.G.: The affect system architecture and operating characteristics. Curr. Dir. Psychol. Sci. **8**(5), 133–137 (1999)
20. Berntson, G.G.: Heart rate variability: origins, methods, and interpretive caveats. Psychophysiology. **34**(6), 623–648 (1997)
21. Billman, G.E.: The LF/HF ratio does not accurately measure cardiac sympatho-vagal balance. Front. Physiol. **4**, 26 (2013). PMC. Web. 29 (Nov 2017)
22. Taylor, S.: Tend and befriend. Biobehavioral bases of affiliation under stress, Curr. Dir. Psychol Sci. **15**(6), 273–277 (2006)

Steady-State Visual Evoked Potential-Based Real-Time BCI for Smart Appliance Control

Noel G. Tavares and R. S. Gad

Abstract Brain–Computer Interface (BCI) provides an alternative way for humans to communicate with the external environment. BCI systems can be of great help to people with severe motor disabilities who cannot perform normal daily activities. In this paper, we introduce a novel steady-state visual evoked potential (SSVEP)-based brain–computer interface system that control home appliances like electric fan, tube light, etc. Designed system aim is to extract the SSVEP signal and then classify them using PCA. We confirmed the generation of SSVEP frequencies in the online analysis using Fast Fourier Transform. The classification of SSVEP signals is done using Principal Component Analysis.

Keywords Brain–computer interface (BCI) · Steady-state visually evoked potentials (SSVEP) · Electroencephalogram (EEG) · Fast fourier transform (FFT) Principal component analysis (PCA)

1 Introduction

Human beings communicate with the external world through the motor and sensory pathways. But damage to these pathways makes it difficult to communicate with the external world. Humans who suffer from neuromuscular diseases like quadriplegic patients, amyotrophic lateral sclerosis cannot move their limbs due to damage to the spinal cord. Brain–computer interface-based systems can be of great help to such type of patients for communication [1]. BCI's have also been used in selective attention based studies, speller systems and prosthetics. There are normally two types, i.e., invasive and non-invasive BCI's. Invasive BCI's include ECoG, neural implants, single cell recording, etc.; whereas non-invasive BCI's include EEG,

N. G. Tavares · R. S. Gad (✉)
Department of Electronics, Goa University, Taleigao Plateau, Goa, India
e-mail: rsgad@unigoa.ac.in

N. G. Tavares
e-mail: elect.noel@unigoa.ac.in

© Springer Nature Singapore Pte Ltd. 2019
P. K. Mallick et al. (eds.), *Cognitive Informatics and Soft Computing*,
Advances in Intelligent Systems and Computing 768,
https://doi.org/10.1007/978-981-13-0617-4_74

MEG, fMRI, etc. Invasive BCI's are more accurate compared to non-invasive BCI, but since it requires surgery it is only used on animals or more recently on patients suffering from severe disabilities or diseases. Electroencephalograph (EEG) recordings are widely used for non-invasive BCI as it has good temporal resolution. EEG's main drawback is its low signal-to-noise ratio and difficulty in source localization. An EEG recording is done by placing an electrode strip or cap on the scalp. There are various non-invasive EEG-based BCI interfaces such as evoked potentials, P300 signals [2], mu rhythm [3], motor imagery [4], etc. The evoked potentials occur due an external stimulus which can be based on vision, sound, or touch (somatosensory) [5, 6] or a combination of them [7]. In P300-based BCI's a positive peak is elicited at a delay of about 300 ms, when an irregular stimulus occurs in after a pattern of regular stimuli. Mu rhythm arises due to imagined motor activity. Visual evoked potentials (VEP) are basically of two types, Transient Visually Evoked Potential (TVEP), and Steady-State Visually Evoked Potential (SSVEP) [8]. In VEP's a stimulus such as flash is used to elicit activity in the occipital lobe which is the visual processing unit of the brain. SSVEP reflect the changes in the cortical activity due to recurrent stimulus presentation. This cortical activity generates electrical signal at the frequency of stimulation and its harmonics. This electrical activity is captured by EEG system which is more prominent in the occipital lobe. The advantages of SSVEP over other BCI's is that the subject requires little or no training, it information transfer rate is high. SSVEP has also been explored for selective attention [9]. Many researchers are exploring research in area of cognitive neuroscience. In one such work a BCI-based speller paradigm has been designed by Movahedi et al. [10]. Chen et al. [11] have built a high information transfer rate speller paradigm with 45 commands. Here in our work we proposed steady-state visual evoked potential-based BCI for smart appliance control. We use non-invasive EEG techniques to read the raw electrical signals using a single channel (a single electrode) system. A stimulus of flickering LED's help produce the steady-state visual evoked potentials, only the potentials taken up from the occipital lobe are amplified and transmitted over a Bluetooth module to a computing system which performs feature extraction (using Fast Fourier Transform (FFT) algorithm) and classification (using Principle Component Analysis (PCA) algorithm) to classify the incoming signal for determined automation control in smart appliances.

2 Methodology

We proposed a BCI system build on SSVEP using B-Alert X-24 device via a Bluetooth interface for controlling home appliances as shown in Fig. 1. The design includes a stimulating platform, EEG signal acquisition unit, signal processing unit, and MATLAB for feature extraction and classification. The stimulating platform generates the frequency based paradigm with help of four light emitting diodes (LEDs) [12].

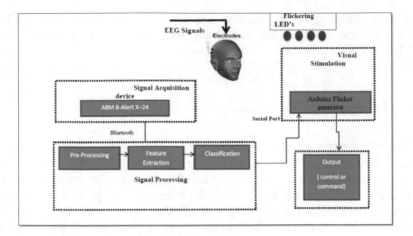

Fig. 1 Block diagram of a SSVEP based BCI system

2.1 Signal Acquisition Unit

Here, B-Alert X-24 series mobile EEG system is used which has of 20 EEG channels that acquire high-quality signals. The International 10–20 system [13] is referred for the locations of scalp electrodes for EEG extraction. We selected O1 channel for processing since SSVEP has eminent magnitude in occipital part of the brain mass.

The B-Alert X-24 series is a Bluetooth-based system which transmits the amplified signal over to a controller where it can be recorded and analyzed. The signals are recorded at a sampling frequency of 256 Hz. The Bluetooth wireless range is about 20 m.

2.2 Subjects in Experiment

Five subjects (all male with mean age of 22 years) participated for the experiment. All the subjects had normal visual acuity. They were comfortably seated in a chair in front of four blinking LED's. The subjects sat at distance of around 40 cm from the target stimuli. Their brain waves were recorded with described EEG system at a 256 Hz sampling frequency in a low decibel dark room. The reference sensor was placed on the right ear lobe and the left ear lobe served as the ground via electrodes. Impedance checking in established to ensure proper contact of the electrodes with the scalp. The subjects had to focus on the flickering LED's over a period of 2 min. The procedure was repeated six times over a frequency. A total of 24 trials, 2 min each were performed on every subject. The raw EEG signals are stored in EDF file format. Preprocessing, feature extraction and classification is established in MATLAB 2017.

2.3 Experimental Paradigm

Individual's seizures occur in brain mass in response to the external stimuli like sound, vision or tactile stimuli, rather than spontaneous occurring seizures [14]. We chose our stimulus such that it does not affect a photosensitive subject. We chose four LED's that flicker at 6, 8, 9, and 10 Hz. Here, lower frequencies are chosen as these frequency regions have stronger amplitude response and higher accuracy. Secondly we have avoided the lower delta range of brain frequency as it may cause dizziness. The microcontroller board is programmed to flicker the LED's at the frequencies discussed above.

2.4 Signal Processing and Feature Extraction

The raw EEG signal is passed through a Finite Impulse Response filter having a lower cut-off at 4 Hz and a higher cut-off at 18 Hz. This range removes all the dc components as well as high frequency and power line noise. The transition width of the filter has been set to 0.2 Hz and filter order is selected to be 30. The filter is tapered to avoid ringing or edge artifacts. Next step is to extract features from the signal. Canonical Correspondence Analysis (CCA) [15, 16] and Fast Fourier Transform (FFT) have been extensively used by researchers [17]. We chose FFT in our work to extract the peak frequencies. The FFT and its inverse are defined as Eqs. (1) and (2), in which the signals are transformed between time domain and frequency domain.

$$X(K) = \sum_{n=0}^{N-1} x(n) e^{j2n\pi k/N} \tag{1}$$

$$x(n) = \sum_{k=0}^{N-1} X(K) e^{-j2n\pi k/N} \tag{2}$$

FFT is digital signal processing method to compute the discrete Fourier Transform (DFT) of a signal. This is used to transform a time domain signal to its frequency domain. The computational complexity of the DFT which is $2N^2$ is reduced to $2N \log_2 N$ in FFT. The sampling frequency for FFT computation was set at 256 Hz to match the EEG sampling frequency.

2.5 Signal Classification

PCA is a statistical attribute extraction method that makes use of a linear transformation to transform a set of probably correlated observations into a set of independent variables called principal components [18]. From the input data linear transformation creates a set of components, arranged according to their variance where the first principal component has the highest viable variance. The brain signal is divided into several independent components due to this variance. PCA projects the input data on a k-dimension Eigen space of k eigenvectors, which are calculated from the covariance matrix Σ of the training data $p = [p_1, p_2, ..., p_n]$. p_i is i-th d-dimension training sample, and n is the number of samples. The covariance matrix ε is computed as

$$\varepsilon = \sum_{i=1}^{n} (p_i - m)(p_i - m)^t \tag{3}$$

where, $m = \frac{1}{n}\sum_{i=1}^{n} p_i$ is the mean vector of the training samples p_i. The covariance matrix Σ is a real and symmetric $d \times d$ matrix, therefore Σ has d eigenvectors and eigenvalues. The most important information contained in the dataset can be found by k eigen vectors which are represented by eigen values. The principal components are the eigenvectors with highest eigen value of the training set p. PCA helps us find those $k1$ largest eigenvector's from k eigenvector's. We extract the feature vector of the test data q from the projection matrix build form the eigenvectors of the training. The k eigenvectors are ordered in ascending order in matrix A, such that the first column contains the eigen vector with the highest eigen value and so on. PCA projects the test data q onto the new subspace and computes the feature vector v, such that

$$v = A^t(q - m), \tag{4}$$

where, $m = \frac{1}{n}\sum_{i=1}^{n} p_i$ is the mean vector of training samples pi. Figure 3 shows the classification of dataset for computed feature vector calculated using PCA. Common feature components are clustered around the region of interest.

3 Results

In SSVEP the main work is to find out the most dominant frequency which corresponds to the frequency of the stimulus from the recorded EEG signals. When we run FFT analysis over the raw EEG signal acquired from the O1 location, we can differentiate between the SSVEP frequencies. Each fundamental frequency has corresponding harmonics at 2f and 3f. These harmonics can also be used to decrease misclassification error. We have chosen the fundamental frequency in such a manner that it does not overlap with the harmonics of other frequencies.

Figure 2a–d depicts the frequency domain representation of a SSVEP signal for stimulus frequencies of 6, 7, 8, and 10 Hz respectively. The FFT output produces dominant frequencies at 6, 7, 8, and 10 Hz which can be used for specific applications. Also one can use PCA feature extraction method to classify the input signals (Ref. Fig. 3) to discriminate the SSVEP signals perceived by subject.

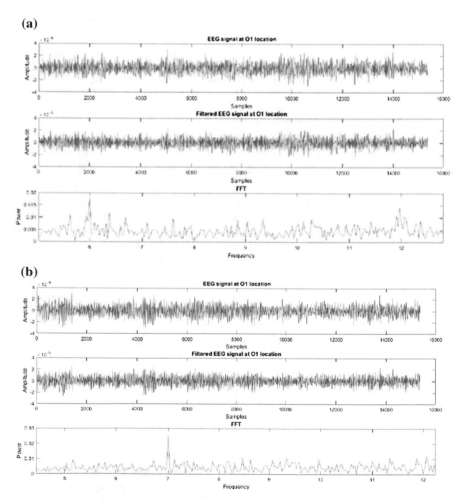

Fig. 2 SSVEP for LED's frequency base paradigm flickering at **a** 6 Hz, **b** 7 Hz, **c** 8 Hz, **d** 10 Hz respectively

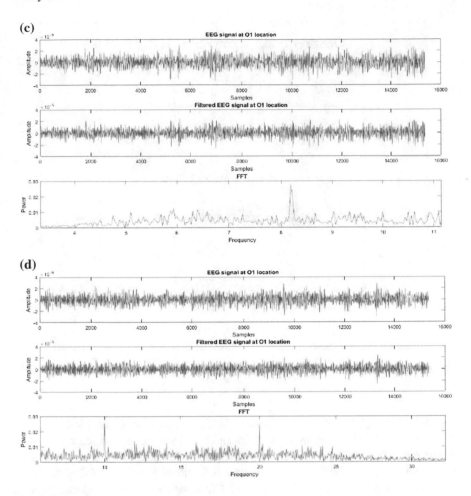

Fig. 2 (continued)

4 Discussion and Conclusions

In this paper, we proposed a BCI-based on SSVEP for home automation with wireless technology. In the proposed system, a subject just gazes a independent flickering target having low frequencies to turn on/off normal day to day electrical gadgets like a bulb or fan. We also discovered that when the flickering frequencies were kept above the delta brain (above 4 Hz) and below beta brain (below 12 Hz) one obtained pretty good detection of frequencies. We also noticed that subject's eyes are tired after some trials. This can be reduced by adjusting the distance of the stimulus LED box for each subject rather than keeping a fixed distance, since each

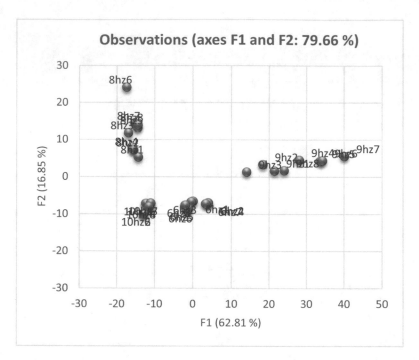

Fig. 3 Classification of the data in clusters of four sets distinguishing the flickering LED's frequencies

user will have different eye sensitivity. In our work, instead of producing the stimulus on a LCD screen we have used LED's which can be carried around and hence portable for the user, thus allowing better flexibility to the user. Another advantage of using a LED's as our stimulus is that we have the flexibility to decide on various frequencies as compared to a LCD screen where the frequencies are a multiple of the monitor refresh frequency rate. One can use a phase-based SSVEP system instead of the current frequency based system to reduce the fatigue caused to the eye retina and to improve performance. At present the frequencies we have selected have one Hz spatial separation between them. Also one can reduce the separation to 0.2 or 0.5 Hz so that we can accommodate more commands within the 4–12 Hz range. The length of each trial can be reduced from 120 s to check how it performs with a shorter trial length in order to increase information transfer rate. Classification algorithms based on ensemble, discriminant or neural networks can be used for to reduce misclassification error.

5 Future Work

Proposed BCI system is designed on a portable B-Alert X-24 system which acquires the raw EEG signals via Bluetooth interface. One can explore a cheaper system such as Neurosky Think gear for acquisition of signals.

Neurosky Think Gear ASIC Module is a single channel (EEG electrode, reference electrode and ground electrode) EEG amplifier with a sampling rate of 512 Hz. The exciting features of this are that, it directly connects to a dry EEG electrode and provides extremely low-level signal detection in the range of 3–100 Hz bandwidth. This module operates at a voltage of 2.97–3.63 V and support a baud of 1200, 9600, and 57,600. Figure 4 gives the basic idea of the acquisition system.

The acquired signal is transmitted over to the Arduino using a Bluetooth module HC-05 which is a slave type. The module supports a baud of 9600, 19,200, 38,400, 57,600, 115,200, 230,400, and 460,800 that can be selected using the AT commands. A master HC-05 Bluetooth module receives the incoming data which is directly connected to the Arduino. In our system we have used the Arduino platform to feature extract and classify the incoming data as shown in Fig. 5. The Arduino UNO R3 is a microcontroller board based on the ATmega328 with a 16 MHz crystal oscillator. It has 14 digital input/output pins (of which 6 can be used as PWM outputs), 6 analogy inputs, a USB connection, a power jack, an ICSP header, and powered using a AC-to-DC adapter or battery. Once the data is classified it is passed over to a relay switch that correspondingly switched the appropriate appliance.

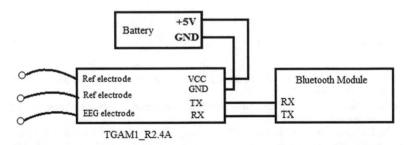

Fig. 4 Acquisition section and transmission

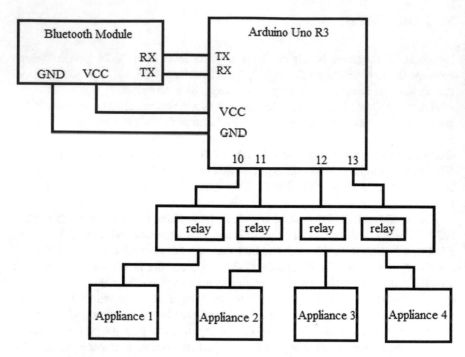

Fig. 5 Reception, preprocessing and classification block diagram

Acknowledgements Authors would like to acknowledge financial assistance from Department of Science and Technology (New Delhi) under Instrument Development Scheme. Author Mr. Noel Tavares would like to thank DST, New Delhi for granting INSPIRE Fellowship to do full time research at Goa University.

References

1. McFarland, D.J., Wolpaw, J.R.: Brain-computer interfaces for communication and control. Commun. ACM **54**(5) (2011)
2. Fazel-Rezai, R., et al.: P300 brain computer interface: current challenges and emerging trends. Frontiers Neuroeng. **5**, 14 (2012)
3. Joshi, R., Saraswat, P., Gajendran, R.: A novel mu rhythm-based brain computer interface design that uses a programmable system on chip. J. Med. Signals Sens. **2**(1), 11 (2012)
4. Neuper, C., Muller-Putz, G.R., Scherer, R.: Motor imagery and EEG-based control of spelling devices and neuroprostheses. Prog. Brain Res. **159**, 393–409 (2006)
5. Muller-Putz, G.R., Scherer, R., Neuper, C.: Steady-state somatosensory evoked potentials: suitable brain signals for brain-computer interfaces? IEEE Trans. Neural Syst. Rehabil. Eng. **14**(1) (2006)
6. Kaufmann, T., Herweg, A., Kübler, A: Toward brain-computer interface based wheelchair control utilizing tactually-evoked event-related potentials. J. NeuroEng. Rehabil. **11**(1), 7 (2014)

7. Kaongoen, N., Jo, S.: A novel hybrid auditory BCI paradigm combining ASSR and P300. J. Neurosci. Methods **279**, 44–51 (2017)
8. Norcia, A.M., et al.: The steady-state visual evoked potential in vision research: a review. J. Vision **15**(6), 4 (2015)
9. Morris, J., Holcomb, P.J.: Effects of spatial selective attention on the steady-state visual evoked potential in the 20–28 Hz range. Brain Res. Cogn. Brain Res. **25**(3) (1998)
10. Movahedi, M.M., Mehdizadeh, A., Alipour, A.: Development of a Brain Computer Interface (BCI) speller system based on SSVEP signals. J. Biomed. Phys. Eng. **3**(3) (2013)
11. Chen, X., Chen, Z.: A high-ITR SSVEP-based BCI speller. Brain-Comput. Interfaces (Taylor & Franics Online) **1** (2014)
12. Zhu, D., Bieger, J.: A survey of stimulation methods used in SSVEP-based BCIs. Comput. Intell. Neurosci. **2010**, Article ID 702357, 12 (2010)
13. Myslobodsky, M.S.; Coppola, R., Bar-Ziv, J., Weinberger, D.R.: Adequacy of the international 10–20 electrode system for computed neurophysiologic topography. J. Clin. Neurophysiol. **7**, 507–518 (1990)
14. Zifkin, B.G., Inoue, Y.: Visual reflex seizures induced by complex stimuli. Epilepsia **45**, 27–29 (2004)
15. Lin, Z., Zhang, C., Wu, W.: Frequency recognition based on canonical correlation analysis for SSVEP-based BCIs. IEEE Trans. Biomed. Eng. **54**(6), 1172–1176 (2007)
16. Nakanishi, M., et al.: A comparison study of canonical correlation analysis based methods for detecting steady-state visual evoked potentials. In: Yao, D. (ed.) PLoSONE **10**(10) (2015)
17. Wu, Z.: SSVEP extraction based on the similarity of background EEG. In: Di Russo, F. (ed.) PLoS ONE **9**(4) (2014)
18. Jolliffe, I.: Principal Component Analysis. Springer, New York (1986)

Author Index

Printed in the United States
By Bookmasters